ACS SYMPOSIUM SERIES **598**

Multidimensional Spectroscopy of Polymers

Vibrational, NMR, and Fluorescence Techniques

Marek W. Urban, EDITOR
North Dakota State University

Theodore Provder, EDITOR
The Glidden Company

Developed from a symposium sponsored
by the Division of Polymeric Materials: Science
and Engineering, Inc., at the 208th National Meeting
of the American Chemical Society,
Washington, D.C.,
August 21–25, 1994

American Chemical Society, Washington, DC 1995

Library of Congress Cataloging-in-Publication Data

Multidimensional spectroscopy of polymers: vibrational, NMR, and fluorescence techniques / Marek W. Urban, editor, Theodore Provder, editor.

p.   cm.—(ACS symposium series, ISSN 0097–6156; 598)

"Developed from a symposium sponsored by the Division of Polymeric Materials: Science and Engineering, Inc., at the 208th National Meeting of the American Chemical Society, Washington, D.C., August 21–25, 1994."

Includes bibliographical references and indexes.

ISBN 0–8412–3262–8

1. Polymers—Analysis—Congresses. 2. Spectrum analysis—Congresses. I. Urban, Marek W., 1953–   . II. Provder, Theodore, 1939–   . III. American Chemical Society. Division of Polymeric Materials: Science and Engineering. IV. Series.

QD139.P6M85   1995
547.7'046—dc20
                                    95–17880
                                        CIP

This book is printed on acid-free, recycled paper.

Foreword

THE ACS SYMPOSIUM SERIES was first published in 1974 to provide a mechanism for publishing symposia quickly in book form. The purpose of this series is to publish comprehensive books developed from symposia, which are usually "snapshots in time" of the current research being done on a topic, plus some review material on the topic. For this reason, it is necessary that the papers be published as quickly as possible.

Before a symposium-based book is put under contract, the proposed table of contents is reviewed for appropriateness to the topic and for comprehensiveness of the collection. Some papers are excluded at this point, and others are added to round out the scope of the volume. In addition, a draft of each paper is peer-reviewed prior to final acceptance or rejection. This anonymous review process is supervised by the organizer(s) of the symposium, who become the editor(s) of the book. The authors then revise their papers according to the recommendations of both the reviewers and the editors, prepare camera-ready copy, and submit the final papers to the editors, who check that all necessary revisions have been made.

As a rule, only original research papers and original review papers are included in the volumes. Verbatim reproductions of previously published papers are not accepted.

M. Joan Comstock
Series Editor

Contents

Preface ix

FOURIER TRANSFORM IR AND RAMAN
SPECTROSCOPY

1. Fourier Transform IR and Raman Spectroscopy of Polymers:
   Section Overview 2
   Peter R. Griffiths and Marek W. Urban

2. Determining the Structure of Polyimide–Metal Interfaces
   Using Fourier Transform IR and Raman Spectroscopy 8
   F. J. Boerio, J. T. Young, and W. W. Zhao

3. Raman Microscopy and Imaging of Polymers 41
   M. Claybourn, A. Luget, and K. P. J. Williams

4. Multimode IR Analysis of Nylon 11 61
   L. J. Fina

5. Rheophotoacoustic Fourier Transform IR Spectroscopy
   To Study Diffusion and Adhesion in Polymers 78
   B. W. Ludwig, B. D. Pennington, and Marek W. Urban

6. Step-Scan Fourier Transform IR Studies of Polymers
   and Liquid Crystals 99
   Richard A. Palmer, Vasilis G. Gregoriou, Akira Fuji,
   Eric Y. Jiang, Susan E. Plunkett, Laura M. Connors,
   Stephane Boccara, and James L. Chao

7. Evaluating the Weatherability of Polyurethane Sealants 117
   Ralph M. Paroli, Kenneth C. Cole, and Ana H. Delgado

8. Diffuse and Specular Reflectance Measurements
   of Polymeric Fibrous Materials 137
   M. Papini

9. Monitoring Polymerization Reactions by Near-IR
   Spectroscopy.. 147
   Shih Ying Chang and Nam Sun Wang

10. UV–Visible Spectroscopy To Determine Free-Volume
    Distributions During Multifunctional Monomer
    Polymerizations 166
    Kristi S. Anseth, Teri A. Walker, and
    Christopher N. Bowman

NMR SPECTROSCOPY

11. Multidimensional NMR Spectroscopy of Polymers:
    Section Overview... 184
    Klaus Schmidt-Rohr

12. Dynamics and Structure of Amorphous Polymers Studied
    by Multidimensional Solid-State NMR Spectroscopy 191
    Klaus Schmidt-Rohr

13. Applications of ^{1}H–^{19}F–^{13}C Triple-Resonance NMR Methods
    to the Characterization of Fluoropolymers.................................... 215
    Peter L. Rinaldi, Lan Li, Dale G. Ray III,
    Gerard S. Hatvany, Hsin-Ta Wang, and H. James Harwood

14. Conformational Disorder and Its Dynamics
    Within the Crystalline Phase of the Form II Polymorph
    of Isotactic Poly(1-butene) 243
    Haskell W. Beckham, Klaus Schmidt-Rohr, and
    H. W. Spiess

15. NMR Study of Penetrant Diffusion and Polymer Segmental
    Motion in Toluene–Polyisobutylene Solutions 254
    Athinodoros Bandis, Paul T. Inglefield, Alan A. Jones,
    and Wen-Yang Wen

16. Miscibility, Phase Separation, and Interdiffusion
    in the Poly(methyl methacrylate)–Poly(vinylidene fluoride)
    System: Solid-State NMR Study... 274
    Werner E. Maas

17. Oxygen Absorption on Aromatic Polymers:
    ^{1}H NMR Relaxation Study 290
    D. Capitani, A. L. Segre, and J. Blicharski

18. **Chain Packing and Chain Dynamics of Polymers with Layer Structures** 311

Hans R. Kricheldorf, Christoph Wutz, Nicolas Probst, Angelika Domschke, and Mihai Gurau

19. **Electron Paramagnetic Resonance and ^{1}H and ^{13}C NMR Study of Paper** 333

D. Attanasio, D. Capitani, C. Federici, M. Paci, and A. L. Segre

FLUORESCENCE SPECTROSCOPY

20. **Applications of Luminescence Spectroscopy in Polymer Science: Section Overview** 356

Ian Soutar

21. **Fluorescence Studies of the Behavior of Poly(dimethylacrylamide) in Dilute Aqueous Solution and at the Solid–Liquid Interface** 363

Ian Soutar, Linda Swanson, S. J. L. Wallace, K. P. Ghiggino, D. J. Haines, and T. A. Smith

22. **Fluorescence Studies of Pyrene-Labeled, Water-Soluble Polymeric Surfactants** 379

Michael C. Kramer, Jamie R. Steger, and Charles L. McCormick

23. **Luminescence Spectroscopic Studies of Water-Soluble Polymers** 388

Ian Soutar and Linda Swanson

24. **Photophysical Study of Thin Films of Polystyrene and Poly(methyl methacrylate)** 410

E. H. Ellison and J. K. Thomas

25. **Photophysical Approaches to Characterization of Guest Sites and Measurement of Diffusion Rates to and from Them in Unstretched and Stretched Low-Density Polyethylene Films** 425

Jawad Naciri, Zhiqiang Hé, Roseann M. Costantino, Liangde Lu, George S. Hammond, and Richard G. Weiss

26. **Photophysical Behavior of Phenylene–Vinylene Polymers as Studied by Polarized Fluorescence Spectroscopy** 446

M. Hennecke and T. Damerau

27. Free-Volume Hole Distribution of Polymers Probed
    by Positron Annihilation Spectroscopy.. 458
    J. Liu, Q. Deng, H. Shi, and Y. C. Jean

28. Monitoring Degree of Cure and Coating Thickness
    of Photocurable Resins Using Fluorescence Probe
    Techniques... 472
    J. C. Song and D. C. Neckers

MULTIDISCIPLINARY APPROACHES

29. Spectroscopic Characterization of Unimer Micelles
    of Hydrophobically Modified Polysulfonates............................... 490
    Yotaro Morishima

30. Fourier Transform IR and NMR Observations
    of Crystalline Polymer Inclusion Compounds.............................. 517
    N. Vasanthan, I. D. Shin, and A. E. Tonelli

31. Solvent-Induced Changes in the Glass Transition Temperature
    of Ethylene–Vinyl Alcohol Copolymer Studied Using
    Fourier Transform IR and Dynamic Mechanical
    Spectroscopy.. 535
    Marsha A. Samus and Giuseppe Rossi

32. Photophysical and Rheological Studies of Amphiphilic
    Polyelectrolytes: Correlation of Polymer Microstructure
    with Associative-Thickening Behavior 551
    Kelly D. Branham and Charles L. McCormick

33. Fourier Transform IR and Fluorescence Spectroscopy
    of Highly Ordered Functional Polymer Langmuir–Blodgett
    Films 568
    Tokuji Miyashita

Author Index 585

Affiliation Index 586

Subject Index... 586

Preface

WHILE CONTINUOUSLY CHANGING ELECTRONICS, computer, and sensor technologies open new opportunities in the development of chemical instrumentation, new concepts in polymer analysis provide a powerful means for expanding analytical resources. This interplay results in the development of more sensitive and powerful techniques and has created a new generation of hyphenated and multidimensional analytical approaches. The hyphenated methods in polymer analysis were documented in *Hyphenated Techniques in Polymer Characterization: Thermal and Other Instrumental Methods,* ACS Symposium Series 581, and this volume focuses on multidimensional spectroscopic approaches in polymer analysis. Multidimensional analysis has been known for a long time, yet it has revived in recent years.

When two or more analytical techniques are tied together, which is commonly referred to as a hyphenated approach, a new level of understanding is achieved, resulting in a synergistic outcome of an experiment. In contrast, if one considers a simple experiment in which spectroscopic analysis is performed as a function of time, concentration, or other additive properties, the output will be multidimensional, and a number of independent variables will determine its dimensions. If such an experiment is conducted by varying spatial coordinates, frequency, or other domains, the situation changes. Such multidimensional experiments will provide an additional wealth of information, further advancing our understanding of structure–property relationships in polymeric materials. In this context, evolution of such spectroscopic probes like Fourier transform IR and Raman, NMR, and fluorescence spectroscopies, with the focus on their multidimensional character and continuously increasing sensitivity, appears to be inevitable.

This volume covers several aspects on multidimensional spectroscopic analysis and focuses on Fourier transform IR, NMR, and fluorescence spectroscopies in the analysis of polymers and multidisciplinary approaches utilized in the analysis of polymers in various environments. The book is divided into four sections describing the most recent developments in each field. This choice was dictated by the importance of these spectroscopic, molecular-level probes and their capabilities to enhance our understanding of structure–property relationships in polymers.

Acknowledgments

We thank all speakers of the International Symposium on Polymer Spectroscopy for their participation, and we thank the contributing authors to this volume for their effectiveness and timely response to the deadlines. A special note of appreciation goes to P. R. Griffiths of the University of Wyoming, Klaus Schmidt-Rohn of the University of Massachusetts, and Ian Sautor of the University of Lancaster, United Kingdom, for contributing introductory remarks for each book section.

The ACS Division of Polymeric Materials: Science and Engineering, Inc., and the Petroleum Research Fund of the American Chemical Society are gratefully acknowledged for the financial support of the symposium.

MAREK W. URBAN
Department of Polymers and Coatings
North Dakota State University
Fargo, ND 58105

THEODORE PROVDER
The Glidden Company
Member of ICI Paints
16651 Sprague Road
Strongsville, OH 44136

January 16, 1995

FOURIER TRANSFORM IR AND RAMAN SPECTROSCOPY

Chapter 1

Fourier Transform IR and Raman Spectroscopy of Polymers

Section Overview

Peter R. Griffiths[1] and Marek W. Urban[2,3]

[1]Department of Chemistry, University of Idaho, Moscow, ID 83843
[2]Department of Polymers and Coatings, North Dakota State University, Fargo, ND 58105

When atoms become attached to each other through a formation of chemical bonds, a molecule or macromolecule is formed. Because all atoms in a molecule possess kinetic energy, they vibrate. Such vibrations can be deconvoluted to so-called normal vibrational modes, and classified into a few classes: some modes may be observed in the Raman spectrum, some in the infrared, and some may or may not be seen in either spectrum. When the molecule possesses a high degree of symmetry, there is a rule of mutual exclusion which states that no vibrational mode may be observed in both the infrared and Raman spectra. This high symmetry is defined by a center of inversion operation. As the symmetry is reduced, and the molecule no longer contains a center of inversion, some vibrational modes may be seen in both the infrared and in the Raman spectra. However, these modes will often have quite different intensity in the two spectra. The quantum mechanical selection rules state that, observation of a vibrational mode in the infrared spectrum requires a change in dipole moment during the vibration. The observation of a vibrational mode in the Raman spectrum requires a change in the electron polarizability resulting from the movement of atoms. Thus, in order for a given vibrational mode to be infrared and Raman active, respectively, the following integrals must be not equal to zero.

The vibration is infrared active if

$$[\mu]_{v',v''} \neq \int \Phi_{v'}(Q_a) \, \mu \, \Phi_{v''}(Q_a) \, dQ_a \qquad (1)$$

Here, $[\mu]_{v',v''}$ is the dipole moment in the electronic ground state, Φ is the vibrational eigenfunction and v' and v'' are the vibrational quantum numbers before and after transition, respectively, and Q_a is the normal coordinate of the vibration.

The vibration is Raman active if

$$[\alpha]_{v',v''} \neq \int \Phi_{v'}(Q_a) \, \alpha \, \Phi_{v''}(Q_a) \, dQ_a \qquad (2)$$

[3]Corresponding author

0097–6156/95/0598–0002$12.00/0

Here, $[\alpha]$ is the polarizability tensor of the vibration, and the remaining parameters are the same as for the infrared activity in equation 1. The apparent differences in the principles governing both effects have led to the development of two physically distinct, yet complimentary, experimental approaches to obtain infrared and Raman spectra. As indicated in equation 2, the detection of Raman scattering involves a completely different assemblage of principles. When monochromatic radiation at frequency v_o strikes a transparent sample, the light is scattered. While most of the scattered light consists of radiation at the frequency of the incident light referred to as the Raleigh scattering, typically 1 out of 10^6 photons are scattered inelastically. This portion of the scattering is referred to as Raman scattering. This inelastically scattered fraction of light, composed of new modified frequencies $(v_o + .v_k)$, is referred to anti-Stokes scattering, and $(v_o - v_k)$ is the Stokes scattering component. Figure 1 illustrates a schematic diagram of the scattering processes leading to IR and Raman spectra. This energy diagram shows that the anti-Stokes scattering requires that the molecules start in an excited vibrational state. Because the easiest way to populate these excited vibrational states is by a thermal excitation, the anti-Stokes intensities will be very temperature dependent and typically quite weak at room temperature. Therefore, Stokes scattering is the most common way to record Raman spectra. Figure 1 also illustrates that the absorption process observed in IR may, and under certain selection rules, correspond to the vibrational energy levels depicted for the Stokes Raman scattering process.

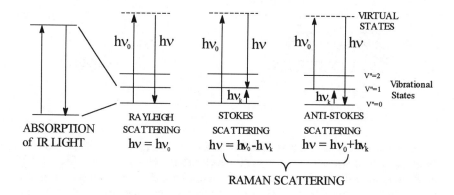

Figure 1. Schematic representation of IR absorption, Rayleight, Stokes, and anti-Stokes scattering processes.

Infrared and Raman spectroscopy have gone through numerous stages of development. At the early days, dispersive instruments dominated the field. When Fourier transform infrared spectrometers utilizing interferometric detection were introduced, numerous developments of sensitive techniques resulted. The sensitivity enhancements of attenuated total reflectance (ATR), reflection-absorption (R-A), diffuse reflectance (DRIFT), photoacoustic (PA), emission, or surface electromagnetic wave (SEW) spectroscopies, and further developments of other experimental approaches were primarily attributed to a higher energy throughput of interferometric instruments. A schematic diagram of selected techniques, along with a brief description is given in Figure 2.

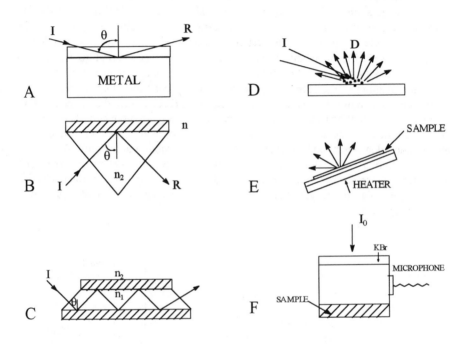

Figure 2. Commonly used surface-sensitive infrared techniques: A - Single reflection-absorption (R-A) setup; incident light (I) penetrates the sample and is reflected (R) by the metal mirror (θ should be between 75 and 89.5°); B - single internal reflection; incident light (I) passes through the internal reflection element and is totally reflected (R) at $\theta > \theta_c$ (n_1 and n_2 are the refractive indices of the sample and the internal element, respectively); C - Multiple-reflection setup in attenuated total reflection (ATR) mode; D - Diffuse reflectance (DRIFT) setup; the incident light (I) is diffusively scattered in all directions (D), collected by hemispherical mirrors, and re-directed to the detector. E - Emission setup; the source of IR light is replaced by a heated sample and emitted light is analyzed by the infrared detector; F - Photoacoustic (PA) setup; the incident modulated light with intensity I_0 impinges upon the sample surface; the light is absorbed, and as a result of re-absorption, heat is released to the surface which, in turn, generates periodic acoustic waves.

With this background in mind, and considering recent instrumental advances in infrared and Raman spectroscopy, let us briefly outline the current developments in vibrational spectroscopy utilized in the analysis of polymers.

Infrared spectroscopy has been used for the study of polymers for about half a century. In about 1980, the Perkin-Elmer Corp., at that time the dominant manufacturer of infrared grating spectrometers, estimated that about 80% of their sales were for polymer-related applications. The advances in Fourier transform infrared (FT-IR) spectroscopy that have been made since that time have given polymer scientists significantly greater sensitivity and/or reduced measurement time. Furthermore, the development of many different sampling accessories for these instruments, such as those for attenuated total reflectance (ATR), specular reflectance (SR), reflection-absorption (R-A) at near-normal and near grazing incidence, diffuse reflectance (DRIFT), and photoacoustic (PA) spectroscopy, has increased the flexibility of FT-IR spectroscopy so that polymers of essentially any morphology can be investigated.

Possibly the most important advance in sampling for infrared spectroscopy over the last 15 years involves the development of microscopes for the mid-infrared region. Not only can transmission spectra of samples as small as 10 μm in diameter be measured, but other techniques such as ATR, SR, R-A, DRIFT, and even PA spectroscopy have been applied to microscopically small domains and inclusions in polymers. In the case of Raman microscopy, an approximate 1 μm spatial resolution can be attained.

Using conventional rapid-scanning FT-IR spectrometers, static and dynamic molecular level information, with time resolution as short as ~50 ms, can be obtained for polymers that are being subjected to a gradually increasing strain. Recently, step-scanning spectrometers have been designed that allow the effect of a low-amplitude modulated strain on the spectra of polymers to be studied. These effects can be caused by reversible changes in the crystallinity or orientation, or even bond angles in the polymer. The phase lag of the infrared signal with respect to the applied strain can be considered to be analogous to the phase angle measured from a stress-strain curve measured by dynamic mechanical analysis (DMA). Since all infrared signals result from vibrations originating from individual functional groups in the polymer, the data obtained in such an experiment may be considered as a type of molecular-level DMA.

The absorptivities of fundamental modes that give rise to bands in the mid-infrared spectrum are usually too high to permit samples much thicker than 100 μm to be investigated. In the situations when thin samples cannot be prepared, the near-infrared (NIR) spectral region, where overtone and combination bands due to C-H, O-H, and N-H groups absorb, becomes of great importance. NIR spectroscopy has, therefore, become tremendously useful for measuring the spectra of thick samples, especially in the field of process analysis where path-lengths less than 1 mm are often difficult to achieve. Diffuse reflectance (DRIFT) is another technique where NIR spectroscopy has proved to be beneficial. Because of the high absorptivities of fundamental vibrational transitions allowed by the selection rules, strong bands in mid-infrared DRIFT spectra of neat specimens either exhibit total absorption or are distorted by the contribution of specular reflection. Samples for DRIFT spectroscopy in mid-infrared are usually in KBr powder prior to the measurement. DRIFT spectra of neat samples can be easily measured in the near-infrared spectral region, since saturation effects are never seen for the weak overtone and combination bands in the NIR.

Raman spectroscopy has also advanced dramatically, with many important instrumental developments made in this past decade. In 1980, measurement of the Raman spectra of most industrial polymers was usually precluded because of the presence of traces of strongly fluorescent impurities. The situation was so bad that even a simple Raman spectrum of transparent poly(vinylidine fluoride) was impossible to obtain. In the mid 1980's, Hirschfeld and Chase demonstrated that Raman spectra could be measured using Nd:YAG laser at 1064 nm to eliminate, or at least significantly reduce, fluorescence, provided that a Fourier transform (FT) spectrometer was used for the measurement. The time interval between the first demonstration of FT-Raman spectroscopy, the commercial introduction of this technique, and its general acceptance was less then 5 years--an amazingly small period. While today acceptable FT-Raman spectra of most polymers can be measured in 1 or 2 minutes, further advances are being made in Raman microscopy which allows mapping polymer surfaces.

The bad news for most polymer scientists who have already purchased an FT-Raman spectrometer is that these instruments are starting to be superseded by polychromators equipped with charge-coupled device (CCD) array detectors and NIR diode laser excitation. These instruments allow spectra to be measured in a few seconds. In this context, the area of continuous interest is resonance Raman and surface enhanced Raman spectroscopy. Resonance Raman (RR) scattering occurs when the exciting line is chosen so that its energy matches the electronic excited state, whereas the surface enhanced Raman scattering (SERS) is induced using a conventional Raman setup when molecules are deposited on rough ·metallic surfaces. While RR spectroscopy is invariably more useful with ultraviolet or visible laser excitation, SERS is equally applicable in any electromagnetic region of radiation. Fiber-optic probes can be interfaced to CCD-Raman spectrometers with greater ease than to FT-Raman instruments. Finally CCD-Raman spectrometers are readily converted for Raman imaging or confocal microscopy with a spatial resolution of 1μm. They should, therefore, prove to be highly beneficial in several areas of polymer science.

Most of the advances described above have been due to advances in optical components that have been made over the past 20 years. Even more spectacular advances in computing technology and data processing algorithms have been made over the same period and these have impacted vibrational spectroscopy as much as, if not more than, the optical advances. Rapid digital data acquisition is required for FT-IR, FT-Raman or CCD-Raman spectroscopy. The raw data obtained from these instruments must always be manipulated before a recognizable spectrum can be displayed. Indeed, the measurement of a spectrum is usually just the start of most contemporary spectroscopic experiments. Both quantitative and qualitative information is available from mid- and near-infrared spectra and Raman spectra. Multicomponent analysis can be achieved for samples containing up to ten components through a variety of multivariate statistical algorithms, such as classical least-squares (CLS), partial least-squares, cross correlation, factor analysis, non-linear methods, and maximum likelihood entropy (MLE) methods. The application of artificial neural networks for spectral interpretation is also starting to assume an important role in infrared and Raman spectroscopy and appears to have the capability of complementing linear multivariate techniques such as CLS. Many other types of data processing can be done to achieve specific molecular information from polymer spectra. For example, the average orientation of individual functional groups relative to the plane of polarization of the radiation can be calculated from the infrared dichroism and trichroism spectra of oriented polymer films.

Several of these advances are described or alluded to in the section of this volume on the vibrational spectroscopy of polymers. It is hoped that these chapters give the reader a taste of the type of measurements that can now be carried out and the molecular information that can be deduced from the measured spectra.

Selected Monographs:

1. J.L.Koenig, *Spectroscopy of Polymers*, American Chemical Society, Washington, DC, 1992.
2. M.W. Urban, "Vibrational Spectroscopy of Molecules and Macromolecules on Surfaces," Wiley & Sons, New York, 1993.
3. J.G.Grasselli and B.J.Bulkin, *Analytical Raman Spectroscopy*, John Wiley & Sons, New York, 1991.
4. M.W.Urban, *Structure-Property Relations in Polymers; Spectroscopy and Performance*, Adv. Chem. Series, 236, American Chemical Society, Washington, DC, 1993.
5. C.D.Craver and T.Provder, *Polymer Characterization; Physical Property, Spectroscopic, and Chromatographic Methods*, Adv. Chem.Series, 227, American Chemical Society, Washington, DC, 1993.
6. D.B.Chase and J.F.Rabolt, *Fourier Transform Spectroscopy from Concept to Experiment*, Academic Press, Inc. Cambridge, MA, 1994.
7. P.B.Coleman, *Practical Sampling Techniques for Infrared Analysis*, CRC Press Inc., Boca Raton, Florida, 1993.
8. P.R. Griffiths, *Adv. Infrared and Raman Spectroscopy*, Vol.9., R.J.H. Clark, R.E. Hester, Eds., Heyden London, 1981.
9. J.R.Ferraro and K. Nakamoto, *Introductory Raman Spectroscopy*, Academic Press, Inc., San Diego, 1994.

RECEIVED March 16, 1995

Chapter 2

Determining the Structure of Polyimide–Metal Interfaces Using Fourier Transform IR and Raman Spectroscopy

F. J. Boerio, J. T. Young, and W. W. Zhao

Department of Materials Science and Engineering, University of Cincinnati, Cincinnati, OH 45221–0012

Reflection-absorption infrared spectroscopy (RAIR) and surface-enhanced Raman scattering (SERS) were used to determine the structure of interphases formed by polyamic acids which were spin-coated onto silver and gold substrates and then heated at elevated temperatures to complete the imidization reaction. Polyimides having relatively rigid backbones, such as those prepared from pyromellitic dianhydride (PMDA) and oxydianiline (ODA), formed films in which the long axes of the molecules were oriented mostly parallel to the substrate surface. Polyimides having relatively flexible backbones, such as those prepared from 4,4'-hexafluoroisopropylidene dianhydride (6FDA) and ODA, mostly formed films in which there was little preferred orientation of the polymer molecules. However, for 6FDA/ODA films having a thickness of only a few tens of angstroms, a preferred orientation was observed in which the C_6H_4 rings of the ODA moieties were oriented parallel to the substrate surface. Regardless of the flexibility of the backbone, the polymers interacted with silver substrates by formation of carboxylate species but there was little evidence of carboxylate formation on gold substrates. Thiol-terminated oligomers such as 4-mercaptophenylphthalimide (4-MPP) formed monolayers when adsorbed onto gold from dilute solutions. However, the orientation of the molecular axes was much different than for PMDA/ODA. Whereas PMDA/ODA was oriented with the long axes of the molecules almost parallel to the gold surface, 4-MPP was oriented with the two-fold symmetry axes of the molecules almost perpendicular to the surface.

The purpose of this paper is to describe the use of reflection-absorption infrared spectroscopy (RAIR) and surface-enhanced Raman spectroscopy (SERS) to

0097–6156/95/0598–0008$15.25/0
© 1995 American Chemical Society

determine the molecular structure and orientation in thin films of polymers and model compounds formed on metal substrates. RAIR has been extensively discussed in the literature (*1-3*). In order to obtain an infrared spectrum of a thin film on a reflecting metal substrate, radiation polarized parallel to the plane of incidence and large, grazing angles of incidence must be used. Under most conditions, the incident and reflected waves combine to form a standing wave that has a node at the metal surface. However, for parallel polarized radiation and grazing angles of incidence, the incident and reflected waves combine to form a wave that has significant electric field amplitude at the metal surface.

The dependence of RAIR absorbance on the angle of incidence and polarization is simulated in Figure 1. In these calculations, it was assumed that the refractive indices of the air, film, and substrate were 1.0, 1.3 - 0.1i, and 3.0 - 30.0i, respectively (*2*). The ratio of film thickness d1 to the wavelength of light was 0.0003. From the figure, it can be observed that absorbance is negligible for radiation polarized perpendicular to the plane of incidence. Absorbance is also small for radiation polarized parallel to the plane of incidence when the angle of incidence is small. However, for parallel polarized radiation, absorbance increases rapidly for angles of incidence greater than about 45° and reaches a maximum at approximately 88°. The absorbance in a properly designed experiment is such that spectra of monolayers of polymers are obtained.

RAIR spectroscopy has several important characteristics. One is that the resultant electric field vector is perpendicular to the metal surface. Therefore, if molecules are adsorbed onto the substrate with a preferred orientation, vibrational modes having transition moments perpendicular to the surface will appear with greater intensity than modes having transition moments parallel to the surface. As a result, RAIR is a powerful technique for determining the orientation of adsorbed molecular species.

Another characteristic of RAIR is that spectra will depend on both the real and complex parts of the refractive index of the absorbing species whereas spectra obtained in transmission depend almost entirely on the complex part. As a result, certain types of "distortions" may appear in RAIR spectra when compared to transmission spectra of the same compound. An example is shown in Figure 2 where optical constants given by Allara et. al. (*3*) for the region of the carbonyl stretching mode of polymethylmethacrylate (PMMA) adsorbed onto a gold substrate were used to simulate the RAIR spectrum corresponding to a transmission spectrum consisting of a single band at 1700 cm^{-1}. It can be seen that for a film having a thickness of 200 Å, the band in RAIR is shifted to higher frequencies by several wavenumbers. When the film thickness increased to 5,500 Å, the band shift increased to about 10 cm^{-1}. However, when the film thickness increased to 14,500 Å, a second component appeared at about 1696 cm^{-1}.

One final comment regarding RAIR spectra is that the absorbance increases almost linearly with thickness for films having thickness less than about 4,000 Å. However, for thicker films the absorbance is a non-linear function of thickness due to interference effects.

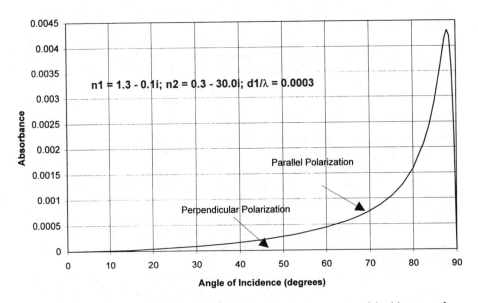

Figure 1. The dependence of RAIR band size on the angle of incidence and polarization. The refractive indices of the ambient, film, and substrate were 1.0, 1.3 - 0.1i, and 3.0 - 30.0i, respectively (2). The ratio of film thickness to the wavelength of light was 0.0003.

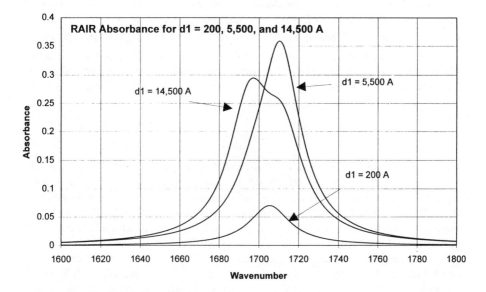

Figure 2. RAIR spectra corresponding to a transmission spectrum consisting of a single band at 1700 cm^{-1}. In this simulation, it was assumed that k_{max} = 0.38, γ = 25 cm^{-1}, ν_0 = 1700 cm^{-1}, n_∞ = 1.38, φ_0 = 82°, and n_2 = 9.5 - 30i. k_{max} and γ were the maximum and full width at half height of the absorption band in k-space while ν_0 was the position of the band. n_∞ was the refractive index of the film away from the absorption band, φ_0 was the angle of incidence, and n_2 was the refractive index of the substrate.

SERS has also been discussed extensively in the literature (*4,5*). The Raman scattering by a monomolecular layer of molecules is usually too weak to be detected by conventional techniques. However, when the monolayer is placed on the roughened surface of certain metals, such as copper, gold, or silver, the intensity of the Raman scattering is enhanced by several orders of magnitude. Two mechanisms seem to be responsible for the enhancement. The first is electromagnetic in origin and is associated with the large electric fields that can be obtained when the roughened surface of a metal is illuminated with electromagnetic radiation. The other is chemical in nature and is associated with a resonance effect involving charge transfer states that are formed when an organic compound is chemisorbed onto the surface of a metal. Electromagnetic enhancement can be as great as 10^4 while chemical enhancement can be as great as 10^2. Both mechanisms are very short-range. The chemical enhancement is only obtained for molecules that are chemisorbed and is therefore restricted to molecules in the "first layer." Electromagnetic enhancement is usually considered to have a form such as

$$\text{Enhancement} \sim (r/r + d)^{12}$$

where r is the radius of an asperity on the metal surface and d is the distance of the scattering molecule from the surface of the asperity (*5*). As a result, molecules adjacent to the surface experience a large enhancement but there is little enhancement for molecules only a few molecular layers away from the surface. Therefore, SERS is *surface-selective* and can be used for non-destructive characterization of polymer/metal interfaces.

Polyimides are widely used as dielectrics in multilevel interconnects. As a result, there has been a great deal of interest in determining the structure of polyimide films on metal substrates. Interest in determining the effect of the substrates on the molecular structure of polyimides and the polyamic acids from which they are usually derived has been especially strong. Much of the work has involved the use of vacuum techniques such as X-ray photoelectron spectroscopy (XPS) to determine the structure of bulk polyimides and their interfaces with metal substrates. However, several applications of RAIR and SERS to the study of polyamic acid and polyimide films on metal substrates have been reported. Burrell et. al. (*6*) utilized RAIR and X-ray photoelectron spectroscopy (XPS) to investigate relatively thick films (1,000-5,000 Å) of PMDA/ODA polyimide prepared by spin-coating polyamic acid onto copper and aluminum substrates and then heating the films at 200°C in air for 30 minutes to promote imidization. Films of polyamic acid on Al substrates were characterized by bands near 1730, 1663, 1540, and 1250 cm^{-1}. These bands disappeared during heating at 200°C and new bands characteristic of imide groups appeared near 1737 and 1780 cm^{-1}, indicating that imidization proceeded readily in films on aluminum. In contrast, no bands related to carboxylic acids were observed in spectra of polyamic acid films on copper substrates. However, bands related to copper/carboxylate complexes which prevented polyamic acid films from curing completely during heat treatment were observed.

Linde (7) used RAIR to determine the structure of thin films (~ 40 Å) formed by spin-coating the polyamic acid of PMDA/ODA onto reactive metals such as copper and silver and unreactive metals including gold, aluminum, and titanium. In the case of silver substrates, polyamic acid was adsorbed mostly through PMDA moieties which were oriented perpendicular to the surface and in which the acid groups were meta to each other. One of the acid groups reacted with the substrate to form a carboxylate salt. The carbonyl group in the other acid group was oriented perpendicular to the substrate. Heating these films to 300°C in nitrogen for 8 minutes destroyed the polymer.

In the case of gold substrates, the band near 1500 cm^{-1}, which was associated with the p-disubstituted benzene rings in the ODA moieties, was very weak. However, the band near 1720 cm^{-1}, which was characteristic of the acid groups in the PMDA moieties, was strong. It was concluded that polyamic acid was adsorbed through a non-polar interaction which tended to align the benzene rings in the ODA moieties parallel to the surface while the rings associated with the PMDA moieties were mostly aligned perpendicular to the surface. This alignment of functional groups was maintained after thermal imidization of the polyamic acid films (7).

Grunze and Lamb (8, 9) used RAIR and XPS to investigate PMDA, ODA, and mixtures of PMDA and ODA vapor-deposited onto polycrystalline silver surfaces. PMDA and ODA were both chemisorbed on the silver surface through oxygen in the PMDA and ODA fragments. Co-deposition of PMDA and ODA followed by heat treatment led to formation of thermally stable polyimide films. Adhesion of polyimide films to silver involved fragments of PMDA and ODA which were initially chemisorbed on the surface.

Pryde (10) used RAIR to determine the structure of films formed by spin-coating solutions of the polyamic acid of benzophenone tetracarboxylic dianhydride (BTDA) and mixtures of m-phenylene diamine (m-PDA) and ODA onto gold and aluminum substrates and then thermally imidizing the films. It was found that the absorbance of the imide out-of-plane bending mode band near 720 cm^{-1}, normalized against the absorbance of the aromatic band near 1500 cm^{-1}, decreased as the thickness of films on aluminum decreased, indicating a strong tendency of the imide groups to orient. This tendency was especially strong when the imidization reaction was carried out at higher temperatures. Pryde reported little tendency for functional groups in the BTDA/PDA/ODA films to orient on gold substrates.

Tsai and co-workers (11) used RAIR to characterize interfaces between polyamic acids and polyimides of PMDA and 2,2 - bis[4-(4-aminophenoxy)phenyl]-hexafluoro-propane (4-BDAF) and silver substrates. They found that the molecules in the bulk of the films were unoriented and well imidized. However, molecules near the polymer/substrate interface tended to orient with the two-fold symmetry axes of the p-disubstituted benzene rings parallel to the substrate surface. Strong evidence for formation of carboxylate species at the interface was also observed.

Very little information has been obtained using Raman spectroscopy to investigate interfaces between polyimides and metal substrates. Perry and Campion (12) used unenhanced Raman spectroscopy to investigate the adsorption of PMDA and ODA onto Ag(110) at 140°K in ultra high vacuum. They found that ODA was physisorbed onto the cold silver substrate but PMDA was dissociatively chemisorbed. Bands which were not observed in the Raman spectra of bulk PMDA appeared near 1414 and 1504 cm^{-1} in the Raman spectra of PMDA adsorbed onto cold silver and were assigned to a bidentate surface carboxylate. Heating a 30 Å thick film of co-dosed PMDA and ODA to 473°K resulted in formation of bands near 1391 and 1788 cm^{-1} that were characteristic of imide groups, indicating polymerization of the monomers.

Young et. al. (13) used SERS to determine the structure of interfaces between the polyamic acid and polyimide of PMDA/ODA and silver substrates. They found that the SERS spectrum of the polyamic acid spin-coated onto silver substrates was similar to the normal Raman spectrum of the polyamic acid and was characterized by a strong band near 1340 cm^{-1} and a weak band near 1574 cm^{-1} that were related to the amide groups and by a strong band near 1623 cm^{-1} that was related to the vibrational mode ν(8a) in the substituted benzene rings. However, a unique band was observed near 1412 cm^{-1} in the SERS spectra of the polyamic acid and assigned to carboxylate species formed by interaction of acid groups in the polyamic acid with silver ions in the substrate. After heat treatment of the films at 200°C for 10 min and immersion in a 1:1 mixture of acetic anhydride and pyridine for 2 hrs, there was little change in the SERS spectra and it was concluded that imidization in the interfacial region was inhibited by formation of the carboxylate species.

Young and co-workers also used SERS to determine the structure of interfaces between the polyamic acid of PMDA/ODA and gold substrates (14). They found little evidence for interaction of the polymers with the substrate and observed that imidization proceeded easily during heat treatment of the films at 200°C or immersion in a 1:1 mixture of acetic anhydride and pyridine.

Experimental

Reflection-absorption infrared (RAIR) spectra of polyamic acid and polyimide films on silver substrates were obtained as follows. The polyamic acids of pyromellitic dianhydride (PMDA) and oxydianiline (ODA) and 4,4'-(hexafluoroisopropylidene)-di-phthalic anhydride (6FDA) and ODA (DuPont Corporation) were spin-coated onto thick silver and gold films which were evaporated onto glass slides. Polyamic acid films deposited on silver and gold substrates were dried at 100°C for 15 minutes to remove the solvent. The films were then thermally imidized by heating at 200°C for an additional 15 minutes.

RAIR spectra of the polyamic acid films before and after imidization were obtained using a Perkin-Elmer Model 1800 Fourier-transform infrared (FTIR) spectrophotometer and external reflection accessories provided by Harrick Scientific Co. In most cases, the RAIR spectra were obtained by averaging 125 scans obtained at an angle of incidence equal to 78° and a resolution of 4 cm^{-1}. However, spectra of the thinnest films were obtained by averaging as many as 750 scans. All RAIR spectra are actually difference spectra obtained by subtracting the spectrum of a metal substrate with no film from the spectrum of a film-covered substrate.

The thickness of the polyamic acid films was measured using a Rudolph Research Model 436 ellipsometer to examine the silver and gold substrates before deposition of the polyamic acid films and after the films were deposited and heated at 100°C for 15 minutes to remove the solvent. The thickness of the films was not expected to differ significantly after imidization.

Transmission infrared spectra of the polyamic acids and corresponding polyimides were obtained using the same spectrophotometer. Samples were prepared by placing a drop of polyamic acid solution onto a potassium bromide disc and then heating the disc at 100°C for 15 minutes to remove the solvent. The resulting polyamic acid films were then thermally imidized by heating at 200°C for an additional 15 minutes.

Samples for SERS investigations were prepared as described below. Glass slides were immersed in 0.1N NaOH and in 0.1N HCl aqueous solutions for one hour. The slides were then rinsed in distilled-deionized water, blown dry with nitrogen, cleaned ultrasonically in absolute ethanol several times, and blown dry with nitrogen again. Clean glass slides were placed in a vacuum chamber which was purged with nitrogen and pumped down to 10^{-6} Torr using sorption, sublimation, and ion pumps. Silver wire was wrapped around a tungsten filament which was resistively heated to evaporate island films onto the glass substrates at a rate of about 1 Å/sec. A quartz crystal oscillator thickness monitor was used to control the thickness of the island films at about 40 Å. Gold substrates were prepared in a similar manner except that the deposition rate was 0.2 Å/sec and the final thickness of the films was about 65 Å.

Polyamic acid films were spin-coated onto the substrates from dilute solutions and then dried by heating at 100°C for 15 minutes. Initially we were concerned that the morphology of the island films would be unstable at the relatively high temperatures (200°C) required to thermally imidize polyamic acids to polyimides and that a significant loss of enhancement would result. Therefore, a combination of thermal and chemical imidization processes was used. Polyamic acid films were thermally imidized by heating at 200°C for 10 minutes and then chemically imidized by immersion in a 1:1 mixture of acetic anhydride and pyridine for two hours. Finally, the specimens were dried by heating at 90°C for an additional 20 minutes.

Subsequently, we found that the morphology of the silver island films was sufficiently stable to enable thermal imidization to be carried out in a nitrogen atmosphere at 200°C. No differences in SERS spectra of polyimides obtained by chemical and thermal imidization or by thermal imidization only was observed.

The thickness of the polyamic acid films deposited on silver and gold island films was again determined by ellipsometry. Thick (several hundred angstroms) films of silver were evaporated onto glass slides. Films of polyamic acid were spun onto silver or gold mirrors from the same solutions and at the same speeds as used to prepare the SERS samples. The thickness of the films was measured by using the ellipsometer to examine the silver substrates before deposition of the polyamic acid films and after the films were deposited and heated at 100°C for 15 minutes to remove the solvent.

SERS spectra were obtained using a spectrometer equipped with a Spex 1401 double monochromator, a Hamamatsu R943-02 photomultiplier tube, Stanford Research Model 400 gated photon counter interfaced to a Hewlett-Packard Vectra computer, and a Spectra-Physics Model 165 argon-ion laser. The slit settings of the monochromator provided a spectral resolution of 10 cm^{-1} for the SERS spectra. The green line of the laser (5145 Å in wavelength) was incident on the sample at an angle of about 65° relative to the normal to the sample surface for SERS experiments and was s-polarized. Scattered light was collected using an f/0.95 collection lens and focused onto the entrance slits of the monochromator. Spectra were obtained using a scan speed of 65 cm^{-1} per minute and time constant of either 2 or 10 seconds. Plasma lines were removed from the spectra by the placement of a narrow-bandpass filter between the laser and sample. In many of the spectra, a considerable background, which was digitally subtracted, was observed.

Normal Raman spectra were obtained from a small amount of powdered sample supported in a glass capillary tube using the instrument described above. Instrumental parameters were the same as used for the SERS spectra except that the slits were set for a spectral width of 5 cm^{-1}.

4-mercaptophenylphthalimide (4-MPP) was synthesized as follows (15). Equal molar amounts of phthalic anhydride (Aldrich Chemical Co.) and 4-aminothiophenol (Lancaster Synthesis) were dissolved in N,N-dimethylacetamide (DMAc) and mixed together. The mixture was refluxed under nitrogen overnight at about 160°C. Upon completion, the mixture was cooled to room temperature and the pale yellow precipitate filtered and rinsed with ether. The precipitate was dried under vacuum at 75°C for several hours to remove residual solvent.

Gold substrates were prepared for SERS investigations as described above. Monolayers of 4-MPP were prepared by immersing a gold substrate into a 1x10^{-3} M solution of 4-MPP in chloroform for 12 hours and then rinsing the substrates thoroughly with chloroform. Similar procedures were used to prepare samples for RAIR and ellipsometry except that thick gold films (~1000 Å) were used as substrates instead of gold island films. SERS, RAIR, and ellipsometry were all carried out as described above except that the 6471 Å line of a Lexel 3000 krypton-ion laser was used to excite the SERS spectra.

Results and Discussion

PMDA/ODA on Silver. Transmission infrared spectra of the polyamic acid and polyimide of PMDA/ODA are shown in Figure 3. The strong bands near 1650, 1540, and 1410 cm^{-1} in the spectrum of the polyamic acid were assigned to modes of the amide groups. The band near 1510 cm^{-1} was assigned to a stretching mode of the C$_6$H$_4$ rings. After imidization, the bands near 1650, 1540, and 1410 cm^{-1} disappeared and new bands which were characteristic of the imide groups appeared near 1720, 1385, and 720 cm^{-1}. The band near 1720 cm^{-1} was assigned to the asymmetric C=O stretching mode while that near 1385 cm^{-1} was assigned to a mode

A

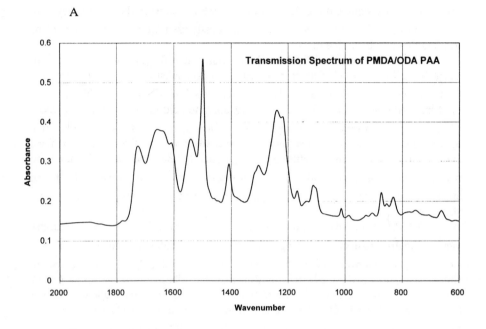

B

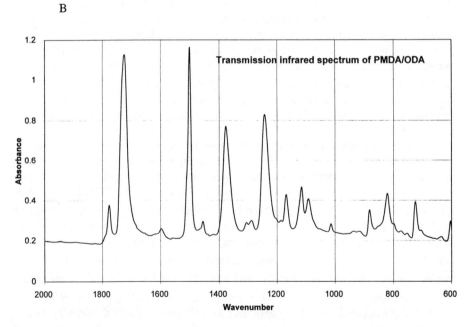

Figure 3. Transmission infrared spectra of the (A) - polyamic acid and (B) - polyimide of PMDA/ODA.

involving mostly symmetric stretching of the C-N bond between the ODA groups and the imide rings. The band near 720 cm^{-1} was mostly associated with an out-of-plane bending mode of the imide rings.

The RAIR spectra of PMDA/ODA polyamic acid films having thicknesses of 1200, 40, and 10 Å which were spin-coated onto silver substrates are shown in Figure 4. The intensity of the band near 1720 cm^{-1} decreased in intensity as the thickness of the films decreased. Since this band was assigned to acid groups and, perhaps, to a few imide groups formed during drying of the films at 100°C, it was concluded that acid groups in the thinner films interacted with silver ions in the substrate to form carboxylate species. This conclusion was supported by the observation that the band near 1410 cm^{-1} became broader as the film thickness decreased, probably due to the appearance of a band near 1400 cm^{-1} that was attributed to the symmetric carboxylate stretching mode.

RAIR spectra obtained after the same polyamic acid films on silver substrates were thermally imidized by heating at 200°C for 30 minutes are shown in Figure 5. Comparing the intensities of the bands near 720 cm^{-1} and 1720 cm^{-1} in the spectra shown in Figures 3A and 3B, it is evident that the orientation of the imide groups changed as the film thickness decreased and that the imide rings tended to orient more nearly perpendicular to the silver surface in the thinner films. It is also evident that the ratio of the intensities of the bands near 1510 and 1385 cm^{-1} relative to that of the band near 1720 cm^{-1} also decreased as the film thickness decreased. Since the transition moment for these bands is mostly parallel to the long axis of the molecules, it was concluded that the molecular axes tended to align more nearly parallel to the substrate surface as the film thickness decreased. Reference to Figure 3C indicates that broad, weak bands were observed near 1600 and 1400 cm^{-1} in spectra of the thinnest films. These bands were assigned to the stretching modes of carboxylate species, supporting the conclusion that the acid groups interacted with silver ions in the substrate.

The SERS spectrum of a film of polyamic acid spin-coated onto a silver island film from a 20% solution in NMP is shown in Figure 6. Results obtained from ellipsometry indicated that the film had a thickness of about 1,500 Å. Significant differences were observed between the normal Raman and SERS spectra of the polyamic acid (13). The strongest band in the SERS spectrum was near 1620 cm^{-1} and was assigned to a combination of the tangential ring stretching mode ν(8a) of the benzene rings and the asymmetric stretching mode of carboxylate groups (16-18). The broad band near 1345 cm^{-1} was attributed to the CN stretching mode of the amide groups. The medium intensity band near 1576 cm^{-1} was assigned to the CNH stretching mode. The sharp, medium intensity band near 1178 cm^{-1} was related to the CH in-plane bending mode ν(9a) of the C_6H_4 rings or to the C-X stretching mode ν(13) of the C_6H_2 ring. Broad bands observed near 1413 and 840 cm^{-1} in SERS spectra of the polyamic acid were assigned to symmetrical stretching and deformation modes of carboxylate groups (16-18). Bands near 1698, 940, and 761 cm^{-1} which

A

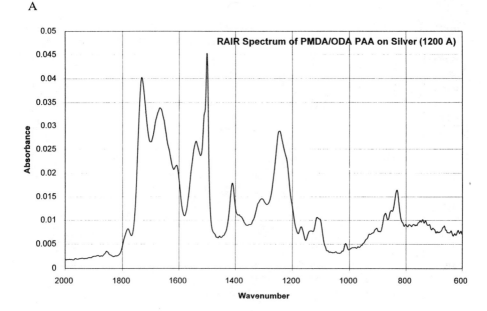

B

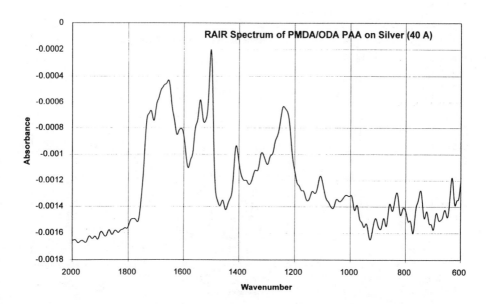

Figure 4. RAIR spectra of thin films of the polyamic acid of PMDA/ODA spin-coated onto silver substrates. The film thicknesses were (A) - 1200, (B) - 40, and (C) - 10 Å. *(Continued on next page.)*

C

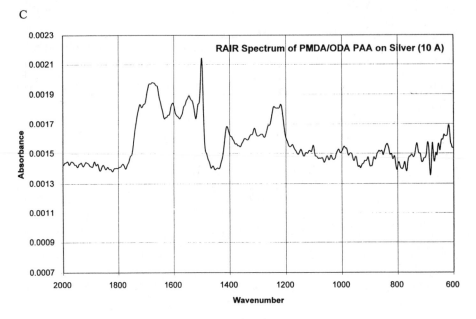

Figure 4. Continued.

A

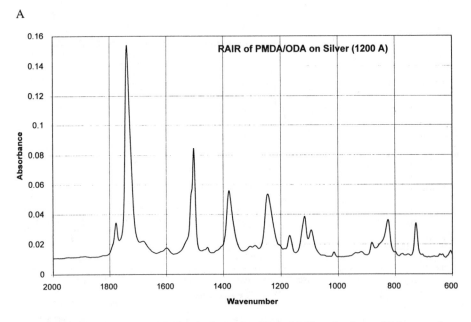

Figure 5. RAIR spectra obtained after polyamic acid films having a thickness of
(A) - 1200, (B) - 40, and (C) - 10 Å were spin-coated onto silver and then
imidized. (Adapted from ref. 13.)

B

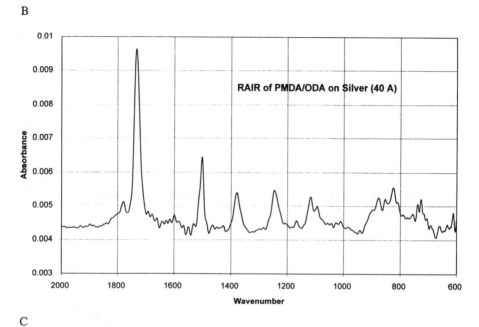

C

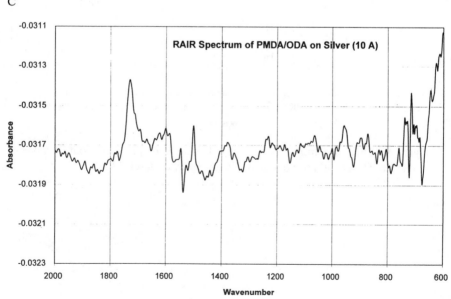

Figure 5. Continued.

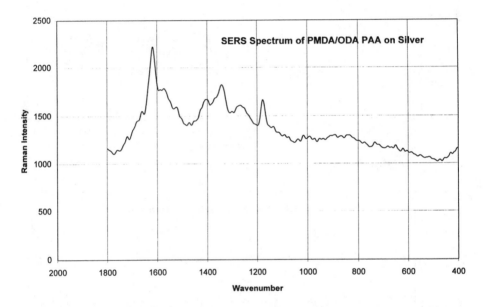

Figure 6. SERS spectrum of a film of polyamic acid spin-coated onto a silver island film from a 20% solution in NMP. (Adapted from ref. 13.)

were assigned to vibrations of the acid groups of the polyamic acid were not observed in the SERS spectra. The appearance of bands due to carboxylate salts in the SERS spectra and the disappearance of bands due to acid groups indicated that the polyamic acid was adsorbed onto silver films to form a carboxylate salt.

SERS spectra of the polyimide obtained by chemically curing films of polyamic acid adsorbed onto silver island films are shown in Figure 7. The polyimide films were obtained by spin-coating films of polyamic acid that were about 1,500, 500, and 120 Å in thickness from 20%, 5%, and 1% solutions in NMP onto silver island films, thermally curing the films at 100°C for 15 minutes and at 200°C in a vacuum oven for an additional 10 minutes, and finally chemically curing the films in a 1:1 mixture of acetic anhydride and pyridine for 2 hours.

There are two important aspects to the spectra shown in Figure 7. The first is that the band intensities in all three spectra were very similar even though the thickness of the polyimide films varied by approximately an order of magnitude. Since the SERS signal was independent of film thickness, it was characteristic of the interface, not the bulk films. The second is that the positions and relative intensities of the bands were more similar to those in SERS spectra of the polyamic acid than to those in normal Raman spectra of the polyimide. For example, the band near 1623 cm^{-1}, which had medium intensity in the normal Raman spectrum, was the strongest band in the SERS spectra of the polyimide and polyamic acid. The band near 1403 cm^{-1}, which was the strongest band in the normal Raman spectrum of the polyimide, was relatively weak in the SERS spectra of the polyimide and the polyamic acid. The band near 1345 cm^{-1} in the SERS spectra of the polyamic acid and polyimide was not observed in the normal Raman spectrum of the polyimide. The band near 1176 cm^{-1} was relatively strong and sharp in SERS spectra of the polyimide and polyamic acid but was very weak in the normal Raman spectrum of the polyimide. Considering the similarities between the SERS spectra of the polyimide and the polyamic acid, it was concluded that imidization of the polyamic acid on the silver surface was inhibited by formation of carboxylate salts with silver ions in the substrate (*13*). Results obtained from SERS were thus consistent with those obtained from RAIR.

PMDA/ODA on Gold. The RAIR spectra obtained from thin films of the polyamic acid of PMDA/ODA spin-coated onto gold substrates are shown in Figure 8. These films had thicknesses of approximately 260, 45, and 15 Å, respectively. Spectra obtained after the same films were thermally imidized are shown in Figure 9. Once again it was observed that the relative intensities of the bands near 1385 and 1720 cm^{-1} and of the bands near 720 and 1720 cm^{-1} decreased as the film thickness decreased, indicating that the molecules tended to align with their long axes parallel to the surface and that the imide rings tended to orient more nearly perpendicular to the surface as the film thickness decreased. However, as expected, spectra of films formed on gold showed little evidence of carboxylate formation which would indicate chemisorption.

A model was used to determine the orientation of the PMDA moieties on the gold surface from the intensities of the bands near 1720, 1385, and 720 cm^{-1} (see

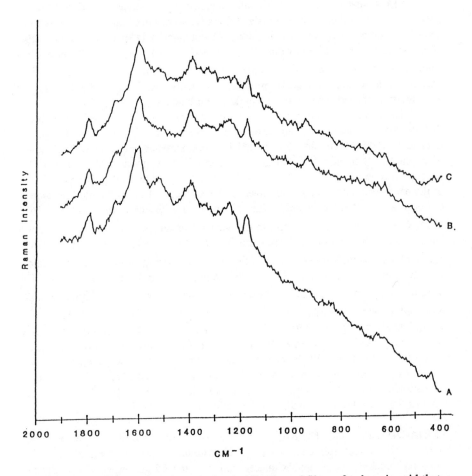

Figure 7. SERS spectra obtained after imidization of films of polyamic acid that were spin-coated onto silver island films from NMP solutions having concentrations of (A) - 20%, (B) - 5%, and (C) - 1%. The films had thicknesses of about 1,500, 500, and 120 Å, respectively. (Reprinted from ref. 13.)

A

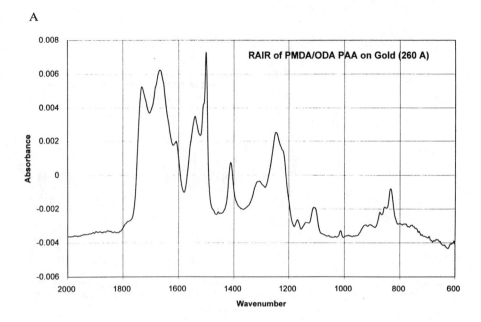

B

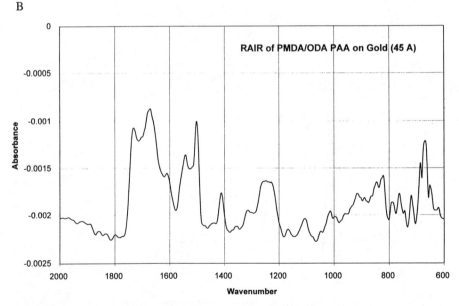

Figure 8.   RAIR spectra obtained from thin films of the polyamic acid of PMDA/ODA spin-coated onto gold substrates. The films had thicknesses of approximately (A) - 260, (B) - 45, and (C) - 15 Å, respectively. (Adapted from ref. 14). *Continued on next page.*

C

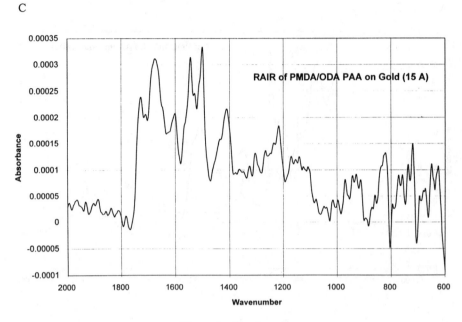

Figure 8. Continued.

A

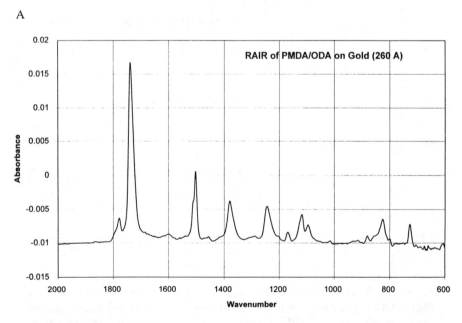

Figure 9. RAIR spectra obtained after imidization of thin films of the polyamic acid of PMDA/ODA on gold substrates.   The films had thicknesses of approximately (A) - 260, (B) - 45, and (C) - 15 Å, respectively. (Adapted from ref. 14.)

B

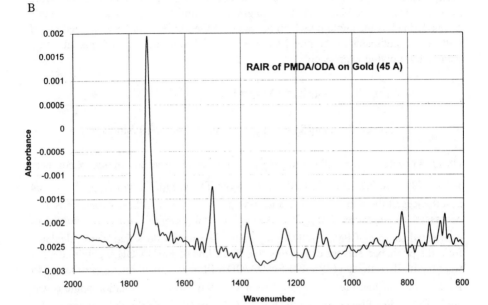

C

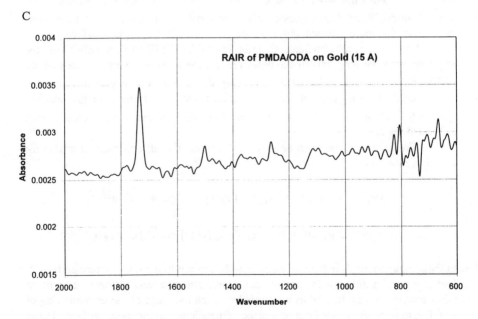

Figure 9. Continued.

Figure 10). The gold surface was defined as the XY plane. The tilt angle θ was defined as the angle between the normal to the gold surface (Z axis) and the 2-fold symmetry axis which bisects the CNC angles of the imide groups. The rotation angle ϕ was defined as the angle by which the plane of the imide groups was rotated out of a plane perpendicular to the metal surface. θ and ϕ were determined quantitatively using the following expressions:

$$A_{1385}(R)/A_{1720}(R) = [A_{1386}(T)/A_{1720}(T)][\cot^2(\theta)/\cos^2(\phi)]$$

$$A_{1385}(R)/A_{720}(R) = [A_{1386}(T)/A_{720}(T)][\cot^2(\theta)/\sin^2(\phi)]$$

where $A_{1385}(R)$, $A_{1720}(R)$, and $A_{720}(R)$ are the absorbances of the bands near 1385, 1720, and 720 cm^{-1} in RAIR and $A_{1385}(T)$, $A_{1720}(T)$, and $A_{720}(T)$ are the absorbances of the same bands in transmission. In order to use these expressions, it is necessary to measure $A_{1385}(T)$, $A_{1720}(T)$, and $A_{720}(T)$ for an amorphous, unoriented sample of PMDA/ODA polyimide. We used values that were calculated from the published spectra of unsupported thin films of PMDA/ODA polyimide that were thermally imidized (19). For PMDA/ODA films having a thickness of about 45 Å, it was found that the tilt angle θ was about 59.6^o while the rotation angle ϕ was about 35.7^o. For films having a thickness of about 260 Å, the tilt and rotation angles were about 56.7^o and 38.6^o, respectively. Thus, polymer molecules in thinner films were somewhat more oriented with their long axes parallel to the gold surface.

Published results indicate that for crystalline PMDA/ODA polyimide, the angle between the long axis of the molecule and the two-fold symmetry axis of the PMDA moieties is about 27° (20). Assuming that a similar situation exists for thin PMDA/ODA films on gold, it can be concluded that the tilt angle for the two-fold symmetry axis of the PMDA moieties is about 63°, which is very close to the calculated value.

Similar expressions can be used to determine the tilt and rotation angles for the ODA moieties:

$$A_{1500}(R)/A_{1117}(R) = [A_{1500}(T)/A_{1117}(T)][\cot^2(\theta)/\sin^2(\phi)]$$

$$A_{1500}(R)/A_{815}(R) = [A_{1500}(T)/A_{815}(T)][\cot^2(\theta)/\cos^2(\phi)]$$

In these expressions, θ is the angle between the normal to the gold surface and the 2-fold axis passing through the oxygen atom, aromatic ring, and nitrogen atom of an ODA moiety. ϕ was defined as the angle by which the plane of the aromatic ring of an ODA moiety was rotated out of a plane perpendicular to the metal surface. Using these expressions and the published spectra of thermally imidized, unsupported thin films of PMDA/ODA (19), it was determined that the tilt and rotation angles for the ODA moieties for the PMDA/ODA film having a thickness of about 45 Å were 58.3^o

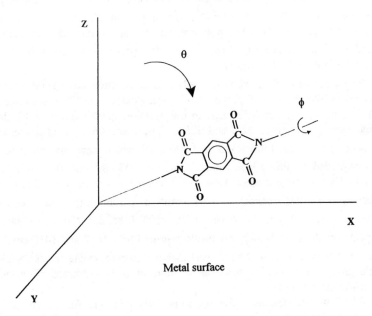

Figure 10. Model used to determine the orientation of PMDA moieties in PMDA/ODA adsorbed onto gold.

and 52.5°, respectively. The tilt angles for the PMDA and ODA moieties were thus similar as expected.

6FDA/ODA on Silver. Transmission infrared spectra of the polyamic acid and polyimide of 6FDA/ODA are shown in Figure 11. The band assignments for both polymers are similar to those for the polyamic acid and polyimide of PMDA/ODA. Thus, strong bands near 1650, 1540, and 1410 cm^{-1} in the spectrum of the polyamic acid were assigned to modes of the amide groups and the band near 1510 cm^{-1} was assigned to a stretching mode of the C_6H_4 rings. After curing, the bands near 1650, 1540, and 1410 cm^{-1} disappeared and new bands characteristic of the imide groups appeared near 1720, 1385, and 720 cm^{-1}. One important difference in the spectra of the 6FDA/ODA and PMDA/ODA polymers was the appearance of bands near 1260 and 1245 cm^{-1} in the spectra of the 6FDA/ODA polymers which were related to modes of the COC and CF_3 groups.

RAIR spectra of 6FDA/ODA polyamic acid films having thicknesses of approximately 320, 60, and 10 Å which were spin-coated onto silver substrates from DMF solutions were generally similar to transmission spectra and varied little with film thickness (see Figures 12A and 12B). These spectra were characterized by strong bands near 1720, 1500, 1385, and 720 cm^{-1} which were characteristic of the imide groups and by strong bands near 1260 and 1245 cm^{-1} which were related to COC and CF_3 stretching modes. However, when RAIR spectra were obtained from films having a nominal thickness of only about 10 Å, some important changes were observed (see Figure 12C). Bands near 1725, 1500, 1385, and 720 cm^{-1} disappeared from spectra of these films and new bands appeared near 1612 and 1410 cm^{-1}. The new bands near 1612 and 1410 cm^{-1} were assigned to carboxylate species, indicating that acid groups in molecules immediately adjacent to the substrate interacted with silver ions in the substrate.

RAIR spectra obtained after the same polyamic acid films were thermally imidized by heating at 200°C for 30 minutes are shown in Figure 13. Spectra of films having thicknesses of about 320 and 60 Å were similar to the transmission spectra, indicating little preferential orientation in the thicker films. However, some interesting features were observed in spectra of the film having thickness of approximately 10 Å. The bands near 1612 and 1410 cm^{-1} were prominent, indicating formation of carboxylate groups by interaction of acid groups with silver ions. Moreover, the band near 1500 cm^{-1}, which was characteristic of the C_6H_4 rings, disappeared upon imidization.

Two possibilities were considered to account for the disappearance of the band near 1500 cm^{-1}. One was that the disappearance of the band near 1500 cm^{-1} upon imidization of the thinnest films was due to an orientation effect. The transition moment for this mode is mostly parallel to the two-fold axes. The very low intensity for the band near 1500 cm^{-1} in spectra of very thin films of 6FDA/ODA spin-coated

A

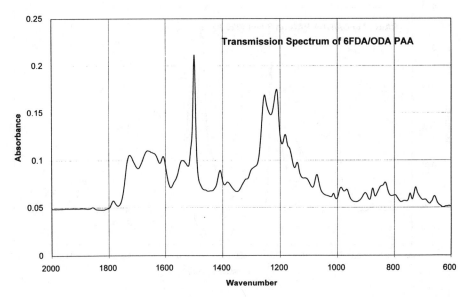

B

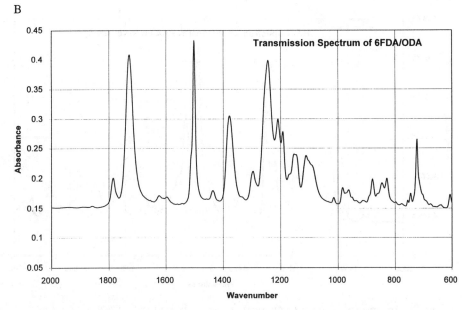

Figure 11. Transmission infrared spectra of the (A) - polyamic acid and (B) - polyimide of 6FDA/ODA.

A

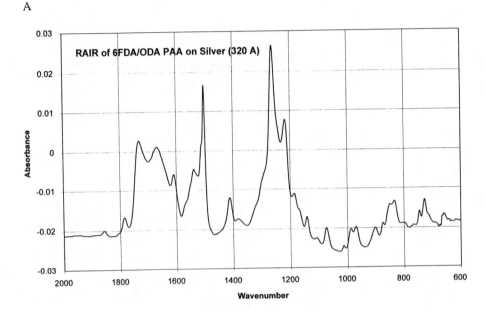

B

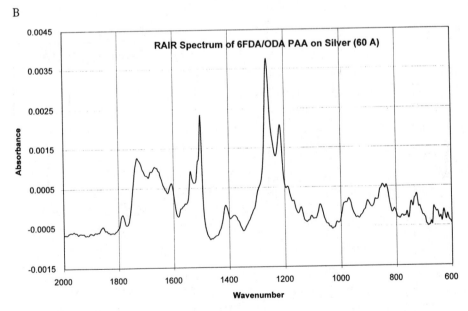

Figure 12. RAIR spectra of 6FDA/ODA polyamic acid films having thicknesses of approximately (A) - 320, (B) - 60, and (C) - 10 Å which were spin-coated onto silver substrates from DMF solutions.

C

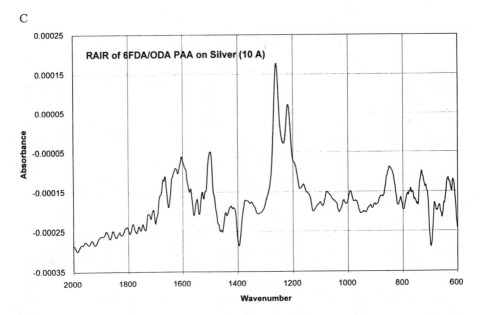

Figure 12. Continued.

A

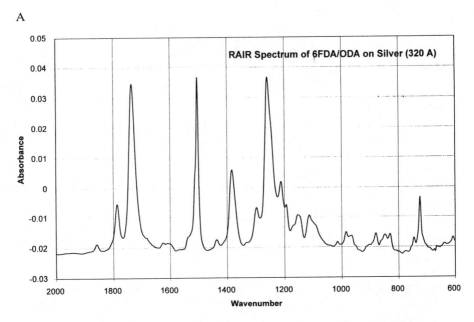

Figure 13. RAIR spectra obtained after imidization of thin films of the polyamic acid of 6FDA/ODA on silver substrates. The films had thicknesses of approximately (A) - 320, (B) - 60, and (C) - 10 Å. *(Continued on next page.)*

B

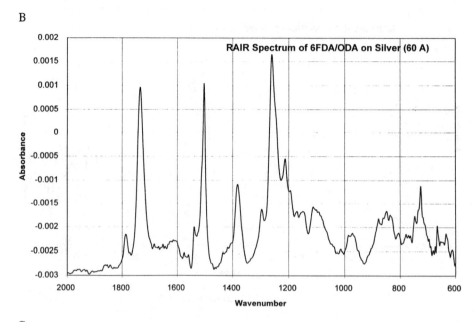

C

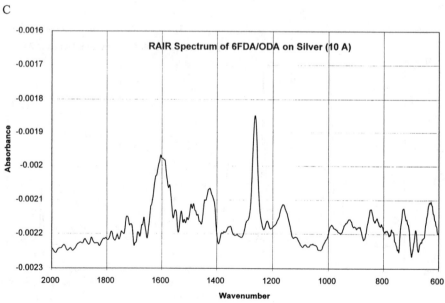

Figure 13. Continued.

onto gold indicated that the two-fold symmetry axes of the C_6H_4 rings were oriented mostly parallel to the substrate surface.

Careful analysis of the RAIR spectrum shown in Figure 13 showed that the band near 1260 cm^{-1} that was characteristic of the COC bonds also disappeared in spectra of the thinnest films of 6FDA/ODA on gold but that the band near 1245 cm^{-1}, which was related to the CF$_3$ groups, appeared with considerable intensity. The very low intensity of the band near 1260 cm^{-1} and the great intensity of the band near 1245 cm^{-1} may indicate that the COC bonds were also oriented parallel to the surface while the CF$_3$ groups had a perpendicular orientation.

The second possibility that was considered to account for the disappearance of the band near 1500 cm^{-1} upon imidization of very thin films of 6FDA/ODA on silver substrates was related to thermal degradation. It is known that silver will catalyze the degradation of thick 6FDA/ODA polyimide films during prolonged heating at temperatures above 200°C (21). Although degradation of films having thicknesses of several hundred angstroms does not occur at 200°C, some degradation of films having thickness of only about 10 Å may occur.

4-Mercaptophenylphthalimide (4-MPP) on Gold. RAIR and SERS have been used to characterize monomolecular films of 4-mercaptophenylphthalimide (4-MPP) adsorbed onto gold (15). The transmission infrared spectrum of 4-MPP was characterized by strong bands near 1710, 1496, 1389, and 717 cm^{-1} (see Figure 14). The bands near 1710, 1389, and 717 cm^{-1} were, of course, characteristic of the imide group while the band near 1496 cm^{-1} was related to a ring stretching mode of the C_6H_4 rings. In addition, a weak band was observed near 2573 cm^{-1} which was related to the SH stretching mode of the thiol groups.

RAIR spectra obtained from 4-MPP adsorbed onto gold from solution in DMF for 12 hours were much different than the transmission spectra (see Figure 15). The band near 2573 cm^{-1} disappeared, indicating that 4-MPP was adsorbed onto gold as a thiolate, as expected. The band near 1389 cm^{-1} was the strongest band in the RAIR spectrum and the bands near 1710 and 715 cm^{-1} were quite weak. Differences in the relative intensities of the bands near 1710, 1389, and 715 cm^{-1} in the transmission and RAIR spectra were attributed to preferential orientation of 4-MPP adsorbed onto gold.

The angle between the 2-fold symmetry axis of 4-MPP and the normal to the gold surface (tilt angle, θ) and the angle by which the plane of the imide groups was rotated out of a plane perpendicular to the metal surface (rotation angle, ϕ) for 4-MPP adsorbed onto gold were determined using the expressions described above. It was found that the tilt angle θ was about 21° while the rotation angle ϕ was about 36°.

Very recently we have carried out a molecular dynamics simulation of 4-MPP adsorbed onto gold (22). Results obtained indicated that the tilt and rotation angles were 25° and 35°, respectively. These values were in excellent agreement with those

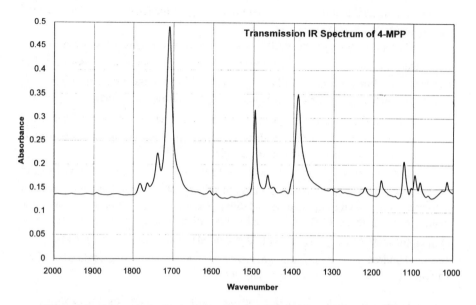

Figure 14. Transmission infrared spectrum of 4-MPP. (Adapted from ref. 15.)

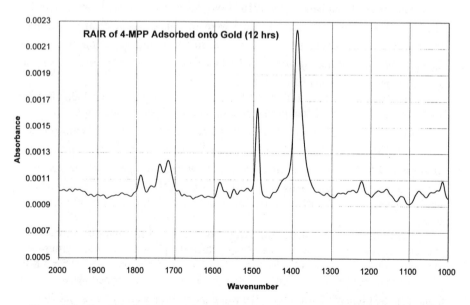

Figure 15. RAIR spectra obtained from 4-MPP adsorbed onto gold from a 10^{-3} M solution in chloroform for 12 hours. (Adapted from ref. 15.)

obtained from RAIR (*15*). To our best knowledge, this is the first time that a polyimide with a nearly vertical orientation (small tilt angle) has been observed.

Miller and coworkers investigated the adsorption onto gold of a series of oligomers of naphthalene - 1,4,5,8 - tetracarboxylic dianhydride and 3,3' - dimethoxybenzidine terminated by thiophenol groups (*23*). They found that the tilt angle was approximately 60° but the rotation angle was 45°, indicating no preferred orientation with respect to rotation about the long axes of the molecules. Miller et. al. also investigated the orientation in monolayers of similar oligomers transferred to gold surfaces by the Langmuir-Blodgett technique (*24*). In that case, a tilt angle of about 58° was found. Differences between the results of Miller et. al. and the results for 4-MPP are undoubtedly related to differences in the structure of the oligomers. Oligomers examined by Miller were much larger and bulkier than those considered here.

The normal Raman spectrum of 4-MPP is shown in Figure 16. Several bands characteristic of the imide group were observed. Thus, the bands near 1789 and 1771 cm^{-1} were assigned to the symmetric and asymmetric stretching modes of the carbonyl bonds, respectively, and the very strong band near 1395 cm^{-1} was assigned to the axial imide CNC stretching mode. Very strong bands near 1604 and 1100 cm^{-1} were assigned to modes $\nu(8a)$ and $\nu(1)$ of the p-disubstituted rings. The band due to the SH stretching mode was observed near 2581 cm^{-1}.

The SERS spectrum of 4-MPP is shown in Figure 17. Strong bands were observed near 1595 and 1085 cm^{-1} due to modes $\nu(8a)$ and $\nu(1)$ of the p-disubstituted rings. A medium intensity band appeared near 1390 cm^{-1} due to the CNC imide axial stretching mode. The band due to SH stretching disappeared from the SERS spectrum, confirming that adsorption had occurred through the thiol group. No bands due to the C=O stretching modes were observed near 1789 and 1771 cm^{-1}.

It is possible to obtain qualitative information regarding the orientation of surface species from the "propensity rules" for Raman scattering which were derived by Moskovits (*17*). According to these rules, modes belonging to the same symmetry species as α_{zz}, where z is perpendicular to the surface, should appear with the greatest intensity, especially if they involve motions perpendicular to the surface. The next strongest lines should belong to the same symmetry species as α_{xz} and α_{yz}.

4-MPP has at most C2v symmetry. The imide CNC axial stretching mode and the symmetric C=O stretching mode both belong to the A1 symmetry species, as does α_{xx}. Assuming that 4-MPP is adsorbed with the long axis of the molecule nearly perpendicular to the surface, the axial CNC stretching mode involves motion perpendicular to the surface while the carbonyl stretching mode involves motion mostly parallel to the surface. That may account for the low intensity of the carbonyl stretching mode relative to that of the CNC axial stretching mode. Modes $\nu(8a)$ and $\nu(1)$ of the p-disubstituted rings both involve considerable motion parallel to the long axis of the molecules. Assuming that 4-MPP is adsorbed with the molecular axes mostly perpendicular to the surface, the corresponding bands should appear with considerable intensity and that is what was observed.

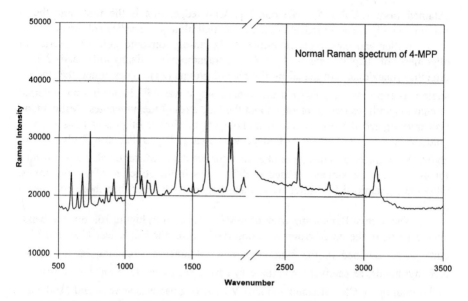

Figure 16. Normal Raman spectrum of 4-MPP. (Adapted from ref. 15.)

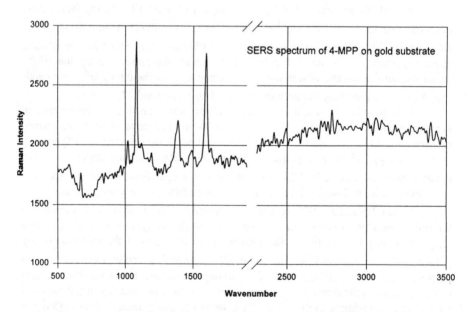

Figure 17. SERS spectrum of 4-MPP adsorbed onto a gold island film from a 10^{-3} M solution in chloroform for twelve hours. (Adapted from ref. 15.)

The results obtained from Raman scattering regarding the orientation of adsorbed 4-MPP are thus qualitatively consistent with those obtained from RAIR. However, the Raman "propensity rules" were derived assuming that the enhancement was strictly electromagnetic and thus do not take account of chemisorption.

Conclusions

Results obtained from reflection-absorption infrared spectroscopy (RAIR) indicate that polyimide molecules interact with silver substrates by formation of carboxylate species but there is little evidence for formation of carboxylate species on gold substrates. Polyimides from pyromellitic dianhydride (PMDA) and oxydianiline (ODA), which have relatively stiff backbones, form films on silver and gold substrates in which the molecules are oriented with their long axes mostly parallel to the surface and the plane of the PMDA moieties nearly perpendicular to the substrate surface. Polyimides having relatively flexible backbones, such as those from 4,4'-hexafluoroisopropylidene dianhydride (6FDA) and ODA, mostly formed films in which there was little preferred orientation of the polymer molecules. However, when 6FDA/ODA films having a thickness of only a few nanometers were deposited onto silver substrates, a preferred orientation was observed in which the C_6H_4 rings of the ODA moieties were oriented parallel to the substrate surface. Results from RAIR also show that thiol-terminated oligoimides such as 4-mercaptophenylphthalimide (4-MPP) interact with gold substrates by formation of thiolate species and that the long axes of these molecules are oriented nearly perpendicular to the metal surface such that the imide carbonyl bonds are nearly parallel to the surface.

In surface-enhanced Raman scattering (SERS), there is a strong enhancement of the Raman scattering by molecules adjacent to the metal surface but little enhancement for molecules located more than a few nanometers from the surface. Therefore, SERS is *surface-selective* and can be used for the non-destructive characterization of polymer/metal interfaces. Results obtained from SERS confirm that PMDA/ODA polyimides interact with silver substrates by forming carboxylate salts and that carboxylate formation inhibits imidization of the polymer near the interface. SERS also confirms that 4-MPP interacts with gold substrates by formation of thiolates.

Acknowledgments

This work was supported in part by a grant from the Office of Naval Research.

Literature Cited

1. Francis, S. A.; Ellison, A. H. *J. Opt. Soc. Am.* **1959**, *49*, 131.
2. Greenler, R. G. *J. Chem. Phys.* **1969, 50**, 1963.
3. Allara, D. L.; Baca, A.; Pryde, C. A. *Macromolecules* **1978**, *14*, 1215.
4. Moskovits, M. *Rev. Mod. Phys.* **1985**, *57*, 783.
5. Otto, A.; Mrozek, I.; Grabhorn, H.; Akemann, W. *J. Phys.: Condens. Matter* **1992**, *4*, 1143.

6. Burrell, M. C.; Codella, P. J.; Fontana, J. A.; McConnell, M. D. *J. Vac. Sci. Technol.* **1989**, *A7*, 55.
7. Linde, H. G. *J. Appl. Polymer Sci.* **1990**, *40*, 2049.
8. Grunze, M.; Lamb, R. N. *Chem. Phys. Lett.* **1987**, *133*, 283.
9. Grunze, M.; Lamb, R. N. *Surface Sci.* **1988**, *204*, 183.
10. Pryde, C. A. *J. Polymer Sci.: Part A: Polymer Chem.* **1993**, *31*, 1045.
11. Tsai, W. H.; Boerio, F. J.; Jackson, K. M. *Langmuir* **1992**, *8*, 1443.
12. Perry, S.; Campion, A. *Surface Science* **1990**, *234*, L275.
13. Young, J. T.; Tsai, W. H.; Boerio, F. J. *Macromolecules* **1992**, *25*, 887.
14. Young, J. T.; Boerio, F. J. *Surf. Interface. Anal.* **1993**, *20*, 341.
15. Young, J. T.; Boerio, F. J.; Zhang, Z.; Beck, T. L. *Langmuir*, submitted for publication, 1994.
16. Shevchenko, L. L. *Russian Chem. Rev.* **1963**, *32*, 201.
17. Moskovits, M. *J. Chem. Phys.* **1982**, *77*, 4408.
18. Russell, T. P.; Gugger, H.; Swalen, J. W. *J. Polymer Sci.: Polymer Phys. Ed.* **1983**, *21*, 1745.
19. Ishida, H.; Wellinghoff, S. T.; Baer, E.; Koenig, J. L. *Macromolecules* **1980**, *13*, 826.
20. Kazaryan, L. G.; Tsvankin, D. Ya.; Ginzburg, B. M.; Tuichiev, Sh.; Korzhavin, L. N.; Frenkel, S. Ya. *Vysokomol. Soed., Ser. A* **1972**, *14*, 1199.
21. Zhao, W. W.; Boerio, F. J., unpublished results.
22. Zhang, Z.; Beck, T. L.; Young, J. T.; Boerio, F. J. *Langmuir*, submitted for publication, 1994.
23. Cammarata, V.; Atanasoska, L.; Miller, L. L.; Kolaskie, C. J.; Stallman, B. J. *Langmuir* **1992**, *8*, 876.
24. Kwan, W. S. V.; Atanasoska, L.; Miller, L. L. *Langmuir* **1991**, *7*, 1419.

RECEIVED February 2, 1995

Chapter 3

Raman Microscopy and Imaging of Polymers

M. Claybourn[1], A. Luget[2], and K. P. J. Williams[3]

[1]ICI Paints, Wexham Road, Slough, Berkshire SL2 5DS, United Kingdom
[2]Trinity Hall College, Cambridge University, Cambridge CB2 1TJ,
United Kingdom
[3]Renishaw Transducer Systems Ltd., Old Town, Wotton-under-Edge,
Gloucestershire GL12 7DH, United Kingdom

Recent developments in Raman spectrometers have led to great improvements in sensitivity, imaging capabilities and instrument robustness. This paper describes a novel design based on a single spectrograph system using CCD detection and holographic Raman rejection filters. The facility of the instrument in the analysis of polymers is demonstrated through a number of examples showing the effectiveness as a microprobe system with high sensitivity. Operation of the instrument in a basic microprobe can be extended very easily with computer controlled optics to function in confocal and imaging modes so that polymer structure in 2 dimensions can be obtained.

Raman spectroscopy is becoming a well established technique in the analysis and characterisation of polymers. The reason is that the vibrational spectrum obtained gives a lot of detailed information about the chemical and morphological structure of the polymer. Raman also provides additional practical advantages of being nondestructive, effectively no restriction on the sample form and size, high spatial resolution (~1μm), mode selection using polarisation for studying oriented polymers, and resonance enhancement by coupling to the electronic structure. There are a number of review articles on the application of Raman spectroscopy to polymers eg *(1-4)*. Figure 1 shows the Raman spectra of four polymers and simply demonstrates the analytical capabilities of the approach; the spectra contain chemical structural information describing functional groups and local bonding specific to that polymer. If the polymer can undergo structural ordering then features in the spectrum may be identified that are characteristic of the physical state of the material. Figure 2 shows two spectra of polyethylene which indicate features relating to crystalline and amorphous structure. For this particular polymer, the technique is quantitative and for these samples the levels

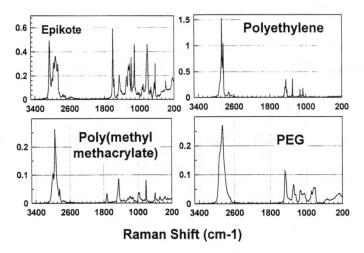

Figure 1 - Raman spectra of four polymers demonstrating the analytical capabilities of the approach; the spectra contain chemical structural information describing functional groups and local bonding specific to that polymer.

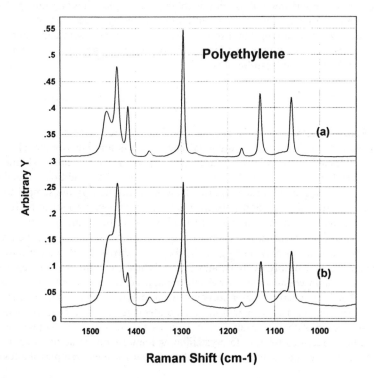

Figure 2 - Raman spectra of two samples of polyethylene show different levels of structural ordering.

of the structural phases are (a) 30% amorphous, 68% crystalline, and (b)70% amorphous, 15% crystalline *(5)*; the remainder in each case has been identified as due to an interfacial phase. With those advantages described above, Raman is clearly an important research and analytical tool in the industrial laboratory.

Coupling a microscope to the Raman spectrometer allows the spatial resolution of the measurement to be taken down to about 1μm. The technique of Raman microscopy was first reported in 1974 *(6,7)*. This design, which has changed little since first demonstrated, comprised a conventional optical microscope coupled to a monochromator with the laser delivered through the microscope and focused onto the sample. Figure 3 shows a schematic diagram of a Raman microscope system. The sample is placed on the microscope stage and is viewed and positioned in white light. Once located, the sample is irradiated with the laser light through the objective lens. The scattered radiation is collected by the microscope objective lens, directed through the transfer optics and onto the slit of the monochromator. The detector system is either single channel (eg photomultiplier tube) or multichannel (diode array detection). With single channel detection, dispersion of the Raman scatter across the exit slit in front of the detector is obtained by rotating the diffraction grating; spectral elements are collected as the grating is scanned. For a multichannel system, the grating is fixed, the exit slit is removed and the whole spectrum is simultaneously measured across the detector. Until recently, these systems have used double or triple monochromators which are relatively inefficient with a large number of optical surfaces.

By using the Raman microscope system as a microprobe, the technique has been used to study polymer fibers eg *(8)* and for single fiber studies, correlation of Raman band shift with tensile strength has been determined *(9)*. The technique also has general analytical capabilities for studying polymer laminates and characterising inclusions, imperfections and contaminants in polymers eg *(10)*.

Recent developments in laser, filter, and detector technologies have led to new instrumental designs that give much higher sensitivities and spatial resolution *(11-14)*. The first major advance has been in the application of holographic optical devices *(15)* as narrow band filters, and beamsplitters. For the Raman measurement this has meant improved rejection of the Rayleigh scatter, and high throughput (about 90%). The result has been a system requiring only a single spectrograph as opposed to a double or triple monochromator to filter the elastically scattered laser radiation *(16)*; apart from the obvious advantage of reducing the cost of such an instrument, this arrangement gives a much higher throughput system. The second major advance has been in the use of the charge-coupled device (CCD) detector. This is a two-dimensional silicon array detector with high quantum efficiency (up to 70%) and very low dark current (generally less than 0.01electrons/pixel/second). This is a major improvement over the 'conventional' single or multichannel detectors *(17,18)*. Currently, for this type of system, a collection efficiency for the Raman scatter of >25% can now be achieved *(19)*; this is compared to <1% for the earlier Raman microscope systems. The improvement in sensitivity gives the advantages of much shorter acquisition times and allows the use of much lower laser powers, hence reducing the potential effect of sample damage under the laser beam.

With the CCD detector operating in a two dimensional mode, the Raman microscope system can be used for imaging allowing the spatial distribution of a specific molecular species to be defined. The Raman band of a particular component in a heterogeneous system can be selected (eg with a tunable filter) and the Raman scatter over an area of the sample defined by the optical setup is collected at the detector. This approach is clearly superior to point-by-point mapping which requires the time consuming acquisition of spectra at many positions on the sample surface followed by data reduction to produce the final image. Raman images showing phase structure in various polymer blends with a spatial resolution of about 1μm have been reported *(20)*.

The CCD detector and Raman microscope system described allows a further type of measurement to be made, namely confocal Raman spectroscopy. In the 'conventional' system, a pinhole is placed in the back focal plane of the microscope. This has the effect of filtering out all the Raman scatter that is not in the plane of focus of the laser in the sample; this greatly enhances the depth resolution of the microscope and collection volumes of the order of about 5μm^3 can be obtained. The factors affecting the depth of resolution are the optical properties of the sample, size of the pinhole, and the numerical aperture, focal length and magnifying power of the objective lens. Depth profiling of thin polymer samples at the micron level have been reported using this approach for confocal Raman (21,22). The system based around the single spectrograph and CCD detector described above achieves the confocal mode by 'binning' the signal at the detector (23). Effectively, this means that the detector area is matched with the illuminated sample area and is achieved by selecting a narrow width of pixels across the array defining the depth of the sampled volume; this is achieved using the instrument software and provides a depth resolution of about 2μm. This approach gets around the difficult problem of aligning pinholes to obtain the confocal arrangement.

A problem that can occur with the Raman microscope systems described above is laser-induced sample fluorescence which can swamp out the Raman scatter. The intensity of the fluorescence is laser wavelength dependent *(24)* so by going to longer wavelength, the effect is weakened. Common lasers used are the Ar+ and He-Ne, both operating in the visible region. For samples displaying fluorescence, near infrared diode lasers operating in the range 750 - 830nm may dispel the problem *(25)* though some materials will still fluoresce in this laser range. FT-Raman systems using a Nd:YAG laser operating at 1.064µm further reduces the probability for fluorescence. FT-Raman microscope spectrometers are available commercially *(26)*. A recent report has demonstrated FT-Raman using the 1.339µm line of the Nd:YAG laser *(27)* which further reduces the possibility of observing fluorescence; however, in this case the instrument operates only in a macro-Raman mode and not as a microprobe. The microscope of an FT-Raman system is of a conventional design with the laser/Raman radiation delivery optics being mirrors or fiber optics. However, the coupling of the microscope to the interferometer remains a non-trivial problem and the efficiency of such systems is poor; for a weakly scattering polymer (eg polyacrylate type), the spectrum collection time can be several hours compared to minutes for a 'conventional' system employing holographic beamsplitter/notch filter, single grating spectrograph and CCD detector *(28)*. There are further inherent problems with the FT-Raman technique in comparison with the CCD/ single spectrograph system described above; these include poor detector sensitivity, poor spatial resolution (about 10µm), weaker Raman scatter (varies with the fourth power of the exciting frequency) and therefore weaker signal. The merits of the CCD/single spectrograph approach over the FT-Raman method have been reported *(29)*.

Experimental
The systems used in this study were a
(a)Bruker FT-Raman microscope based on the Bruker IFS66 FTIR and FRA106 Raman module with the microscope optically connected via fiber optics. The signal was detected using a liquid nitrogen cooled Ge photodiode detector. The laser was a 350mW Nd:YAG operating at 1.064microns. The laser energy at the sample was about 35mW with a spot size of about 10µm. The spectral range for the instrument is ~100 - 3500cm-1. The Raman spectra were not corrected for instrumental response because it is almost flat over the spectral range 3300cm^{-1} to the Rayleigh line. The system used here has been previously described *(30)*.
(b)Renishaw Ramascope (details may be found in *(31)*) using a low-powered (25mW) air-cooled laser (He-Ne or diode laser), a single spectrograph (250mm focal length) for spectroscopy mode, a set of angle tunable bandpass filters for imaging mode, an Olympus microscope, a Peltier cooled CCD detector (576x384 pixels), a holographic notch filter used as a beamsplitter. The holographic filter reflects >90% of the laser energy into the microscope and is designed to reject the back-scattered Rayleigh component while permitting the Raman scatter to be transmitted into the spectrometer. This arrangement gives an efficiency of about 25%. The laser spot size at the sample was 1µm and the laser energy was varied between <1mw to full power (about 7mW at the sample) using ND filters. This

system was also used in confocal and imaging modes; the experimental setup for each of these is described below.

Results and Discussion
Microprobe Measurements. Raman microscopy can play a significant role in analytical applications such as identifying microscopic unknowns (eg inclusions in polymers) *(10)*. The major defect types found in coatings are pinholes and craters that can be very difficult to analyze using FTIR microspectrometry since the defect may be too small to characterise or the spectrum of a contaminant may be completely obscured by other components in the sampled area. However, Raman microscopy is now providing an important tool for characterising contaminants in these types of coating defects. Most pinholes we find in coatings are beyond the diffraction limit of FTIR (ie <10microns). Figure 4 shows a schematic diagram of a possible pinhole; the substrate is likely to be another coating such as a basecoat. These types of defects are often met in automotive coatings where the environment of application may be contaminated by extraneous material such as a silicone oil or poorly miscible additives which are not correctly dispersed and cause defects such as craters or pinholes on curing.

Many of these defects are caused by siloxanes; Figure 5 shows the Raman and FTIR spectra for polydimethyl siloxane. The FTIR spectrum is dominated by the bands at about 1020 and 1100cm^{-1} due to Si-O-Si stretching modes with two sharp bands either side at 800cm^{-1} (Si-C stretching) and about 1260cm^{-1} (CH$_3$ deformation of Si-CH$_3$). The Raman spectrum is dominated by the C-H symmetric and asymmetric modes of CH$_3$ in the 2900-3000cm-1 region. The bands characteristic of the siloxane are at 488 and   710cm^{-1}. Clearly both techniques could provide accurate analytical methods for characterising contaminants found in coating defects. The FTIR spectrum obtained for material found at the bottom of a pinhole that was about 5 microns in diameter at its base, is shown in Figure 6; the typical structure of such a defect is shown in Figure 4. An immediate problem can be identified since the wavelength of the incident light is of the same dimensions of the bottom of the defect. To obtain sufficient spatial resolution in FTIR microscope, generally some form of aperture or pinhole is used (eg 0.3mm aperture with x15 objective gives a spatial resolution of about 8µm - as in this measurement). Any attempt to obtain a spatial resolution beyond the wavelength of the incident light using apertures is prevented by the diffraction limit - in other words, an IR response outside the aperture-defined region could be diffracted at the aperture and reach the detector. The result shown in Figure 6 combines this diffraction limitation on spatial resolution below 8µm (corresponding to 1250cm^{-1} in the spectrum) and also, a response from the coating at the edge of the defect, as defined by the aperture, is also detected. Although the FTIR measurement may indicate the possible presence of a silicone; the spectrum is too poor for any great confidence in this interpretation. Figure 7 shows the Raman spectrum for the same

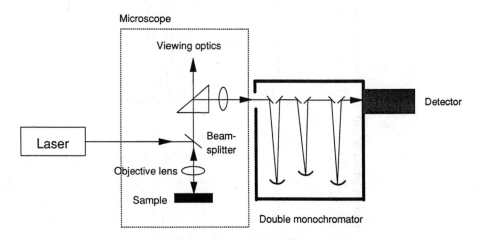

Figure 3 - A schematic layout of a Raman microscope system.

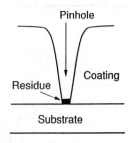

Figure 4 - A schematic diagram of a pinhole; the substrate is likely to be another coating such as a basocyte.

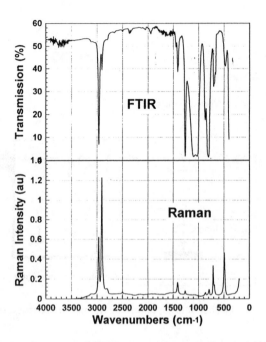

Figure 5 - Raman and FTIR spectra for polydimethylsiloxane.

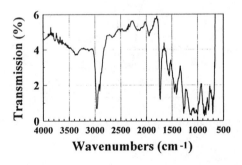

Figure 6 - FTIR spectrum obtained for material found at the bottom of a pinhole that
was about 5 μm in diameter at its base. The sample spot size was about 8μm.

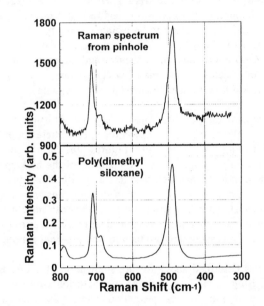

Figure 7 - Raman spectrum for the same pinhole as in Figure 6. This was obtained using the Renishaw Raman system using a He-Ne laser with a laser spot size of 1µm; the scan time was about 2mins.

defect with a laser spot size was about 1μm. Due to the improved spatial resolution, the Raman spectrum now clearly shows the material to be polydimethylsiloxane - bands at 488 and 710cm^{-1} *(32)*; the origin of this component was found to be an antiflow agent that was phase separating during film formation. FT-Raman microscopy measurements with a spatial resolution of about 5μm were also made on this sample but was found to cause heating due to low levels of carbon (<1%) used in the coating; this resulted in sample burning.

Confocal Measurements. An important area of polymer defect analysis where use of the confocal mode of operation can be made is in the characterisation of inclusions. Figure 8(a) shows the spectrum of an inclusion embedded in a poly(ethylene terephthalate) (PET) preform used in bottle production. Use of the confocal mode allowed the inclusion to be isolated spatially from the bulk polymer and a spectrum obtained. Figure 8(b) shows a spectrum of the bulk polymer. The differences in the C=O band width at 1735cm^{-1}, and increase in the intensity of the 1096 and 859cm^{-1} bands are associated with structural ordering in the polymer *(33)*; the C=O bandwidth is thought to correlate directly with crystallinity *(34)*.

A similar type of problem was successfully solved using the same approach. In this case there was an inclusion of a gel particle about 2μm in diameter embedded in a polyethylene film. Figures 9(a) and (b) show spectra for the inclusion and bulk film respectively. Comparison of the bulk spectrum with that for the inclusion indicate that defect is less crystalline than the material found in the bulk - this can be seen by the reduction in the intensity of the band at about 1416cm^{-1} (5). Modifications in the structural properties can have a major effect on the final product behaviour. These types of analyses make significant headway in the understanding of processing and production line problem solving.

The examples described above for the application of confocal operation simply require focusing of the laser light on the inclusion with careful 'binning' at the CCD detector to give the vertical resolution. The confocal mode can also be used for profiling in the z-direction by focusing through the sample. Figure 10 shows an example of this approach for a 2μm polyethylene film on a polypropylene plaque. The measurement employed a 50x microscope objective, and a CCD active area of 600 (spectral elements) x 4 (depth resolution) pixels. Figure 10(a) shows the spectrum with the laser focused on the top surface of the polyethylene and indicates that the spatial resolution is not perfect since features due to the polypropylene appear. Figures 10(b) through 10(d) are Raman spectra taken by focusing through the polyethylene layer and into the polypropylene. Clearly as the point of focus of the laser is changed then the different polymers may be isolated in the Raman spectrum. Figure 10(e) shows a polypropylene spectrum for reference. Using a 100x microscope objective, an improvement in the spatial resolution to better than the required 2μm is observed; this is demonstrated

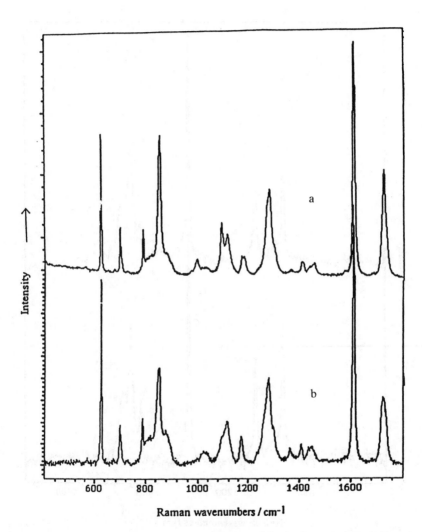

Figure 8 - Raman spectra from (a) an inclusion in a PET preform and (b) bulk polymer. The spectra were obtained in 40secs using a He-Ne laser operating at 5mW.

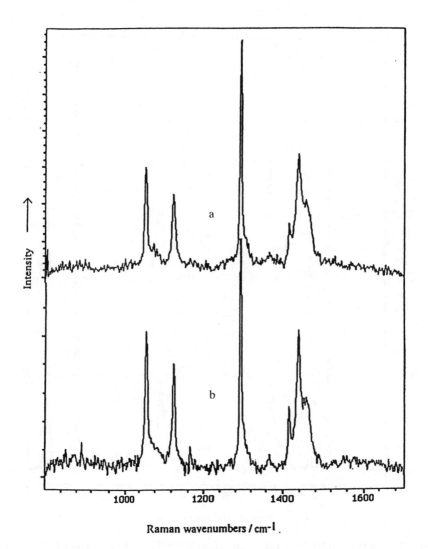

Figure 9 - Raman spectra from (a) 2μm gel particle in a polyethylene film and (b) bulk polymer. The spectra were obtained in 80secs using a He-Ne lasers operating at 5mW.

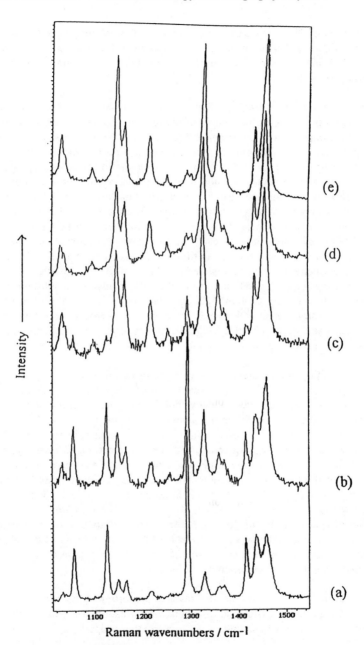

Figure 10 - Raman spectra obtained by focusing through a 2μm polyethylene film on a polypropylene placque: (a)on the surface of the polyethylene film, (b-d) through the polyethylene film and into the polypropylene, and (e) polypropylene reference spectrum. This was obtained using the Renishaw Raman system using a He-Ne laser at 5mW, 50x microscope objective and an accumulation time of about 4mins.

in Figure 11(a) - there is clearly no evidence for the polypropylene bands in the spectrum. Figure 11(b) shows the same measurement but using a 36 pixel height on the CCD; the polypropylene bands begin to appear demonstrating the loss in confocal operation.

Raman Mapping and Imaging. Raman mapping requires spectra to be obtained across a grid of points on the sample of interest. To obtain the chemical map of any of the components in the material, the data must be processed and reduced to the appropriate form. This type of measurement can take many hours to build up a chemical distribution map of the sample. An example of a Raman map is shown in Figure 12 for a large white inclusion found in a clear styrenated-acrylic copolymer film. The map was produced using data obtained on the Bruker FT-Raman system. Spectra were obtained at 100μm intervals on a grid of dimensions 500 x 500μm with the sample located manually using the vernier scale on the microscope stage. The measurement time for each spectrum was about 30mins to obtain the required signal/noise. Spectra for the film and inclusion are shown in Figure 13. From these spectra the inclusion is clearly a lump of TiO_2. The map was obtained by normalising the integrated intensity of the TiO_2 bands in the 400-600cm^{-1} range against the styrene band at about 1000cm^{-1}. The time required to obtain the map is very clearly restrictive on making this a routine measurement. An improved method as described in the Introduction is the use of a 2-D detection system giving a direct chemical image across a sample with resolution down to about 1μm.

The imaging measurements were made using the Renishaw system which employs a tunable filter to select a wavelength of interest so that this can be imaged across a sample surface. An example of imaging as an important tool for characterising different phases in a polymer is given in Figure 14. The image shows the distribution of the ester band at 1730cm^{-1} of a polyester/epoxy blended coating where clearly the ester component (white indicates high intensity, black indicates low intensity) is spherulitic. The dimensions of the spherulites in the image are of the order of 1-2microns. For these comparatively weak Raman scatterers, the image took about 20mins to accumulate. Figure 15 shows a similar coating that had undergone a different processing and in the case, the ester component appears to be localised at the interfaces of large, epoxy-rich domains. The white light image gives a poorly resolved distribution - this demonstrates that the Raman image gives very much more significant information about phase separation and chemical distribution.

For these types of results, the physical properties of the coatings can be correlated with detailed chemical distribution information at the micron level and can give a clearer understanding of the final coating properties.

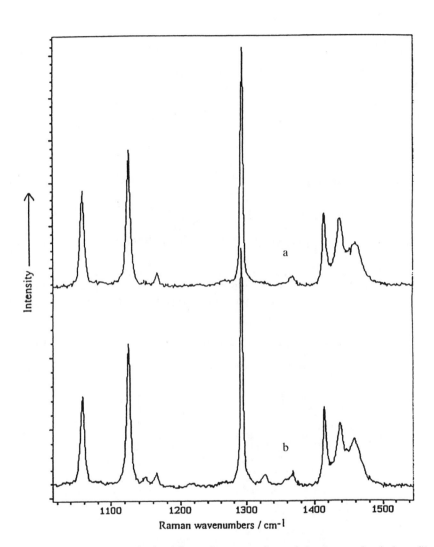

Figure 11 - Raman spectra obtained from the top surface of the 2μm polyethylene film on the polypropylene plaque obtained as in Figure 10 but using the 100x microscope objective and (a) with 4 pixels height on the CCD and (b)36 pixels height on the CCD.

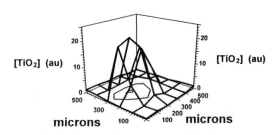

Figure 12 -   Raman map for the region around a large white inclusion found in a clear styrenated-acrylic copolymer film.   The map was produced using data obtained on the Bruker FT-Raman system (Nd:YAG laser, 35mW laser power, 10x microscope objective).

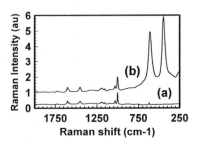

Figure 13 - Spectra for the film (a) and inclusion (b) of Figure 12.

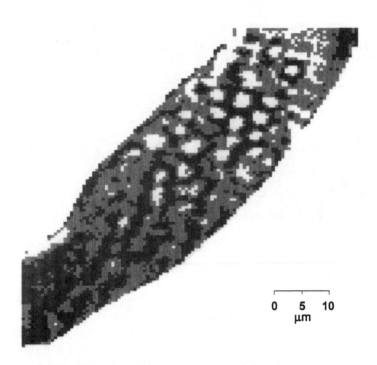

0   5   10
μm

Figure 14 - Raman image at 1735cm⁻¹ of a bis-phenol A epoxy/poly(ethylene terephthalate) blended polymer showing the distribution of the ester component.

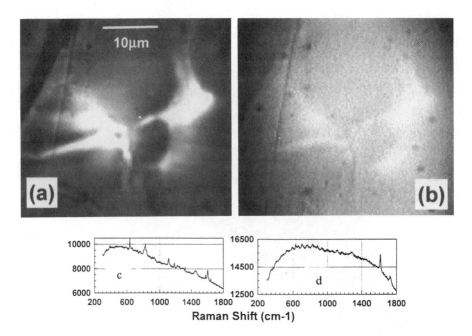

Figure 15 - (a)Raman image at 1735cm-1 for a similar coating as in Figure 14 but that had undergone a different processing and in the case (b) white light image - the phase structure can only just be identified, (c) and (d) Raman spectra taken in the dark (epoxy-rich) and light (ester-rich) regions shown in the Raman image.

Conclusions

The discussion presented above illustrates the applications currently accessible with a commercial Raman system optimised for microprobe and imaging capabilities. Comparison is made with an FT-Raman microscope system which is clearly not sufficiently sensitive to provide any sort of routine applications with a reasonable signal/noise. Consequently, the work has focused on the applications generated using the Renishaw Raman system which can give high quality spectra on the micron dimension within a few minutes. This microscope can operate in a confocal mode by 'binning' the signal at the CCD giving a spatial resolution of 1µm laterally and 2µm vertically. Examples of applications of this system operating in a confocal configuration have been demonstrated and can be achieved with simple instrument control under the system software. The instrument can also operate in an imaging mode with a spatial resolution of 1µm; chemical distribution maps can be obtained within a few minutes.

Acknowledgments
We wish to express our thanks to Drs G D Pitt, A Whitely and Mr C Dyer of Renishaw Transducers Ltd., to Prof. D Batchelder of Leeds University and Dr N Everall of ICI, Wilton for giving us access to the Renishaw Raman Microprobe and Imaging systems for many measurements. We would also like to thank Dr N M Dixon of BP helpful discussions on Raman microscopy. Finally, we would like to thank ICI for allowing publication of this work.

Literature Cited
1. Grasselli, J. G.; Snavely, M. K.; Bulkin, B. J. *Phys. Rep.*1980, *65*, 231.
2. Gerrard, D. L.; Bowley, H. J. *Anal. Chem.* 1986, *58*, 6R.
3. Gerrard, D. L.; Maddams, W. F. *Appl. Spectrosc. Reviews* 1986, *22*, 251.
4. Schlotter, N. E. *Compr. Polymer Sci. 1 Polymer Charact.* 1989, 469.
5. Strobl, G. R.; Hagedorn, W. *J. Polym. Sci.: Polym. Phys.* 1978, *16* 1181.
6. Rosasco, G. J.; Etz, E. S.; Cassatt, W. A. *IVth International Conference on Raman Spectroscopy*, Brunswick, ME, 1974.
7. Delahaye, M.; Dhamelincourt, P. *IVth International Conference on Raman Spectroscopy*, Brunswick, ME, 1974.
8. Pastor, J. M. *Makromol. Chem., Makromol. Symp.*, 1991, *52*, 57.
9. van Eijk, C. P.; Meier R. J.; Kip, B. J. *J. Polym. Sci.:B:Polym. Phys.*, 1991, *29*, 99.
10. Adar, F.; Leclerq, M.; Grayzel, *R. E. Int. Lab.*, 1982, 82.
11. Batchelder, D. N.; Cheng, C.; Muller, W.; Smith, B. J. E. *Makronol. Chem., Makromol. Symp.* 1991, *46*, 171.
12. Liu, K.-L. K.; Chen, L.-H.; Sheng, R.-S.; Morris, M. D. *Appl. Spectrosc.* 1991, *45*, 1717.

13. Greeve, J.; Puppels, G. J.; Otto, C. *Proceedings of the 13th International Conference on Raman Spectroscopy*, Eds.: Kiefer, W.; Cardona, M.; Schaak, G.; Schneider, F. W.; Schroet W., (J. Wiley, Chichester, 1992), p21.
14. Treado, P. J.; Levin, I. W.; Lewis, E. N. *Appl. Spectrosc.* **1992**, *46*, 1211.
15. Tedesco, J. M.; Owen, H.; Pallister, D. M.; Morris, M. D. *Anal Chem*, **1993**, *65*, 441A.
16. Mason, S. M.; Conroy, N.; Dixon, N. M.; Williams, K. P. J. *Spectrochimica Acta*, **1993**, *49A*, 633.
17. Murray, C. A.; Dierker, S. B.; *J. Opt. Soc. Am.*, **1986**, *A3*, 2151.
18. Falkin, D.; Vosloo, M., *Spectroscopy Europe*, **1993**, *5*, 16.
19. Williams, K. P. J.; Batchelder, D. N. *Spectroscopy Europe*, **1994**, *6*, 19.
20. Garton, A.; Batchelder, D. N.; Cheng, C. *Appl. Spectrosc.*, **1993**, *47*, 922.
21. Tabaksblat, R., *Spectroscopy Europe*, **1992**, *4*, 22.
22. Tabaksblat, R.; Meier, R. J.; Kip, B. J., *Appl. Spectrosc.*, **1992**, *46*, 60.
23. Williams, K. P. J.; Pitt, G. D.; Batchelder, D. N.; Kip, B. J. *Appl. Spectrosc.* **1994**, *48* 232.
24. Williams, K. P. J.; Gerrard, D. L. *Optics & Laser Technol.*, **1985**, 245.
25. Ferris, N. S.; Bilhorn, R. B., *Spectrochimica Acta*, **1991**, *47*, 1149.
26. Sawatski, J. *Fresenius J Anal Chem.* **1991**, *339*, 267.
27. Asselin, K. J.; Chase, B. *Appl. Spectrosc.* **1994**, *48*, 699.
28. Claybourn, M. unreported results
29. Mason, S. M.; Conroy, N.; Dixon, N.M; Williams, K. P. J., *Spectrochimica Acta*, **1993**, *49*, 633.
30. Claybourn, M.; Agbenyega, J. K.; Hendra, P. J.; Ellis, G., in *Structure-Property Relations in Polymers, Advances in Chemistry Series No. 236*, Eds: M W Urban, C D Craver, ACS, Washington, **1993,** p443
31. Williams, K. P. J.; Pitt, G. D.; Smith, B. J. E.; Whitley, A.; Batchelder, D. N.; Hayward, I. P., *J. Raman Spectrosc.*, **1994**, *25*, 131.
32. Schrader, B. *Raman/Infrared Atlas of Organic Compounds, Second Ed.*, VCH, Weinheim, **1989**.
33. Claybourn, M.;Turner, P. H., in *Structure-Property Relations in Polymers, Advances in Chemistry Series No. 236*, Eds: M W Urban, C D Craver, ACS, Washington, **1993,** p407
34. Melveger, A. J., *J. Polym. Sci.*, **1972**, *10*, 317.

RECEIVED March 17, 1995

Chapter 4

Multimode IR Analysis of Nylon 11

L. J. Fina

Department of Mechanics and Materials Science, College of Engineering, Rutgers University, Piscataway, NJ 08855

This chapter summarizes some of the recent work on the structure and morphology of a particular polyamide, nylon 11, as viewed with infrared spectroscopy. Attenuated total reflectance spectroscopy is used to study the development of dipole orientation and the existence of structural gradients from the surface in draw and electric field poled films and bilaminates. The new technique of two-dimensional infrared spectroscopy is used to study the dynamic stretching properties of melt-crystallized nylon 11. The complex absorption behavior of the amide I region is addressed. Lastly, the technique of trichroic infrared spectroscopy is used to elucidate the three dimensional absorption behavior of nylon 11 films that are subjected to one-way drawing, annealing and poling treatments.

In the group of polymers known as polyamides, difference in the number of carbon atoms between the amide groups results in a wide range of physical and mechanical properties. Nylons are often partially crystalline and properties are controlled to a large extent by the amount of crystallinity, the average crystallite size, the order, or lack of it in the amorphous phase, and the orientation in the crystalline and amorphous phases. When a polyamide sample is drawn, the crystallites act as cross-links, as do the amide group hydrogen bonds and to a lesser extent the van der Waals interactions in the amorphous phase. These influences lead to a high degree of elastic recovery in drawn materials.

Many of the physical properties of nylons such as the high melting point and the physical toughness result from the hydrogen bonding between adjacent amide groups. Since polyamides often exist with crystal structures

0097–6156/95/0598–0061$12.00/0
© 1995 American Chemical Society

that maximize hydrogen bond formation(*1*), the chain arrangement in the crystals can be predicted. In the A_n type of polyamides, the sense or direction of the chains in the crystal is important to the final structure. This is not the case for the $(AB)_n$ type where a center of symmetry along the chain removes the directional property. Broadly speaking, x-ray studies of polyamides have revealed that two different categories of crystal forms can exist. The form adopted depends on the type of polyamide and the method of preparation. In the first category are the original triclinic α-like crystal structures of Bunn and Garner(*2*) for nylon 6 and nylon 610. They consist of fully extended planar zigzag chains which are completely hydrogen bonded and the chains are either parallel or antiparallel within the hydrogen bonded sheets. Additionally, the hydrogen bonded sheets can show either progressive or staggered shifts with respect to one another. In the second category are γ- (or δ-) like crystals. The c-axis is slightly contracted from that of the fully extended planar zigzag due to internal bond rotation. The distances between hydrogen bonded sheets and between chains within hydrogen bonded sheets are approximately equal, forming a pseudohexagonal lattice.

Nylon 11 has been found to have four and possibly five crystal forms. Two are triclinic and two(or three) are pseudohexagonal. The differences within each type are subtle. Nylon 11 falls into a unique group of polyamides in that the unit cell can be prepared with a net polarization. The odd nylons have this characteristic because the number of methylene units between the amide groups renders the net electric dipoles associated with the amides groups as parallel. When the crystals of a sample can be oriented in three dimensions, i.e., when the amide planes of the crystals are contained in the same plane and in the same direction in that plane, the sample has a net bulk polarization. Therefore, the odd nylons fall into the same polar group of polymers as the much more highly studied poly(vinylidene fluoride) and can be useful in piezoelectric and pyroelectric applications.

In this work the orientation of the amide planes and the methylene segments in response to orientation, thermal treatments and electric field poling is explored. The influence of the surface of thin films in the orientation of the amide planes is studied with attenuated total reflectance spectroscopy. The three dimensional orientation of the amide groups is also studied with tilted film spectroscopy. Band resolution of the complex amide I region of the infrared spectrum is improved with two dimensional infrared spectroscopy, i.e., by the hyphenation of dynamic mechanical analysis and step-scanning interferometry. Remnant electric field analyses are conducted to provide information on the structural changes associated with electric field poling.

FTIR-ATR for Surface Gradients in Nylon 11 and Nylon 11/PVF$_2$ Bilaminates

A considerable amount of work has been done on the microstructural changes associated with the drawing process and electric field poling in poly(vinylidene fluoride) (PVF$_2$). Several techniques have been used to characterize the distribution of polarization as a function of distance from the film surface in poled films. Past work has demonstrated that films poled at room temperature have a low degree of polarization at the electrode-film interfaces and the polarization reaches a maximum in the center of the film(*3,4*). On the other hand, for PVF$_2$ that has been annealed or weakly oriented, the polarization maximum is no longer in the center of the film, but concentrated at the polymer film-positive electrode interface(*4,5*). Unlike the plethora of work that has appeared on PVF$_2$, very little work has been done on the distribution of polarization in nylon 11. In this section variable-angle FTIR-ATR spectroscopy is applied in a depth profiling mode to the surface of nylon 11 films that have been one-way drawn and electric field poled, both as single films and as co-melted pressed films of nylon 11 and PVF$_2$.

The attenuated total reflection (ATR) spectra used in this section are collected with a variable-angle reflection accessory in the angular range of 41-60°. An hemispherical zinc selenide crystal is used to produce an optically simple single reflection. A wire grid polarizer is used to give polarized light. Infrared scanning is done after the sample had been clamped for one hour to eliminate any effects of sample aging. Figure 1 shows the axis system where the axes are defined with respect to the polymer film. The X-direction is the draw direction, the Y-direction is perpendicular to the draw direction and in the plane of the film, and the Z-direction is perpendicular to the film plane. In order to use the treatment of Flournoy and Schaffers(*6*) (which yields the directional absorption coefficients), the position of the polymer is fixed on the ATR crystal, and one set of *s*- and *p*-polarized spectra is collected. The sample is rotated by 90° with respect to the crystal and another set is collected. This procedure yields the three directional absorption coefficients k_x, k_y and k_z, where the absorption coefficient k is defined by the complex refractive index as $\hat{n} = n - ik$. With a standard definition of the effective depth of penetration d_e(*7*), the surface of polymers can be semiquantitatively depth profiled. A large amount of work has been done to characterize polymer surfaces using this treatment (*8-11*). At this juncture it is relevant to point out that the effective depth is different for *s*- and *p*-polarized light collected at the same angle of incident light, and it is not actually a distance but a measure of the strength of interaction of the evanescent wave within the probed material. A measure of the dipole distribution and orientation can be obtained by using the variable-angle ATR intensities in an inverse Laplace transform as shown previously(*12-14*). However, such a treatment is not appropriate for the analysis given here since the band intensities are too large. In fact, the accuracy of the d_e treatment for depth profiling purposes also breaks down when the band intensities are high, albeit not as quickly. In the application given here, past work has established the limits of the d_e treatment(15).

In this study, efforts are concentrated on the 3300 cm^{-1} infrared absorption band which is assigned solely to N-H stretching vibrations of many polyamides(16,17). In nylon 11 the amide planes respond to the external treatments of drawing and poling and therefore, the 3300 cm^{-1} band is ideal to follow microstructural changes. The amide I mode (~1645 cm^{-1}) can also be very informative, although it is too intense in the ATR spectra to apply semiquantitative methods. Figure 2 shows the ATR-derived absorption coefficients in the three spacial directions for variously treated samples of nylon 11 as calculated with the equations of Flournoy and Schaffers(6) using a refractive index value of 1.53 as found from the interference fringe pattern of a thin film infrared transmission measurement. The depth of penetration for these conditions is in the range of one to three microns. Figure 2 contains the depth-dependent coefficients of nylon 11 for a drawn single film, a drawn bilaminate film (co-melt pressed and drawn with PVF$_2$), a drawn plus poled bilaminate film and a drawn plus poled single film. In the bilaminates, the polymer-polymer interface is examined after separation of the two polymers. The inverse Laplace treatment referred to earlier was applied to the ATR intensity data in the direction parallel to the draw direction (X-direction in Figure 1) where the intensities fall within Beer's Law. This serves as a quantitative check of the accuracy of the effective depth analysis shown in Figure 2. The result of the treatment is shown in Figure 2 as dashed lines. The close agreement between the dashed lines and the open circles attest to the validity of the methods.

In one-way drawn polymers it is common for uniaxial or fiber symmetry to develop, i.e., isotropy exists in the plane perpendicular to the draw direction (transverse plane). This is not the case in PVF$_2$ as shown previously(15,18). Figure 2a show the orientational properties of the transverse plane in one-way drawn nylon 11 by the filled circles and the open boxes. The large difference between the two sets of data points establishes that the amide planes strongly align in the plane of the film surface. The slight convergence of the two sets as the effective depth is increased suggests that the amide plane alignment is a surface induced effect that decreases in the bulk. The same trend is seen in the drawn bilaminate of Figure 2b where the amide plane alignment is a result of the nylon 11/PVF$_2$ interface. The orientation of nylon 11 at both the air-polymer (Fig. 2a) and polymer-polymer (Fig. 2b) interfaces is much the same.

It is commonly postulated that electric field poling of nylon 11 results in the switching of the electric dipoles associated with the amide group into the poling field direction. A comparison of the N-H stretching data of Figure 2a (single film, one-way drawn) and 2d (single film, one-way drawn and poled at 1.60 MV/cm) shows that very little amide dipole

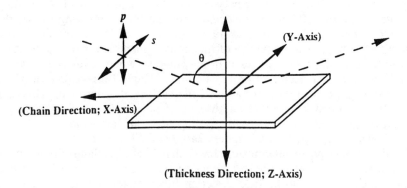

Figure 1. Experimental coordinate system for an ATR experiment.

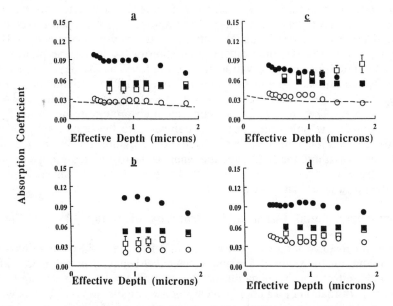

Figure 2. Absorption coefficients of the 3300 cm^{-1} peak of nylon 11 obtained from differently treated films as a function of effective depth. a) drawn single film; b) drawn bilaminate film (polymer-polymer interface); c) drawn and poled bilaminate film (polymer-polymer interface); d) drawn and poled single film (polymer-electrode interface). Open circles - k_x; filled circles - k_y; open boxes - k_z; filled boxes - isotropic equivalent value; dashes - inverse Laplace prediction of k_x.

reorientation occurs at the air-polymer interface as a result of the poling. Since no poling effects are detected with the N-H stretching mode, the amide I mode is also examined for the presence of frequency shifts as a result of poling. The frequency of the amide I mode is related to the ratio of ordered to disordered material in polyamides, where a lower frequency indicates a higher degree of order. Poling of single films of drawn nylon 11 results in no changes in the amide I frequencies of absorption. This data can be explained in terms of an absence of polarization buildup near the polymer-electrode interface when poling at room temperature. On the other hand, a comparison of Figures 2a (single film, one-way drawn) and 2c (drawn and poled bilaminate) shows that N-H stretching dipoles have switched into the electric field direction, i.e., the absorption coefficient intensities increase in the field direction (open boxes) in Figure 2c at the expense of the dipoles in the film plane direction (filled circles). The dipole polarization in the transverse plane after poling is depth dependent as seen in Figure 2c by the opposite slopes of the data sets (open boxes and filled circles). It can further be seen from Figure 2c that after poling, the absorption coefficient intensities in the transverse plane (open boxes and filled circles) are nearly equal, outside of the depth dependence. This suggests a nearly isotropic dipole orientation in the transverse plane as a result of poling, an observation that is contrary to the development of substantial remnant polarization in these nylon 11/PVF$_2$ bilaminates. The apparent dilemma can be understood with a consideration of the amide I frequencies. They show a lower frequency of absorption in the poling field direction in the poled bilaminates as compared to the drawn single film. This observation indicates that in well poled nylons the ordered amide groups switch into the field direction whereas the disordered groups do not. Therefore, the origin of the nearly equal intensities in the transverse plane in the poled bilaminates is explained.

Two-Dimensional Infrared Spectroscopy of Nylon 11

Two-dimensional infrared spectroscopy (2D-IR) is a rapidly emerging new technique for the study of dynamic mechanical behavior in polymers. 2D-IR is based on the hyphenation of step-scanning interferometry and any cyclic perturbation of a sample which produces dipole motion. In this work the cyclic perturbation is dynamic mechanical oscillation. The primary benefits of the technique come from the ability to distinguish the phase of the movement of chemical subgroups in polymer chains, i.o.w., a determination of which subgroups respond in-phase and out-of-phase with the dynamic mechanical stretch. Additionally, through a correlation analysis(19,20) it becomes apparent which subgroups are in-phase with each other during the oscillation and which are out-of-phase. The latter characteristic give 2D-IR spectroscopy a greatly increased spectral resolution. For example, bands in the infrared spectrum from two different chemical groups on a polymer chain may absorb near the same frequency such that in the normal transmission spectrum they are unresolvable. If the two groups have a different phase relationship to an applied dynamic perturbation, often a splitting of the two groups can be seen in the correlation spectra. The splittings are usually more apparent in the asynchronous correlation spectra.

In the recent past 2D-IR spectroscopy has been used to elucidate the mechanical behavior in a variety of polymeric systems. The reorientational behavior in polystyrene of the methylene groups in the main chain and the phenyl ring side groups was separated(*20*). In atactic poly(methyl methacrylate) three type of methyl and methylene groups were resolved in the CH_2/CH_3 stretching region(*21,22*). Significantly different reorientational rates between polymers were observed in a bilaminate of poly(γ–benzyl-L-glutamate) and polypropylene(23). The in-phase dynamic spectra of polypropylene were further analyzed in terms of dynamic mechanical frequency, orientation and static polarization axis(*24*). In this chapter 2D-IR spectroscopy is applied to melt-crystallized samples of nylon 11. The intricacies of the complex amide I region are addressed.

Figure 3 shows the experimental set-up for conducting a 2D-IR experiment with a step-scanning FT-IR spectrometer. In this set-up the beam is brought outside of the main and external sample compartments by parabolic and flat plate mirrors. The beam is passed through an optical filter which filters out either the lower or upper half of the mid-infrared frequency range, and is used to decrease the frequency of the laser sampling, which decreases the scan time. The beam is next passed through the sample film that is dynamically stretched at a frequency of 11 Hz, and finally through a static linear polarizer in front of the detector. The interferometer is stepped through the optical retardation range at 1 step/4 sec while the "dither" of the stepping mirror is set at 400 Hz. Two lock-in amplifiers are used to sort out the electrical signal output from the detector. The first acts to remove the carrier "dither" signal while the second removes the dynamic mechanical frequency. Further details can be found elsewhere(*25*).

In this chapter the 2D-IR spectra are used in order to more closely define the sub-bands within the amide I region of the infrared spectrum of nylon 11. The information is used to give support to the curve-fitting procedures used in the next section of the chapter. In Figure 4 are shown the amide I, amide II and methylene bending regions of the normal absorbance, in-phase and quadrature spectra of a thin film of nylon 11 melt-crystallized on a Teflon substrate. The amide I mode is primarily composed of a carbonyl stretching motion with smaller contributions from C-N stretches and in-plane N-H bends(*26*). The in-phase spectrum shows a bisignate feature associated with both the amide I and amide II modes and can in part be attributed to different directional reorientations associated with the oscillatory strain. As expected, the quadrature signal shows a considerably smaller response as compared to the in-phase signal.

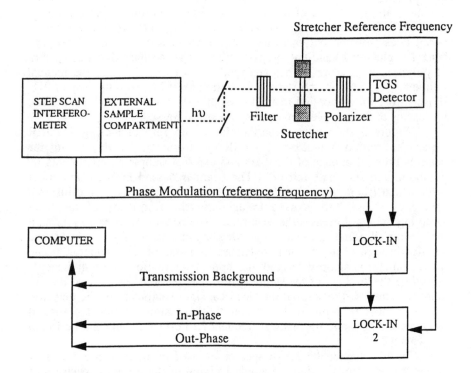

Figure 3. Optical and electronic set-up of the 2D-IR double-modulation-demodulation experiment.

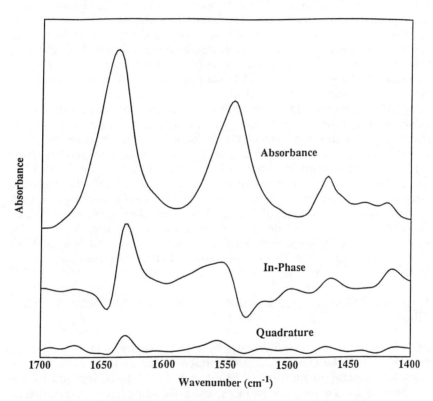

Figure 4. Normal IR absorbance spectrum and the in-phase and quadrature dynamic IR spectra of a thin film of nylon 11 melt-crystallized on a Teflon substrate in the amide I-amide II region.

According to the developments of Noda(*19,20*), the in-phase and quadrature intensities can be processed in a correlation analysis, which has the effect of highlighting the in- and out-of-phase motions between sets of two dipoles. Figure 5 shows the synchronous correlation plot in the amide I/amide II region which results from Figure 4. The main features of the plot are a strong autopeak for the 1636 cm^{-1} part of the amide I band, occurring along the diagonal, and a cross-peak between the 1636 and the 1560 cm^{-1}, the latter of which is a component of the amide II band. Since the sign of the cross-peak is positive, i.e., an intensity which comes up from the page, the reorientation of the dipoles associated with the 1636 and the 1560 cm^{-1} bands is parallel. This can also be observed directly in Figure 4 from the positive-going peaks in the in-phase spectrum (at 1636 and the 1560 cm^{-1}). Figure 4 further establishes that the two dipoles respond in-phase with the applied strain oscillation.

The asynchronous correlation spectra are generally more informative than the synchronous and this also is the case in nylon 11. Figure 6 shows the asynchronous plot in the amide I/amide II region. The amide I region of the spectrum covers the range from about 1610 to 1690 cm^{-1} and the peak splits into four bands at about 1620, 1636, 1648 and 1678 cm^{-1}. The location of the cross peaks in this region indicates that the peak assignable to ordered or crystalline species at 1636 cm^{-1} is out of phase with the three other components of the amide I band. The speed relationships that can be derived from the sign of the asynchronous correlation peaks indicates that the 1636 cm^{-1} reorients faster that the less well-ordered 1648 cm^{-1} peak, in concurrence with other studies. Further interpretations of the 2D-IR spectra of nylon 11 can be found elsewhere(*25,27*).

Trichroic Infrared Transmission Spectroscopy of Nylon 11 Thin Films

As alluded to earlier, the polyamides have an unusual response to one-way drawing in that the planar structures formed by the amide groups tend to align in the plane of the film. The effect is found to be induced by the presence of the air-polymer surfaces, since the alignment decreases as a function of distance from the surface, and by the propensity of the polymers to form hydrogen bonded sheets. A three-dimensional alignment was also observed in one-way drawn poly(vinylidene fluoride)(*18*) for reasons that are not yet fully understood. Trichroic infrared spectroscopy based on a sample tilting is an informative technique to fully characterize samples which display overall symmetry which is greater that uniaxial. Such is the case in one-way drawn, annealed one-way drawn and electric field poled nylon 11.

Trichroic infrared spectroscopy is conducted in order to predict the absorption spectrum in the direction parallel to the electromagnetic radiation, a direction that is normally lost in transmission measurements. It is based on the fact that the intensities in three dimensions can be

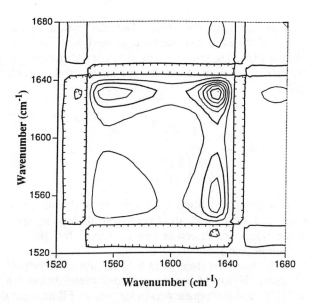

Figure 5. Synchronous 2D-IR correlation spectrum of melt-crystallized nylon 11 in the amide I-amide II region. Cross-hatching indicates negative intensities.

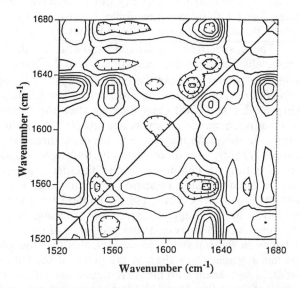

Figure 6. Asynchronous 2D-IR correlation spectrum of melt-crystallized nylon 11 in the amide I-amide II region. Cross-hatching indicates negative intensities.

represented by an intensity ellipsoid. The absorption in any direction within one plane of the ellipsoid can be predicted when at least two other absorptions in the plane are known. The spectrum of the sample parallel to the electromagnetic radiation, a.k.a. the thickness or A_z direction, can be found from(28):

$$
A_z = \frac{A_{\alpha,\beta,\chi}\left(1 - \dfrac{\sin^2\alpha}{n^2}\right)^{1/2} - A_y}{\left(\dfrac{\sin\alpha}{n}\right)^2} + A_y \tag{1}
$$

where $A_{\alpha,\beta,\chi}$ is a polarized tilted spectrum, α is the tilt angle, n is the refractive index and A_y is an untilted polarized absorption spectrum. The spectrum in the thickness direction can be found by tilting around the parallel or perpendicular axes.

All nylon 11 films referred to in this section were initially prepared in the same fashion. Nylon 11 powder was melt-pressed between aluminum foil sheets at 210°C and melt-quenched in ice water. Films were then drawn at a final draw ratio of 3:1 at room temperature. Subsequent annealing treatments were done at 180°C at constant strain. For electric field poling treatments, gold electrodes were evaporated onto both surfaces to a thickness of ~200Å. The field used is 1.50 MV/cm. All poled films were cycled six times from +1.50 to -1.50 MV/cm to remove initial cycling effects. Figures 7 and 8 show the three-dimensional spectra for a one-way drawn film where the A_{TH} spectra are calculated from $A_{\parallel}$ and $A_{\perp}$ and a tilted film spectrum. The $A_{\parallel}$ spectrum is polarized parallel to the draw direction, while $A_{\perp}$ is perpendicular to the draw direction and in the plane of the film. The large difference between $A_{\parallel}$ and the other two spectra is indicative of the chain alignment during the drawing procedure. The frequencies of the N-H stretching modes at ~3300 cm^{-1} indicate that the hydrogen bond strength between adjacent amide groups is anisotropic ($\parallel$ = 3296, $\perp$ =3298, TH = 3301 cm^{-1}). In the plane transverse to the draw direction (transverse plane) the amide groups are more strongly hydrogen bonded in the perpendicular direction as compared to the thickness direction. The anisotropy in the amide group orientation and character is apparent in Figure 8 where the lowest peak position of the amide I band at ~1650 cm^{-1}, which corresponds to the highest degree of ordering, occurs in the perpendicular direction. An analysis of the intensities of the N-H stretching and amide I bands in the transverse plane indicates that the plane of the amide groups has a tendency to align in the plane of the film during the drawing process, in agreement with the ATR studies shown earlier.

Annealing of the drawn nylon 11 films causes the expected effects of peak width narrowing. The intensities of the N-H stretching and amide I modes dramatically increase in the perpendicular direction and decrease in the thickness direction. When the direction of the transition moments of

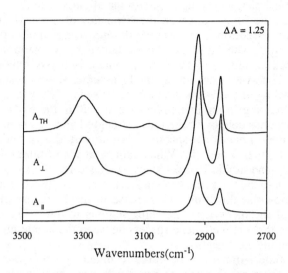

Figure 7. Parallel, perpendicular, and calculated thickness direction infrared absorption spectra for a one-way drawn nylon 11 film in the region 3500 to 2700 cm⁻¹.

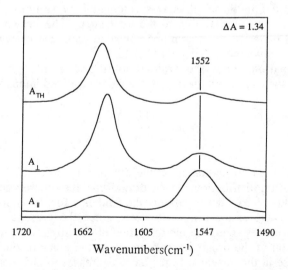

Figure 8. Same as Figure 7 in the region 1720 to 1490 cm⁻¹.

these modes with respect to the amide plane is considered, the intensity changes establish that annealing improves the alignment of the amide planes in the plane of the film. However, the frequency of the N-H modes in the perpendicular direction increases with annealing, indicating weaker hydrogen bonds. The latter can be explained by a examination of the methylene rocking region of the spectra, which is crystal phase sensitive. As a result of one-way drawing, nylon 11 assumes the γ-crystal structure, whereas annealing of γ-crystals produces α-crystals. The γ to α conversion yields a 6% increase in the interchain spacing within the hydrogen bonded sheets(29), which accounts for the lower strength hydrogen bonds.

The three-dimensional spectra of a one-way drawn and poled film are shown in Figures 9 and 10. When compared with Figures 7 and 8, the N-H stretching and amide I peaks show evidence for the switching of the amide planes into the thickness (poling field) direction. Both peaks decrease in the perpendicular direction and increase in the thickness direction. A comparison of the methylene stretching regions in the poled and unpoled films suggests that the methylene spacers between the amide groups are not affected by poling treatments.

The amide I mode can be curve-resolved into four components of order and disorder based partly on previously used procedures(30) and the 2D-IR results shown earlier in this chapter. A Gaussian band shape was chosen because all preliminary trials where the band shape was a variable showed over 90% Gaussian character. The parallel, perpendicular and thickness direction spectra of all samples were curve-fit. The relative amount of order in these films was calculated by summing the curve-resolved areas and normalizing to the total area. The drawn films have 32% ordered material, the drawn and annealed 45%, the drawn and poled 38%, and the drawn, annealed and poled 45%.

Information about the orientation of the amide groups in the transverse plane can be obtained from the curve-resolved intensities using:

$$\theta = \tan^{-1}\left(\sqrt{\frac{A_{TH}}{A_{\perp}}}\right) \tag{2}$$

where θ contains information about the dipole orientation and the breadth of the orientation distribution. In the draw films, the ordered amide groups show a θ value of 36°, demonstrating the tendency for the amide planes to align in the plane of the film. However, since the presence of the air-polymer interface is central to the formation of the alignment, the center of the distribution of the ordered amide groups is the plane of the film. This is also the case in the annealed films, but θ decreases to 28°, demonstrating the tendency to further align in the plane of the film. In the poled films the center of the orientation distribution becomes the thickness direction and θ assumes a value of 59°. All disordered amide I curve-resolved peaks show

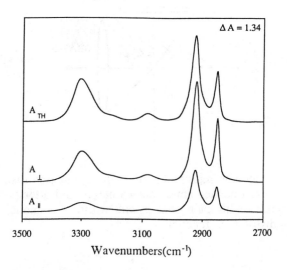

Figure 9. Three-dimensional spectra of a drawn plus poled nylon 11 film in the region 3500 to 2700 cm⁻¹.

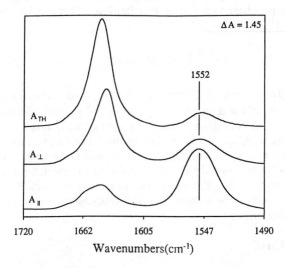

Figure 10. Same as Figure 9 in the region 1720 to 1490 cm⁻¹.

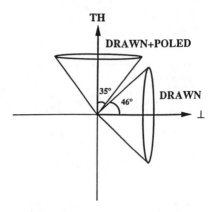

Figure 11. Schematic representation of the orientation distribution of ordered hydrogen bonded amide planes before and after poling. The angle associated with each cone is the width of a model Gaussian orientation distribution function.

near random order in the transverse plane as do the methylene stretching peaks in these remnant polarization studies.

Further analysis of the polarization properties in these materials can be realized by a separation of the effects on the infrared intensities of the width and center of the orientation distribution. This can be accomplished by the application of:

$$A_i = \iint f(\alpha, \theta) M_i^2 d\alpha d\theta \tag{3}$$

to model the intensities. A Gaussian distribution function for f where the width and center of the distribution can be varied is used. Simulated values of θ (as defined in eqn. 2) are generated with eqns. 2 and 3 and compared with the θ values calculated from the spectra. Following this procedure, a 90° rather than a 60° switching mechanism in the ordered regions is considered to be the primary mode of polarization switching in nylon 11. A schematic of the poling process in the transverse plane is shown in Figure 11 where the cones represent the infrared-determined widths of the orientation distributions.

Literature Cited

(1)   Trifan, D. S.; Terenzi, J. F. *J. Polym. Sci.* **1958**, *28*, 443.
(2)   Bunn, C. W.; Garner, E. V. *Proc. Roy. Soc. A* **1947**, *189*, 39.
(3)   Lang, S. B.; Das-Gupta, D. K. *Ferroelectrics* **1984**, *60*, 23.
(4)   Bihler, E.; Holdik, K.; Eisenmenger, W. *IEEE Trans. Elec. Ins.* **1989**, *24*, 541.

(5)   Wübbenhorst, M.; Wünsche, P. *Prog. Colloid Polym. Sci.* **1991**, *85*, 23.
(6)   Flournoy, P. A.; Schaffers, W. J. *Spectrochim. Acta* **1966**, *22*, 5.
(7)   Harrick, N. J. *Internal Reflection Spectroscopy*; Harrick Scientific Corporation: Ossining, New York, 1987.
(8)   Flournoy, P. A. *Spectrochim. Acta* **1966**, *22*, 15.
(9)   Hobbs, J. P.; Sung, C. S. P.; Krishnan, K.; Hill, S. *Macromol.* **1983**, *16*, 193.
(10)   Yuan, P.; Sung, C. S. P. *Macromol.* **1991**, *24*, 6095.
(11)   Kaito, A.; Nakayama, K. *Macromol.* **1992**, *25*, 4882.
(12)   Fina, L. J.; Tung, Y. S. *Appl. Spectrosc.* **1991**, *46*, 986.
(13)   Fina, L. J. In *Advances in Chemistry Series, Structure-Property Relations in Polymers: Spectroscopy and Performance*; C. D. Craver and M. W. Urban, Ed.; American Chemical Society: Washington, DC, 1993; pp 289-303.
(14)   Chen, G. C.; Fina, L. J. *J. Appl. Polym. Sci.* **1993**, *48*, 1229.
(15)   Chen, G. C.; Su, J.; Fina, L. J. *J. Polym. Sci.: Polym. Phys.* **1994**, *32*, in press.
(16)   Jakes, J.; S., K. *Spectrochim. Acta* **1971**, *27A*, 35.
(17)   Moore, W. H.; Krimm, S. *Biopolym.* **1976**, *15*, 2439.
(18)   Fina, L. J.; Koenig, J. L. *J. Polym. Sci.: Polym. Phys.* **1986**, *24*, 2541.
(19)   Noda, I. *Appl. Spectrosc.* **1993**, *47*, 1329.
(20)   Noda, I. *Appl. Spectrosc.* **1990**, *44*, 550.
(21)   Noda, I. *Polym. Prep. Japan* **1990**, *39*, 1620.
(22)   Noda, I.; Dowrey, A. E.; Marcott, C. *Polym. Prep.* **1990**, *31*, 576.
(23)   Palmer, R. A.; Manning, C. J.; Chao, J. L.; Noda, I.; Dowrey, A. E.; Marcott, C. *Appl. Spectrosc.* **1991**, *45*, 12.
(24)   Budevska, B. O.; Manning, C. J.; Griffiths, P. R.; Roginski, R. T. *Appl. Spectrosc.* **1993**, *47*, 1843.
(25)   Singhal, A.; Fina, L. J. *Appl. Spectrosc.* **1994**, in review.
(26)   Fina, L. J.; Yu, H. H. *J. Polym. Sci.: Polym. Phys.* **1992**, *30*, 1073.
(27)   Singhal, A.; Fina, L. J. *J. Polym. Sci.: Polym. Phys.* **1994**, in preparation.
(28)   Fina, L. J.; Koenig, J. L. *J. Polym. Sci.: Polym. Phys.* **1986**, *24*, 2509.
(29)   Mathias, L. J.; Powell, D. G.; Autran, J. P.; Porter, R. S. *Macromol.* **1990**, *23*, 963.
(30)   Skrovanek, D. J.; Painter, P. C.; Coleman, M. M. *Macromol.* **1986**, *19*, 699.

RECEIVED February 2, 1995

Chapter 5

Rheophotoacoustic Fourier Transform IR Spectroscopy To Study Diffusion and Adhesion in Polymers

B. W. Ludwig, B. D. Pennington, and Marek W. Urban[1]

Department of Polymers and Coatings, North Dakota State University, Fargo, ND 58105

The utilization of photoacoustic FT-IR (PA FT-IR) spectroscopy in the studies of polymeric materials and processes has numerous advantages. In this chapter, we discuss selected applications of PA FT-IR spectroscopy applied to diffusion and adhesion studies of polymeric systems. It is demonstrated that, with a proper PA FT-IR experimental set up and theoretical foundations, it is possible to quantify diffusion of small molecules through polymer networks and elucidate crosslinking kinetics of polymer network formation. Furthermore, the utility of rheo-photoacoustic (RPA) FT-IR will be disclosed in its application to quantitatively determine the work of adhesion.

Photoacoustic Fourier transform infrared (PA FT-IR) spectroscopy provides several advantages in the characterization of materials over conventional infrared spectroscopy techniques, adding a multi-dimensional character to the analytical process. For example, while the ability to obtain molecular level information without altering a specimen is one of the appealing features, determination of the stratification and degradation processes in polymers *(1,2)*, non-equilibria processes *(3,4)*, or crosslinking of thermosetting polymers *(5,6)* add another dimension.

A schematic diagram of the basic PA FT-IR experimental setup is presented in Figure 1A. In this experiment, modulated infrared light enters an acoustically isolated photoacoustic cell, and those wavelengths of incident modulated radiation which correspond to vibrational modes of chemical bonds are absorbed. As a result of reabsorption, energy is given off in the form of heat. As heat reaches the surface, pressure variations in the coupling gas over the sample are induced at the frequency of modulated light. These pressure fluctuations are detected by the use of a sensitive microphone, and the signal is Fourier transformed into an infrared spectrum *(7)*.

[1]Corresponding author

0097–6156/95/0598–0078$12.25/0

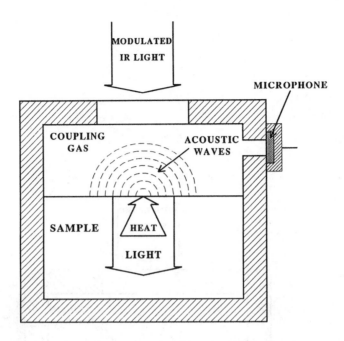

Figure 1A. A schematic diagram of the PA FT-IR cell.

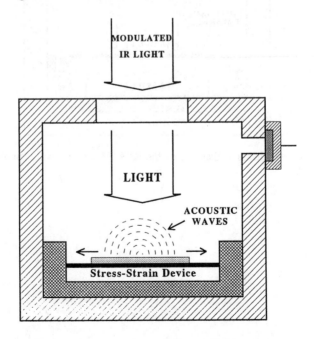

Figure 1B. A schematic diagram of the rheo-photoacoustic (RPA) FT-IR cell.

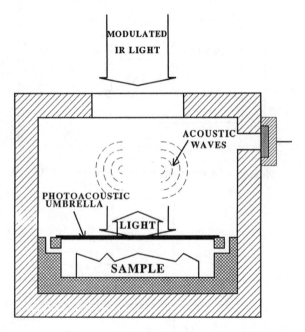

Figure 1C. A schematic diagram of the RPA FT-IR cell with "photoacoustic umbrella".

In an effort to further expand the versatility of the PA FT-IR method, our research group has introduced two new concepts to the basic PA experiment. The first involved the development of the rheo-photoacoustic (RPA) technique *(8)*, which enables collection of the PA spectra from uniaxially elongated samples held at fixed elongations during spectral acquisition. A schematic diagram of the experimental setup is illustrated in Figure 1B, and appears to be sensitive to the interchain bonding of poly(p-phenylene terephthalamide) (PPTA) fibers. Other applications were also demonstrated *(8,9)*. More significantly, this approach enables the determination of the strength of molecular level interactions which contribute to adhesion between polymeric materials *(9-11)*.

The RPA FT-IR setup also opened the opportunity to monitor diffusion of small molecules from polymeric materials *(12-14)*. In this approach, an aluminum "photoacoustic umbrella" was utilized, which allows determination of the concentration of gas-phase species over a solid sample. This is depicted in Figure 1C. Because theoretical *(15)* and experimental details *(13,16)* were presented elsewhere, the following sections will highlight selected applications relevant to polymer morphology and diffusion, followed by molecular level quantitative approaches to polymer-polymer adhesion.

RPA FT-IR Spectroscopy and Diffusion

Poly(vinylidene fluoride). When external forces are applied to semi-crystalline polymers, interactions of tie molecules with crystals, which are generally considered to be impermeable to organic diffusants, inhibit lateral contraction. As crystallites are drawn apart, amorphous regions widen and their volume increases, giving rise to an increase of fractional free volume (FFV). This, in turn, leads to the enhancement of the transport properties through the network *(17,18)*. At higher elongations, it is anticipated that the diffusion coefficient may decrease due to the formation of microfibrillar structures in semi-crystalline polymers. Tie molecules stretched tightly between the crystallites compress the amorphous phase, causing a decrease of the FFV and therefore, permeability through the polymer network *(19)*.

With these considerations in mind, we examined a poly(vinylidenefluoride) (PVDF) network, and the behavior of ethyl acetate (EtAc) molecules migrating through this polymer, using a combination of the rheo-photoacoustic and photoacoustic umbrella techniques shown in Figures 1B and 1C. The mobility of ethyl acetate in PVDF is of particular interest because EtAc is known *(20)* to plasticize PVDF. Furthermore, when PVDF is elongated, the film morphology changes from spherulitic to fibrillar structures, and the monoclinic α phase undergoes transformation to the orthorhombic β phase *(21,22)*, resulting in a modification of the film's transport properties. A portion of the film, which has undergone plastic deformation to the fibrous β form, is referred to as the "neck" of the film.

With this background in mind, let us go back to the experimental setup depicted in Figure 1C, saturate PVDF with EtAc, and use the C=O stretching band of EtAc to monitor its migration to the gas phase. The absorption band at 1768 cm^{-1} attributed to the carbonyl normal vibration of EtAc in the vapor phase, can be used to quantify the amount of EtAc present in the photoacoustic cell by calibrating the cell with a known amount of EtAc. It appears that a linear response of the photoacoustic intensity can be obtained when a photoacoustic umbrella is used to eliminate contributions of EtAc present in the polymer *(12)*.

The effect of elastic deformations on the permeability can be examined by monitoring EtAc diffusing out of the polymer, as a function of strain for films containing various amounts of EtAc. When PVDF films are elongated to induce elastic elongation, the concentration of EtAc in the vapor phase initially increases, followed by its steady decrease. As seen in Figure 2, a maximum rate of diffusion occurs at 1.7% elongation for samples with a high EtAc concentration. Specimens treated in the same manner, but which have a low EtAc concentration, do not exhibit a maximum diffusion rate until 4.2% strain is reached. Based on these observations two competing processes which determine the amount of EtAc diffusing from the film and entering the vapor phase can be identified. Since the PVDF samples are initially saturated with EtAc, the decreasing concentration of EtAc within the film as the experiment proceeds tends to decrease the amount of EtAc diffusing out of the network. However, as the film is elongated, an increase of the FFV of the amorphous phase results in an increase in the rate of evaporation, followed by a rapid decline of EtAc diffusion due to a limited supply of EtAc.

When a polymer film is elongated beyond the range of elastic deformation, plastic deformation is encountered. In this elongation range, PVDF experiences so-called neck formation, which can be envisioned as a narrowing of the center of the specimen as a result of elongation. The increase of EtAc exudating during the initial elongation stage, before the neck begins to form, is attributed to the opening of spherulitic structures and subsequent increase of FFV. Because above 4.2% strain the FFV changes in the regions not undergoing neck formation cease, subsequent deformations occur in the film regions that undergo transformation to fibrous morphology. Elongations up to 5% result in a whitening of the film due to the void formation between spherulites. The voids' effect on the transport properties of PVDF is demonstrated in Figure 3. When elongation reaches 6.7%, a completion of the void formation in regions that form the initial neck is achieved. The voids provide less obstructed diffusion pathways, greatly enhancing transport properties of the network. Thus, the increasing intensity of the 1768 cm^{-1} band due to EtAc in the gas phase is attributed to the opening of the spherulitic structures, which increase FFV and void formation. These data indicate that the permeability of the films reaches a maximum at 6.7% elongation. Extended strains induce no change in the polymer morphology, which could further increase the rate of diffusion, making the amount of EtAc diffusing out of the network dependent upon the EtAc concentration.

Quantitative Analysis.   RPA FT-IR experimental setup makes it possible to obtain quantitative data, which may yield not only information about the diffusion rates, but

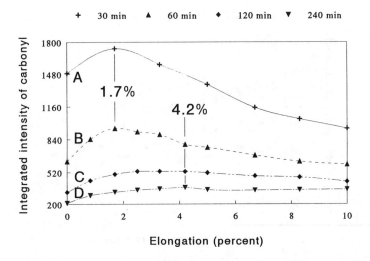

Figure 2. Diffusion monitored in PVDF film undergoing elastic deformation: Integrated intensity of the EtAc carbonyl band in the vapor phase (1768 cm^{-1}) for films strained 0-10%, and allowed to dry prior to experiment for: (A) 30 min.; (B) 60 min.; (C) 120 min.; (D) 240 min. (Reproduced with permission from ref. 12. Copyright 1992 Butterworth-Heinemann Ltd.)

MULTIDIMENSIONAL SPECTROSCOPY OF POLYMERS

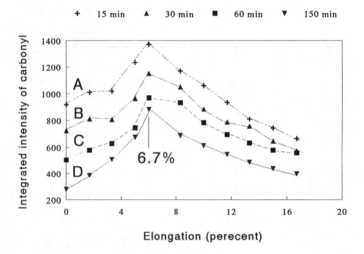

Figure 3. Diffusion monitored in PVDF undergoing plastic deformation: Integrated intensity of the EtAc carbonyl band in the vapor phase (1768 cm^{-1}) for films strained 0-10%, and allowed to dry prior to the experiment for: (A) 15 min.; (B) 30 min.; (C) 60 min.; (D) 150 min. (Reproduced with permission from ref. 12. Copyright 1992 Butterworth-Heinemann Ltd.)

also the changes in polymer morphology. It has been recognized for a long time that the rate of diffusion of small molecules in polymer networks is exponentially related to the diffusant concentration *(19)*. In our case, the following relationship can be used to relate diffusion to the intensity of a photoacoustic signal

$$PA = A \exp[x(-t^{1/2})] \tag{1}$$

where: PA is the photoacoustic intensity of the IR band of the diffusant in the vapor phase, t is the time at which the spectrum was collected (for a saturated sample, $t = t_0$), A is a pre-exponential term representing the log of the integrated photoacoustic FTIR intensity at $t = 0$, and x is the diffusion parameter, which is related to an independently determined desorption diffusion coefficient, D_d, given by equation 3. The natural log of the intensity plotted as a function of the square root of time results in a slope equal to x, a parameter that depends on polymer morphology. The x values obtained for stressed and non-stressed PVDF are listed in Table I. It is apparent from this data that, for PVDF films in a spherulitic form, application of stresses results in the increased rate of diffusion. This is illustrated by the higher x values obtained for the strained films. In contrast, stress decreases the rate of diffusion from the specimens with a fibrillar structure, which is demonstrated by low x values for strained fibrous PVDF.

Diffusion and Polymer Crystallinity. While a qualitative assessment provides one dimension in the analytical process, quantitative analysis becomes essential in multi-component systems. In order to quantitatively measure diffusion processes in PVDF, it is desirable to examine a series of specimens with a crystalline content ranging from 22 to 43% *(13)*. While X-ray diffraction measurements can be used to quantify the amount of crystalline phase, the desorption diffusion coefficient can be determined *(23,24)* using equation 2

$$D_d = 0.05 (L^2 / t_{1/2}) \tag{2}$$

where: D_d is the desorption diffusion coefficient, L is the sample thickness, and $t_{1/2}$ is the desorption half-time.

Figure 4 represents a plot of diffusion coefficients, obtained from equation 2, plotted as a function of x, obtained for the same specimens using equation 1. The relationship between x and D_d was found to be

$$D_d \times 10^6 = A + Bx \tag{3}$$

where: $A = -3.34$ and $B = 21.78$ are system dependent experimental constants. Since x can be determined from RPA FT-IR experiments, equation 3 allows the determination of D_d by measuring the elution of vapor in the RPA FT-IR cell. Such an approach makes RPA FT-IR spectroscopic evaluation of D_d in PVDF, as well as in other polymers, possible. The previously reported method of determining D_d using transmission spectroscopy *(25)* is limited in that respect.

It has been reported in the literature *(26)* that the diffusion rates in polymers are inversely related to the content of crystalline phase. In order to establish the relationship between the diffusion coefficient D_d and the crystalline fraction (α_c) for the PVDF/EtAc system, the diffusion coefficients determined from the relation shown in Figure 4 were plotted as a function of α_c. This is shown in Figure 5. The correlation coefficient between D and α_c is better than 0.995, and this relationship formulates the basis for the assessment that the diffusion coefficient of EtAc in PVDF is proportional to the concentration of the amorphous fraction in the film.

In contrast to other polymers, poly(tetrafluoroethylene) (PTFE) exhibits unusual behavior; it thickens upon uniaxial deformation. This phenomenon is commonly referred to as a negative Poisson's ration (NPR). In an effort to determine the morphological changes, RPA FT-IR measurements of permeability as a function of elongation were conducted concurrently with X-ray diffraction experiments *(14)*. The permeant used as a probe of PTFE morphology was perfluoromethylcyclohexane (PFMCH). In this case, the integrated intensity of the 976 cm^{-1} band, attributed to a C-F deformation of the fluorine group attached to a cyclohexane ring *(27)*, was used to determine the concentration of PFMCH within the RPA cell, allowing the determination of the diffusion rate of PFMCH in PTFE.

Curves A and B of Figure 6 present the relative permeability of PTFE films at various elongations which were determined by monitoring the desorption of PFMCH from samples held at constant strains. Curve C represents the integrated intensity of the X-ray peak at 17.7° 2Θ attributed to the 1010 plane in the PTFE crystal *(28)*. Because this plane will only contribute to the diffractogram when the crystal plains are horizontal, it is a sensitive measure of the proportion of crystals which are parallel to the film surface.

In order to understand the origin of two minima in the diffusion rate curves, and eliminate concerns resulting from the influence of rapidly changing concentration of the diffusant molecules in the early stages of evaporation, the data in curve B of Figure 6 was collected. This data represents specimens which were not saturated with PFMCH. Because saturated (curve A) and unsaturated (curve B) specimens exhibit transitions at the same elongations, kinetic effects due to the decreased concentration of PFMCH in the film as a function of time are therefore eliminated.

Although the initial, rapid decrease of the diffusion rates up to 4% strain depicted in curves A and B of Figure 6 is somewhat surprising, this behavior is likely attributed to intercrystalline amorphous phase chains being pulled taut, limiting the mobility of other amorphous phase chains, and thereby, causing a decrease of the PTFE free volume. However, another possibility is that the decrease of the diffusion rate with increasing strain is due to a reorganization of the polymer morphology, such that more crystals are oriented parallel to the plane of the film. As more impermeable crystals align parallel to the plane of the film, diffusants take longer paths around them in order to reach the film surface. As shown in Figure 6, curve C, the intensity of the X-ray peak at 17.7° 2Θ increases when going from 0 to 4 % strain, indicating the increased number of crystallites

Table I.
Values for the diffusion parameter (x), obtained from
Equation 1 for PVDF samples under various conditions

PVDF Sample	Diffusion time			
	0-140 min.		140-300 min.	
	x^a	$\ln A^b$	x	$\ln A$
Spherulitic form				
0% elongation	0.198	8.70	0.105	7.60
1.7% elongation	0.214	8.86	0.108	7.60
5% elongation	0.232	8.99	0.122	7.73
Fibrous form				
0% elongation	0.168	8.09	0.099	7.26
1.7% elongation	0.167	8.02	0.087	7.09

[a]Exponent in equation 1
[b]Intercept of the integrated intensity axis at time = 0

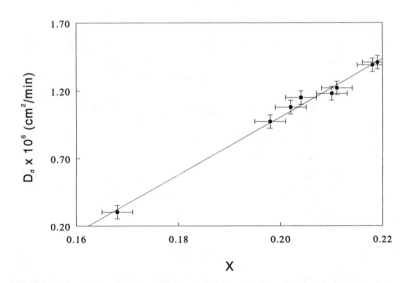

Figure 4. Diffusion coefficients, D_d, determined from equation 1, plotted as a function of x. (Reproduced with permission from ref. 13. Copyright 1993 Butterworth-Heinemann Ltd.)

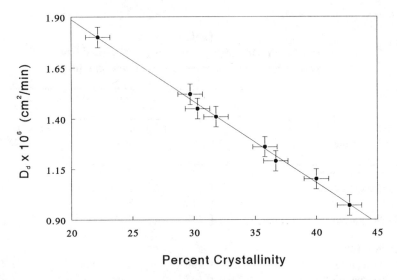

Figure 5. Diffusion coefficients, D_d, determined from the relation in Figure 4, plotted as a function of percent crystallinity. (Reproduced with permission from ref. 13. Copyright 1993 Butterworth-Heinemann Ltd.)

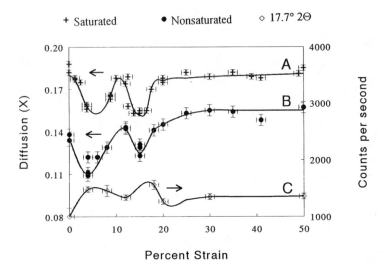

Figure 6. Relative diffusion rates recorded using RPA FT-IR, plotted as a function of x obtained from equation 1. (left Y axes): A-saturated PTFE; B-unsaturated PTFE, exposed for 30 min. before measurements; C-Integrated intensity of the 17.7° 2Θ peak (right Y axis). (Reproduced with permission from ref. 14. Copyright 1994 Butterworth-Heinemann Ltd.)

lying horizontal to the film plane. Increased order in the amorphous regions, as well as an enhanced tortuosity of the diffusion path due to the alignment of crystallites, account for the slower diffusion rates during the first stage of deformation.

Further analysis of the results presented in Figure 6 indicates that, between 4 and 12% strain, a rapid increase of the diffusion rate is observed. This observation corresponds well with the decrease of the $17.7°$ 2Θ peak, and is attributed to the restructuring of the polymer morphology as more tie molecules are pulled taut. A significant fraction of crystals, not aligned parallel to the film surface, rotate perpendicular to the film surface, as they move to assume orientations preferentially parallel to the draw direction. The disruption caused by these rotating crystals is demonstrated in Figure 6, curve C, showing a decrease of the $17.7°$ 2Θ peak, which indicates that a portion of the crystals which were parallel to the draw direction, are displaced from their parallel alignments. The rotation of the crystals perpendicular to the draw direction will increase the thickness of the film, resulting in a NPR.

A third transition in the diffusion and X-ray diffraction data is detected between 12 and 17% strain. As illustrated in Figure 6, permeability decreases rapidly during this stage of deformation. The corresponding increase of the $17.7°$ 2Θ intensity is also seen in curve C of Figure 6, indicating that a greater number of crystals oriented parallel to the plane exists. With more crystals now oriented parallel to the draw direction, the diffusion paths are more tortuous, and with a significant fraction of the crystalline rotations completed, there are fewer disruptions of the amorphous phase. As a result, an overall increase of the density occurs. The increased density, which corresponds to a decreased free volume, in addition to the greater overall orientation of the crystalline phase, results in decreased diffusion rates observed at approximately 17% strain.

When the degree of strain reaches 17%, there are regions within the network which have attained such a high degree of alignment that crystal slippage and chain rupture are the means available of stress relaxation. These mechanisms result in the formation of microvoids within the film and a decrease in the overall order, as illustrated by the decreased intensity of the diffraction peak at $17.7°$ 2Θ (Figure 6, curve C). It is the microvoid formation that provides unhindered pathways through the network, greatly enhancing diffusion rates. The overall order of the film increases between 20 and 40% strain.

Quantitative Analysis of Crosslinking Processes. In our previous studies *(29)*, we demonstrated that the photoacoustic effect is sensitive to crosslinking reactions, such as the formation of a polydimethylsiloxane elastomer network. While network formation reactions lead to thermal property changes affecting intensity of the PA FT-IR signal, side reactions may also occur. For example, during reactions leading to the formation of polyurethane networks, side reactions, such as a reaction between water and isocyanate are not uncommon. This reaction, yielding an unstable carbamic acid which dissociates to form carbon dioxide and an amine, is shown below *(30,31)*:

$$\text{R—N}{=}\text{C}{=}\text{O} + \text{H}_2\text{O} \longrightarrow \left[\text{R}{-}\underset{\text{H}}{\text{N}}{-}\underset{\text{O}}{\text{C}}{-}\text{O}{-}\text{H} \right] \xrightarrow{-\text{CO}_2} \text{R—NH}_2$$

Because the evolution of gaseous products within the urethane film may not be desired, it is important not only to be able to detect, but also to quantify the amount of CO_2 generated in polyurethane formation as a function of time. The first step in this development was to construct a calibration curve. Figure 7 illustrates the relationship between PA FT-IR intensity of the 667 cm^{-1} band due to C = O bending mode in CO_2, plotted as a function of the CO_2 concentration introduced into the cell. Using this calibration curve, the amount of CO_2 evolved at any time during the polyurethane formation can be quantified. An example of the amount of CO_2 evolving from a crosslinking polyurethane as a function of time is depicted in Figure 8, and was obtained using the experimental setup depicted in Figure 1C. A correlation of the band intensity with the calibration curve allows a quantitative assessment of the amount of CO_2 produced during the reaction.

RPA FT-IR Spectroscopy and Adhesion

Measurement of adhesion in a bilayer system is strongly influenced by physico-chemical changes at polymer interfaces. Although RPA FT-IR spectroscopy has been utilized in the studies of polymer-polymer interfacial interactions, no correlations have been made between spectroscopic data and the work of adhesion. Using the experimental setup shown in Figure 1B, one can induce interfacial stresses in a bilayer polymer system and determine the work of adhesion. Before such correlations can be made, it is necessary to develop theoretical foundations leading to assessment of the work of adhesion.

Theoretical Considerations of Vibrational Energy Changes and Work of Adhesion.   A typical bilayer experiment is designed in such a way that a substrate is stretched and an adhered top film resists deformation. Therefore, shear stresses are induced at the polymer-polymer interface. If the interfacial interactions are significant, the potential energy of bonds involved in adhesion will change. If this is the case, vibrational frequencies of chemical bonds involved will also change, and the correlation between the wavenumber changes resulting from stresses induced at the interface and the work of adhesion can be made.

The total potential energy for a bilayer system can be expressed as a sum of the following interactions

$$V_{Total} = V_{11} + (V_{int})_{12} + V_{22} \qquad (4)$$

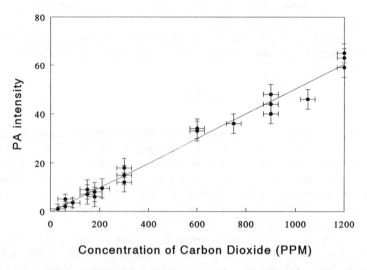

Figure 7. Calibration curve: PA FT-IR intensity of the 667 cm^{-1} band plotted as a function of CO_2 concentration.

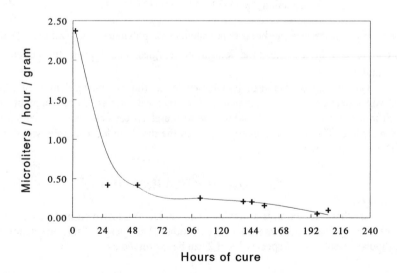

Figure 8. The amount of CO_2 generated from a polyurethane system plotted as a function of time.

where: $(V_{int})_{12}$ is the interfacial interaction potential between a polymer substrate [1] and a top polymer layer [2], and V_{11} and V_{22} are the interaction potentials between respective molecular layers.

The total force at the interface experienced by the bilayer can be determined by differentiating the total potential energy with respect to the change in bond distance to give

$$(\partial V_{Total} / \partial r_1) = (\partial V_{11} / \partial r_1) + (\partial (V_{int})_{12} / \partial r_1) \qquad (5)$$

The total force is equal to the force in the unperturbed substrate molecules and the force resulting from 1-2 interactions. For example, the intermolecular potential for an outer diatomic segment of a substrate will change as a result of interactions with a top film layer. Therefore, a vibrational frequency perturbation is anticipated. Because dispersive, repulsive, dipolar orientational, and inductive interactions have a significant effect on the net potential energy, let us estimate each of the energies and establish their contributions to the work of adhesion.

The dispersion forces are created by the Coulombic interactions of electrons and nuclei in atoms. This motion produces an instantaneous dipole that induces a secondary dipole in the neighboring atoms which can be estimated using the London relationship *(32)*

$$V_{dispersion} = (\alpha_1 \alpha_2 / R_{12}^6) (I_1 I_2 / I_1 + I_2) \qquad (6)$$

where: α_1 and α_2 are the bond polarizabilities of polymers 1 and 2 in the bilayer system, I_1 and I_2 are the bond ionization energies, and R_{12} is the "effective" interbond distance.

The interaction between two molecules at the polymer-polymer interface, one with a permanent dipole moment and another with a zero dipole moment, will result in a secondary (induced) dipole due to the electrical field existing around the first molecule. The attraction energy between the dipole-induced dipole was shown by Debye *(33)* to be

$$V_{induction} = (\mu_1^2 \alpha_2 + \mu_2^2 \alpha_1) / (R_{12}^6) \qquad (7)$$

Another interaction is a dipole-dipole interaction, in which molecules with a permanent dipole moment interact with each other. The energy of this interaction for the dipolar orientation of species 1 and 2 can be expressed as

$$V_{dipolar\ orientation} = (\mu_1 \mu_2 / R_{12}^3) \, \varphi \qquad (8)$$

where: μ_1 and μ_2 is the bond dipole moments, and φ is the orientation factor.

The repulsion component between the electrostatic charges in two interacting molecules can be determined from

$$V_{repulsion} = a \exp(-b\ R_{12}) \qquad (9)$$

where: a and b are empirical constants. The repulsive component results from the force due to the overlap of the electronic orbital structure between the two molecules.

The force for each type of interaction can be determined by taking the derivative of each potential energy term with respect to r_1, giving the overall force

$$F = F_{att} + F_{rep} = (V'_{dis} + V'_{or} + V'_{ind}) + V'_{rep} \qquad (10)$$

As a result of perturbation, the vibrational frequency will change. At the same time, however, the bond distance, as well as the force constant, will also change. To account for these changes, one can adopt an approach similar to that proposed for solute-solvent interactions. It is assumed that the substrate frequency shift produced by the attractive and repulsive forces of the surrounding medium are proportional to the changes in the bond length *(34,35)*

$$\Delta v / v_0 = -a\ (r_1 / r_e) \qquad (11)$$

$$a = r_e\ [-3/2(g/f) + (G/F)] \qquad (12)$$

where: G is the ensemble average for the cubic force over the perturbation coordinates, and g is the cubic force constant. The proportionality constant, *a*, depends on the properties of the two interacting molecules. The g/f term can be determined by utilizing Badger's rule *(36)*, whereas the G/F term can be evaluated using empirical correlation from Oxtoby *(37)* and Herzfeld *(38)*.

From the above relationships, ΔW will be determined by integrating the force of interaction, F, over the limits of equilibrium bond distance to ∞, giving the work of adhesion

$$\Delta W = -\int F\ dr = -\int F_{att} + F_{rep}\ dr$$
$$= -\int V'_{dis} + V'_{or} + V'_{ind} + V'_{rep}\ dr \qquad (13)$$

where: F is represented as the ensemble average of the quadratic force over the perturbation.

Experimental Results of RPA Analysis.   In order to set the stage for spectroscopic measurement of adhesion and correlation of spectral changes with the theoretical derivations shown above, the photoacoustic response to tensile elongation was examined. Figure 9 illustrates PA FT-IR spectra of acrylic-coated

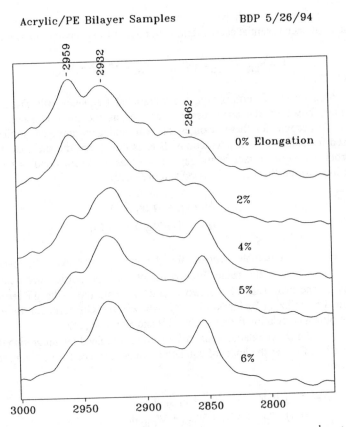

Figure 9. Acrylic / PE RPA FT-IR spectra in the 3000 to 2750 cm^{-1} region.

polyethylene in the 3000-2750 cm^{-1} region elongated from 0 to 6%. The bands at 2930 and 2855 cm^{-1} are due to the polyethylene C-H asymmetric and symmetric vibrational stretching modes, respectively. The spectral features at 0% elongation illustrate the initial conditions of the unperturbed state, while the spectra recorded at various increasing elongations exhibit the development of polyethylene bands. As a tensile force is applied to the substrate, the bilayer system is being stretched and becomes thinner. Should the interfacial attractive forces be strong enough, the acrylic coating will follow the deformation of the polyethylene substrate, causing the thickness of the top layer to decrease. A linear increase in the polyethylene bands would be expected from an uniform elongation, since the sampling depth has remained constant. However, a significant increase of the polyethylene bands at 2930 and 2855 cm^{-1} with elongations exceeding 5% occur, while the C-H stretching band due to acrylic at 2959 cm^{-1} is diminished. A sudden increase of the PE bands is attributed to the presence of interfacial voids which provide a means for heat to escape without passing through the top-coat. This will result in a substantial increase of the substrate band intensities at the point where interfacial failure occurs.

As the bilayer system is elongated, in addition to the intensity changes, vibrational energies of specific bands also change. As seen in Figure 9, as the acrylic-PE bilayer system is elongated, a change of the substrate vibrational energy is reflected in a shift of the bands at 2930 and 2855 cm^{-1}. The CH_2 symmetric band is shifted from 2857 to 2855 cm^{-1}, while the CH_2 asymmetric frequency changes from 2932 to 2930 cm^{-1} for elongations up to 6%. It appears that the magnitude of the wavenumber shift of the C-H stretch is affected by the degree of elongation of the substrate, adhesive properties of the bilayer, and the nature of interactions. For example, shifts up to 16 cm^{-1} were detected for a PDMS-PE bilayer *(11)*. The apparent overlap with the acrylic C-H stretching bands is resolved by correlating the reduced intensity of the 2959 cm^{-1} band due to the thinning process. Since our previous studies *(8)* have shown that elongation of the substrate alone does not produce any changes in the vibrational energy *(11)*, the effect of the interfacial forces accounts for this shift. This information can be used to quantify the data in conjunction with the theoretical considerations presented above (equations 4-13).

Although studies of several bilayer systems showed similar trends, and the substrate bands were detected at various elongations, the full extent of interfacial failure was detected at the point where the substrate band intensity remains unchanged. In the case of epoxy-PP, this elongation occurs at 16%, epoxy-PE at 12%, acrylic-PE at 5%, and acrylic-PP at 6%.

Measurement of Bilayer Interactions with RPA FT-IR. The effect of different intermolecular forces on the work of adhesion is of our central focus. As changes in the substrate bond length produced by elongation occur, vibrational energy changes are induced. At the same time, the potential energy changes with the bond distance. Because the most outer segments of a molecule will be affected by the surroundings, the C-H stretching vibrational energy is expected to be sensitive to environmental changes. Indeed, in all systems we studied, the C-H asymmetric stretching mode is

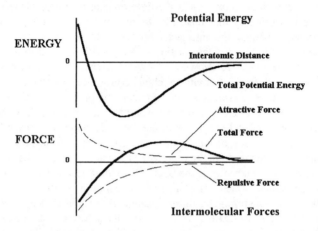

Figure 10. Interlayer energy and intermolecular forces between two molecular layers plotted as a function of distance.

Table II.
Calculated Work of Adhesion Values

	V(dis) ergs * 10^{16}	V(rep) ergs * 10^{10}	V(or) ergs * 10^{18}	V(ind) ergs * 10^{15}	W_A(total) J/m^2 * 10^4
ACRYLIC/PE	1.13	1.41	0	3.02	9.82
ACRYLIC/PP	1.10	1.39	4.17	3.77	8.56
EPOXY/PE	1.12	1.45	0	2.98	7.49
EPOXY/PP	1.09	1.44	4.14	3.71	4.68

often the most sensitive to interfacial stresses. Similar observations were accounted in the studies of solute-solvent interactions (*34*). If one envisions that the repulsive and attractive components of the overall work of adhesion are in a state of equilibrium, the intermolecular distance will determine to what extent each component dominates. This is shown in Figure 10. For an equilibrium situation, the attractive and repulsive forces will be equal. As external forces are imposed on the system, the bond distance decreases, and the repulsive component will dominate. In contrast, as the bond distance increases from the initial equilibrium bond distance, the attractive forces will prevail.

The experimentally measured frequency shifts can be used in equation 11 with the constant a to determine the change in the C-H bond distance from equilibrium. The ratio of the cubic force constant to the quadratic force constant (g/f) can be determined by Badger's rule, with the corresponding forces (G/F) from the empirical correlation expressed by equation 12, to give the proportionality constant, a. With the corresponding potential energy, the quadratic force for each contributing component to the total potential energy function can be determined from equation 10. In view of the above considerations, and according to equations 6-8, there are dispersive, dipolar orientational, and inductive intermolecular forces. However, these attractive forces are not the only ones that are responsible for interfacial interactions. If substrate and film molecules approach each other, the interfacial distance (r) decreases, and at some point, the repulsive forces will counteract attractive forces. Having established each potential energy component contributing to the interfacial interactions, the actual work of adhesion can be calculated from the frequency shifts of vibrational bonds sensitive to interfacial interactions. Using the results from equations 6-9 and combining with equation 13, the estimated value of the work of adhesion for each bilayer system can be determined. Table II summarizes work of adhesion values obtained for selected systems. Since polymeric materials are not expected to strongly adhere, the work of adhesion values are expected to be small in relation to other systems, for example polymer-metal interfaces.

In summary, RPA FT-IR spectroscopy, with a proper experimental setup, can be utilized to monitor diffusion processes in polymers, morphology changes, and interfacial interactions induced in bilayer polymer systems. With proper calibration curves, RPA FT-IR can be used for quantitative analysis of processes at the ppm level. Monitoring changes on the molecular level during substrate elongation allows attractive and repulsive interfacial forces to be related to changes in the substrate bond distance and utilized in the measurement of work of adhesion.

Literature Cited

1. Salazar-Rojas, E.M.; Urban, M.W. *J. Polym. Sci.: Poly. Chem. Ed.* **1990**, *28*, 1593.
2. Urban, M.W. *Prog. Org. Coat.* **1989**, *16*, 371.
3. Huang, J.B.; Urban, M.W. *J. Chem. Phys.*, **1993**, *98*, 5259.
4. Huang, J.B.; Urban, M.W. *J. Chem. Phys.*, **1994**, *100*, 4509.

5. Urban, M.W.; Gaboury, S.R. *Macromolecules* **1989**, *22*, 1486.
6. Gaboury, S.R.; Urban, M.W. *ACS Proc. PMSE* **1989**, *60*, 875.
7. Urban, M.W. In *Structure-Property Relationships in Polymers;* Urban, M.W.; Craver, C.D. Eds.; Am. Chem. Soc.: Washington, D.C., 1993.
8. McDonald, W.F.; Goettler, H.; Urban, M.W. *Appl. Spectrosc.* **1989**, *43*, 1387.
9. Urban, M.W.; Goettler, H. U.S. Patent 5,036,708, **1991**.
10. McDonald, W.F.; Urban, M.W. *J. Adhesion Sci. Technology* **1990**, *4*, 751.
11. Pennington, B.D.; Urban, M.W. *J. Adhesion Sci. Technology* **1995**, in press.
12. Ludwig, B.W.; Urban, M.W. *Polymer* **1992**, *33*, 3343.
13. Ludwig, B.W.; Urban, M.W. *Polymer* **1993**, *34*, 3376.
14. Ludwig, B.W.; Urban, M.W. *Polymer* **1994**, *35*, 5130.
15. Huang, J.B.; Urban, M.W. *J. Chem. Phys.* **1993**, *98*, 5259.
16. Ludwig, B.W.; Urban, M.W. *J. Coatings Technology* **1994**, *66*, 59.
17. Peterlin, A.; Phillips, J.C. *Poly. Eng. Sci.* **1983**, *23*, 734.
18. Peterlin, A. *J. Macromol. Sci.: Phys.* **1975**, *1*, 57.
19. Vittoria, V., et. al, *J. Poly. Sci.: Poly. Phys. Ed.* **1986**, *24*, 1009.
20. Phillips, J.C., et. al, *Poly. Mater. Sci. and Eng.* **1983**, *49*, 555.
21. Siesler, H.W. *J. Poly. Sci.: Poly. Phys. Ed.* **1985**, *23*, 2413.
22. Matsushige, K.; Takemura, T. *J. Poly. Sci.:Poly. Phys. Ed.* **1978**, *16*, 921.
23. Crank J.; Park, G.S. *Diffusion in Polymers;* Acedemic Press: New York, NY, 1969.
24. Mirkin, M.A. *Poly. Sci. U.S.S.R.* **1989**, *31*, 447.
25. Phillips, J.C.; Peterlin, A. *J. Poly. Sci.: Poly. Phys. Ed.* **1981**, *19*, 789.
26. Phillips, J.C.; Peterlin, A. *Poly. Eng. Sci.* **1983**, *23*, 734.
27. Thompson, H.W.; Temple, R.B. *Trans. Faraday Soc.* **1945**, *41*, 236.
28. Wecker, S.M. PhD. Thesis, Northwestern University, Evanston, IL, USA, 1973.
29. Urban M.W.; Gaboury S.R. *Macromolecules* **1989**, *22*, 1486.
30. Wicks, Z.W.; Jones, F.N.; Pappas, S.P. *Organic Coatings, Science and Technology;* Wiley: New York, NY, 1992.
31. Robinson, G.N.; Alderman, J.F.; Johnson, H.W. *J. Coatings Technology,* **1993**, *65*, 51.
33. London, F. *Trans. Far. Soc.* **1937**, *33*, 8.
33. Debye, P.J.W. *Phys. Z.* **1920**, *21*, 178.
34. Zakin, M.R.; Herschbach, D.R. *J. Chem. Phys.* **1985**, *83*, 6540.
35. Zakin, M.R.; Herschbach, D.R. *J. Chem. Phys.* **1986**, *85*, 2376.
36. Herschbach, D.R.; Laurie, V.W. *J. Chem. Phys.* **1961**, *35*, 458.
37. Oxtoby, D.W. *J. Chem. Phys.* **1979**, *70*, 2605.
38. Schwartz, R.N.; Slawsky, Z.I.; Herzfeld, K.F. *J. Chem. Phys.* **1952**, *20*, 1591.

RECEIVED February 2, 1995

Chapter 6

Step-Scan Fourier Transform IR Studies of Polymers and Liquid Crystals

Richard A. Palmer[1], Vasilis G. Gregoriou[1,3], Akira Fuji[1,4], Eric Y. Jiang[1], Susan E. Plunkett[1], Laura M. Connors[1,5], Stephane Boccara[1,6], and James L. Chao[2]

[1]Department of Chemistry, Duke University, Box 90346, Durham, NC 27708–0346
[2]IBM Corporation, Research Triangle Park, NC 27709

The paper reviews the application of dynamic infrared spectroscopy to the study of the molecular and sub-molecular (functional group) origins of the macroscopic rheological properties of organic polymers and liquid crystals and particularly the responses of polymer films to modulated mechanical fields and of liquid crystals to modulated (AC) or pulsed (DC) electric fields. The emphasis is on the use of step-scan FT-IR spectroscopy for these measurements. Two examples of applications, one to dynamic mechanical analysis of a tri-block co-polymer and the other to the study of the electro-reorientation dynamics of a low molecular weight nematic liquid crystal, are presented to illustrate the power of the step-scan method.

Dynamic spectroscopy can conveniently be divided into experiments that use the impulse-response technique (time-resolved, or time domain, spectroscopy) and those that use modulation-demodulation techniques (phase-resolved, or frequency domain, spectroscopy). In the impulse-response mode the response of the sample is measured as an explicit function of time, whereas in the modulation-demodulation mode, it is the phase of the sample response with respect to that of the perturbation that is detected. For dynamic spectroscopy in the near-IR, visible and UV regions, the high intensity of sources, and the high speed of detectors allow great flexibility for dynamic measurements. However, in the mid-infrared the relative weakness of broad band sources and slowness of detectors dictate that virtually all *broad spectral range* measurements, both static and dynamic, use the multiplex and throughput advantages of Fourier transform infrared (FT-IR) methods. Conventional continuous-scan, or rapid-scan, FT-IR is remarkably effective for static measurements, but its applications to dynamic spectroscopy are limited generally to time domain measurements with $t_{1/2}$ ≥ 1 s and frequency domain measurements with f ≥ 1 kHz (although some faster time domain measurements have recently been demonstrated by both synchronous and

[3]Current address: Polaroid Corporation, 1265 Main Street, W4–1D, Waltham, MA 02254
[4]Current address: Kao Corporation, 2–1–3 Bunka Sumida-ku, Tokyo 131, Japan
[5]Current address: Department of Chemistry, Roanoke College, Salem, VA 24153
[6]Current address: Ecole Supérieure d'Ingenieurs de Marseille, IMT–Technopole de Chateau-Gombert, Marseille 13451, France

0097–6156/95/0598–0099$12.00/0
© 1995 American Chemical Society

asynchronous sampling). These limitations are basically a consequence of the time dependence of the spectral multiplexing (inherent in the constant velocity operation of the interferometer mirror), which must be separated from the time dependence of the sample signal, and by the practical limitations on mirror velocities. This is especially a problem for frequency domain measurements when the modulation frequencies are in the range of the natural Fourier frequencies of the continuous-scan interferometer, as well as for certain ranges of repetition rates and time resolutions in time domain experiments. Although dramatic progress has been made in recent years in dynamic IR measurements by means of pump/probe laser methods; in terms of both time and spectral resolution, as well as sensitivity, FT-IR retains an important advantage in cost, ease of operation, and when the ms-ns dynamic response of the sample over a wide range of wavelengths must be measured.

Step-Scan FT-IR. The limitations of the continuous-scan FT-IR method for dynamic spectroscopy are avoided by the use of the step-scan mode of data collection (*1,2*). In step-scan FT-IR, data are collected while the retardation is held constant (or while it is oscillated about a fixed value). Thus the time dependence of the spectral multiplexing is removed. For modulation-demodulation experiments the step-scan mode is particularly powerful, especially when modulation frequencies below 1 kHz must be used. The advantages of step-scan operation include the ability to apply virtually any modulation frequency to the IR radiation and to carry out multiple modulation frequency experiments. In addition, the use of lock-in amplifiers (or of digital signal processors (DSPs))(*3*) provides a high degree of noise rejection, analogous to the Fourier filtering effective in the continuous-scan mode. Another advantage is the easy retrieval of the signal phase, which is possible because the in-phase and in-quadrature components of the signal are easily obtained as outputs of a two-phase lock-in amplifier (or DSP), and because both components have the same instrument (beamsplitter) phase. These vector components can either be used individually or used to calculate the phase spectrum (as in photoacoustic spectral depth profiling) or to calculate a frequency correlation function (as in the 2D FT-IR analysis of dynamic polymer rheology and liquid crystal electro-optical data described in this paper).

The capabilities of step-scan FT-IR are particularly appropriate for the frequency domain study of the molecular reorientations associated with mechanical and electrical field perturbations in polymers and liquid crystals since the modulation frequencies used are typically below 100 Hz. The IR spectrum offers information on the microscopic (molecular, sub-molecular, and functional group) origins of macroscopic rheological behavior, and *FT*-IR allows high sensitivity with high spectral resolution and a broad spectral range for these measurements. Finally, *step-scan* FT-IR provides access to the real-time dynamics of these processes. The analysis of these data by frequency correlation methods can be used to create 2D FT-IR spectra (*4,5*).

In addition to its use in frequency domain measurements, the step-scan mode of FT-IR can also be used very effectively and flexibly for time domain (impulse-response) experiments. Whereas both synchronous (stroboscopic) and asynchronous sampling in the continuous-scan mode are constrained, with respect to usable repetition rates, by the available range of mirror velocities, the step-scan mode is not. Since the retardation can, in principle, be maintained constant indefinitely in the step-scan mode, data may be coadded from any desired number of repeated pulse sequences at any desired repetition rate. In the conventional time domain configuration with the temporal response to the perturbation pulse monitored by the IR emission or absorption, the time resolution is limited only by the speed of the detector, the speed and capacity of the data collection and processing electronics and the strength of the signal. With respect to detectors, standard (photoconductive) mercury-cadmium telluride (MCT) detectors allow practical time resolution of ~1 μs, whereas photovoltaic MCT's are available with rise times of < 10 ns. The limit of current standard 16 bit analog/digital converters (ADC's) is ~5 μs; faster digitizers (e.g., 8 bit) are required in order to take advantage of the speed of the faster detectors. Time resolution below 1 ns does not seem likely in either step-scan or continuous-scan FT-IR without the introduction of pump/probe techniques and the use of a ps wide-band probe pulse, such as recently demonstrated in asynchronously sampled continuous-scan FT-Raman by Sakamoto et al. (6).

2D FT-IR. The concept of 2D correlation spectroscopy was originated by I. Noda (7). In the 2D analysis, the in-phase and in-quadrature responses of a sample to a synchronously modulated perturbation (as monitored, for example, by the change in the IR spectrum) are used in a cross-correlation analysis analogous in some respects to that of 2D nuclear magnetic resonance techniques. The original 2D IR work was based on spectra obtained over limited wavelength ranges by use of dispersive instrumentation and was applied to the study of polymer rheology through a combination of infrared spectroscopy and dynamic mechanical analysis (DMA) (8).

The first 2D *FT*-IR data were published in 1991 (9). FT-IR is especially valuable for 2D analysis when making correlations over a wide wavelength range because of the technique's well-known multiplex and throughput advantages; however, as mentioned above, the step-scan method of FT-IR data collection is particularly important for the low-modulation frequencies appropriate for studies of molecular reorientation. As pointed out by Noda, this type of analysis can be used on a wide variety of modulation-demodulation dynamic measurements in correlating responses to either the same stimulus or two different stimuli (4).

The specific rules for analysis of the 2D IR plots have been detailed by Noda (4). Briefly, in the experiments described here, the cross peaks in the synchronous correlation indicate which pairs of transition dipoles (and thus which pairs of functional groups) are reorienting in phase with each other in response to the perturbation and whether this reorientation is mutually parallel or perpendicular. In contrast, cross peaks in the asynchronous correlation indicate mutually out of phase reorientation and the relative rates of reorientation of different transition dipoles, or functional groups. In effect, these 2D plots accentuate the relative phases of the responses of those functional groups associated with the various bands in the spectrum and thus provide a view of the relative sub-molecular dynamics of the molecular reorientation process.

Experimental

Instrumentation. The FT-IR instrumentation developed for the original polymer and liquid crystal sub-molecular rheology work in our laboratory has been described previously (10). In this instrument (an IBM-Bruker IR-44 converted to optional step-scan operation) phase modulation (path difference modulation) is used with dual lock-in feed-back loops both to control the retardation and to provide for accurate stepping. As a result, this instrument is particularly adapted to frequency domain measurements, although somewhat limited for time domain experiments. The frequency domain techniques used with this instrument for the modulated mechanical field studies of polymer films (1,11—13) and the modulated electrical field studies of liquid crystals (1,15) have also been described. In both types of experiment, sample modulation frequencies in the range of 10-30 Hz are used. In order to take advantage of the sensitivity of the MCT detector, a higher frequency, usually on the order of 400 Hz, is used as a carrier. For reasons previously discussed (10), this carrier frequency is best provided by path difference modulation (phase modulation) of the interferometer.

In the usual step-scan modulation/demodulation experiment the data are demodulated sequentially. The first stage demodulation at the 400 Hz carrier frequency yields the static absorption spectrum of the sample. This output is also split to the second stage lock-in amplifier, where two-phase demodulation (at the frequency of the mechanical or electrical field modulation) yields the in-phase and in-quadrature components of the dynamic response to the perturbation. Although high-frequency polarization modulation can also be added (4), this was not used in the examples included in this paper. However, the signal is enhanced by prior orientation of the sample and the use of static polarization of the IR beam. The dynamic in-phase and in-quadrature single beam transmission signals are normalized to the static transmission spectrum and then converted to the corresponding components of the dynamic absorption spectrum. It is these data which are used to calculate the 2D FT-IR frequency correlation maps described above. A schematic of the experimental setup is given in Figure 1.

For the time domain experiments the commercial step-scan modification of the Bruker Instruments IFS 88 spectrometer was employed. The general design and operation specifications of this instrument have been described (15). For the results presented here, the instrumental set-up is illustrated in Figure 2. The time-resolved data for these relatively slow molecular reorientation processes were collected by use of the standard 16-bit ADC, although a dual 8-bit ADC internal transient digitizer circuit, which has the capability of 5 ns/200 MHz speed, is available when higher time resolution is needed.

The master clock for the experiment is the SRS Digital Delay/Pulse Generator Model DG 535, which pretriggers the bench (TTL) to start data collection before the perturbation is applied. In this way the reference interferogram is collected as pre-pulse time slices alternating with the response (post-pulse) time slices at each step of the interferometer. The DG 535 also is used to set the voltage (perturbation) pulse profile and repetition rate by triggering the Data Pulse 100 pulse generator after a suitable delay time for reference point collection. The Data Pulse 100 provides either a short pulse (typically 10 V) followed by a longer delay (0 V) during which the sample recovers. Alternatively, a symmetrical or unsymmetrical square wave can be applied, with or without a DC offset, depending on the nature of the sample response. This process is repeated as many times as required to achieve the desired signal-to-noise ratio, at each data collection point of the interferogram. The free spectral range and spectral resolution are determined, as in the continuous-scan mode, by the spacing of the data points and the total change of retardation (length of the interferogram), respectively. After each step, a mirror settling time of 20-50 ms is allowed before the reference data are collected.

Polymer Rheo-Optics. For the polymer rheo-optical studies, films, typically ~50 μm thick, are subjected to a sinusoidally modulated tensile strain of amplitude ~50 μm and frequency between 10 and 50 Hz. In this combination of infrared spectroscopy and dynamic mechanical analysis the amplitude of the deformation is typically less than 0.5 % of the length of the sample and is confirmed to produce only a linear response. In order to enhance the dynamic signal due to the reorientation, the IR light transmitted by the sample is polarized parallel to the direction of the modulated deformation. The micro-rheometer has been described previously (14) and is schematically represented in Figure 3.

Liquid Crystal Electro-Reorientation. In the liquid crystal electro-reorientation studies, liquid crystal films, typically ~10-20 μm thick, are placed between germanium plates which have been surface-treated so as to induce as nearly as possible, a single, homogeneous phase. The surface treatment involves coating the surface of the plate with a 0.1% aqueous solution of polyvinyl alcohol (PVA) and then rubbing the residual PVA film unidirectionally with a cotton swab. The two surface-treated plates are placed face-to-face, anti-parallel, and separated by a ~10 μm spacer, in the controlled-temperature cell. The pathlength between the plates is determined by the use of the infrared interference fringe pattern produced by the empty cell. The liquid crystal is injected between the plates, where it is drawn and held by capillarity. The desired degree of orientation of the homogeneous phase, that is, that a single domain has been formed, is verified by measuring the static polarization ratio of characteristic bands in the spectrum which are indicative of orientation of the director. For example, a well-aligned cyanophenyl mesophase sample will have a ratio $A_{\parallel}/A_{\perp}$ of ~3 for the $C \equiv N$ stretching band. (The parallel direction is the rubbing direction of the plates and

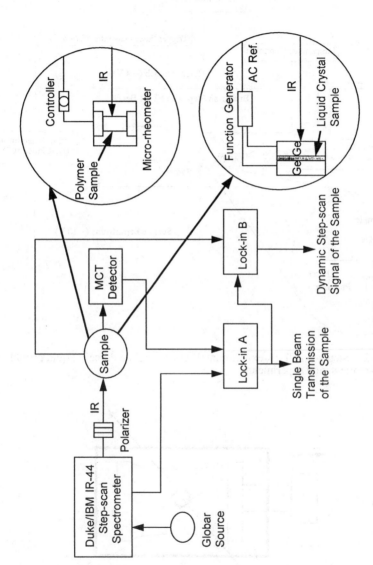

Figure 1. Step-scan FT-IR experimental setup for frequency domain dynamic polymer rheo-optic and liquid crystal electro-reorientation experiments.

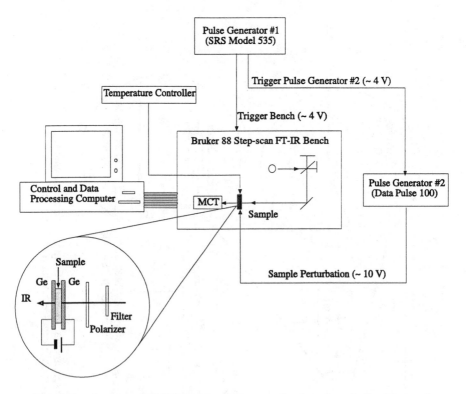

Figure 2.   Step-scan FT-IR experimental set-up for time domain liquid crystal electro-reorientation experiments.

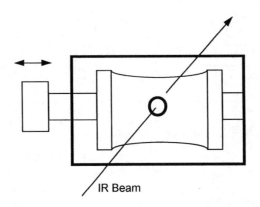

Figure 3.   Micro-rheometer schematic.

is also the orientation of the polarizer during the time-resolved measurements.) A schematic representation of the variable temperature liquid crystal dynamic FT-IR cell used in the work reported here is shown in Figure 4. It is a modified Harrick controlled temperature cell and is capable of $\pm 0.2°C$ control between 25 and 85°C.

In the frequency domain studies the periodic transition to the homeotropic phase is achieved by application of an AC potential, typically 5-10 V, across the germanium plates. In some cases a DC off-set is used in order to compensate for a threshold voltage effect and thus achieve the maximum response with the minimum AC potential. However, the AC potential may also be applied without an off-set. The zero off-set technique is useful for mesophases which are sensitive to DC fields because of ionic or other impurities. Because of the dependence of the director orientation on the square of the electric field, the signal may be usefully detected at the first harmonic of the modulation frequency in these instances. For nematic mesophases the AC frequency used is on the order of 5-15 Hz. The molecular and sub-molecular dynamics of the reorientation process are monitored by the same techniques as in the polymer rheology experiments (above), and the normalized in-phase and in-quadrature signals used to calculate analogous 2D frequency correlation maps.

For the time domain studies the same sample preparation techniques and temperature-controlled cell as used for the frequency domain experiments are used. The appropriate voltages are determined by preliminary continuous-scan spectra measured as a function of applied DC voltage, as in the frequency domain studies, and are typically on the order of 10-20 V. The appropriate pulse profile and repetition rate are determined from initial trials in which the approximate voltage-on and voltage-off response times are observed with the aid of an oscilloscope.

For the nematic liquid crystal experiments a typical pulse sequence is as follows: after the settling time, 250-400 data are collected at 1 ms intervals, 20 reference data before the perturbation pulse is applied and then 230-380, during and after the pulse, sufficient to follow the entire response cycle. The voltage pulse may be short with respect to the response time of the mesophase or may be maintained until the sample has reached a reorientation maximum (homeotropic state). Typically, there is a hysteresis in the homogeneous/homeotropic transition, and the voltage-on pulse is followed by a much longer voltage-off period, during which the sample relaxes to the homogeneous state.

For the time domain nematic liquid crystal results presented here, the step-scan data were collected at 4 cm^{-1} spectral resolution with an undersampling ratio of 4 (data point spacing of $2\lambda_{HeNe}$; free spectral range of 3950 cm^{-1}), with a phase resolution of 64 cm^{-1}, Mertz phase correction, Blackman-Harris apodization and 4 orders of zero-filling. A 3950 cm^{-1} low-pass optical filter was used to prevent aliasing. The < 1 μs-rise time liquid nitrogen-cooled MCT detector was DC-coupled and the offset was carefully zeroed before each scan.

The packed interferogram comprised 1888 retardation points, each consisting of 250 temporal points at 1 ms intervals, which were each the average of the response to 50 voltage pulses. The packed interferogram was unpacked and each temporal component was transformed by use of the standard Bruker (Opus) software. The transforms (single beam transmission spectra) were averaged in groups of 10 to further increase the SNR, and then those collected after the initial pulse-on were ratioed to the average of those acquired before (reference slices) to yield $\Delta A(\bar{v},t)$ spectra at 10 ms intervals (where $-\log[I(\bar{v},t)/I_0(\bar{v})] = \Delta A(\bar{v},t)$). The sequence of $\Delta A(\bar{v},t)$ spectra follow the entire homogeneous-homeotropic-homogeneous transition process.

Results and Discussion

Modulated Mechanical Field Responses of Polymer Films. The combination of dynamic mechanical analysis and infrared spectroscopy is a well established technique of polymer rheology (*16*). The analysis of DMA/IR data to

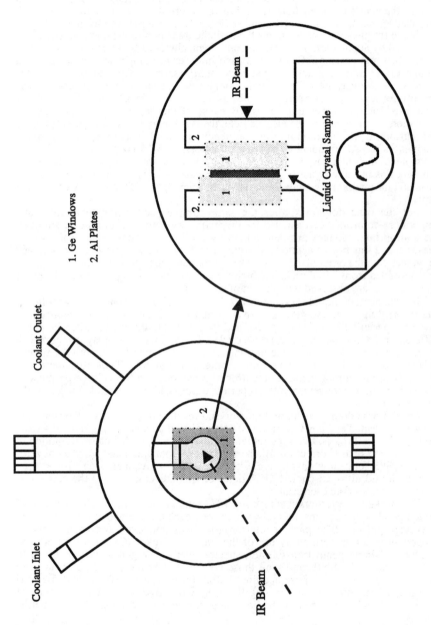

Figure 4.   Variable temperature liquid crystal IR cell.

produce 2D IR spectra was introduced by Noda (*4,7*). The power and utility of the method has now been illustrated in numerous studies of polymer films by both dispersive and FT instrumentation (*1,8,9,11-13,17—21*). As mentioned above, the ability to apply FT-IR instrumentation to these investigations has been linked to the development of reliable and practical methods of step-scan data collection.

As an example of the application of step-scan FT-IR to dynamic polymer film rheology, Figures 5a and 5b show the synchronous and asynchronous 8 cm^{-1} resolution 2D FT-IR correlation spectra of a film of the micro phase-separated tri-block copolymer Kraton D1102 (styrene-butadiene-styrene; butadiene:styrene = 72:28 weight %), solvent-cast on a Teflon supporting substrate. These 2D spectra are calculated from the changes in the polarized transmission spectrum that were in-phase and in-quadrature with a 23.5 Hz sinusoidally modulated uniaxial tensile strain applied by the micro-rheometer. The data were first demodulated at the 400 Hz phase modulation (carrier) frequency. The output of this lock-in amplifier was split, one part to be transformed to the reference transmission spectrum and the other to provide input to the second lock-in, which was referenced to the micro-rheometer frequency. The outputs of the second lock-in amplifier, i.e., the I and Q interferograms, were transformed and then ratioed to the (400 Hz) transmission spectrum. The resulting dynamic absorption vector components were then used to produce the 2D spectra shown. Only the region of the asymmetric CH$_2$ stretching frequency is shown since this is the region of most significant dynamic response.

In the CH stretching region (3200-2800 cm^{-1}) of the static spectrum of the Kraton three sub-regions can be distinguished (*22*). The aromatic CH stretching modes of the polystyrene (PS), as well as the vinyl CH stretch of the polybutadiene (PB) are localized between 3100 and 3000 cm^{-1}. The CH$_2$ stretches of both PB and PS are very closely overlapping, with the symmetric modes occurring at 2840 and 2855 cm^{-1}, respectively, and the asymmetric stretches at 2918 and 2932 cm^{-1}, respectively. In both PS and PB the asymmetric CH$_2$ mode is the most intense of the CH stretching bands. However, the 72:28 PB:PS weight ratio in the tri-block copolymer, combined with the structures of the two components, predicts an approximately 10:1 dominance of the CH$_2$ region in the static spectrum of the copolymer by the vibrations of the PB. The 2:1 ratio of vinyl protons to aromatic protons in the copolymer suggests a more equal sharing of the intensity in the region above 3000 cm^{-1} between the CH modes of the two components.

In the dynamic 2D maps virtually all the intensity is confined to the region of the asymmetric CH$_2$ modes, a region, as pointed out above, dominated by absorption in the PB phase in the static spectrum. Curiously, neither the region of the symmetric CH$_2$ modes nor that of the vinyl/aromatic CH modes shows any detectable diagonal peaks in the synchronous 2D map. The only suggestion of dynamic response in these modes is the appearance of weak off-diagonal peaks coupling them to the strong asymmetric CH$_2$ modes in both the synchronous and asynchronous maps. The fact that at room temperature PS is below its glass transition temperature, whereas PB is not, could explain the absence of dynamic response in the aromatic CH region. This would suggest that the deformation of the phase-separated block copolymer is concentrated essentially in the rubbery PB phase, while the glassy micro-domains of PS remain essentially rigid bodies with little internal reorganization. However, the relative lack of response from the vinyl CH and symmetric CH$_2$ modes (also concentrated in the PB phase) is puzzling.

The most intriguing feature in the dynamic spectra is the existence of the multiple bisignate peaks in the region of the asymmetric methylene CH$_2$ stretching band centered at 1920 cm^{-1}. The 2D FT-IR synchronous and asynchronous correlation maps in Figure 5 enhance the resolution of this structure and clearly suggest the existence of sub-structures in the PB phase which have different temporal response to the modulation of the tensile strain. The synchronous correlation (Figure 5a) shows a positive cross peak between components at 2935 and 2918 cm^{-1}, and the lower half of the asynchronous correlation (Figure 5b) shows negative peaks between

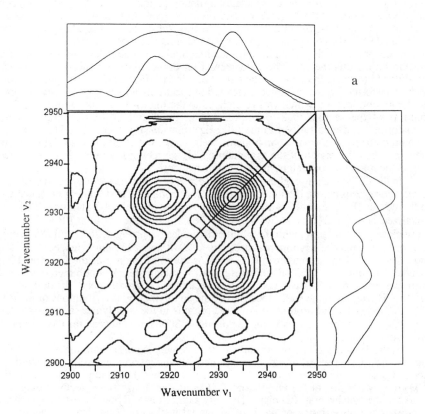

Figure 5. Step-scan 2-D FT-IR spectra of Kraton film response to sinusoidally modulated tensile strain.   a. Synchronous correlation; b. Asynchronous correlation. Shaded peaks are negative. Reference spectra along the top and side of each correlation map represent the static absorbance and the dynamic magnitude x60 (structured spectrum).

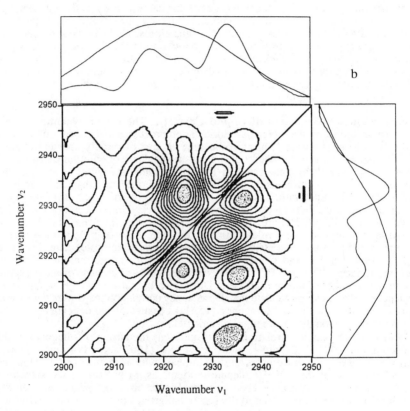

Figure 5. Continued.

components at 2935 and 2918, between 2935 and 2932 and between 2924 and 2918 cm^{-1}, as well as a positive cross peak between components at 2932 and 2924 cm^{-1}. No structure is evident in the corresponding static (1D) spectra (shown at the top and right of each 2D spectrum), nor could any be resolved by Fourier self-deconvolution of the band centered at 2920 cm^{-1}. (The *structured* spectrum also shown at the top and side of the 2D correlation maps is the *dynamic magnitude*, M = (I^2 + Q^2)$^{1/2}$.)

These data thus illustrate the two most important aspects of the enhancement provided by the 2D IR analysis. The dramatic difference in the intensity of the peaks for PS and PB in the dynamic in-phase and in-quadrature spectra (that is, that the PS bands appear to be essentially null), suggests that the two major components of the copolymer are virtually independent in their rheological response. And the appearance within the otherwise unresolved (predominantly PB) CH$_2$ asymmetric stretching band centered at 1920, of a complex pattern of bisignate cross peaks in the asynchronous 2D spectrum suggests the presence of microdomains in the PB component which respond out of phase with each other to the modulated strain. It is likely that these domains are differentiated with respect to their degree of association with the surfaces of the micro phase-separated, glassy and relatively unresponsive PS domains within the structure of the copolymer.

Electric Field Reorientation of Liquid Crystals. The use of dynamic infrared measurements to study the molecular reorientation processes associated with electric field effects in liquid crystals has also become an area of rapidly growing interest *(22,23)*. An example of the study of the sub-molecular reorientation dynamics of nematic liquid crystals by step-scan 2D FT-IR is illustrated in Figures 6a and 6b. The synchronous and asynchronous 2-D FT-IR frequency correlation maps derived from the in-phase and in-quadrature IR responses, respectively, of a 12.5 μm film of the nematic mesophase 5PCH [1-(4-cyanophenyl)-4-pentylcyclohexane] to a ± 8 V, 7.5 Hz AC field at 42 ± 0.2°C are shown. The 2-D spectra show the correlation of transition dipole responses related to the the sub-molecular dynamics of the homogeneous-homeotropic transition, as the liquid crystal molecules "flip" back and forth between parallel and perpendicular alignment to the germanium window-electrode surfaces. The zero DC off-set technique was used because of the observed long-term instability of this mesophase to DC fields. In this method the in-phase and in-quadrature data used to calculate the 2D frequency correlation maps were detected at the first harmonic of the AC frequency (15 Hz) because of the dependence of the director orientation on the square of the electric field vector.

In the synchronous correlation map (Figure 6a) the strong off-diagonal peaks connecting the C≡N (2227 cm^{-1}) and phenyl C-C (1607 and 1494 cm^{-1}) modes confirm that these transition dipoles reorient in-phase with each other and thus that the cyanophenyl part of the 5PCH molecule reorients as a rigid rod in the switching process. However, in the asynchronous map (Figure 6b) cross peaks between the vibrational bands that correspond to the functional groups of the cyanophenyl group and those of the rest of the molecule show that the rigid and flexible parts of the molecule reorient out of phase with each other in response to the AC field. The signs of the cross peaks in the asynchronous map, specifically those connecting the CH$_2$ deformation peak at 1467 cm^{-1} with the C≡N and phenyl C-C stretching bands at 2227 and 1607/1494 cm^{-1}, respectively, indicate that the pentyl "tail" of 5PCH leads the rigid portion of the molecule in some part of the reorientation process. This is consistent with the results of similar and related experiments on the reorientational dynamics of 5CB (4-pentyl-4'-cyanobiphenyl) *(14,22,24)* and with preliminary results on 5PCH *(1,25)*.

As noted in our earlier report *(1)*, this appearance of cross peaks in the asynchronous 2D FT-IR map of 5PCH is a clear indication in itself of out of phase response of the rigid and flexible portions of the molecule, and the sign information does indicate that the flexible part responds first. However, the 2D results do not allow any conclusion as to whether the flexible "tail" responds first as the voltage is *increased*, or as it is *decreased*, during the sinusoidal modulation. Similar ambiguity

a

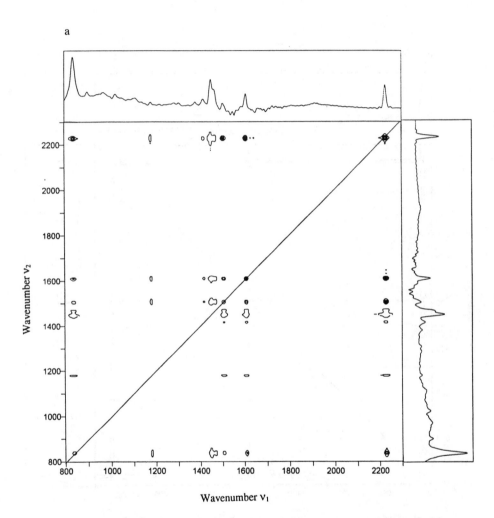

Figure 6.   Step-scan 2-D FT-IR spectra of nematic liquid crystal 5PCH response to sinusoidally modulated (AC) electric field.   a. Synchronous correlation; b. Asynchronous correlation.   Reference spectra along the top and side of each correlation map represent the static absorbance.   *(Continued on next page.)*

b

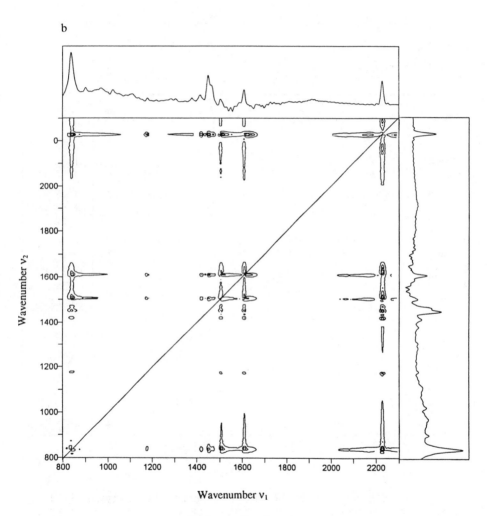

Figure 6. Continued.

is encountered in the analogous dynamic polymer rheology experiments, such as that described above. However, in contrast to the polymer stretching experiment, since the liquid crystal reorientation process essentially "has no (macroscopic) moving parts", it is equally as easy to carry out a pulsed, time domain, experiment as to do the synchronously modulated, frequency domain, experiment. In the time domain experiment, not only is it possible to separate the "rise" dynamics from the "decay" dynamics, but induction times can be measured, quantitative response rates determined and there is the possibility of detecting complex responses involving intermediate forms. Preliminary time domain measurements by use of the stroboscopic continuous scan technique allowed an inference that the differential response occurs in the voltage-off part of the response (*25*). However, the constraints of the stroboscopic method as regards repetition rate and pulse structure make the step-scan mode more flexible for investigating the complete cycle of voltage response.

The results of time domain step-scan FT-IR investigation of electro-reorientation dynamics of 5PCH are summarized in Figure 7 and Table I. In Figure 7 the voltage pulse profile for the experiment is correlated with the intensity response curves for the various bands. As described above, each data point represents the averaged value of $\Delta A(\bar{v}, t)$ for 10 consecutive 1 ms time slices. It should be noted that all of the data for each sample were collected in a single scan, that is, with *simultaneous temporal and spectral multiplexing*. The total data collection time was approximately 6.5 hours.

As seen in Table I, 5PCH responds with a $t_{1/2}$ of < 2 ms when the voltage pulse is applied. On the other hand, the recovery (voltage off) process, *after* the 30 ms, 11 V pulse, is much slower, requiring almost 200 ms. However, the most interesting result is that the $t_{1/2}$ for the response of the pentyl CH_2 deformation mode at 1467 cm^{-1} in the voltage off, homeotropic-to-homogeneous, transition is approximately 3 times shorter than the corresponding response times for the modes associated with the rigid part of the molecule ($t_{1/2}$ = 6.3 ms *vs.* ≥ 15 ms). At the time resolution and voltage of these experiments it does not appear that a comparable differential response occurs in the voltage-on process (nor, incidentally, does there appear to be any significant induction time after the pulse is applied before the reorientation begins).

Thus the ambiguity of the frequency domain measurements, that is, in what part of the electro-reorientation process does the pentyl "tail" get out of phase with the rigid part of the molecule, is clearly resolved by the time domain results. Although both parts of these low molecular weight liquid crystal molecules respond together ($t_{1/2}$ < 2 ms) when the voltage is applied, the flexible "tails" of the molecules reorient faster than the rigid part as the mesophase returns (relatively slowly) to its homogeneous state after the electric field is removed. Analogous experiments on the biphenyl analogue of 5PCH, 5CB (4-pentyl-4'-cyanobiphenyl), indicate that, although the relaxation (voltage off) response rate of 5CB is approximately 3 times slower than that of 5PCH for all segments of the molecule, the flexible (pentyl) tail leads the rigid part of the molecule by roughly the same factor as for 5PCH in the return to the homogeneous state (A. Fuji, unpublished data).

Previous dynamic infrared studies of 5CB have also detected a differential response between the rigid and flexible parts of the molecule to an applied electric field. In the first such work the stroboscopic continuous-scan FT-IR method was applied (*24*). This was followed by the first report of the use of the step-scan FT-IR technique to study liquid crystal electro-reorientation dynamics, which also focused on 5CB (*14*). In that work it was confirmed that there is a differential response between the rigid and flexible segments of the molecules, and that the flexible pentyl "tail" leads the response of the rigid segment of the molecule at some phase of the reorientation cycle. Subsequently in a more extensive investigation of 5CB by both stroboscopic continuous-scan and by step-scan FT-IR (*22*) evidence was found that in the recovery, or voltage-off, homeotropic-to-homogeneous, process the pentyl chain undergoes a fast local motion in addition to the rotational relaxation of the entire molecule. This conclusion was drawn primarily from careful analysis of the dynamic response in the

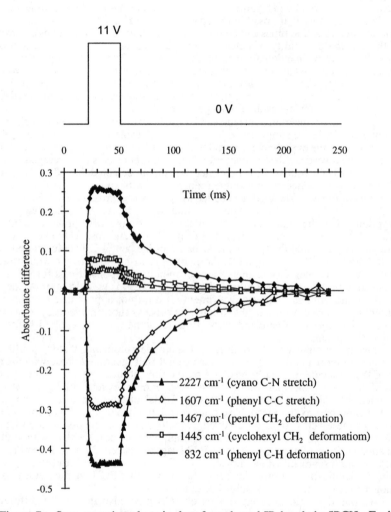

Figure 7. Step-scan time domain data for selected IR bands in 5PCH. Each point represents an average of 10 1-ms $\Delta A(\bar{\nu},t)$ data, each from 50 co-added pulse sequences (see text). Corresponding voltage profile for each pulse sequence is given at top of the figure.

Table I.   Step-scan time domain electro-reorientation half-lives for selected IR bands in 5PCH

Wavenumber (cm^{-1})/ Assignment	$t_{1/2}$, voltage-on (ms)	$t_{1/2}$, voltage-off (ms)
2227/CN stretch	1.5	14.9
1607/phenyl CC stretch	1.4	18.4
1503/phenyl CC stretch	1.4	15.7
1467/pentyl CH_2 deformation	1.9	6.3
1445/cyclohexyl CH_2 deformation	1.5	10.3
832/phenyl CH deformation	1.5	18.1

C-H stretching region of the spectrum. Although the differential $t_{1/2}$ values for the various transition dipole responses were not given, it would appear that under the specific conditions of that study, the relaxation period of the rigid segment of the molecule observed in that study is somewhat shorter than has been observed in our laboratory, but this may be due to the use in that study of a pulse width of only 10 ms, compared to the 30 ms used in our experiments on both 5CB and 5PCH. The difference in the voltage used (10 V *vs.* 11 V) is not considered significant. However, from numerous experiments with both 5CB and 5PCH, as well as with other low molecular weight liquid crystals, it is clear that exact values of threshold voltages, induction times and reorientation rates are highly dependent on the details of sample preparation, as well as pulse voltages and widths.

Such differences in sample preparation may explain the particularly divergent results recently reported for 5CB in which band intensity changes several orders of magnitude smaller than those illustrated and discussed above were observed by use of time-resolved dispersive IR techniques (*26*). However, in common with the time-resolved FT-IR results, the dispersive data show a clear difference in segmental response and suggest that it has its origin in the relaxation (voltage-off) process.

That the more flexible segments of these mesogenic molecules should return to their equilibrium (homogeneous) state more rapidly than the rigid parts correlates with the relative degrees of conformational freedom available to the different segments, which promotes a more rapid thermal relaxation of the pentyl "tail" compared to the rigid segment. This suggests that for 5PCH, the cyclohexyl group, with an intermediate degree of conformational freedom between that of the pentyl tail and the cyanophenyl group, might show an intermediate relaxation rate. In fact, this appears to be the case, as seen in the values of $t_{1/2}$ in Table I. In addition, there is preliminary evidence that calculation of the relative phase spectrum, $\phi(\bar{v}) = -\arctan[Q(\bar{v})/I(\bar{v})]$, from the frequency domain in-phase $I(\bar{v})$ and in-quadrature $Q(\bar{v})$ data may provide the resolution necessary to make such a distinction even more clearly. If this result can be verified and shown to be general, it will provide even stronger support for the complementary relationship of frequency- and time-domain techniques for these liquid crystal dynamic electro-reorientation studies.

Acknowledgments

The authors acknowledge helpful discussions with I. Noda and E. T. Samulski, as well as contribution of equipment from the Bruker Instrument Company and the IBM

Corporation. Financial support from the Bruker Instrument Company, from Kao Corporation and from Duke University is also gratefully acknowledged.

Literature Cited

1. Palmer, R. A.; Chao, J. L.; Dittmar, R. M.; Gregoriou, V. G. and Plunkett, S. E. *Appl. Spectrosc.* **1993**, *47*, 1297.
2. Palmer, R. A. *Spectroscopy* **1993**, *8*(2), 26.
3. Manning, C. J. and Griffiths, P. R. *Appl. Spectrosc.* **1993**, *47*, 1345.
4. Noda, I. *J. Am. Chem. Soc.* **1989**, *111*, 8116; *Appl. Spectrosc.* **1990**, *44*, 550 and *Appl. Spectrosc.* **1993**, *47*, 1329.
5. Marcott, C.; Dowrey, A. E. and Noda, I. *Anal. Chem.* **1994**, *66*, 1065A.
6. Sakamoto, A.; Furukawa, Y.; Tasumi, M. and Masutani, K. *Appl. Spectrosc.* **1993**, *47*, 1457.
7. Noda, I. *Bull. Am. Phys. Soc.* **1986**, *31*, 520.
8. Noda, I.; Dowrey, A. E. and Marcott, C. *Mikrochim. Acta [Wein]* **1988**, *1*, 101.
9. Palmer, R. A.; Manning, C. J.; Chao, J. L.; Noda, I.; Dowrey, A. E. and Marcott, C. *Appl. Spectrosc.* **1991**, *45*, 12.
10. Smith, M. J.; Manning, C. J.; Palmer, R. A. and Chao, J. L. *Appl. Spectrosc.* **1988**, *42*, 546 and Manning, C. J.; Palmer, R. A. and Chao, J. L. *Rev. Sci. Instrum.* **1991**, *62*, 1219.
11. Dittmar, R. M.; Gregoriou, V. G.; Chao, J. L. and Palmer, R. A. *Polym. Prepr.* **1991**, *32*, 673.
12. Palmer, R. A.; Gregoriou, V. G. and Chao, J. L. *Polym. Prepr.* **1992**, *33*, 1222.
13. Gregoriou, V. G.; Noda, I.; Dowrey, A. E.; Marcott, C.; Chao, J. L. and Palmer, R. A. *J. Polym. Sci. B. Polym. Phys.* **1993**, *31*, 1769.
14. Gregoriou, V. G.; Chao, J. L.; Toriumi, H. and Palmer, R. A. *Chem. Phys. Lett.* **1991**, *179*, 491.
15. Hartland, G. V.; Xie, W.; Dai, H.-L.; Simon, A. and Anderson, M. J. *Rev. Sci. Instrum.* **1992**, *63*, 3261.
16. Siesler, H. W. "Rheo-Optical FT-IR Spectroscopy of Macromolecules" in *Advances in Applied FT-IR Spectroscopy*, Ed. M. W. Mackinsie (John Wiley, Chichester, 1988).
17. Noda, I.; Dowrey, A. E. and Marcott, C. *Appl. Spectrosc.* **1988**, *42*, 203.
18. Noda, I.; Dowrey, A. E. and Marcott, C. *J. Mol. Struct.* **1990**, *224*, 265.
19. Chase, B. and Ikeda, R. *Appl. Spectrosc.* **1993**, *47*, 1350.
20. Budevska, B. O.; Manning, C. J.; Griffiths, P. R. and Roginski, R. T. *Appl. Spectrosc.* **1993**, *47*, 1843.
21. Saijo, K.; Suehiro, S.; Hashimoto, T. and Noda, I. *Polym. Prepr. Jpn.* **1989**, *38*, 4212.
22. Nakano, T.; Yokoyama, T. and Toriumi, H. *Appl. Spectrosc.* **1993**, *47*, 1354.
23. Kohri, S.; Kobayashi, J.; Tahata, S.; Kita, S.; Karino, I. and Yokoyama, T. *Appl. Spectrosc.* **1993**, *47*, 1367; Masutani, K.; Yokota, A.; Furukawa, Y.; Tasumi, M. and Yoshizawa, A. *ibid.*, p.1370; Czarnecki, M. A.; Katayama, N.; Ozaki, Y.; Satoh, M.; Yoshio, K.; Watanabe, T. and Yanagi, T. *ibid.*, p.1382; Hasegawa, R.; Sakamoto, M. and Sasaki, H. *ibid.*, p.1390.
24. Hatta, A. *Mol. Cryst. Liquid Cryst.* **1981**, *74*, 195 and Toriumi, H.; Sugisana, H. and Watanabe, H. *Japan. J. Appl. Phys.* **1988**, *27*, L279.
25. Gregoriou, V.G., PhD Dissertation, Duke University, 1993.
26. Urano, T. I. and Hamaguchi, H.-O. *Appl. Spectrosc.* **1993**, *47*, 2108.

RECEIVED February 2, 1995

Chapter 7

Evaluating the Weatherability of Polyurethane Sealants

Ralph M. Paroli[1], Kenneth C. Cole[2], and Ana H. Delgado[1]

[1]Institute for Research in Construction, National Research Council of Canada, Ottawa, Ontario K1A 0R6, Canada
[2]Industrial Materials Institute, National Research Council of Canada, Boucherville, Quebec J4B 6Y4, Canada

Photoacoustic-FTIR (PAS-FTIR) and thermogravimetry (TG) have been used to study the effect of UV-radiation and/or water exposure for up to 8000 hours on three polyurethane-based sealants. The three sealants were found to contain calcium carbonate and/or titanium dioxide and different types of urethane polymer. One material appears to be chemically less stable than the others; under water exposure significant break-down occurs. It undergoes substantial changes in less than 1000 hours. PAS-FTIR results indicated that, for this type of exposure, sealant PU1 was the least stable chemically. Water exposure was found to have induced significant breakdown of the urethane groups. Combined UV/water exposure caused significant changes to occur in less than 1000 hours. PU2 and PU3 were found to undergo small changes after water exposure. The most noticeable observation was an increase in hydroxyl absorption. Substantial change in chemical composition was observed to occur between 1000 and 2000 hours of combined UV/water exposure.

Polyurethane sealants have been in use since the early 1970's. In 1994, they are the second most used high-performance sealants in the building industry. Polyurethane sealants are produced by reacting an isocyanate group with a polyol (see scheme I). They are available as one- or two- component materials and as high or low-

$$R-N=C=O \quad + \quad R'-OH \quad \rightarrow \quad R'-O-\overset{\overset{\displaystyle O}{\|}}{C}-\overset{\overset{\displaystyle H}{|}}{N}-R \qquad \text{Scheme I}$$

modulus formulations. A typical formulation for a urethane sealant is shown in Table I. It is relatively complex since fillers, plasticizers, and other minor additives are present in addition to the polymer. Generally, urethane sealants exhibit excellent elasticity, good flexibility (down to -54 °C), excellent chemical resistance, good recovery (~90%), and good weathering properties. The disadvantages include poor water immersion resistance (i.e., adhesion failure), yellowing when exposed to UV-light and fading of light colours *(1-2)*.

Table I. Typical Polyurethane Sealant Formulation

Component	Amount (%)
Polymer	35-45
Fillers	30-40
Plasticizers	15-25
Solvent	1-4
Pigments	2-3
Adhesion Additives	1-3
Thixotropic Agents	1-2

The weatherability of polyurethane sealants, will be affected by the pigments, plasticizers, etc. Different methods can be used to analyze the weathering effects of construction materials (e.g., tensile strength, elongation, color changes, spectroscopy, etc.) *(2-9)*. Obviously, after weathering one can easily look for visual changes such as surface crazing and discolouration. Visual changes, however, may not affect the actual service performance of the material but may constitute an aesthetic failure. The performance of the materials can be verified by studying the mechanical properties. Changes in the mechanical properties will inform the manufacturer and user that the material has changed. However, the mechanical data will not help in explaining why the changes took place. Changes at the micro level may not necessarily be observed upon visual inspection yet can dramatically affect the performance of the material.

Durability or service-life of construction products is of concern to designers, owners and users. Upon weathering, polyurethane sealants may, due to the synergistic effects of solvent evaporation, ozone attack, migration of plasticizers, ultraviolet radiation, water immersion, etc., undergo crosslinking or loss of plasticizer. This may lead to embrittlement, thus reducing movement capability and causing the sealant to fail. It is therefore essential to evaluate the weatherability of sealants. Natural weathering is an option but requires much patience. Manufacturers are not able to wait 15 years to evaluate the weatherability of a product. Artificial weathering is an alternative requiring less time. Unfortunately, the artificial weathering may not replicate the actual degradation of the material in question since other parameters, e.g., physical aging, may also be a factor. FTIR spectroscopy allows for the monitoring of the material degradation at a molecular level. It is possible to compare natural and accelerated weathering and evaluate the relevancy of the weathering programme. However, construction materials, such as polyurethane sealants, cannot be characterized by FTIR transmission or reflection techniques because they are black, gray or opaque and thick. ATR is not suitable because after aging there is no intimate contact between the sealant and the crystal. Photoacoustic-FTIR (PAS-FTIR), however, is capable of providing spectral information for this type of material with minimal, if any, sample preparation.

In recent years, there has been a growing interest in photoacoustic-FTIR spectroscopy as a tool to study samples of irregular morphology. Moreover, carbon black, a highly absorbing filler used in many construction products, does not present problems (if less than 15% is present) *(10)*. Photoacoustic spectroscopy makes possible the simultaneous study of sample surface and bulk *(11)*. It has been demonstrated to be a useful technique for obtaining mid-infrared spectra of solid samples with minimal preparation *(12-15)*. PAS-FTIR is particularly useful for measuring the spectra of compounds that suffer from structural alteration *(16,17)* arising from sample preparation. The applications of this technique are quite widespread and include the evaluation of accelerated weathering for polyester-urethane coatings *(18)*, cured paint media and weathering of alkyd paints *(19)*.

Another family of techniques useful in the characterization of weathered materials is thermal analysis *(20)*. This includes the classical methods such as thermogravimetry (TG), differential thermal analysis (DTA) and differential scanning calorimetry (DSC), as well as the modern thermoanalytical methods, e.g., dynamic mechanical analysis (DMA) and thermomechanical analysis (TMA). TG provides information on any reaction involving mass change as a function of temperature and/or time; thus information on the heat stability and composition of the sample can be obtained. DTA provides information on the temperature difference between a reference and a sample while they are subjected to a common temperature program. The main application of DTA is in establishing thermal events associated with chemical changes such as oxidative stability, chemical reactions, phase transitions, enthalpies, melting point, etc. Today, simultaneous thermal analysis (STA or TG/DTA or TG/DSC) is popular because data from two thermoanalytical techniques, TG and DTA (or DSC), are acquired simultaneously, by the same instrument on a single specimen (i.e., physical and chemical changes are recorded simultaneously). Simultaneous measurement helps to circumvent problems related to the inhomogeneity in building materials.

In this paper, PAS-FTIR and TG are used to evaluate the weatherability of polyurethane-based sealants. More specifically, the chemical changes that occur during UV-radiation and/or water exposure up to 8000 hours are monitored.

Experimental

Artificial Weathering. Three commercially-available sealants were weathered in a xenon arc Weather-o-meter up to 8000 hours. One series of samples was exposed to water and the other series was exposed to water and light. A piece of each sealant was taken every 1000 hours. Samples were weathered under the following conditions:

Irradiance:	$0.37\ W\ m^{-2}\ nm$
Black panel temperature:	63 °C
Relative humidity arc on:	50%
Light cycle:	3.5 hours on, 0.5 hours off
Specimen spray cycle:	When arc was off

Photoacoustic-Fourier transform infrared (PAS-FTIR) spectroscopy. The weathered sealants were analyzed by PAS-FTIR spectroscopy. Infrared spectra were measured on a Nicolet 170SX instrument equipped with an MTEC Model 200 photoacoustic detector. For sampling, disks of 8 mm diameter were cut with a cork borer. In a few cases, where samples were too thick, the thickness was reduced by slicing with a scalpel. Conditions used for recording the spectra were: resolution 8 cm^{-1}; mirror velocity setting 4; number of scans 512; helium purge; Happ-Genzel apodization, standard phase correction. Along with the weathered samples, two specimens of each material were analyzed before exposure and used as controls. All spectra were ratioed against a background recorded earlier the same day with carbon lampblack (Fisher Scientific, Cat. No. C-198) or with a carbon black membrane standard reference.

Spectra were recorded for control and samples exposed for 1000, 2000, 4000, 6000 and 8000 hours. However, only those that showed significant changes are discussed.

Thermal analysis. The weathered sealants at 6000 hours were analyzed by Simultaneous Thermal Analysis (STA). The analysis was performed using the Seiko Simultaneous Thermal Analyzer (STA) model TG/DTA320 and data acquired with the Seiko SSC5200H disk station. Approximately 7-8 mg of sample was heated under a nitrogen atmosphere (100 mL/min.) from 40 °C (no weight loss observed before 40 °C) to 900 °C at 10 °C/min. Unweathered sealants were also analyzed as controls. All samples were run in duplicate.

Results and Discussion

In general, the sealants showed four weight losses. The first weight loss occurred in the 140 to approximately 340 °C region and can be attributed to the loss of solvent and plasticizer. The second weight loss is in the 340-500 °C region and is probably due to the decomposition of the polyurethane polymer (i.e., breakage of the urethane linkages). A small weight loss in the 500-600 °C region may result from the release of other additives. The last major weight loss occurs above 600 °C region and is due to the decomposition of calcium carbonate. The weight losses are summarized in Table II.

PU1. The spectrum of the unexposed material is given in Fig. 1. The pigment used for this sealant is probably titanium dioxide, giving the broad absorption from 700-500 cm^{-1}. The polymer appears to be a polyether urethane, for which some of the bands may be assigned as follows: N-H stretching at 3300 cm^{-1}; C-H stretching at 2970, 2928, 2857 cm^{-1}; urethane carbonyl at 1729 cm^{-1}; aromatic ring stretching at 1589 cm^{-1}; amide II (C-N stretch plus N-H deformation) at 1539 cm^{-1}; polyether C-O-C at 1113 cm^{-1}.

Spectra of samples exposed to water only for 2000 and 8000 hours are shown in Fig. 1 (2000W and 8000W). Small but progressive changes occur, notably: an increase in OH absorption (3600-3200 cm^{-1}); a decrease in the amide II band at 1539 cm^{-1}; an increase in the band at 1435 cm^{-1}; a change in the relative intensities

Table II. Weight Loss for Polyurethane Sealants in Nitrogen

Sample	Weathering (hr)		Weight Loss (%)				
	*L/W**	*W*	*140-325 °C*	*325-405 °C*	*405-600 °C*	*600-900 °C*	*Total*
	0	0	42.7	28.3	7.9	4.4	83.8
PU1	6000		-	68.0	9.7	3.6	81.9
		6000	-	69.8	7.4	4.0	81.7

Sample	Weathering (hr)		Weight Loss (%)				
	*L/W**	*W*	*160-340 °C*	*340-500 °C*	*500-600 °C*	*600-900 °C*	*Total*
	0	0	39.0	23.5	2.6	11.5	77.1
PU2	6000`		37.4	18.9	4.8	10.2	71.8
		6000	37.6	22.7	3.2	11.7	75.7
	0	0	18.6	47.4	3.1	12.6	82.3
PU3	6000		16.9	34.0	1.6	21.0	74.7
		6000	18.5	41.0	1.2	16.7	78.4

*L/W = light and water; W = water

of the bands at 1146 and 1113 cm^{-1}. It is clear that some changes in chemical composition are occurring, involving breakage of urethane crosslinks (see Ref. 4 for possible mechanisms).

The spectra of samples exposed to light and water also undergo change (Fig. 1 1000L, 2000L and 8000L). After only 1000 hours, substantial change in the chemical composition has occurred. The changes are greater than after 8000 hours exposure to water only. The shape of the 1800-1500 cm^{-1} region has changed completely. After 2000 hours, further change has occurred, as can be seen particularly from the decrease of the band at 1100 cm^{-1}. Beyond this point, change in chemical composition continues (increase of absorption at 1700-1600 cm^{-1}) at a more gradual rate. Overall, the change in chemical composition appears to be more drastic than for the other polyurethane samples.

The PU1 sealant series showed the first weight loss around 140-405 °C. The unexposed sample (Fig. 2) showed two unresolved weight losses. The first one (~42%) occurred in the temperature range of 140 to ~325 °C and the second (29%) between 325 and 405 °C. The derivative thermogravimetry (DTG) curves for exposed samples showed only one major weight loss in the same region; however, the DTG showed more than one weight loss occurring at slightly lower temperatures than the major weight loss (~300-308 °C). This is more clearly seen in the DTG curve of the sample exposed to water only (Fig. 3). The series shows two more weight losses, one in the 405-600 °C region and the other in the 600-900 °C. region. The sample exposed to UV/water (Fig. 4) showed a ~10% weight loss in the former region whereas the unexposed sample and the sample exposed only to water showed similar weight losses (7%). The total weight loss for the series is ~2% higher for the unexposed which implies that some components may have been lost during weathering.

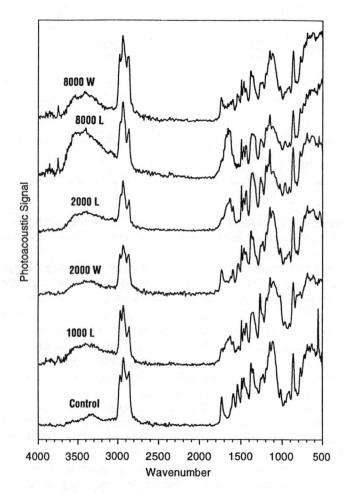

Figure 1. PAS-FTIR spectra of polyurethane sealant PU1: control, 1000L (UV and water exposure), 2000W (water exposure only), 2000L (UV and water exposure), 8000L (UV and water exposure) and 8000W (water exposure only).

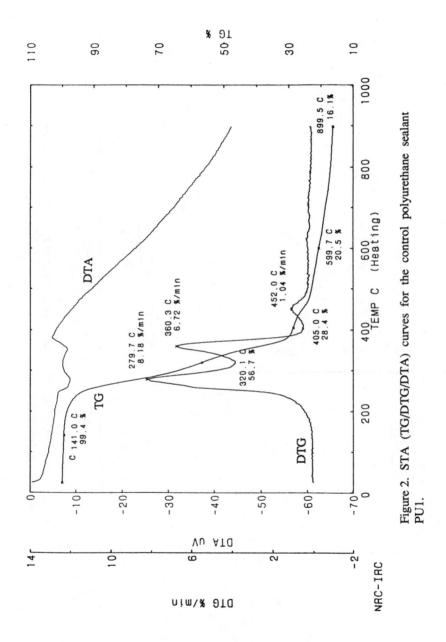

Figure 2. STA (TG/DTG/DTA) curves for the control polyurethane sealant PU1.

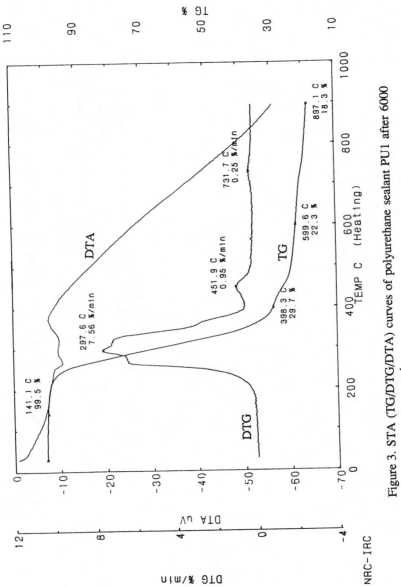

Figure 3. STA (TG/DTG/DTA) curves of polyurethane sealant PU1 after 6000 hours water exposure only.

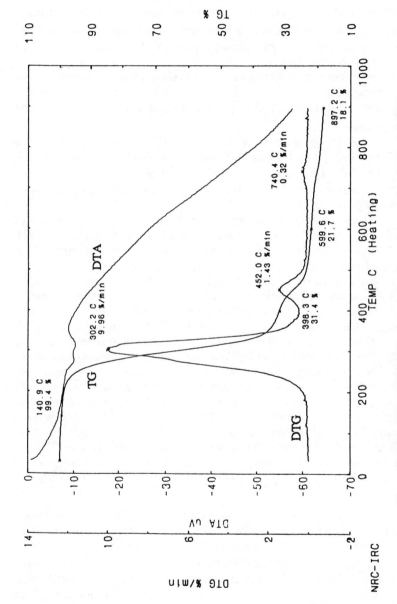

Figure 4. STA (TG/DTG/DTA) curves of polyurethane sealant PU1 after 6000 hours UV and water exposure.

PU2. The spectrum of the unexposed sample is given in Fig. 5. It is different from the previous case. The main pigment seems to be titanium dioxide (700-500 cm^{-1}), but there is probably a small amount of calcium carbonate (weak peaks at 2513, 1794, 876 cm^{-1}). The strong ester band at 1733 cm^{-1} suggests a polyester urethane, but the ether band near 1100 cm^{-1} is also strong. The sharp peak at 1634 cm^{-1} indicates polyurea linkages from moisture curing. The N-H band at 3321 cm^{-1} is stronger than in the previous cases.

Exposure to water up to 8000 hours has little effect on the general appearance of the spectrum (Fig. 5). There is some variation in relative band intensities (e.g. 1733/1634 cm^{-1}, polyurethane/TiO$_2$), but not much change in the nature of the bands themselves.

Exposure to light and water has a greater effect (Fig. 5). Although the changes are small after 1000 hours, between 1000 and 2000 hours there is a drastic change. A strong broad OH band appears (3600-3200 cm^{-1}) and a strong band appears at 1632 cm^{-1}. However, as in the case PU1, change is much more gradual between 2000 and 8000 hours.

The PU2 series start to lose weight at ~140 °C (Figs. 6-8). The series showed four weight losses. The first one, not completely resolved, occurs in the 100-335 °C region. The weight loss in this region is almost the same for the unexposed (39%) and exposed samples (39 and 38%). The DTG curves for the unexposed sample (Fig. 6) showed a second unresolved weight loss in the first peak of the DTG curve. This unresolved peak becomes less apparent in the exposed samples (Figs. 7-8). The weight losses in the first and second temperature regions showed similar peak-heights for the unexposed sample and the sample exposed to UV/water. However, a large increase is observed in the sample exposed to water only. This is probably due to the breaking of urethane linkages by the hydroxyl group and formation of new products. The second distinctive weight loss of the series takes place in the 335-500 °C region. In this region, the unexposed samples showed 2% higher weight loss than the sample exposed to UV/water and about 5% more than the sample exposed to water only. The third weight loss in the 500-600 °C region is quite small for all the samples (between 3 and 5%). The last weight loss (600-900 °C) remains almost the same for the unexposed and exposed samples. The total weight loss for the series ranges from 77% (unexposed), 72% (exposed to UV/water) to 76% (exposed to water only). This series is more heat resistant than PU1.

PU3. The spectrum of the control sealant is given in Fig. 9. It contains calcium carbonate but little if any titanium dioxide. The strong band at 1114 cm^{-1} suggests a polyether urethane. Exposure to water has little effect after 2000 hours and only a slight effect after 8000 hours, namely a broader band at 1640 cm^{-1} and a weaker band at 1014 cm^{-1} (Fig. 9). Exposure to light and water brings about a decrease in polyurethane absorption compared to carbonate. After 1000 hours, changes are small, but between 1000 and 2000 hours a strong band appears at 1646 cm^{-1}.

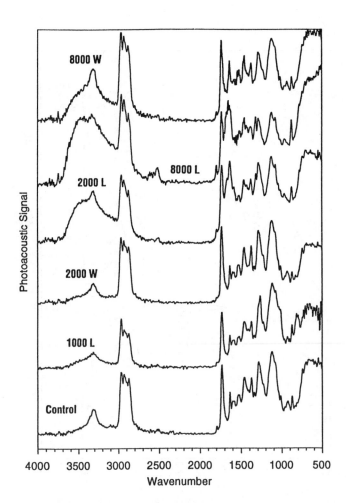

Figure 5. PAS-FTIR spectra of polyurethane sealant PU2: control, 1000L (UV and water exposure), 2000W (water exposure only), 2000L (UV and water exposure), 8000L (UV and water exposure) and 8000W (water exposure only).

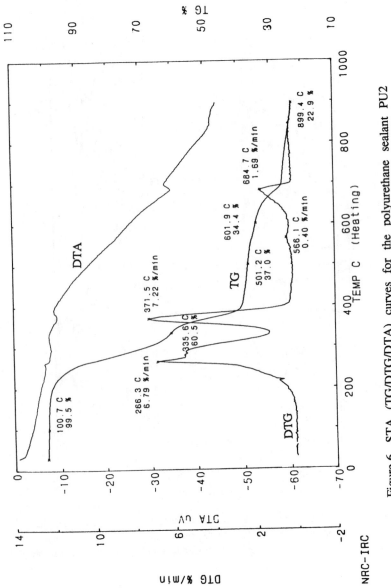

Figure 6. STA (TG/DTG/DTA) curves for the polyurethane sealant PU2 (control).

NRC-IRC

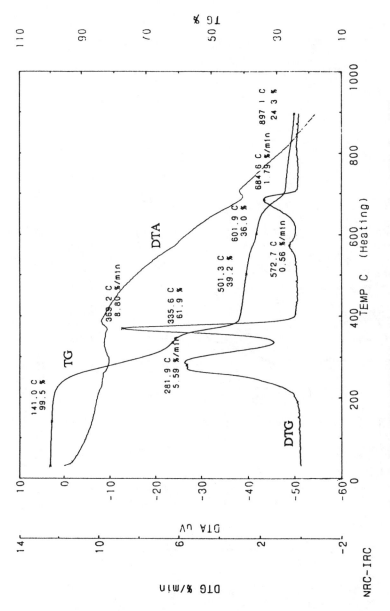

Figure 7. STA (TG/DTG/DTA) curves of polyurethane sealant PU2 after 6000 hours water exposure only.

NRC-IRC

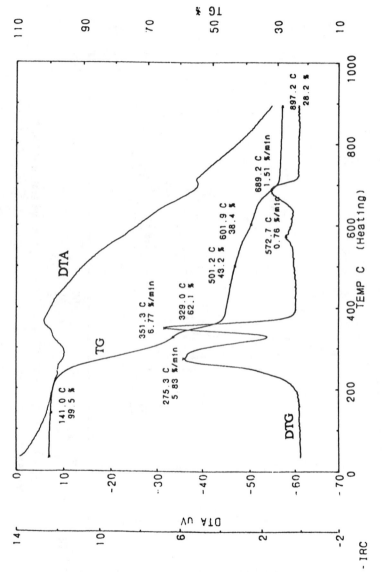

Figure 8. STA (TG/DTG/DTA) curves of polyurethane sealant PU2 after 6000 hours UV and water exposure.

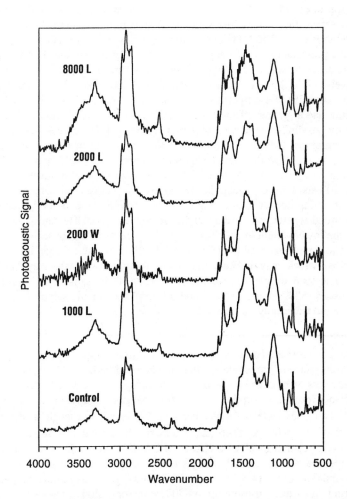

Figure 9. PAS-FTIR spectra of polyurethane sealant PU3: control, 1000L (UV and water exposure), 2000W (water exposure only), 2000L (UV and water exposure), 8000L (UV and water exposure) and 8000W (water exposure only).

Beyond 2000 hours, changes are more gradual. No carbonate bands were present in the spectrum of the sample exposed for 4000 hours; however, the bands did appear after 8000 hours of weathering. This is probably due to a variation in composition rather than a degradation effect.

The specimens of the PU3 series start to lose weight at ~160 °C. As in the previous series, the weight losses in the first two temperature regions are not completely resolved. The DTG curves for the series showed smaller weight losses than in the previous cases for the first two losses. The unexposed (Fig. 10) and the sample exposed to water only (Fig. 11) showed a similar weight loss (19 and 20% respectively) in the 140-340 °C region. The sample exposed to UV/water (Fig. 12), however, lost ~2% less weight than the other two. This implies that the UV/water weathering had a greater effect on these specimens than on specimens exposed to water only. A similar trend was observed from the DTG curves for the largest second weight loss (340-500 °C region). In this region, the unexposed sample loses ~5% more weight than the sample exposed to water only (42%) and ~ 13% more than the one exposed to UV/water. The series also showed a small weight loss between 500 and 600 °C. Again, the weight loss for the exposed samples is ~1/2 of that of the unexposed. A fourth weight loss occurred in the 600-900 °C region. An opposite trend is observed for the weight loss in this region. The unexposed sample showed the lower weight loss (13%) followed by the sample exposed to water only (16%). The sample exposed to UV/water showed the greatest weight loss (21%) in this region, which may indicate that more calcium carbonate may have been present. The total weight loss for the series showed the same trend as individual weight losses: 82% for unexposed, and 75 and 78% for the samples exposed to UV/water and water only, respectively.

Summary

The three polyurethane-based materials contain different types of urethane polymer, along with calcium carbonate and/or titanium dioxide. PU1, appears to be chemically less stable. The most prominent change in the FTIR spectra was an increase in hydroxyl absorption. The other changes included: a decrease in the amide II band; an increase in the band at $1435 \, cm^{-1}$; a change in the relative intensities of the bands at 1146 and $1113 \, cm^{-1}$. It is clear that some changes in chemical composition involved breakage of urethane crosslinks. Under light/water spray exposure, the substantial changes observed between take place in less than 1000 hours. The other two (PU2 and PU3) show generally similar behaviour on weathering. For water exposure up to 8000 hours, only slight changes are observed in the spectra. The same is true for light/water spray exposure up to 1000 hours, but between 1000 and 2000 hours substantial change in chemical composition occurs (as determined by IR). Between 2000 and 8000 hours, change in chemical composition continues but at a slower rate. It is interesting to note that PU1, the only sealant which contained no calcium carbonate, was the most sensitive to weathering.

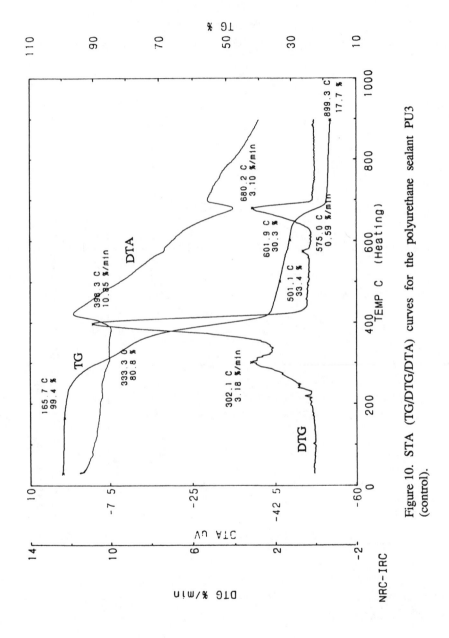

Figure 10. STA (TG/DTG/DTA) curves for the polyurethane sealant PU3 (control).

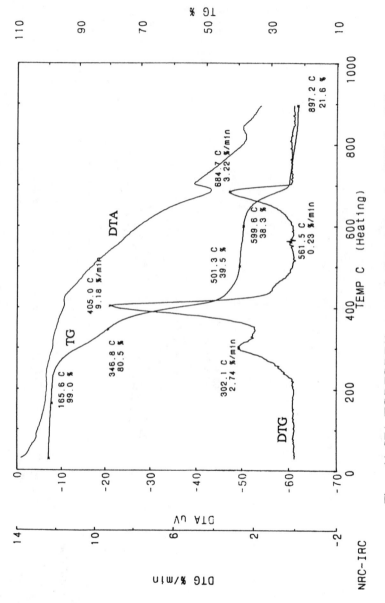

Figure 11. STA (TG/DTG/DTA) curves of polyurethane sealant PU3 after 6000 hours water exposure only.

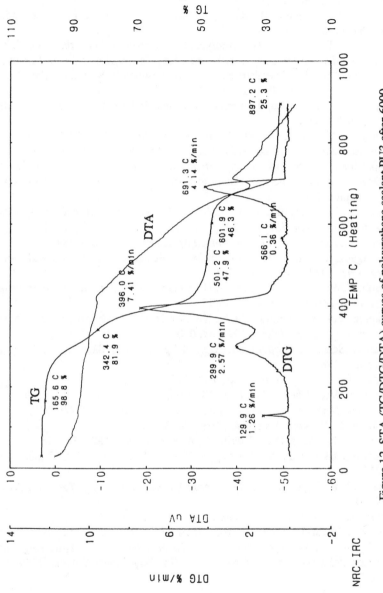

Figure 12. STA (TG/DTG/DTA) curves of polyurethane sealant PU3 after 6000 hours UV and water exposure.

Literature Cited

1. Panek, J.R.; Cook, J.P. Construction Sealants and Adhesives, 2nd Edition, John Wiley and Sons, New York, NY, 1991.
2. Feldman, D. Polymeric Building Materials, Elsevier Science Publishers Ltd., New York, NY, 1989.
3. Leonard, R.G. In *The Analytical Laboratory in the Specialty Sealants/Adhesive Industry:* Smyth, M.R. Ed., *Chemical Analysis in Complex Matrices,* Analytical Chemistry Series, Ellis Horwood PTR Prentice Hall, New York, NY, 1992, pp. 149-191.
4. Wypych, J. Weathering Handbook, Chemtec Publishing, Toronto, Canada 1990, pp. 222-266.
5. Paroli, R.M.; Delgado, A.H. In *Applications of TG-FTIR in the Characterization of Weathered Sealants,* Provder, T.; Urban, M. *ACS Symposium Series Volume: Hyphenated Techniques in Polymer Characterization,* American Chemical Society, Washington, DC, **1994,** Vol. 581 Chapter 10.
6. Paroli, R.M.; Delgado, A.H.; Cole, K.C. *Canadian Journal of Applied Spectroscopy,* **1994,** *39,* pp. 7-14.
7. Penn, J.J.; Paroli, R.M. *Thermochimica Acta,* **1994,** *226,* pp. 77-84.
8. Rodriguez, I.; Dutt, O.; Paroli, R.M.; Mailvaganam, N.P. *Materials and Structures,* **1994,** *160,* pp. 355-361.
9. Paroli, R.M.; Dutt, O.; Delgado, A.H.; Stenman, H.K. *Journal of Materials in Civil Engineering,* **1993,** *5,* pp. 83-95.
10. Paputa Peck, M.C.; Samus, M.A.; Killgoar, Jr. P.C.; Carter III, R. O. *Rubber Chemistry and Technology,* **1991,** *64,* 610.
11. Harrick Scientific Corporation, "Optical Spectroscopy Sampling Techniques Manual", 1987.
12. Busse, G.; Bulemer, B. *Infrared Phys.,* **1978,** *18,* 631
13. Rockley, M.G. *Chem. Phys. Lett.,* **1979,** *68,* 455.
14. Rockley, M.G. Davis, D.M.; Richardson, H.H. *Science,* **1980,** *210,* 918.
15. Vidrine, D. W. *Appl. Spectrosc.,* 1980, *34,* 314.
16. Lowry, S.R.; Mead, D.G.; Vidrine, D.W. *Anal. Chem.,* **1982,** *54,* 546.
17. Renugopalakrishnan, V.; Bhatnagar, R.S. *J. Am. Chem. Soc.,* **1980,** *106,* 2217.
18. Bauer, D.R.; Paputa Peck, M.C.; Carter III, R.O. *J. Coatings Technol.,* **1987,** *59,* 103.
19. Hodson, J.; Landert, J.A. *Polymer,* **1987,** *28,* pp. 251-256.
20. Flynn, J.H. In *Thermal Analysis:* Mark H.F.; Bikales, N.M.; Overberger, C.G.; Menges, G.; *Encyclopedia of Polymer Science and Engineering,* 2nd Edition, John Wiley & Sons: New York, NY, Supplement Volume; 1989; pp. 690-723.

RECEIVED February 13, 1995

Chapter 8

Diffuse and Specular Reflectance Measurements of Polymeric Fibrous Materials

M. Papini

Institut Universitaire des Systèmes Thermiques Industriels, Unité Associée au Centre National de la Recherche Scientifique 1168, Université d'Aix-Marseille I, Centre de Saint-Jérôme, 13397 Marseille Cedex 20, France

Polymeric fibrous materials generally present a rough surface. Measurements of the total and diffuse near-normal hemispherical reflectances over two wavelength ranges (0.25-2.5 µm and 2.5-20 µm) have been performed with two spectrophotometers fitted with integrating spheres. The results obtained on commonly used polymer fibers show that radiative properties vary with the interrelated following factors a) the wavelength: the obtained spectrum allows a non destructive analysis of the sample, it represents its spectral fingerprint, b) the orientation of the fiber: this produces a variable pathlength and then, the insulating property or the radiative heat transfer will be angular dependent, c) the size of the fiber: absorption and scattering are function of particle size.

Vibrational spectroscopy appears as an important, very useful and destruction free tool providing for information on the chemical composition and physical structure of polymer surfaces. Fibrous materials are commonly used as fabrics or insulating materials. The knowledge of the optical characteristics of materials represents a large interest especially for applications involving radiative heat transfer. Fibers have a large surface-to-volume ratio and their radiative properties depend on parameters such as the surface structure or the aspect ratio of the fibers. Heat transfer can occur by radiation, convection or conduction. Only radiative transfer through fibrous polymeric materials is considered in the present paper.

When an incident beam impinges on a layer, radiation undergoes specular reflection, scattering, refraction and absorption in varying proportions before exiting the surface. These effects attenuate the energy of each spectral component of the resulting radiation and concern each fiber in the case of fibrous materials. The reflected radiation contains both specular and diffuse components; nevertheless, some rays can be transmitted directly through the voids between the fibers if not too closely arranged.

The resulting beam contains information on the chemical composition, the surface roughness and the particle size. The shape and intensity of the obtained

0097–6156/95/0598–0137$12.00/0

spectra depend in a wide range on the composition of the material. The specular component of the reflected beam is small for a matte surface while the diffuse component is high. Integrating spheres are useful in the case of highly textured materials, the resulting radiation being reflected towards random directions (1-2). The reflectance is predominantly diffuse for currently used materials like fabrics and insulating materials, each fiber represents a scattering center for.

Experimental Measurements

Spherical particles scatter radiation into all directions spanning the 4 π steradian solid angle while cylindrical fibers scatter radiation along a diffusion angle ϕ equal to the angle between the incident beam and the fiber axis: the scattering is anisotropic in this case, it propagates along a cone surface whose half apex angle is $\phi(3)$.

Fibers can be modelled as a collection of discrete cylinders arranged in specific or random directions in space. Several diffuse and specular reflectance studies have been performed in the two wavelength ranges: 0.25-2.5 μm and 2.5-20 μm to learn out the properties of polymer fibers.

Measurements of the near-normal hemispherical spectral reflectance were accomplished with two kinds of spectrometers. For the first wavelength range, a Beckman UV 5240 dispersive spectrophotometer fitted with an integrating sphere coated on the inside with a diffuse rough barium sulfate white paint was used; the sample was located in the center of the sphere. For the second wavelength range we used a Digilab Fourier transform infrared (FTIR) spectrophotometer fitted with an integrating sphere whose inner surface is gold-coated and diffuse; the sample was placed on the sphere wall in this case. In the first wavelength range, two detectors are used: a photomultiplier for the visible wavelengths and a lead sulfide detector for the near infrared (NIR) wavelengths. Measurements of the near-normal hemispherical spectral reflectance for the 2.5-20 μm region (MIR) was performed using a liquid air cooled mercury cadmium telluride (MCT) detector. The reflected rays undergo numerous reflections inside the sphere before reaching the detectors, and then, no information concerning the sample is lost. In spite of the low efficiency of the integrating sphere, the method is a convenient means for determining radiative characteristics of materials presenting various surface states with little or no sample preparation. Atmospheric carbon dioxide and water vapor are reduced by an appropriate purging of the compressed air circulating inside the experimental set up. In order to eliminate the specular radiation, a light trap is placed in the direction of the specular component. Total and diffuse radiations are measured and the specular component is deduced from the difference between the two spectral data. The black trap eliminates not only the specular radiation but also the scattered rays resulting from successive reflections on the surfaces of the fibers and transmissions through the fibers and arriving to the trap with the same direction as the specular ray: in this direction, it is impossible to discriminate between the true specular and the diffuse components.

The transmitted rays contain structural information about the sample while the true specular radiation provides for little information about it. Fabrics were studied without any preparation and fibers loaded in cells constituted successively by a KBr window, ten polytetrafluoroethylene (PTFE) rings 1 mm thick and a black non reflecting surface for the bottom. 256 scans were signal-averaged in order to improve the signal-to-noise ratio and a resolution of 4 cm^{-1} was used. All the spectra were run at room temperature.

Reflectance values **R** for the solar radiation spectrum in the 0.25-2.5 μm wavelength range are evaluated by integrating the product of the experimental spectral values R by the blackbody curve. We then ratio the calculated values to the integrated blackbody curve over the chosen spectral region. The blackbody is considered at 5800 K, which is the surface temperature of the sun.

Results and Discussion

A few examples are given hereunder to illustrate how radiative properties vary with the interrelated following factors: wavelength, surface state of the sample, orientation and diameter of the fibers. Assignments of main bands were accomplished through published data in the NIR (*4-9*) and in the MIR (*10-13*) wavelength ranges. The calculated **R** values are representative of the specific arrangement of the fibers in the studied sample. They may be considered as indicative for the sample with its arrangement.

Wavelength. Radiative properties are spectrally dependent: absorption bands are observed mainly in the NIR and MIR regions.

Absorption bands are due to the vibration of chemical bonds: they correspond to fundamental vibrational modes occurring in the MIR region and to overtone and combination bands of fundamental vibrational modes in the NIR region. The absorption peaks are broad and weaker in the NIR than in the MIR and they involve only vibrations of light atoms entering in strong hydrogenic bonds such as C-H, N-H and O-H bonds. The spectrum is the fingerprint of the sample. It reflects all its physical and chemical characteristics with a minimal sample preparation. Figure 1 illustrates the variation of the near-normal hemispherical spectral reflectance of a commonly used manufactured acrylic fabric, mainly composed of polyacrylonitrile units but also containing other constituents, here polymethylmethacrylate. Different assignments can be made concerning the C-H, $C{\equiv}N$ and C=O stretching bands at 2939.6, 2243.3 and 1933.6 cm^{-1} respectively while CH_2 bending band appears at 1453.6 cm^{-1} for example. Combination and overtone bands for the C-H and C=O bonds appear in the NIR spectrum of the acrylic sample drawn in Figure 2. Assignments of principal bands were accomplished with published data. The calculated **R** value is 0.29.

Some fabrics like tweed are made of blends between natural and synthetic polymers; let us take the case of a wool/polyester blend. The obtained spectra give the radiative property of the sample as shown in Figures 3 and 4 where spectra of natural wool and polyester (PET) are also drawn to show that absorption bands of the two constituents are found for the blend in the NIR and MIR wavelength range. The shape of the spectra in the UV and visible region depends on the colour of the sample. In the MIR spectra: for PET, the C-H, C=O, C-O-C stretching vibrations for instance are respectively present at 2968, 1743.5 and around 1340 cm^{-1}; for wool, a large absorption area is found between about 3600 and 2900 cm^{-1} due to the N-H and C-H stretching vibrations; other absorption bands at 1685 and 1565 cm^{-1} are assigned to the C=O stretching and NH bending vibrations. It is clear that for the blend, the large absorption area for wool between 3600 and 2900 cm^{-1} softens off the spectrum of the blend in this region, masking the peaks of the PET, while peaks appear again when the reflectance of wool is higher. In the NIR wavelength range, the absorption bands in the spectra also constitute a fingerprint of the samples and can be assigned to the different bonds existing in the samples. They agree with the published data. The calculated reflectance values **R** are 0.59, 0.28 and 0.31 for wool, PET fibers and a wool/PET blend respectively.

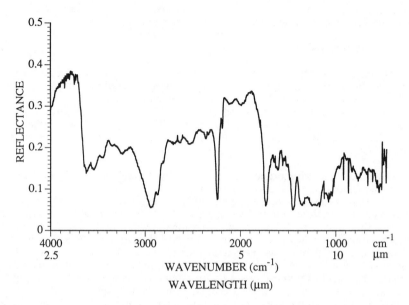

Figure 1. Total near-normal hemispherical spectral reflectance R versus wavenumbers (cm^{-1}) of a fabric composed of acrylonitrile and methylmethacrylate units.

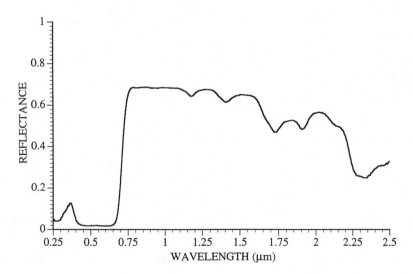

Figure 2. Total near-normal hemispherical spectral reflectance R versus wavelength (μm) of the same fabric as in Figure 1.

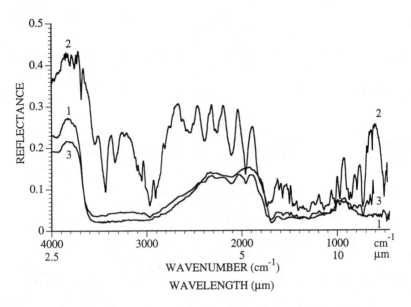

Figure 3. Total near-normal hemispherical spectral reflectance R versus wavenumbers (cm^{-1}) of fabrics composed of: curve 1, wool; curve 2, PET; curve 3, wool/PET blend.

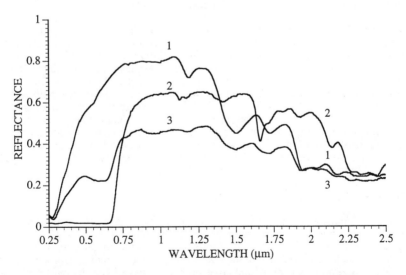

Figure 4. Total near-normal hemispherical spectral reflectance R versus wavelength (μm) of fabrics composed of: curve 1, wool; curve 2, PET; curve 3, wool/PET blend.

Surface State. It has been shown that reflectances vary with size for spherical particles. This behavior is wavelength dependent: reflectances increase with particle size in the NIR wavelength range while they decrease in the MIR (*14*). The limit case corresponds to the bulk material having a plane smooth surface with large peaks in the NIR and none in the MIR region as shown in Figures 5 and 6 for a polypropylene sample. The behavior of a bulk material differs then from that of particles: the surface state appears as another factor influencing reflectances of materials.

Orientation. Fibrous materials can be oriented in specific directions: parallel or perpendicular to cell boundaries or randomly oriented in a plane or in space. The cross-section varies with the orientation of the fibers in the medium. Radiative properties were then shown to be angular dependent. Backscattering is important for fibers oriented parallel to cell boundaries and small for fibers oriented perpendicular to the cell boundaries (*15*). Then, the radiative energy transport is high for this last orientation and low for the parallel orientation. The scatter of measured values is significant for different orientations of the fibers as shown in Figures 5 and 6 for polypropylene fibers. The calculated **R** values are 0.44 and 0.31 for horizontally and perpendicularly oriented fibers respectively. The **R** value for the bulk material is 0.42.

Absorption and scattering affect radiative properties. Fabrics composed of the same material can present different surface states and therefore spectra have different intensities. This effect is illustrated in Figure 7 for three samples of polyester fabrics, a) fibers woven horizontally in a plane, b) a velvet with obliquely oriented long fibers, and c) a velvet with near-perpendicular short fibers. Reflectances are higher in the case of horizontal woven fibers. The evolution is the same in the NIR region. The calculated **R** values are: 0.44, 0.29 and 0.15 respectively. The effect of the orientation can be observed when rotating a sample through different angles, maintaining it in the same plane with respect to the incident radiation. Variations are also observed and the obtained spectra are within the limits presented in Figures 8 and 9 for the two wavelengths regions. The calculated **R** values are 0.26 and 0.29.

Diameter. Short polyamide fibers have been used to describe the influence of the diameter in the case of randomly oriented fibers. Polyamide threads 0.108 and 0.215 mm in diameter d were used. They were cut regularly at the following lengths l = 1.27, 1.95 and 2.32 mm using a special tool built in our laboratory. The intensities of the absorption peaks were found to change with the diameter of the fibers. This seems to have, in this case, a greater effect on the reflectance than their length has, as illustrated in Figure 10 where curves are displayed in two groups corresponding to the diameter but not to the length of the fibers. The following explanation can be given: though fibers were statistically randomly dispersed, the KBr window influences the orientation of the small thin fibers near the surface; in fact the arrangement of the fibers was not completely random. The calculated **R** values vary from 0.34 (d = 0.215 mm, l = 2.32 mm) to 0.40 (d = 0.108 mm, l = 1.27 mm).

Conclusion

The results obtained on commonly used polymer fibers have shown that radiative properties vary with different parameters. The obtained spectrum allows a non destructive analysis of the sample, it represents its spectral fingerprint. The surface state, the orientation and the size of the fiber greatly influence riadiative properties: the angular and size dependence of radiative properties have been demonstrated. The choice of a given material will be determined knowing the variations of its properties with the above-mentioned factors.

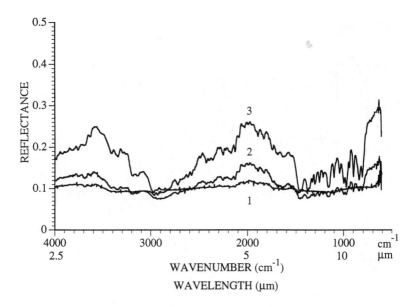

Figure 5. Total near-normal hemispherical spectral reflectance R versus wavenumbers (cm⁻¹) for a bulk sample of polypropylene (curve 1); fibers perpendicular (curve 2) and parallel (curve 3) to the cell boundaries. (Reproduced with permission from ref. 15. Copyright 1994 Society for Applied Spectroscopy.)

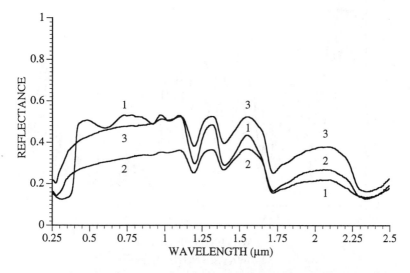

Figure 6. Total near-normal hemispherical spectral reflectance R versus wavelength (μm) for a bulk sample of polypropylene (curve 1); fibers perpendicular (curve 2) and parallel (curve 3) to the cell boundaries. (Reproduced with permission from ref. 15. Copyright 1994 Society for Applied Spectroscopy.)

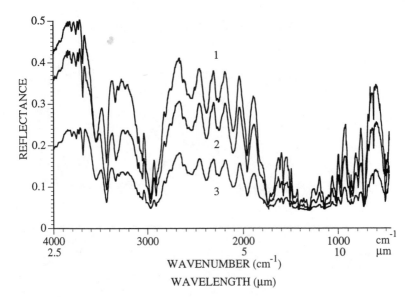

Figure 7. Total near-normal hemispherical spectral reflectance R versus wavenumbers (cm^{-1}) of polyester fabrics: curve 1, fibers woven horizontally in a plane; curve 2, velvet with long fibers and curve 3, velvet with near-perpendicular short fibers.

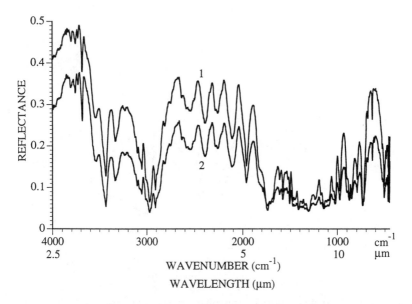

Figure 8. Total near-normal hemispherical spectral reflectance R versus wavenumbers (cm^{-1}) for a bundle of polyester fibers: curves 1 and 2 are obtained for two perpendicular orientations of the fibers maintained in the same plane.

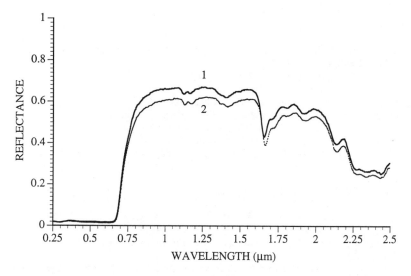

Figure 9. Total near-normal hemispherical spectral reflectance R versus wavelength (μm) for a bundle of polyester fibers: curves 1 and 2 are obtained for two perpendicular orientations of fibers maintained in the same plane.

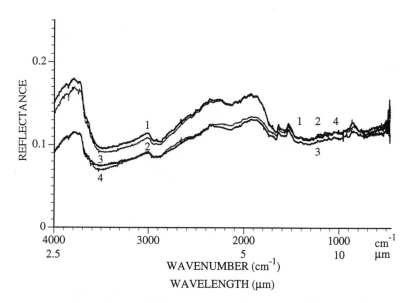

Figure 10. Total near-normal hemispherical spectral reflectance R versus wavenumbers (cm^{-1}) for polyamide fibers: curves 1 and 2 are for fibers with a diameter d = 0.108 mm and a length l = 1.27 and 2.32 mm respectively; curves 3 and 4 are for fibers with d = 0.215 mm and l = 2.32 and 1.95 mm respectively.

References

1. Sheperd, R. A. SPIE, *Characterization, Propagation and Simulation of Infrared Scenes* **1990,** *volume 1311*, page 55.
2. Richter, W. *Appl. Spectrosc.* **1983,** *volume 37,* page 32.
3. Lee, S. C. J. *Quant. Spectrosc. Radiat. Transfer* **1986,** *volume 36,* page 253.
4. Martin, K. A. *Appl. Spectrosc. Rev.* **1992,** *volume 27,* page 325.
5. Elliott, A. *British J. of Appl. Phys.* **1954,** *volume 5,* page 377.
6. Miller, C. E. *Appl. Spectrosc. Rev.* **1991,** *volume 26,* page 275.
7. Wheeler, O. H. *Chem. Rev.* **1959,** *volume 59,* page 629.
8. Miller, R. G.; Willis, H. A. *J. Appl. Chem.* **1956,** *volume 6,* page 385.
9. Roy, R. B.; Kradjel, C. *J. of Polym. Sc.: Part A: Polymer Chemistry,* **1988,** *volume 26,* page 1733.
10. Tungol, M. W.; Bartick, E. G.; Montaser, A. *J. of Forensic Sc., JFSCA* **1991,** *volume 36,* page 1027.
11. Chalmers, J. M.; Mackenzie, M. W.; Willis, H. A.; Edwards, H. G. M.; Lees, J. S.; Long D. A. *Spectrochimica Acta* **1991,** *volume 47 A,* page 1677.
12. Skaare, L. E.; Klaeboe, P.; Nielsen, C. J. *Vibrational Spectroscopy* **1992,** *volume 3,* page 23.
13. Silverstein, R. M.; Bassler, G. C. *Identification Spectrométrique des Composés Organiques;* Masson Ed., Paris, 1968, Chapter 3, pp 35-95.
14. Papini M. *Infrared Phys.* **1993,** *volume 34,* page 607.
15. Papini M. *J. Appl. Spectrosc.* **1994,** *volume 48,* page 472.

RECEIVED February 2, 1995

Chapter 9

Monitoring Polymerization Reactions by Near-IR Spectroscopy

Shih Ying Chang and Nam Sun Wang

Department of Chemical Engineering, University of Maryland, College Park, MD 20742

The following overview is an investigative survey on the feasibility of near-infrared (NIR) spectroscopy as an on-line process analytical technique for control and optimization. The review focuses on current developments in the field of polymerization monitoring. A brief background description on instrumentation and methods of data analysis leads to suggestions for further investigation. After NIR's introduction in the early 1950s, it remained unnoticed by polymer chemists and engineers through the 1960s and 1970s. NIR recaptured chemical engineers' attention in the late 1980s for two major reasons: (1) the development of advanced optical fiber materials and solid state detectors and (2) the incorporation of recent progress on multivariate data analysis algorithms. A review of the current trends in reactor monitoring strategies comprises the last section.

NIR spectra, like the mid-infrared spectra, are the result of light absorption by certain chemical bonds in molecules. In the NIR region of the spectrum, which covers the range from 780 nm (12,820 cm^{-1}) to 2500 nm (4,000 cm^{-1}), the absorption bands are the result of overtones or combinations originating in the fundamental mid-infrared region of the spectrum (4,000~600 cm^{-1}). Because of energy considerations, the majority of the overtone peaks seen in the NIR region arise from the R-H stretching modes with the fundamental vibrations located around 3000 cm^{-1} (O-H, C-H, N-H, S-H, etc.). Other fundamental vibrations manifest in the NIR region only as higher-order overtones that are generally too weak to provide useful analytical information. Another characteristic of NIR spectra is the large number of overlapping bands. For a given molecule, many active overtone and combination bands are typically present in a narrow NIR region and force the peaks to overlap significantly.

Because of the aforementioned weak absorptivity and highly overlapping peaks in the NIR region, NIR analysis, in comparison to its mid-IR counterpart, was largely

0097–6156/95/0598–0147$12.00/0

neglected by spectroscopists before the 1960's. After the 1960's, as NIR spectroscopy was emerging as a rapid, nondestructive method in the agricultural field, relatively few investigations in NIR applications dealt with polymers. Probable reasons for the lack of interest included difficulties in resolving overlapping bands and complex band assignments in the NIR region and the establishment of mid-IR spectroscopy as a standard and reliable polymer analysis technique.

Currently, agricultural analyses comprise the majority of established NIR applications. Other applications are found in pharmaceutical, biomedical, textile, and petrochemical industries. For a pertinent review, some references are available (1-6).

INSTRUMENTATION AND DATA ANALYSIS

NIR Instrumentation

Recent advances in NIR instrumentation have undoubtedly contributed to its increased use in quality analysis and process monitoring. Since the first NIR spectrophotometer of Dickey-John and Neotec in 1971, tremendous progress has been made in several aspects of instrument design towards a high signal-to-noise ratio and a stable spectral measurement. Improved design features include better optical configuration, faster scan rate, superior detector, implementation of sample averaging, and simplified sample presentation. These features are summarized in Table I (7).

Figure 1 displays the primary building blocks of a typical NIR spectrophotometer. It is a predispersive, single monochromator-based instrument in which light is dispersed prior to striking the sample. It employs a tungsten-halogen lamp as the light source, a single monochromator with a holographic diffraction grating, and an uncooled lead sulfide (PbS) detector. Interchangeable gratings allow experimentation with different holographic diffraction to achieve wavelengths from 400 to 2500 nm. The wavelength resolution is approximately 1~10 nm. Both transmittance and reflectance modes are offered in most instruments. Because of different requirements, some contemporary spectrophotometers incorporate advanced diode array detectors, NIR emitting diode sources, Hadamard mask exit slits, acousto-optical tunable filters (AOTF), ultrafast spinning interference filter wheels, tunable laser sources, and interferometers with no moving parts (7).

Modern NIR-based on-line process monitoring often benefits from advances in optical fiber techniques. Several types of optical fibers that can efficiently transmit light throughout the entire NIR region are presently available. As a result, remote sampling is now feasible with an NIR spectrophotometer coupled with optical fibers. A specific type of measurement configuration is called *transflectance*. In this configuration, light travels through a relatively thin layer of sample where a reflecting surface such as a ceramic disc or a metallic mirror forms the opposite physical boundary. After reflection, light subsequently transmits through the sample back toward the detector. This type of measurement is generally suitable for liquid samples or slurries.

Spectral Pretreatment

In the measurement process, a spectrophotometer records as raw data the fraction of the transmitted or reflected energy of the sample. The signal level is generally referenced to

Table I. Design characteristics of NIR instrumentation

1. **Optical configurations**
   A. Interference filters
   B. Moving diffraction grating(s)
   C. Prism(s)
   D. Near-infrared emitting diodes (NIR-EDs)
   E. Interferometer (Michelson, etc)
   F. Acoustooptical tunable filters (AOTF)
   G. Fixed grating
   H. Diode array detector (fixed grating)
2. **Scan rate**
   A. Slow (60-90 sec, or longer, for full range)
   B. Medium (10-60 sec for full range)
   C. Fast (0.1-10 sec for full range)
3. **Source type**
   A. Infrared (Globar ,Nernst glower, Nichrome wire, tunable laser)
   B. Near-infrared (tungsten-halogen monofilament, NIR-ED)
4. **Detector type**
   A. Infrared thermal type (thermocouples, thermistors, etc)
   B. Infrared photon detectors (semiconductors such as InAs, PbS, InSb)
   C. Near-infrared photon detectors (such as InGaAs, Pbs, and InAs)
5. **Sample averaging technique**
   A. Spinning or rotating cup
   B. Stepwise averaging of long-path cell
   C. Integrating sphere
6. **Sample presentation technique**
   A. Noncontact reflectance (telescoping optics)
   B. Near-infrared transmittance (NIRT)
   C. Near-infrared reflectance (NIRR)
   D. Reflectance and/or transmittance (NIRR/NIRT)

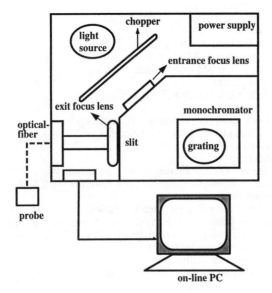

Figure 1 Diagram of a predispersive grating spectrophotometer

an appropriate standard. Before the calibration step, raw data may require pretreatment to linearize the optical response or to enhance the signal-to-noise ratio. Either an electronic device or a mathematical treatment program accomplishes the pretreatment task. Various algorithms appear in several excellent review articles (*2,8,9*). Below, we briefly summarize the main features of some common ones.

In transmission spectroscopy of homogeneous and nonscattering samples, the Beer-Lambert law describes the theoretical relationship between concentration c and absorbance A quite well (*10*)

$$c \propto A = log\left(\frac{I_0}{I}\right) = log\left(\frac{1}{T}\right) \tag{1}$$

where I is the light intensity measured by a spectrophotometer's detector, and I_0 is the intensity of the reference signal. Together, they yield transmittance, T. This linearization treatment of the directly measured light intensity via logarithmic transformation is universally accepted.

For diffuse-reflectance measurements, the most common choice of data pretreatment is the apparent absorption defined by $log(1/R)$. The concentration of the absorber should closely correlate to the reflectance.

$$c \propto log\left(\frac{1}{R}\right) \tag{2}$$

Unlike the Beer-Lambert Law governing light transmission, the above argument lacks theoretical rigor because the functional relationship between c and $log(1/R)$ remains to be developed. Nevertheless, the use of the above empirical relation is popular in practical applications.

Another widely accepted expression for the theoretical description of diffuse-reflectance signal is the Kubelka-Munk function (K-M function) (*11*):

$$f(R) = \frac{(1 - R_\infty)^2}{2R_\infty} = \frac{K}{S} \tag{3}$$

in which the ratio of an absorption coefficient, K, and a scatter coefficient, S, is given in terms of R_∞, where R_∞ is the reflectance from an infinitely thick sample. With strong variation in the light scatter, the K-M function after scatter correction seems to perform better than the $log(1/R)$ transformation (*12*).

The following operation based on spectral intensities at three consecutive wavelengths (λ_0, λ_1, and λ_2) generates the numerical approximation to the corresponding second-derivative.

$$Y'_{\lambda_1} = Y_{\lambda_2} - 2Y_{\lambda_1} + Y_{\lambda_0} \tag{4}$$

By repeating the process at each wavelength with a moving window that covers three wavelengths, a derivative spectrum results, wherein, the derivative operation eliminates baseline shifts and helps resolve overlapping peaks. The gap width, i.e., $\Delta\lambda = \lambda_2 - \lambda_1 = \lambda_1 - \lambda_0$, may be adjusted for optimal performance.

Further references to mathematical treatments can be found in the literature (*12,13*) for special sample systems, e.g., multiplicative scatter correction for a highly scattering medium.

Data Analysis

Progress in spectral analysis is perhaps the major reason behind improvements and extended applications of NIR spectroscopy. Modern multivariate regression methods play a critical role in both qualitative and quantitative spectral analyses. One can refer to many particularly pertinent statistical textbooks for details on some common regression methods, such as multiple linear regression (MLR) and principal component regression (PCR) (*14,15*).

Partial least squares (PLS) regression is presented here as a member of the bilinear class of regression methods. The aforementioned PCR is another popular example of these bilinear methods. These methods are powerful and flexible approaches to multivariate calibration. The PLS differs from PCR in that it actively utilizes Y (dependent variables, e.g., concentrations or material properties to be predicted) during the data compression of X (independent variables, usually spectral intensities). By balancing information contained in X and Y, the method reduces the impact of large, but irrelevant X-variations in the calibration model.

A simplified PLS model consists of a regression between the scores for both the X and Y data blocks. The model consists of two outer relations of X and Y blocks and an inner relation that links the two data blocks. The outer relation for the X block, which decompose the original data into h significant components (similar to PCA) is as follows:

$$X = TP' + E = \sum_{h} t_h p'_h + E \qquad (5)$$

where T is the score, P is the loading, and E is the residual not explained during the decomposition process. The outer relation for the Y block is similarly constructed:

$$Y = UQ' + F = \sum_{h} u_h q'_h + F \qquad (6)$$

We must accurately describe Y by forcing the error term F to be as small as possible. At the same time, we also shall establish a good correlation equation between X and Y by regressing each component of U, the score of the Y block, against the corresponding component of T, the score of the X block. The simplest model for such a relation is a linear one:

$$u_h = b_h t_h \qquad (7)$$

where $b_h = u'_h t_h / t'_h t_h$ is simply the coefficients from 1-dimensional least squares regression.

Various models of PLS have been developed since its inception in 1975 by Herman Wold as a practical solution to concrete data-analytic problems in econometric and social sciences (*16*). During the late seventies, S. Wold and H. Martens' groups pioneered the PLS method for chemical applications. The PLS model described above is basically the algorithm published by Geladi and Kowalski in 1986 (*17*).

Finally, there is significant progress in specialized algorithms for tackling nonlinear systems. These include the locally weighted regression (*18*), nonlinear PLS models (*19,20*), and neural networks (*21*). In an attempt to assign NIR absorbance to specific contributing bonds or functional groups, several researchers have recently studied the cross-correlation of two different spectroscopic methods, e.g., NIR-Raman (*22*) or NIR-IR (*23*). Furthermore, new studies investigate the potential of NIR image analysis for classification (Y. R. Chen, USDA-ISL, personal communication, 1994).

POLYMER CHARACTERIZATION

Material Characterization

NIR is a newly emerging method of characterizing polymer materials. Earlier applications include unsaturation content determination (*24*), ethylene/propylene ratio measurement for ethylene-propylene elastomers (*25,26*), monomer residue in poly(methyl methacrylate) (*27*), and hydroxyl number determination (*28*). Recent applications include polyurethane (*29-31*), high density and low density polyethylene (*32*), polyethylene tetraphthlate and polycarbonate (*33*), and epoxy resins (*34*). A couple of review articles (*9,35*) contain details pertaining to this field of application.

Polymerization Monitoring

A long-recognized weak link in the polymer reactor control scheme is undoubtedly on-line instrumentation. Since feedback control of the polymer quality is not possible without some form of on-line polymer characterization/monitoring, there have been numerous attempts to achieve it. Such attempts utilize the material's various chemical/physical properties: densitometry (*36-39*), refraction and dielectric constants (*40*), light scattering and turbidity (*41*), Raman and light scattering (*42-44*), and fluorescence (*45*). Various unresolved problems such as sampling loop clogging, measurement delay, and limited number of monitored variables continue to challenge researchers. Some estimation methods, like Kalman filter, have been applied to solve the problem of unavailable or delayed measurements (*46,47*); however, the lack of adequate kinetic models with valid kinetic parameters and the limited filter observability remain problematic. Reference (*48*) addresses this topic.

Until recently, there have been few reported NIR applications on polymerization monitoring. Powell et al. (*49*) followed the bulk polymerization of styrene in a 2.54 cm pathlength cell and noted an increase in the transmission at 1211 nm as the reaction proceeded. Significant work by Miller and Willis dealt with the composition ratio in butadiene/styrene copolymers and the determination of methyl methacrylate monomer residue in its polymer counterparts (*27*). Without presenting quantitative results, they also entertained the possibility of following methyl methacrylate reaction and measuring the molecular weight of polyethylene oxide. Crandall et al. (*50*) reported monitoring the imidization reactions involving a series of polyimides following an amide band around 2300 nm and an amine band around 1950 nm. Crandall and Jagtap (*51*) tabulated the absorption bands for the most common step-reaction polymers. Pomerantseva et al. (*52*) recently monitored various copolymerizations of olefins. Based on the terminal epoxide C-H band around 2207 nm, Qaderi et al. (*53*) followed the second-order curing reaction of epoxy in the presence of amine and acid groups, as well as solvent and other species. Most reports successfully demonstrated feasibility for monitoring reactions with NIR; however, there is room for more vigorous quantitative treatments.

MODERN APPLICATIONS

This section reviews the latest applications of NIR to polymerization studies over the last decade. We will compare the methods of collecting NIR spectra and the algorithms of quantifying the reaction contents.

The renaissance of NIR in polymerization reaction monitoring and control applications began with Buback's group in the kinetic study of ethylene polymerization (54-56). In his study of thermal, laser, and chemically initiated high-pressure and high-temperature ethylene polymerizations, Buback used a specially designed transmission optical cell. The cell with two sapphire windows was machined from stainless steel for operation up to 350°C and 3300 bar. The optical pathlength was approximately 12 mm. The temperature variation within one polymerization experiment was less than 0.5°C. The reaction vessel was removed from the irradiation assembly and introduced into the spectrophotometer before NIR or UV spectra were recorded.

A typical NIR spectrum is shown in Figure 2. The peaks around 1115 nm (8960 cm^{-1}) and 1144 nm (8730 cm^{-1}), which were attributed to C-H stretching of monomeric ethylene, decreased over the course of polymerization, while the formation of polyethylene was responsible for the increase in the 1211 nm (8260 cm^{-1}) peak. Another set of peaks around 1629 nm and 1761 nm were used as well, the latter region being the first overtone of C-H stretching. Ethylene concentration was determined from the integrated absorbance of the short wavelength half-band, which is defined as:

$$\int A(\lambda)d\lambda = C_E \cdot l \cdot \int \epsilon(\lambda)d\lambda = C_E \cdot l \cdot B_{\frac{1}{2}} \qquad (8)$$

where $A(\lambda)$ is the absorbance at wavelength λ, C_E is the ethylene concentration, l is the optical pathlength, and $B_{\frac{1}{2}}$ is the so called half-band molar integrated intensity. Because the empirical constant $B_{\frac{1}{2}}$ had to be determined in separate experiments, the absolute accuracy of the ethylene concentration, C_E, thus determined was estimated to be ±3%. Another significant contribution of Buback's work was the possibility of obtaining structural information about polyethylene from the difference between consecutive absorbance measurements ΔA. The component around 1190 nm, which was assigned to methyl groups, indicates the formation of polymer materials either with a lower molecular weight or with a higher degree of short chain branching.

Laurent et al. at British Petroleum Chemicals patented a process of NIR reaction monitoring (57). Primarily aimed at polyolefins, this invention covered extending an FTIR spectrophotometer to operate in the 1100~2500 nm NIR range and an appropriate sampling loop to separate samples from the reaction mixture. They recorded absorbance at 2~20 separate wavelengths and correlated the measured signal to polymer properties with multivariate regression. Transmission, reflection, and a combination thereof were listed in the patent as optional measurement methods. Laurent et al. suggested that polymer properties, including composition, density, viscosity, and molecular weight, might be successfully analyzed by the apparatus.

Aldridge et al. reported the use of short-wavelength NIR to monitor bulk polymerization of methyl methacrylate (58). A special sample cell with a retroreflective array, much like that used in traffic signs, was constructed to reduce the lensing effect caused by considerable changes in the refractive index during polymerization. The cell consisted of a microassembly of corner cube reflectors that broke the incident light beams into an array of subbeams. Being able to cajole the subbeams to travel through the sample by traversing exactly the same path backward, this device counteracted the beam-bending effect caused by local variations in the refractive index. With the distortion largely eliminated, the spectra were treated with a second derivative transformation and referenced to the initial monomer spectrum. Figure 3 illustrates second derivative difference spectra that were obtained by subtracting the initial second derivative

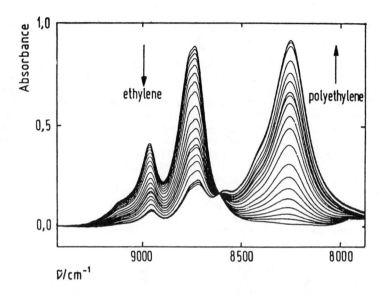

Figure 2 Near-infrared absorbance of a laser-induced high-pressure ethylene polymerization. (Reproduced with permission from ref. 54. Copyright 1986, Hüthig & Wepf Verlag.)

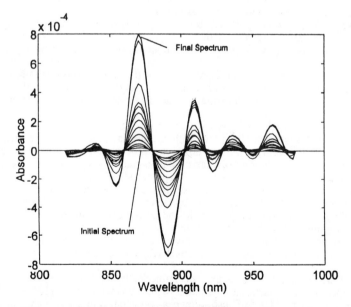

Figure 3 NIR second-derivative spectra of methyl methacrylate bulk polymerization using a photodiode array spectrophotometer. (Reproduced from ref. 58. Copyright 1994 American Chemical Society.)

spectrum from all the remaining spectra. Two intensities at 868 nm and 922 nm were selected to build up a multilinear regression model.

Gossen et al. applied UV-Visible and NIR spectroscopy to analyze a copolymer latex system of styrene and methyl methacrylate (59). For experimental tractability, they designed the latex batches to achieve variations in particle sizes, polymer fractions with either high or low molecular weight, and different copolymer ratios. Allowing the reagents to convert completely (i.e., 100% conversion), they were able to follow emulsion polymerization from pre-designed input charges and simplify the characterization of the latex products. The NIR spectra were obtained with an optical fiber probe that was directly immersed in the samples. The probe had a reflective mirror tip and a 3 mm gap, which effectively yielded a 6 mm pathlength. They recorded absorbance spectra and employed principal component analysis to group the spectra into an initial stage and a final stage along the reaction trajectory. Absorbance readings in the entire UV-Vis and NIR regions were used to develop separate PLS calibration models for each reaction stage. Difficulties in eliminating the interference from scattering latex particles led to relatively poor performance during the final stage compared to the good predictions achieved during the initial reaction stage.

Long et al. utilized a fiber optical probe in the range 1000~2500 nm to monitor in real-time the reaction kinetics of anionic polymerization of styrene and isoprene in polar and nonpolar solvents (60). The probe tip was equipped with a rhodium reflective surface that provided a pathlength of approximately 1 cm. They assigned the absorption peak of two pairs of vinyl carbon-hydrogen bonds at 1624 nm. In the absence of independent experiments, this assignment is probably based on comparison with the literature. A semilog plot of peak absorbance at this wavelength versus reaction time confirmed the reaction rate to be first-order. They did not disclose the details on concentration determination. They also monitored copolymerization reaction with NIR; however, a time plot of peak absorbance, rather than concentration, reflected their difficulties in quantitative determination. They emphasized that the successful application of on-line spectroscopic polymerization monitoring required clear separation of the absorbance bands of the monomer, polymer, and solvent. This requirement, though preferable, seems unrealistic, especially in view of the highly overlapping NIR spectra and the complex environment in which industrial polymerization normally proceeds. Nevertheless, methods of modern multivariate analysis can successfully tackle this problem.

Finally, through several studies in the authors' own laboratory, we have successfully demonstrated the feasibility of NIR as an on-line process analyzer for polymerization. In two types of batch solution polymerizations, namely methyl methacrylate in ethyl acetate and styrene in toluene, we were able to predict simultaneously the concentrations of the monomer and its polymer counterpart based on NIR spectra taken with a 1 mm transmission optical cell (61). In one study, samples were taken out from the reactor periodically and were measured by NIR and referenced to independent gravimetry methods. Figures 4 and 5 show the difference spectra for methyl methacrylate and styrene polymerization reactions, respectively, obtained by subtracting out the pure solvent spectrum under the same conditions during the course of reaction.

The method of partial least squares was numerically applied to spectral deconvolution. Compared to the classical least-squares regression methods, PLS has the following advantages: (1) the problems of absorption peak broadening and shifting resulting from changes in the sample compostition (unrelated to instrumentation drift/malfunction)

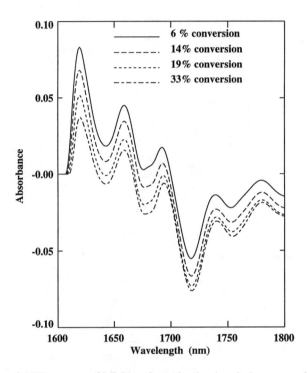

Figure 4 NIR spectra of MMA polymerization in ethyl acetate solution

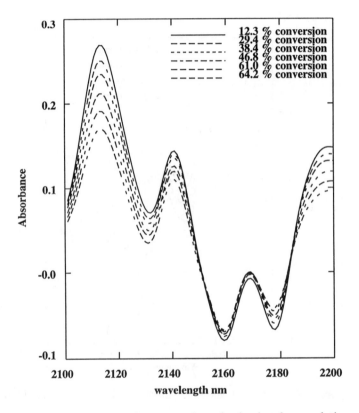

Figure 5 NIR spectra of styrene polymerization in toluene solution

can be addressed; (2) redundant but relevant absorbance readings help reduce over-all measurement errors. The C-H stretching of methylene and vinyl carbon bonds at 1670~1770 nm and the combination bands of C-H at 2100~2300 nm yielded the best model. The prediction accuracies during the two polymerization processes, in terms of the relative mean squared errors based on independent primary off-line measurements, were 95% and 98% for methyl methacrylate and poly(methyl methacrylate), respectively; and 94% and 87% for styrene and polystyrene, respectively. The scatter plots of Figures 6 and 7 illustrate the prediction accuracies. Other runs resulted in similar error levels; thus, the calibration model seemed quite stable. Alternatively, a PLS model was built from data taken in a previous run and subsequently predicted another run conducted under the same set of conditions. It gave conversion predictions that were as accurate as those from a PLS model calibrated with completely independent measurements.

Further investigations into the characteristics of the spectra showed that a highly linear relationship between absorbance and concentration generally led to a reliable PLS model. We found high linearity between the raw spectra and the PLS components in the selected regions around 1700 nm and 2200 nm. In addition, absorbance in these two spectral regions also exhibits much sensitivity to concentration variation. Linearity and sensitivity together form the foundation of a successful PLS calibration model.

In another one of our polymerization studies, we collected spectra with a fiber op-tical probe inserted directly into the reaction media. Figure 8 shows the setup for this on-line measurement. NIR served not so much as an analytical instrument for polymer characterization but as an on-line process analyzer for process control and optimization. This approach demands speed, accuracy, and reliability. From the operational stand point, we prefer a sensor system without a sampling loop, as most polymerization media are highly viscous. Spectra were taken every 5~10 minutes during the polymerization reaction. We developed a method of temperature compensation to account for the dif-ference in temperature between the on-line reaction samples and calibration samples. In addition, we formulated a weighting algorithm that automatically selected the optimum spectral regions to build up a calibration model. Figure 9, which shows satisfactory prediction of various methyl methacrylate polymerizations, confirms the feasibility of the method. Applications to heterogeneous reaction systems are in progress.

Using a similar probe configuration, DeThomas et al. (*62*) also collected spectra directly from the processes media. They attached an optical probe (1 cm pathlength) to the process stream of a polyurethane reactor and monitored the isocyanate concentration. They formulated a correlation equation based on only one second-derivative absorbance at 1648 nm and reported good prediction results.

CONCLUSIONS

Through NIR spectroscopy and appropriate spectral analysis, rugged and fast on-line polymerization monitoring can provide chemical information in-situ. This real-time information helps an engineer control and optimize a polymerization process. The benefits of NIR spectroscopy include increased manufacturing yield, better product quality control, and greater savings. Various examples cited in this article firmly support the notion that NIR is an excellent on-line analyzer for the polymer industry.

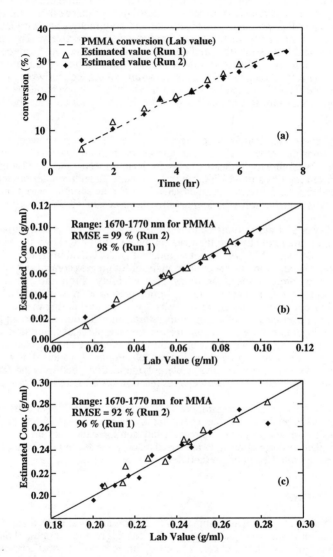

Figure 6 NIR measurement of MMA conversion during polymerization

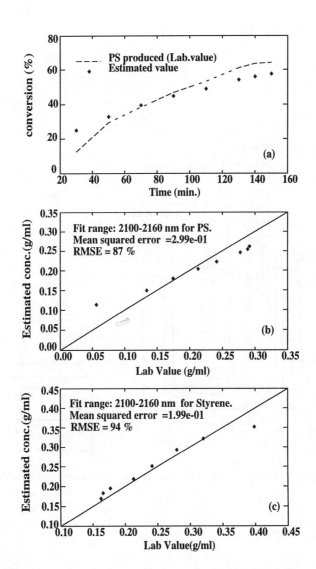

Figure 7 NIR measurement of polystyrene conversion in toluene solution polymerization

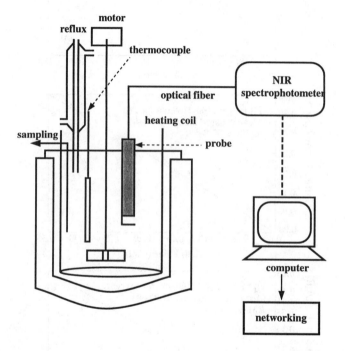

Figure 8 Setup of on-line polymerization measurement with NIR

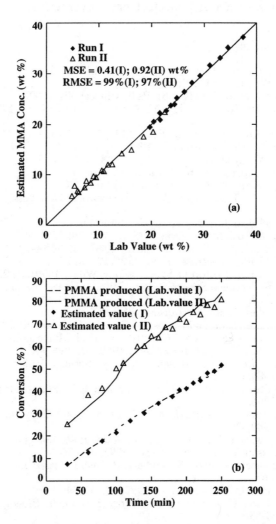

Figure 9 NIR measurement of MMA conversion in ethyl acetate solution polymerization

ACKNOWLEDGEMENT

This work is partially supported by a grant from the Maryland Industrial Partnerships Program (MIPS) through the collaboration of the Engineering Research Center at the University of Maryland, College Park, and NIRSystems, Silver Spring, Maryland. We would like to thank Mr. Karl H. Norris for his help in our studies.

LITERATURE CITED

1. *Near-Infrared Technology in the Agricultural and Food Industries;* Williams, P. C.; Norris K. Eds.; American Association of Cereal Chemists: St. Paul, Minneapolis, **1987**.
2. Stark, E.; Luchter, K.; Margoshes, M. *Appl. Spectrosc. Rev.*, **1986,** *22*, 335-399.
3. Martin, K. A. *Appl. Spectrosc. Rev.*, **1992,** *27*, 325-383.
4. Whitfield, R. G., *Pharm. Manuf.*, **1986,** April, 31-34.
5. Ciurczak, E. W., *Appl. Spectrosc. Rev.*, **1987,** *23*, 147-163.
6. *Near-Infrared Spectroscopy: Bridging the Gap between Data Analysis and NIR Applications;* Hildrum, K. I.; Isaksson, Næs, T.; Tandberg, A. Eds.; Ellis Horwood: New York, **1992**; pp319.
7. *Handbook of Near-Infrared Analysis;* Burns, D. A.; Ciurczak, E. W. Eds.; Marcel Dekker: New York, **1992**; Ch. 4.
8. Enke, C. G.; Nieman, T. H. *Anal. Chem.*, **1976,** *48*, 705A-709A.
9. Miller, C. E. *Appl. Spectrosc. Rev.*, **1991,** *26*, 277-339.
10. Smith, A. L. *Applied Infrared Spectroscopy*, Wiley: **1979;** pp 220.
11. Kortum, G. *Reflectance Spectroscopy;* Springer-Verlag: Berlin; New York, **1969**.
12. Martens, H.; Næs, T. in *Near-Infrared Technology in the Agricultural and Food Industries;* Williams, P. C.; Norris K. Eds.; American Association of Cereal Chemists: St. Paul, Minneapolis, **1987**.
13. Miller, C. E.; Næs, T. *Appl. Spectrosc.*, **1990,** *44*, 895-902.
14. Jolliffe, I. T. *Principal Component Analysis;* Springer-Verlag: New York, **1986**; Ch. 10.
15. Wold, H. in *Multivariate Analysis*, Krishnaiah, P. R., Ed., Academic Press: New York, **1966**.
16. Martens, H.; Næs, T. *Multivariate Calibration;* John Wiley: Chichester; New York, **1989**; Ch. 3.
17. Geladi, P.; Kowalski, B. R. *Anal. Chim. Acta*, **1986,** *185*, 1-17.
18. Næs, T.; Isaksson, T. *Appl. Spectrosc.*, **1992,** *46*, 34-42.
19. Wold, S., Kettaneh-Wold, N.; Skagerberg, B. *Chem. Intell. Lab. Sys.*, **1989,** *7*, 53-65.
20. Wold, S. *Chem. Intell. Lab. Sys.*, **1992,** *14*, 71-84.
21. McAvoy, T. J.; Su, H. T.; Wang, N. S.; Ming, H. *Biotech. Bioeng.*, **1992,** *40*, 53-63.
22. Miller, C. E. *Spectrochimica Acta*, **1993,** *49A* , 621-632.
23. Barton, F. E. (II); Himmelsbach, D. S.; Duckworth, J.H. *Appl. Spectrosc.*, **1992,** *46*, 420-427.
24. Goddu, R. F. *Anal. Chem.*, **1957,** *29*, 1790-1794.
25. Bly, R. M.; Kiener, P. E.; Fries, B. A. *Anal. Chem.*, **1986,** *38*, 217-220.
26. Skaare, L. E.; Klaeboe, P.; Nielsen, C. J. *Vibrational Spectrosc.*, **1992,** *3*, 23-33.

27. Miller, R. G. J.; Willis, H. A. *J. Appl. Chem.*, **1956**, *6*, 385-390.
28. Hilton, C. L. *Anal. Chem.*, **1959**, *31*, 1610-1612.
29. Miller, C. E. *J. Appl. Polym. Sci.*, **1991**, *42*, 2169-2190.
30. Miller, C. E. *Appl. Spectrosc.*, **1990**, *44*, 576-579.
31. Miller, C. E.; Eichinger, B. E. *Appl. Spectrosc.*, **1990**, *44*, 887-894.
32. Skagerberg, B.; MacGregor, J. F.; Kiparissides, C. *Chem. Intell. Lab. Sys.*, **1992**, *14*, 341-356.
33. Zhu, C; Hieftje, G. M. *Appl. Spectrosc.*, **1992**, *46*, 69-72.
34. Dannenberg, H. *Soc. Plast. Engineering*, Trans., **1963**, *3*, 78-88.
35. Lee, K. A. B. *Appl. Spectrosc. Rev.*, **1993**, *28*, 231-284.
36. Schork, F. J.; Ray, W. H. *Am. Chem. Soc. Ser.*, **1981**, *165*, 505-514.
37. Abbey, K. J. *Am. Chem. Soc. Ser.*, **1981**, *165*, 345-356.
38. Ponnuswamy, S.; Shah, S. L. *J. Appl. Polym. Sci.*, **1986**, *32*, 3239-3253.
39. Canegallo, S.; Storti, G.; Morbidelli, M.,; Carra, S. *J. Appl. Polym. Sci.*, **1993**, *47*, 961-979.
40. Schmidt, A. D.; Ray, W. H. *Chem. Eng. Sci.*, **1981**, *36*, 1401-1410.
41. Kiparissides, C.; MacGregor, J. F.; Singh, S.,; Hamielec, A. E. *Can. J. Chem. Engin.*, **1980**, *58*, 65-71.
42. Chu, B.; Lee, D. *Macromolecules*, **1984**, *17*, 926-937.
43. Damoun, S.; Papin, R.; Ripault G.,; Rousseau, M. *J. Raman Spectrosc.*, **1992**, *23*, 385-389.
44. Gulari, E.; Mackeigue, K.; Ng, K. Y. S. *Macromolecules*, **1984**, *17*, 1822-1825.
45. Wang, F. W.; Lowry, R. E.; Fanconi, B. M. *Polymer*, **1984**, *25*, 690-692.
46. Adebekun, D. K.; Schork, F. J. *Ind. Eng. Chem. Res.*, **1989**, *28*, 1846-1861.
47. Dimitratos, J. *Chem. Eng. Sci.*, **1991**, *46*, 3203-3218.
48. Chien, D. C. H.; Penlidis, A. *Macromol. Chem. Phys.*, **1990**, *C30*, 1-42.
49. Evans, A.; Hibbard, R. R.; Powell, A. S. *Anal. Chem.* **1951**, *23*, 1604-1610.
50. Crandall, E. W.; Johnson, E. L.; Smith, C. H. *J. Appl. Polym. Sci.*, **1975**, *19*, 897-903.
51. Crandall, E.W.; Jagtap, A.N. *J. Appl. Polym. Sci.*, **1977**, *21*, 449-454.
52. Pomerantseva, E. G.; Tsareva, L. A.; Chervyakova, G. N.; Nozrina, F. D. and Fokeeva, N. V.,*Plast. Massy*, **1989**, *12*, 66-70.
53. Qaderi, S. B. A.; Paputa Peck, M. C.; Bauer, D. R. *J. Appl. Polym. Sci.*, **1987**, *34*, 2313-2323.
54. Brackemann, H.; Buback, M.; Vögele, H.-P. *Makromol. Chem.*, **1986**, *187*, 1977-1992.
55. Buback, M.; Lendle, H. *Makromol. Chem.*, **1983**, *184*, 193-206.
56. Buback, M. *Angewandte Chemie*, **1991**, *30*, 641-653.
57. Laurent, J.; Martens, A.; Vidal, J. L.; *PCT Int. Appl.* **1989**, WO 89/06244.
58. Aldridge, P. K.; Kelly, J. J.; Callis, J. B. *Anal. Chem.*, **1993**, *65*, 3581-3585.
59. Gossen, P. D.; MacGregor, J. F.; Pelton, R. H. *Appl, Spectrosc.*, **1993**, *47*, 1852-1870.
60. Long, T. E.; Liu, H. Y.; Schell, B. A.; Teegarden, D. M.; Uerz, D. S. *Macromolecules*, **1993**, *26*, 6237-6242.
61. Chang, S. Y.; Wang, N. S. AIChE annual Meeting, Material Science and Engineering, paper 202n, San Francisco, Nov. 1994.
62. DeThomas, F. A.; Hall, J. W.; and Monfre, S. L. *Talanta*, **1994**, *41*, 425-431.

RECEIVED February 6, 1995

Chapter 10

UV–Visible Spectroscopy To Determine Free-Volume Distributions During Multifunctional Monomer Polymerizations

Kristi S. Anseth, Teri A. Walker, and Christopher N. Bowman

Department of Chemical Engineering, University of Colorado,
Boulder, CO 80309–0424

Efforts have focused on developing a spectroscopic technique to characterize the microstructural heterogeneity and free volume distribution during the polymerization of multifunctional monomers. The technique uses UV-Vis spectroscopy to monitor the trans to cis conformational changes in photochromic probes. Since the absorbance spectra of the two conformations are significantly different, UV-Vis spectroscopy can be used to monitor the isomerization rate of the probe by measuring the absorbance at certain wavelengths. The photoinduced forward isomerization of the photochromic probe is a sensitive measure of the local microstructure and free volume. Therefore, this experimental technique which combines UV-Vis spectroscopy and photochromic probes provides a method to monitor the free volume distributions *in situ* during the polymerization. The analysis has been discussed in detail, and results are presented for a series of multiethylene glycol dimethacrylate polymerizations. The photochromic probes studied were azobenzene, m-azotoluene, stilbene, and 4,4'-diphenylstilbene.

This work develops a technique using UV-Visible (UV-Vis) spectroscopy to characterize the microstructural evolution and volume distributions of a polymer network by examining trans to cis conformational changes in photochromic probes during the polymerization. Since the absorbance spectra of the two probe conformations are substantially different, UV-Vis spectroscopy can be used to monitor the isomerization rate of the probe by measuring the absorbance at certain wavelengths. The photoinduced forward reaction of photochromic probes (trans to cis isomerization) has been found to be a sensitive measure of the local free volume and microstructure and has been used to establish volume distributions by others studying physical aging in linear polymer systems (*1-7*). Unlike the previous studies, this experimental technique provides a method for monitoring the free volume distributions *in situ* during the polymerization by monitoring the absorbance changes of the probe with a UV-Vis spectrophotometer while simultaneously photopolymerizing the system (*8*).

Photopolymerizations have been extensively used in the coatings industry to rapidly cure liquid, multifunctional monomers (typically diacrylates) at room temperature into highly crosslinked, polymer networks. Other commercial applications for light induced polymerizations include dental restorative materials, aspherical lenses , and optical fiber coatings (*9-11*). Depending on the structure and functionality of the monomer as well as the reaction conditions, the crosslinking density and the microstructure of the system will vary. This crosslinking density and local microstructure of the polymer control many of the macroscopic material properties, such as the dimensional and thermal stability and the swelling (or resistance to swelling) of the polymer. Thus, it is important to develop techniques to characterize the structural evolution of the polymer and to determine the effects of reaction conditions and monomer type on this evolution. These formation-structure relationships provide insight into many of the macroscopic properties of the network.

While the macroscopic properties of the polymer are strongly dependent on the microstructure of the network, the polymerization behavior of multifunctional monomers is extremely complex, and the resulting network structures are very difficult to characterize either experimentally or theoretically. The primary reason for this difficulty is the structural heterogeneity that develops (*9*). The inhomogeneities that occur during the polymerization of multifunctional monomers are microgels of highly crosslinked material formed within regions of a much lower crosslinking density (*12-14*). The heterogeneity results from unequal reactivity of functional groups in the reaction medium. In general, this reactivity is not only unequal, but changes with conversion in the system. As the reaction begins, the high concentration of pendant double bonds in the vicinity of the active radical leads to their higher reactivity. This higher reactivity produces increased crosslinking and cyclization and the formation of the microgel regions.

Near the end of the reaction, the pendant double bonds lose their enhanced reactivity as the unreacted pendant bonds (formed earlier in the reaction) are shielded from the active radicals by the already formed polymer network. Then, the relative reactivity of the monomeric double bonds increases. Kinetic gelation simulation results (*15,16*) indicate that pendant double bonds can be up to 50 times more reactive than monomeric double bonds early in the reaction because of their proximity to the radical location. This enhanced reactivity leads to the formation of microgel regions which can significantly decrease the mechanical properties of the polymer network when compared to a homogeneous network. Hence, controlling the reaction conditions that minimize the heterogeneity in the polymer network structure will greatly enhance the ability to improve and predict the final network properties.

As a result of the inhomogeneous nature of the polymerizations of multifunctional monomers, a free volume distribution develops in the polymer network. Therefore, many of the free volume based theories for linear polymerizations that assume a homogeneous distribution of free volume throughout the system do not accurately describe or predict the free volume in crosslinking polymerizations. Since the reaction behavior and the material properties are strongly influenced and controlled by the local free volume, it is desirable to develop techniques to measure and quantify the local free volume distribution.

Electron spin resonance spectroscopy (ESR) (17,18), positron annihilation lifetime spectroscopy (PALS) (19-21), and x-ray scattering (SAXS and WAXS) (22,23) have all been used to study the microstructure and volume distributions in polymers. While these techniques have been successfully used in studying effects such as aging on the polymer microstructure, the time scale of these experimental techniques makes them improbable for studying the microstructural evolution during a photopolymerization. For example, PALS experimental accumulation times range from several hours to days, while a photopolymerization is complete in seconds.

Alternatively, Torkelson and coworkers (1-4) have developed a technique based on photochromic probes which quantifies volume distributions during physical aging in linear polymer systems. Some of the first studies of photochromic probes in polymer systems were done by Gardlund (24) and Gegiou et al. (25) in the late 1960's. In particular, Gegiou (25) reported a dependence on the amount of cis isomer at the photostationary state related to the polymer viscosity. It was suggested that for the probe to isomerize, the local environment around the probe must have enough mobility, or alternatively, a minimum volume. Since these early studies, several researchers have investigated the isomerization process of photochromic probes in polymer systems. Several studies have focused on the thermal back reaction (cis to trans isomerization) as well as the photoinduced forward reaction (trans to cis isomerization). In general, it appears that the photoinduced forward reaction is more easily interpreted and related to the behavior of the polymer than the thermal back reaction.

Sung and coworkers (5-7,26,27) have studied the trans to cis isomerization of azo-probes and azo-labeled polymers in solution, as well as in the rubbery and glassy state. In solution, they found the isomerization to exhibit first order kinetics, but in the rubbery or glassy state, the isomerization was described by the sum of two first order processes. The first process exhibited a characteristically fast isomerization of the probe in the liquid-like environment, and the second process was a slow isomerization dependent on the local viscosity. They found the first order decay constant for the slow isomerization to be nearly 100 times smaller than the decay constant for the fast isomerization. Sung and coworkers (5-7,26,27) also studied the mobility of chain ends, side chains, and backbone polymer by site-specific labeling with azobenzene chromophores. Horie et al. (28) and Mita et al. (29) developed a kinetic based model for the isomerization process based on three different environments in the polymer. The three environments were identified as follows: liquid-like allowing isomerization as easily as in solution, isomerization controlled by the polymer mobility, and no isomerization. They also found a dependence of the probe size on the mobility or ability of the probe to isomerize in the polymer. Torkelson and coworkers (1-4) have greatly expanded on this work by incorporating several probes into different linear polymer systems (e.g. polystyrene, polycarbonate, poly(methyl methacrylate)). Furthermore, they quantified the minimum critical volume required by the probes to isomerize. By monitoring the trans to cis photoisomerization process, they also found the extent of isomerization to depend on the size of the probe. Additionally, they studied the effects of physical aging on the mobility of the probe and the cumulative volume distribution of the polymer. Interestingly, the volume distributions obtained by Torkelson through the photochromic technique were found to have a similar shape when compared to those obtained by PALS (*19*).

Expanding on the previously described techniques, we have used photochromic probes to study the microstructure and volume distributions during a polymerization (*8*). In particular, this work uses UV-Vis spectroscopy coupled with a series of photochromic probes to establish free volume distributions during photopolymerizations. This technique allows the determination of the degree of heterogeneity and the free volume distribution as a function of polymerization conditions, e.g. temperature, number of double bonds, initiation rate, monomer properties and polymerization time. Because of the known heterogeneous nature of polymerizations of multifunctional monomers (*9,12-14*), i.e. monomers with multiple double bonds per monomer, our research has focused on using this technique to characterize free volume distributions in these polymerizations.

Experimental

A series of multiethylene glycol dimethacrylate monomers was chosen for this study: ethylene glycol dimethacrylate (EGDMA), diethylene glycol dimethacrylate (DEGDMA), triethylene glycol dimethacrylate (TEGDMA), poly (ethylene glycol 200) dimethacrylate (PEG(200)DMA), and poly (ethylene glycol 600) dimethacrylate (PEG(600)DMA) (Polysciences Inc., Warrington, PA). All monomers were used as received. Polymerizations were initiated with the photoinitiator 2,2-dimethoxy-2-phenylacetophenone (DMPA; Ciba Geigy, Hawthorn, NY). The photochromic probes chosen for these studies were azobenzene (AB; Aldrich, Milwaukee, WI), m-azotoluene (AT; ICN Biomedical Inc., Aurora, OH), stilbene (SB; Aldrich, Milwaukee, WI), and 4,4'-diphenylstilbene (DPS; Aldrich, Milwaukee, WI). Torkelson and coworkers (*1-4*) have previously calculated the critical volume required for the isomerization of these probes. Azobenzene, m-azotoluene, stilbene, and 4,4'-diphenylstilbene require a minimum of 127Å^3, 202Å^3, 224Å^3, and 575Å^3 to isomerize, respectively.

By choosing to study the multiethylene glycol dimethacrylate monomer series, the effects of crosslinking density and monomer mobility on the microstructural evolution of the network was studied. By changing the number of ethylene glycol units between the methacrylate double bonds, the concentration of double bonds in the system as well as the relative reactivity of the pendant functional group changes. This reactivity of the pendant functional group will strongly influence the formation of microgels in the network. In addition to the effects of monomer type on the polymer structure, the effects of double bond conversion and rate of polymerization were also studied. The rate of polymerization can be controlled by such parameters as the initiator concentration, temperature, and light intensity.

Samples were prepared by dissolving 0.1 wt% of a given photochromic probe and 1.0 wt% DMPA in each monomer. To prepare thin films of the liquid mixture for analysis, a cover slip was mounted on a quartz microscope slide with a tape spacer (approximately 30 μm) between the two. The monomer mixture was drawn by capillary action between the slide and the cover slip creating a thin film. This slide was then place in a temperature controlling device that circulated a reservoir of water behind the slide. The reservoir was sealed on one side by the microscope slide and on the opposite side by a quartz window, providing a path for the UV-Vis beam to pass through the sample.

The thin film of monomer-probe-initiator was then cured using a 365 nm 6 W UV light source (Cole-Parmer, Chicago, IL). The UV light also served to induce the trans to cis conformational changes in the photochromic probes. The polymerization was performed in the UV-Vis spectrophotometer where the absorbance was monitored *in situ* during the polymerization. Thus, the absorbance in the polymerizing system as a function of time was obtained. In addition to this curve, the absorbance of the probe as a function of time in a monomer system, without polymerization but with isomerization, was measured. This measurement was obtained by dissolving the probe in the appropriate monomer without initiator and exposing the system to the UV light to monitor the isomerization. The wavelength that was monitored was chosen near the peak absorbance of the trans isomer (320 nm for AB, 324 nm for AT, 308 nm for SB, and 340 nm for PS). These two curves provide the information required to determine the fraction of probe that is mobile as a function of polymerization time. When identical experiments are repeated, the fraction of mobile probes is generally reproduced within one percent at short times but may deviate by up to several percent as the isomerization time increases. The increased error results from the inherent difficulties in calculating the slope of the nonpolymerizing system as the slope approaches zero.

To determine the probe mobility as a function of conversion, the polymerization was monitored with a differential scanning calorimeter with a photocalorimetric accessory (Perkin-Elmer DSC-DPA 7; Norwalk, CT) under identical reaction conditions (temperature and light intensity). From the rate profiles, the double bond conversion as a function of polymerization time was determined. The theoretical heat evolved per methacrylate double bond was taken as 13.1 kcal/mol (*30-32*).

Analysis

Establishing a free volume distribution within a polymerizing sample is dependent on determining the state of a given photochromic probe molecule. The four possible states for any given photochromic probe are the trans and mobile (or free) state, the trans and immobile (or bound) state, the cis and mobile state, and the cis and immobile state. Here, mobile refers to a molecule surrounded by a local volume that is greater than the volume the molecule requires to isomerize. A molecule is considered immobile when the local volume is smaller than this critical volume to isomerize and the probe becomes locked in a single conformation.

The probe molecule may undergo any of the following reactions during the polymerization:

$$t_f \underset{k_2}{\overset{k_1}{\rightleftharpoons}} c_f \tag{1}$$

$$t_f \xrightarrow{k_3} t_b \tag{2}$$

$$c_f \xrightarrow{k_4} c_b \tag{3}$$

Here, t represents the trans state, c represents the cis state, f refers to the free or mobile state, and b is the bound or immobile state. If all the probes molecules are mobile in the liquid monomer, then reactions 2 and 3 would not exist in the absence of polymerization. During polymerization, all of the probe reactions 1-3 are occurring in parallel with the polymerization reaction.

For a monomer system (i.e., a system containing only monomer and probe but no initiator), the transformation of a probe molecule from the trans to the cis state may be monitored by UV-Vis spectroscopy by measuring the system absorbance as a function of time. The system is exposed to the curing light in the absence of initiator (no polymerization occurs), and the rate of trans to cis isomerization is monitored at a wavelength for which the probe has a high absorbance. The absorbance of the probe in the monomer state is then related to the trans and cis concentrations by

$$A_m = \varepsilon_t l t_f + \varepsilon_c l c_f \tag{4}$$

Here, ε is the molar absorption coefficient of the respective state, l is the sample thickness, m refers to the monomer system, and A is the absorbance.

In a polymerizing sample, a similar equation exists for the absorbance as a function of the state of the probe molecule. In a polymerizing system not all of the probe molecules are mobile, and the fraction of probe that is mobile changes with polymerization time (or conversion). Thus,

$$A_p = \varepsilon_t l(t_f + t_b) + \varepsilon_c l(c_f + c_b) + A_i \tag{5}$$

where p refers to the polymer system and A_i is the absorbance of the photoinitiator. The initiator absorbance as a function of the polymerization time was measured by polymerizing a sample without the photochromic probe. This absorbance was then subtracted from the absorbance of the polymerizing probe system to isolate the effects of the probe conformations on the polymer absorbance. Generally, A_i was not significant above 300 nm for the photoinitiator concentrations used.

Assuming first order reactions for the first three equations, the differential balances on each probe state may be written for a monomer and a polymerizing system. For the polymerizing system, the kinetic constants in equation 1 were assumed to be independent of double bond conversion in the sample. This assumption will be addressed in the following paragraph. From physical principles, the kinetic constants for equations 2 and 3 should be the same and represent the rate of conversion of any mobile state to a bound state. Now, defining a variable z which is the fraction of probe molecules that are mobile

$$z = \frac{[P]_f}{[P]} = \frac{c_f + t_f}{[P]} \tag{6}$$

where the total concentration of the probe in the sample is [P]. Then,

$$k_3 = k_4 = -\frac{\partial \ln z}{\partial t} \tag{7}$$

The above analysis reduces to a system of 8 equations and 8 unknowns: t_f, c_f, t_b, c_b, k_1, k_2, k_3, and z; however, a simple, analytical solution for z in terms of experimentally measured quantities provides that

$$z = \frac{\partial A_p / \partial t}{\partial A_m / \partial t}\bigg|_{t=\text{constant}} \tag{8}$$

To address the assumption that the free probes in a polymerizing sample will behave as those in a monomer sample (i.e. k_1 and k_2 are not functions of conversion in the sample), a series of experiments were conducted. Samples were prepared by dissolving 0.1 wt% photochromic probe and 1.0 wt% thermal initiator in a monomer. Samples were thermally cured for 0 minutes, 10 minutes, 30 minutes, and 8 hours. Thermally curing the sample insured that the probe remained in the trans state until UV-Vis analysis, and the complication of simultaneous polymerization and isomerization present in photopolymerizations was eliminated. The absorbance versus time was monitored for each sample (i.e., the probe isomerization rate was monitored at different polymerizations times or double bond conversions). Assuming first order kinetics of the isomerization, the decay time of the probe in each sample was independent of conversion (less than 1% variation in the decay time for samples polymerized different times).

This conclusion might appear contrary to the earlier works of Sung and coworkers (5-7,26,27), Horie et al. (28), and Mita (29) et al. , but not if the time scale of our experiments is considered. Essentially, we are monitoring and distinguishing between the probe molecules that are in a liquid-like environment from those that are in a region of restricted mobility. We do not distinguish between probe molecules that have completely restricted motion and those that isomerize dependent on the polymer mobility. This second type of molecule is considered immobile during the time scale of our experiments (typically less than 5 minutes).

Results and Discussion

Figure 1 shows typical experimental results for the absorbance versus time in systems with and without photoinitiator. The absorbance is measured at the wavelength corresponding to the peak absorbance of the trans probe in all of the following time dependent experiments. At low reaction times the conversion in the polymerizing sample is close to zero, so the probes are freely mobile in both the monomer and polymerizing sample. Therefore, early in the reaction the slopes for each curve are equal, and z is one (equation 8). With continued UV exposure, the polymerization begins to autoaccelerate and conversion increases. As conversion increases, some probe molecules become locked in the trans state which leads to a higher absorbance in the polymerizing system. Finally, when the polymerization is complete (after 300s in Figure 1), the slope of the polymer curve reduces to the same exponential decay as the slope of the monomer curve, but the polymerizing sample has a different preexponential factor which accounts for the trapped probe molecules. Thus, z, which is a ratio of the two slopes, reduces to a constant value that represents the fraction of probe molecules that remain mobile in the polymer system.

In Figure 2, the absorbance spectrum of the probe stilbene is plotted as a function of wavelength. Three curves are drawn which represent the absorbance of the probe in a monomer system before isomerization, the absorbance of the probe in an unpolymerized system isomerized for 10 min., and the absorbance of the probe in a system polymerized and isomerized for 10 minutes. Comparing equal isomerization times in the monomer and polymerized sample, it is clear that a significant amount of probe is locked in the trans state in the polymerized sample. The higher absorbance in the polymerized sample is indicative of the higher concentration of trans probe (which is immobile) in the polymer.

To determine the fraction of probe that is mobile in a system that remains the same throughout the isomerization (e.g. a preformed polymer), the extent of isomerization in the polymer system and a fully mobile system (e.g. a monomer or solvent) must be known. In both of these systems, the extent of isomerization is calculated by

$$Y = \frac{1 - \dfrac{A}{A_{trans}}}{1 - \dfrac{\varepsilon_{cis}}{\varepsilon_{trans}}} \tag{9}$$

where A is the absorbance at the photostationary state in the system (e.g., fully mobile or polymer), A_{trans} is the absorbance of the trans probe before isomerization, and ε_{cis} and ε_{trans} are the molar extinction coefficients at the wavelength that the absorbances were measured. The fraction of mobile probe is the extent of isomerization in the polymer system divided by the extent of isomerization in the fully mobile system.

This analysis can be applied to thermal polymerization (where isomerization is not coupled with polymerization), but is not applicable to photopolymerizations. The disadvantage of this method in studying the photochromic probe mobility during a polymerization is that several separate experiments are required. Several samples must be polymerized to different conversions, the polymerization quenched, and the probe subsequently isomerized. From each sample, z is determined for a single conversion. In contrast to this technique, we photopolymerize and isomerize the sample simultaneously. By monitoring the absorbance with time, we are then able to determine z as a continuous function of conversion with a single experiment.

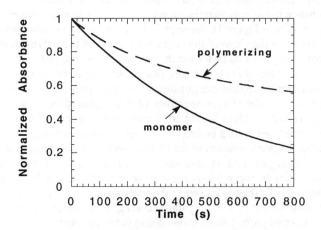

Figure 1. A typical absorbance versus time curves for a monomer system and a polymerizing system.

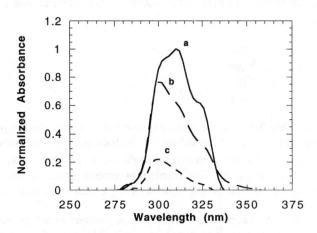

Figure 2. Absorbance of stilbene versus wavelength in (a) an unpolymerized system before isomerization, (b) a system polymerized and isomerized for 10 minutes, and (c) an unpolymerized system after 10 minutes of isomerization.

In Figure 3, z is plotted as a function of double bond conversion for AB, SB, AT, and DPS during the polymerization of DEGDMA at 25°C with 1 wt% I651 and 365 nm UV light at an intensity of 0.2 mW/cm^2. Even at low conversion, the fraction of probe molecules that are mobile is decreasing for all the probes. A plausible explanation for this decrease at such low conversions is that some probe molecules are trapped in the microgel regions. Recall that microgels are regions of a much higher crosslinking density and lower free volume as compared to the macroscopic system. As the polymerization continues and conversion increases, the average free volume of the entire system decreases, and an increasing number of probe molecules become trapped. The result is a more rapid decrease in z with increasing conversion. Finally, the free volume in the system becomes so low that the mobility of the reacting specie limits further polymerization, and maximum attainable double bond conversions are reached (*33-35*).

When the fraction of mobile probes in AB, AT, SB, and DPS is compared for a given double bond conversion, AB is the most mobile at all conversion followed by AT, SB, and DPS, respectively. The final fraction of mobile probe in the polymer structure is 0.75 for AB, 0.57 for AT, 0.46 for SB, and 0.10 for PS. This mobility is related to the critical volume (i.e., the minimum volume required) for each probe to isomerize which is 127 Å^3 for AB, 202 Å^3 for AT, 224 Å^3 for SB, and 575 Å^3 for DPS. By incorporating several probes that require different free volumes to isomerize, the volume distribution in the polymerizing sample as a function of conversion may be determined as discussed in detail later.

In addition to studying the effect of critical isomerization volume on the probe mobility in the polymer, the effect of polymerization temperature on this mobility was studied. Figure 4 provides z as a function of conversion for AB and SB during the photopolymerization of DEGDMA at 25°C and at 40°C while the light intensity and initiator concentration remained constant at 0.20 mW/cm^2 and 1.0 wt%, respectively. As the temperature of the system was increased, the average free volume and mobility in the system increased. These effects were seen as the mobile fraction of both AB and SB increased with higher temperature at all conversions.

In this temperature range, the increase in mobile probe fraction for SB was much greater than the increase for AB. The small critical free volume required for AB to isomerize might suggest that these molecules become immobile primarily in the densely crosslinked microgel regions. Thus, increasing the temperature from 25 to 40°C has little effect on the free volume and mobility in these regions. At much higher double bond conversions when the average free volume of the system is greatly reduced, a slight increase in the mobility of the AB probe is seen where some probe might be locked in the macroscopic network (e.g. in the interstitial regions between the microgels). In contrast, the increase in z for SB is observably greater as early as 10% conversion. One possible explanation for this observation is the larger critical free volume required for SB to isomerize. At a much lower double bond conversion, SB may become immobile in the network outside the microgel regions. The system has reached macroscopic gelation, but a significant amount of unreacted double bonds and monomer remain. As the temperature is increased, the mobility of the loosely crosslinked polymer (swollen in unreacted monomer) will be enhanced much more than that in the highly crosslinked microgel regions.

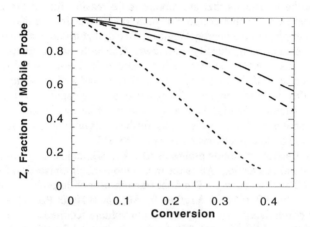

Figure 3. Fraction of probes that are mobile as a function of double bond conversion of DEGDMA polymerized with 0.2 mW/cm^2 of UV light and 1 wt% DMPA. (——, AB), (— — —, AT), (- - - -, SB), (- - - - -, PS)

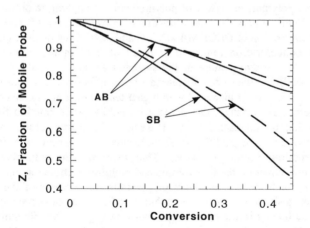

Figure 4. Fraction of probes that are mobile as a function of double bond conversion of DEGDMA polymerized with 0.2 mW/cm^2 of UV light and 1 wt% DMPA at two different temperatures. (——, 25°C), (— — —, 40°C)

The effects of monomer size and crosslinking density on the mobility of the probe were also studied. These results are shown in Figure 5 where z is plotted versus conversion during the polymerization of DEGDMA, PEG(200)DMA, and PEG(600)DMA. The polymerizations were conducted at 25°C and initiated with 1.0 wt% DMPA and 0.2 mW/cm^2 of UV light. The results are shown for the SB probe. As the number of ethylene glycol units between the dimethacrylate double bonds is increased from 2 (DEGDMA) to ~14 (PEG(600)DMA), the average crosslinking density is decreased and mobility in the system is increased. This effect is observed macroscopically as PEG(600)DMA is a rubbery network after curing, while DEGDMA is glassy.

Comparing the mobility of SB in each of these systems, z is the highest in PEG(600)DMA, followed by PEG(200)DMA, and the lowest in DEGDMA for a given conversion. This difference in mobility was primarily attributed to the crosslinking density of the network. The general shape of z indicates a slower decrease in z early in the reaction followed by a steeper decrease as higher conversions are reached. Again, the initial decrease is attributed to trapping of probe molecules in the microgel regions followed by further trapping in the macroscopic network as the free volume of the system dramatically decreases with conversion.

Comparing the initial decrease from 0 to 10% conversion in these systems, the fraction of mobile SB in DEGDMA has decreased much more than that of PEG(200)DMA and PEG(600)DMA. If this decrease is related to the formation of microgels in the polymer, then DEGDMA might appear to be more heterogeneous with a greater tendency to form more highly crosslinked microgel regions than PEG(200)DMA or PEG(600)DMA. Recall that the formation of microgels is related to the enhanced reactivity of pendant double bonds near the vicinity of the active radical. Therefore, if the pendant functional group reactivity of DEGDMA is considerably greater than that of PEG(600)DMA , the tendency to form microgels in DEGDMA will be correspondingly greater, and the system will have a higher degree of heterogeneity.

The preceding discussions have considered the effects of double bond conversion, polymerization conditions, and crosslinking density of the polymer network on the mobility of the probe molecules. Once the mobility of the probe molecules is characterized as a function of conversion in the system, then the distribution of volume in the polymer network can be elucidated. These volume distributions are closely related to the distribution of free volume in the polymer (the probe volume distribution should be symmetric with the free volume distribution). Thus, if the mean free volume of the system is known, then the free volume distribution can be easily calculated.

In Figure 6, the fraction of probe molecules that are immobile in the system (i.e., 1-z) is plotted as a function of the critical free volume of the probe at various conversions. This data is taken from Figure 3 for the polymerization of DEGDMA. Each set of points represents the cumulative distribution of free volume in the system at that conversion. The distribution is cumulative since the critical free volume of the probe represents a minimum volume required for the probe to isomerize. As conversion is increased from 0.10 to 0.40, the fraction of trapped probe molecules increases at all free volume sizes. If several probes were studied, a generalized

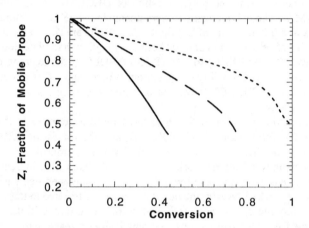

Figure 5. Fraction of probes that are mobile as a function of double bond conversion for DEGDMA (———), PEG(200)DMA (— — —), and PEG(600)DMA (– – – –) polymerized with 0.2 mW/cm^2 of UV light and 1 wt% DMPA.

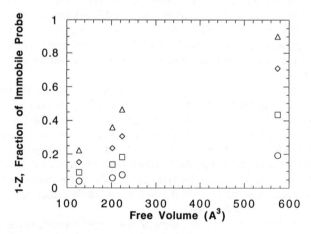

Figure 6. Fraction of probes that are immobile as a function of free volume for DEGDMA at several different double bond conversions: o (10%), ◻ (20%), ◊ (30%), and Δ (40%).

curve could be fit to the points at a given conversion, and the derivative of this curve would represent the free volume distribution. Since these results are for only 4 probes, a Gaussian distribution of the free volume was assumed and the average free volume and standard deviation were calculated from these points at a given conversion. The results are shown in Figures 7 and 8 for DEGDMA and PEG(600)DMA polymerized under identical conditions.

Figure 7 shows the changes in the average free volume as a function of double bond conversion for DEGDMA and PEG(600)DMA. DEGDMA is a considerably smaller molecule as compared to PEG(600)DMA, and the viscosity of DEGDMA is correspondingly much lower. Thus, at the beginning of the reaction, DEGDMA has a much higher free volume compared to PEG(600)DMA. As polymerization continues, the network formed by DEGDMA is much more highly crosslinked than PEG(600)DMA. Additionally, the concentration of double bonds in DEGDMA is nearly 3 times as great as that in PEG(600)DMA. Thus, in a given volume of monomer at 10% conversion of double bonds, nearly 3 times as many double bonds have reacted in DEGDMA as in PEG(600)DMA at the same conversion. This concentration difference leads to the sharp decrease in the average free volume of DEGDMA with conversion.

Around 20% conversion, a crossover is seen in the two averages. The amount of free volume that has been consumed up to this point in the polymerizing DEGDMA leads to a reduction in the average free volume below that of PEG(600)DMA. As polymerization continues and higher conversions are reached, the average free volume continually decreases in both systems, but the average free volume in DEGDMA drops further below that of PEG(600)DMA. Eventually, the free volume in DEGDMA drops so low that mobility of the reacting species is severely limited and polymerization stops. In contrast, the free volume of PEG(600)DMA remains higher, and these systems reach nearly 100% conversion of double bonds.

In Figure 8, the complementary results related to the standard deviation calculated from the Gaussian distribution are shown. The standard deviation divided by the average free volume is plotted as a function of conversion for DEGDMA and PEG(600)DMA. In comparing these two systems, DEGDMA has a broader distribution of free volume at all conversion relative to PEG(600)DMA. The broader distribution is indicative of the higher degree of heterogeneity in DEGDMA. Additionally, as the conversion in both of the systems is increased, the breath of the distribution increases, and the difference between the two polymerizing samples also increases.

Conclusions

A technique has been developed to monitor changes in free volume distributions and microstructural heterogeneity during polymerizations. The technique uses UV-Vis spectroscopy to monitor the trans to cis isomerization rates of a variety of photochromic probes. These probes each require that a certain minimum free volume be available to allow isomerization. By comparing rates of isomerization in samples with and without photoinitiator, it is possible to determine the fraction of

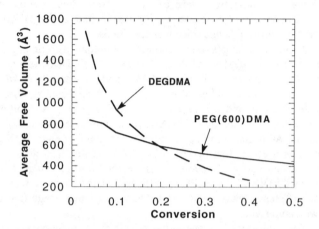

Figure 7. The average free volume (for a Gaussian distribution) as a function of double bond conversion in DEGDMA (— — —) and PEG(600)DMA (———).

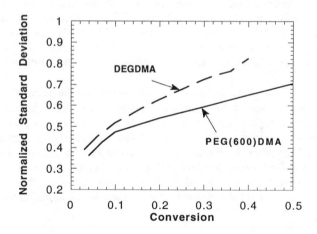

Figure 8. The normalized standard deviation (for a Gaussian distribution) as a function of double bond conversion in DEGDMA (— — —) and PEG(600)DMA (———). The normalized standard deviation is defined as the standard deviation of the distribution divided by the average free volume.

probes which are able to isomerize as a continuous function of conversion. These experiments provide an *in situ* method for quantifying free volume distributions during photopolymerizations, and in this work, were applied in a preliminary study of photopolymerizations of multiethylene glycol dimethacrylates.

Acknowledgments

The authors thank the National Science Foundation for their support of this research in the form of a grant (CTS-9209899) and a fellowship to KSA. We also thank the University of Colorado's Office of the Dean of the Graduate School for a Dean's Small Grant Award.

Literature Cited

1.   Victor, J.G.; Torkelson, J.M. *Macromolecules* **1987**, *20*, 2241.
2.   Victor,J.G.; Torkelson, J.M. *Macromolecules* **1988**, *21*, 3490.
3.   Royal, J.S.; Victor, J.G.; Torkelson, J.M. *Macromolecules* **1992**, *25*, 729.
4.   Royal, J.S.; Torkelson, J.M. *Macromolecules* **1992**, *25*, 4792.
5.   Yu, W.-C.; Sung, C.S.P. *Macromolecules* **1988**,*21*, 365.
6.   Yu, W.-C.; Sung, C.S.P.; Robertson, R.E. *Macromolecules* **1988**, *21*, 355.
7.   Lamarre, L.; Sung, C.S.P. *Macromolecules* **1983**, *16*, 1729.
8.   Anseth, K.S.; Rothenberg, M.D.; Bowman, C.N. *Macromolecules* **1994**, *27*, 2890.
9.   Kloosterboer, J.G. *Adv. Poly. Sci.* **1988**, *84*,1.
10. Zwiers, R.J.M.; Dortant, G.C.M. *Appl Opt.* **1985**, *24*, 4483.
11. Watts, D.C.   In *Materials Science and Technology: A Comprehensive Treatment*; Williams, D.F., Ed.; VCH Publishers: New York, 1992; p 209.
12. Bastide, J.; Leibler, L. *Macromolecules* **1988**, *21*, 2647.
13. Funke, W. *Br. Polym. J.* **1989**, *21*, 107.
14. Matsumoto, A.; Matsuo, H.; Ando, H.; Oiwa, M. *Eur. Polym. J.* **1989**, *25*, 237.
15. Bowman, C.N.; Peppas, N.A. *Chem. Engng. Sci.* **1992**, *47*, 1411.
16. Anseth, K.S.; Bowman, C.N. *Chem. Engng. Sci.* **1994**, *49*, 2207.
17. Zhu, S.; Tian, Y.; Hamielec, A.E.; Eaton, D.R. *Macromolecules* **1990**, *23*, 1144.
18. Tian, Y.; Zhu, S.; Hamielec, A.E.; Fulton, D.B.; Eaton, D.R. *Polymer* **1992**, *33*, 384.
19. Liu, J.; Deng, Q.; Jean, Y.C. *Macromolecules* **1993**, *26*, 7149.
20. Jean, Y.C.; Deng, Q. *J. Polym Sci.: Polym. Phys.* **1992**, *30*, 1359.
21. Jeffrey, K.; Pethrick, R.A. *Eur. Polym. J.* **1994**, *30*, 153.
22. Roe, R.-J.; Curro, J.J. *Macromolecules* **1983**, *16*, 428.
23. Song, H.H.; Roe, R.-J. *Macromolecules* **1987**, *20*, 2723.
24. Gardlund, Z.G. *J. Polym. Sci., Polym. Lett. Ed.* **1968**, *B6*, 57.
25. Gegiou, D.; Muszkat, K.A.; Fischer, E. *J. Am. Chem. Soc.* **1968**, *90*, 12.
26. Sung, C.S.P.; Gould, I.R.; Turro, N.J. *Macromolecules* **1984**, *17*, 1447.
27. Sung, C.S.P.; Lamarre, L.; Chung, K.H. *Macromolecules* **1981**, *14*, 1839.
28. Horie, K.; Hirao, K.; Kenmochi, N.; Mita, I. *Makromol. Chem.* **1988**, *9*, 267.
29. Mita, I.; Horie, K.; Hirao, K. *Macromolecules* **1989**, *22*, 558.

30. Miyazaki, K.; Horibe, T.J. *J. Biomed. Mater. Res.* **1988**, *22*, 1011.
31. Horie, K.; Otagawa, A.; Muraoka, M.; Mita, I. *J. Polm. Sci.* **1975**, *13*, 445.
32. Moore, J.E. In *Chemistry and Properties of Crosslinked Polymers*; Labana, S.S., Ed.; Academic: New York, 1977; p 535.
33. Simon, G.; Allen, P.; Bennett, D.; Williams, D.; Williams, E. *Macromolecules* **1989**, *22*, 3555.
34. Cook, W.D. *J. Polym Sci., Part A: Polym. Chem.* **1993**, *31*, 1053.
35. Kloosterboer, J.; Lijten, G.; Boots, H. *Makromol. Chem., Macromol. Symp.* **1989**, *24*, 223.

RECEIVED February 2, 1995

NMR SPECTROSCOPY

Chapter 11

Multidimensional NMR Spectroscopy of Polymers

Section Overview

Klaus Schmidt-Rohr

Department of Polymer Science and Engineering, University of Massachusetts, Amherst, MA 01003

Nuclear Magnetic Resonance (NMR) provides powerful techniques for characterizing molecular and supermolecular properties of polymers. It uses the magnetic and electric interactions of nuclear spins to analyze structure, orientation, and mobility of well-defined chemical moieties. The nuclear magnetic moments are excellent probes that do not affect the state of the system, as NMR energies are only 0.1 kJ/mol or less. Since many polymers are insoluble, or exhibit their most interesting properties in the solid state, NMR research of polymers is dominated by solid-state NMR studies. Nevertheless, solution NMR methods play an important complementary role by determining composition and configuration (1). The following provides a brief introductory overview of the large field of NMR of polymers, including multidimensional techniques. More complete treatments can be found in recent monographs (2-6).

Basics of Pulsed NMR

The nuclear magnetization (which is proportional the expectation value of the nuclear spin) is aligned with the external B_0 field in thermal equilibrium. It can be manipulated, e.g. rotated to be transverse to the B_0 field, by radio frequency (rf) pulses. These are applied to the sample by a coil 4-20 mm in diameter which is part of a resonant circuit. The coil generates B_1 fields, perpendicular to the B_0 field, which oscillate with a frequency that is resonant with the nuclear Larmor frequency

$$\omega_L = -\gamma B_0 \tag{1}$$

Here, γ is the magnetogyric ratio of the specific nucleus. With typical B_0 field strengths of 5-18 Tesla, proton resonance frequencies are 200-750 MHz. Those of other nuclei of interest in polymers are 2.5 to 10 times smaller. The transverse nuclear magnetization precesses around the B_0 field with a frequency ω_L, inducing a voltage in the NMR coil. The resulting free induction decay (FID) is detected in a phase-sensitive detector and the radio frequency is subtracted from the signal. This is equivalent to the transition into

0097–6156/95/0598–0184$12.00/0

the rotating frame, which is a general concept in magnetic resonance that greatly simplifies the analysis of the action of rf pulses and the behavior of the magnetization. Fourier transformation of the digitized FID produces the 1D NMR spectrum. Usually, between tens and thousands of FIDs (scans) are added in the computer memory to improve the signal-to-noise ratio, since the noise in different scans partially cancels. In spite of the significant improvements by signal averaging and the Fourier technique, the sensitivity of NMR is still a limiting factor and makes it difficult to obtain spectra from isolated thin films or from surfaces in low-surface-area materials.

Dipolar Decoupling. In solids, due to dipolar interactions between nuclear spins, dipolar decoupling is a prerequisite for obtaining highly resolved NMR spectra. High-power irradiation of the proton resonance achieves decoupling of heteronuclear dipolar interactions of protons with the nuclear spins of interest. This heteronuclear, as well as the more complex multiple-pulse homonuclear decoupling, can be explained stringently by the average-Hamiltonian theory, which considers the average Hamiltonian in an interaction frame "toggling" under the action of the radio-frequency pulses (*7,2*). The homonuclear dipolar couplings of protons, generating line broadenings of 30-75 kHz, are so strong that even with advanced multiple-pulse decoupling sequences (*7*), ^1H spectra of solid polymers attain a resolution of only 0.3 - 2 ppm, much less than the chemical-shift resolution of solution NMR. Therefore, ^{13}C rather than ^1H spectroscopy is used most often in solid-state NMR.

NMR Interactions, Magic-Angle Spinning, Spin Diffusion and Cross Polarization. The versatility, but also the complexity, of solid-state NMR arises in part from the variety of interactions, or local fields, to which the nuclear spins are subject. The *chemical shift*, resulting from the shielding of the B_0 field by the electrons in the molecule, is characteristic of the electronic, i.e. chemical, environment of the nucleus. Its orientation-independent part, the isotropic chemical shift, is used extensively in both solution and solid-state NMR spectroscopy for chemical identification. To obtain highly-resolved chemical-shift spectra in solids, dipolar decoupling must be applied and the broadening by the *chemical-shift anisotropy* has to be removed by fast magic angle sample spinning (*MAS*), around an axis making an angle of 54.74°, the root of $3\cos^2\theta_r - 1$, with the B_0 field (*8*).

In other cases, the anisotropy of NMR interactions, i.e. the dependence of the NMR frequency on the orientation of the molecular segment relative to the B_0 field, can be of great value, since it allows to measure segmental orientations and reorientations. The *quadrupole coupling* of deuterons has proven a particularly useful 'molecular protractor', which is also sensitive to segmental reorientations (*9*). The quadrupolar coupling probes the angle θ between the C-^2H bond direction and the B_0 field, according to

$$\omega(\theta) = \omega(0°) \ (3\cos^2\theta - 1)/2. \tag{2}$$

Dipolar spin-pair couplings, which probe the direction of the internuclear vector and the B_0 field, can provide similar orientational (*10*) and dynamical (*11*) information. In addition, they can establish couplings between spins, which is useful for probing proximities of segments (*12,13*), as well as for characterizing domain sizes by a multi-step magnetization-transfer process known as *spin diffusion* (*14-16*).

    Cross polarization from protons to rare nuclei, a technique widely used for signal enhancement in solids, also relies on dipolar couplings, which bring about heteronuclear magnetization transfer in the presence of radio-frequency fields of matched strength (*17*). Dipolar decoupling and cross polarization require double-resonance equipment for simultaneous rf irradiation on protons and the rare nuclei of interest. In recent years, an increasing number of triple-resonance experiments has also been developed, which are applied to systems containing fluoropolymers (^{1}H-^{19}F-^{13}C), partially deuterated polymers (^{1}H-^{13}C-^{2}H) (*12*), or biopolymers (^{1}H-^{13}C-^{15}N, ^{1}H-^{13}C-^{31}P) (*13*). Several of the following papers give a flavor of these developments.

Two- and Multi-Dimensional NMR

The introduction of two-dimensional (2D) NMR (*18,19*) has greatly enhanced the potential of NMR and helped to make it one of the most versatile and powerful spectroscopic techniques available today. In such 2D NMR spectra, a *correlation* between peaks can be established that provide detailed structural or dynamical information; spectral overlap can be eliminated by *separation* of spectral patterns in the second dimension; and two-dimensional *exchange* intensity patterns can be obtained that characterize segmental dynamics in unprecented detail. Figure 1 shows the basic building blocks of a 2D experiment. By systematic incrementation of the *evolution* time t_1 and acquistion of the NMR signal during the *detection* period t_2, a two-dimensional time signal $s(t_1,t_2)$ is obtained. The evolution of the magnetization or coherence of the spin system with a frequency ω_1 is thus recorded indirectly through the modulation of the amplitude or phase of the actually detected NMR signal oscillating with the frequency ω_2. The 2D spectrum, which is obtained by Fourier transformation of the 2D time signal $s(t_1,t_2)$, will then exhibit a signal at (ω_1,ω_2). In a 2D *exchange* spectrum, off-diagonal intensity, at $\omega_2 \neq \omega_1$, provides detailed information about the *mixing* process, which may be chemical exchange, molecular reorientation, or dipolar exchange. In 2D *separation* spectra, one dimension separates the signals by their well-resolved isotropic chemical shifts, while a broadline spectrum of the same site, reflecting dipolar couplings, chemical-shift anisotropy, quadrupolar couplings, or the like, is displayed along the other dimension.

    Three-dimensional (3D) NMR is a logical extension of the concept of 2D NMR. The third dimension can be used to separate overlapping 2D spectra (*20,21*), to correlate a further anisotropic interaction (*22*), or to study details of a complex exchange process (*23,2*). Due to the large size of the 3D data fields, the development of 3D NMR has been coupled with the increased availability of sufficient computer storage capacity.

NMR in Structural Studies of Polymers. In the last two decades, NMR has seen the steady development of new or improved methods, many of them two-dimensional, for the investigation of polymer structure on many levels, from the Ångström to the centimeter length scale:

- Solution NMR and solid-state magic-angle spinning experiments characterize polymers in terms of chemical groups   (0.1-1 nm). *(1,3)*

- The nuclear Overhauser effect in solution and dipolar splittings in solids yield inter- and intramolecular distances   (0.1-1 nm). *(1,12,13,24)*

- Anisotropic interactions measure orientation distributions of various segments in both crystalline and amorphous regions of polymer fibers and films   (0.2-2 nm). *(9,22)*

- Conformation dependent chemical shifts, in solids for instance based on the γ-gauche effect, provide information on torsion angles along a chain   (0.3-0.9 nm). *(1,5)*

- Packing-induced chemical-shift differences confirm occurrence of various sites in the crystal structure   (0.3-0.7 nm). *(5)*

- Solution NMR yields information about polymer configurational statistics   (0.3-1.5 nm). *(1,5)*

- Relaxation times and lineshapes characterize molecular mobility in various phases   (0.5-500 nm). *(9, 25)*

- Spin and noble-gas diffusion allow for estimating the sizes of supramolecular domains in semicrystalline polymers, polymer blends, and block copolymers   (0.5-5000 nm). *(14,15,26,27,2)*

- Magnetic-resonance imaging techniques can provide spatially resolved spectroscopic information   (20 μm-1 cm). *(28)*

Polymer Dynamics and Multidimensional NMR. Next to structural studies, the second large area of interest for NMR in polymer science is the investigation of molecular dynamics. By relaxation measurements, line-shape studies, and two-dimensional exchange experiments, correlation times between 10^{-10}-10^{-2}, 10^{-5}-10^{-1}, and 10^{-3}-10^{2} seconds can be determined, respectively. While structural NMR studies often have to compete with similarly powerful scattering techniques, multidimensional exchange NMR in solids is without rival in providing details about polymer dynamics on a molecular level.

Two-dimensional exchange spectra without MAS exhibit ridge patterns which are characteristic of the geometry of the reorientation occurring during the mixing time. In ^{2}H 2D NMR, a reorientation by a specific jump angle β gives rise to a pair of identical elliptical patterns *(29,2)*. The dependence of the exchange patterns on β is shown in Figure 2. The ratio of the lengths a and b of the semi-axes of each ellipse is related to the jump angle β according to

$$|\tan\beta| = b/a. \qquad (3)$$

Diffusive or ill-defined motions, with a wide reorientation-angle distribution, exhibit featureless 2D exchange patterns *(29,30)*. 3D exchange spectroscopy probes the orientation of a molecular segment three times, making it possible to investigate orientational memory and to determine the number of energy minima that are available to a reorienting molecular segment *(2,23.)*. A reduced 4D experiment has proved useful for distinguishing heterogeneous and homogeneous distributions of correlation times *(30)*.

The contributions to the following NMR section include a variety of dynamical studies.   They concern the nature of the α relaxations in semicrystalline and amorphous polymers, the β relaxation in poly(methacrylates), sidegroup dynamics, phenyl ring libration, motion in

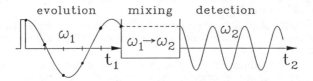

Figure 1. Schematic representation of the generic two-dimensional NMR experiment. During the evolution time t_1, the spin system evolves with frequency ω_1. After the mixing period, in detection time t_2, the signal with frequency ω_2 is detected. The two-dimensional time signal obtained by systematically incrementing t_1 is Fourier transformed to yield the two-dimensional NMR spectrum.

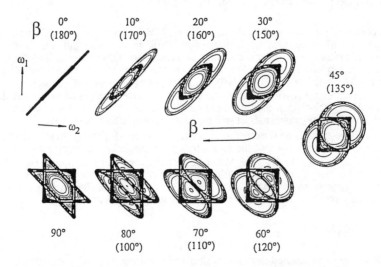

Figure 2. Contour plots of exchange patterns in ^2H 2D NMR spectra. The figure shows the dependence on the reorientation angle β, which is the angle between the $C-^2H$ bond orientations before and after the molecular reorientation. The spectra exhibit elliptical ridges which are characteristic of β.

liquid-crystalline polymers, and the diffusion of small molecules in a polymer solution.

Segment Identification by NMR

One of the great strengths of NMR is the assignment of structural or dynamical features to specific moieties or phases in the material. As mentioned above, for complex polymers this is achieved in 2D or 3D experiments that separate the signals of different groups according to the corresponding isotropic chemical shifts.

An alternative, chemical, route to obtaining high structural resolution is provided by selective isotopic labelling. While most routine applications of NMR observe the signals of ^{13}C, ^{1}H, ^{31}P, ^{29}Si, etc. in natural abundance, many specific NMR studies rely on synthetic chemistry for isotopic labelling. In the following articles, systematic deuteration as well as ^{13}C labelling are encountered. For NMR studies of proteins, ^{15}N labelling by biosynthetic methods is also employed.

In addition to achieving selectivity, isotopic enrichment also serves to overcome the sensitivity problem of NMR, and of solid-state NMR in particular. While rare nuclei or spin pairs can often be observed without enrichment in solution NMR, they are not detectable in the solid state because the peak heights in solids are smaller, as the linewidths are larger by orders of magnitude.

As NMR techniques and technologies are developed further, problems currently solved by isotopic enrichment will become accessible in unlabelled samples, with the necessary segment identification achieved by chemical-shift separation in a multi-dimensional experiment. Much of the technical progress will rely on enhancements of the signal-to-noise ratio, for instance by further increases in magnetic field strengths and by line-narrowing techniques. Even then, many interesting, ever more complex, scientific questions will be accessible only by isotopic enrichment.

Outlook

In the future, multidimensional NMR will certainly continue to provide unprecedented insights into polymer dynamics. We may also expect more 2D NMR investigations of the chain structure in amorphous and crystalline phases of synthetic as well as biological polymers. Spin-diffusion NMR studies will be applied to all kinds of heterogeneous polymer systems. These developments will make NMR comparable and complementary to the scattering methods that have been applied to these problems so far. In summary, it is to be expected that NMR will establish itself firmly as a set of techniques for studying polymers not only on a segmental level, but actually on a wide range of length scales.

Literature Cited

(*1*) Bovey, F. A. *Chain Structure and Conformation of Macromolecules*; Academic Press: New York, 1982.

(*2*) Schmidt-Rohr, K; Spiess, H. W. *Multidimensional Solid-State NMR and Polymers*; Academic Press: San Diego, 1994.

(3) McBrierty, V. J.; Packer, K. J. *Nuclear Magnetic Resonance in Solid Polymers*; Cambridge University Press: Cambridge, 1993.

(4) Koenig, J. L. *Spectroscopy of Polymers*; ACS Reference Book Series: Washington, D.C., 1992.

(5) Tonelli, A. E. *NMR Spectroscopy and Polymer Microstructure: The Conformational Connection*; VCH Publishers: New York, 1989.

(6) Komoroski, R. A., Ed. *High Resolution NMR Spectroscopy of Synthetic Polymers in Bulk*; VCH Publishers: Deerfield Beach, Fla., 1986.

(7) Haeberlen, U. *Advances in Magnetic Resonance. Supplement 1*; Academic Press: New York, 1976.

(8) Schaefer, J.; Stejskal, E. O. *J. Am. Chem. Soc.* **1976**, *98*, 1031.

(9) Spiess, H. W. *Adv. Polym. Sci.* **1985**, 66, 23.

(10) Opella, S. J. *Ann. Rev. Phys. Chem.* **1994**, *45*, 659.

(11) Schaefer, J.; Stejskal, E. O.; McKay, R. A.; Dixon, W. T. *Macromolecules* **1984**, *17*, 1749.

(12) Pan, Y.; Gullion, T.; Schaefer, J. *J. Magn. Reson.* **1990**, *90*, 330.

(13) Schmidt, A.; Kowalewski, T.; Schaefer, J. *Macromolecules* **1993**, 26, 1729.

(14) Caravatti, P.; Neuenschwander, P.; Ernst, R. R. *Macromolecules* **1985**, *18*, 119.

(15) VanderHart, D. L. *J. Magn. Reson.* **1987**, 72, 13.

(16) Clauss, J.; Schmidt-Rohr, K.; Spiess, H. W. *Acta Polym.* **1993**, *44*, 1.

(17) Pines, A.; Gibby, M. G.; Waugh, J. S. *J. Chem. Phys.* **1973**, *59*, 569.

(18) Jeener, J.; Meier, B. H.; Bachmann, P.; Ernst, R. R. *J. Chem. Phys.* **1979**, 71, 4546.

(19) Ernst, R. R.; Bodenhausen, G.; Wokaun, A. *Principles of Nuclear Magnetic Resonance in One and Two Dimensions*; Clarendon: Oxford, 1987.

(20) Nakai, T.; Ashida, J.; Terao, T. *J. Chem. Phys.* **1988**, *88*, 6049.

(21) Lee, Y. K.; Emsley, L.; Larsen, R. G.; Schmidt-Rohr, K.; Hong, M.; Frydman, L.; Chingas, G. C.; Pines, A. *J. Chem. Phys.* **1994**, *101*, 1852.

(22) Chmelka, B. F.; Schmidt-Rohr, K.; Spiess, H. W. *Macromolecules* **1993**, 26, 2282.

(23) Schmidt-Rohr, K.; Kulik, A. S.; Beckham, H. W.; Ohlemacher, A.; Pawelzik, U.; Boeffel, C.; Spiess, H. W. *Macromolecules* **1994**, 27, 4733.

(24) Raleigh, D. P.; Creuzet, F.; Das Gupta, S. K.; Levitt, M. H.; Griffin, R. G. *J. Am.. Chem.. Soc.* **1989**, *111*, 4502.

(25) Schmidt-Rohr, K. Clauss, J.; Spiess, H. W. *Macromolecules* **1992**, *25*, 3273.

(26) Colombo, M. G.; Meier, B. H.; Ernst, R. R. *Chem. Phys. Lett.* **1988**, *146*, 189.

(27) Tomaselli, M.; Meier, B. H.; Robyr, P. Suter, U. W., Ernst, R. R. *Chem. Phys. Lett.* **1993**, *205*, 145.

(28) Blümich, B.; Kuhn, W., Eds. *Magnetic Resonance Microscopy*, VCH Publishers: Weinheim, 1992.

(29) Schmidt, C.; Blümich, B.; Spiess, H. W. *J. Magn. Reson.* **1988**, *79*, 269.

(30) Schmidt-Rohr, K.; Spiess, H. W. *Phys. Rev. Lett.* **1991**, *66*, 3020.

RECEIVED March 15, 1995

Chapter 12

Dynamics and Structure of Amorphous Polymers Studied by Multidimensional Solid-State NMR Spectroscopy

Klaus Schmidt-Rohr[1]

Chemistry Department, University of California, Berkeley, CA 94720
and
Max-Planck-Institut für Polymerforschung, Postfach 3148, D–55021
Mainz, Germany

Multidimensional exchange NMR techniques have been used to study dynamics, and conformational structure of amorphous polymers. The molecular motions underlying the β relaxations in poly(methyl methacrylate) (PMMA) and poly(ethyl methacrylate) (PEMA) were elucidated by two-dimensional (2D) and three-dimensional (3D) ^{13}C exchange NMR of the carboxyl moiety. 3D NMR proves a relatively well-defined motion between two potential-energy minima for the ca. 50% of the sidegroups that are mobile. 2D spectra show that the sidegroup OCO planes undergo 180°(±20°) flips which are accompanied by a main-chain rotation about the local chain axis, with a 20° rms amplitude. In PEMA, the merging of the β and the α process has been investigated. At 20 K above the glass transition temperature T_g, most sidegroups undergo 180° flips. While the flip-angle imprecision remains similar to that of the pure β process, the amplitude of the rotations around the main-chain axis increases strongly with temperature according to the α-process WLF curve. Nevertheless, the reorientations of the chain axes themselves remain restricted up to more than 60 K above T_g. This anisotropic glass-transition dynamics is in contrast to the predominantly isotropic behavior of most other amorphous polymers investigated so far. Conformational structure and dynamics have been investigated in amorphous aliphatic polymers by magic-angle spinning NMR, exploiting the γ-gauche effect on the isotropic chemical shifts of CH_2 units. In amorphous poly(propylene), trans and gauche are the preferred but not the only accessible conformations. Major conformational dynamics are found to set in only above T_g, with correla[1] tion times comparable to those of segmental reorientations. Nevertheless, the absence of orientational memory in 3D NMR shows that dynamics on a slightly distorted diamond lattice is not a suitable model for polymer dynamics above T_g.

[1]Current address: Department of Polymer Science and Engineering, University of Massachusetts, Amherst, MA 01003

0097–6156/95/0598–0191$13.00/0

Solid-state NMR has proven to provide powerful techniques for investigating the dynamics, structure and order of polymers. Multidimensional NMR spectroscopy in particular (1) has yielded ample molecular-scale information on reorientational and translational dynamics in semicrystalline and amorphous polymers (2-7), on their chemical and phase structure (8-10), and on orientational order (11-14). In this paper, results obtained recently on amorphous polymers will be in focus: the elucidation of the molecular nature of the β relaxation in methyl methacrylates (15) , and its merging with the α relaxation above the glass-transition temperature T_g (16), isotropic and anisotropic chain dynamics above T_g (16), as well as the investigation of conformational structure and dynamics in amorphous aliphatic polymers (17). Major parts of this article have been adapted from references (15-17) .

Experimental

The ^{13}C spectra of PMMA and PEMA were acquired with cross polarization (contact time of 2 ms) and proton decoupling (18) in a variable-temperature ^1H-^{13}C probehead on a Bruker MSL-300 spectrometer operating at a ^{13}C resonance frequency of 75.47 MHz, with 90° pulse lengths of 5 μs and recycle delays of 2 sec. 1D and 2D MAS spectra were acquired in a Bruker variable-temperature MAS probehead, with 90° pulse lengths of 3.5 μs. All 2D spectra were measured off-resonance, so that only a single-phase (cosine) dataset had to be acquired (1,2). Measuring times for a 2D spectrum ranged between 8 and 24 hours.

NMR Background

This section gives a short introduction to the aspects of solid-state ^{13}C NMR spectra that are relevant in this study. More detailed descriptions can be found in references (1) and (2).

Angle-Dependent NMR Frequencies. Solid-state NMR methods can yield information on molecular reorientations due to the angle dependence of anisotropic NMR interactions. The NMR frequency reflects the orientation of a given molecular unit relative to the externally applied magnetic field B_0 according to (1,19)

$$\omega(\theta,\phi) = \omega_{11}(\cos\phi \, \sin\theta)^2 + \omega_{22}(\sin\phi \, \sin\theta)^2 + \omega_{33}(\cos\theta)^2. \qquad (1)$$

Here, ω_{11}, ω_{22}, and ω_{33} are the principal values of the chemical-shift interaction tensor and (θ,ϕ) are the polar angles of B_0 in the principal-axes system of the ^{13}C chemical-shift, whose orientation with respect to the molecular unit is an inherent property of every type of functional group and reflects its local symmetry [cf. Fig. 1(a)]. In carboxyl groups, for symmetry reasons one principal axis of the ^{13}C shift tensor must be perpendicular to the OCO plane; it corresponds to ω_{33}, the most upfield tensor principal value. (19) The ω_{22} axis in esters is usually close to the C=O bond. (19) The simulated 2D exchange patterns shown below were obtained with an angle of 3° between the C=O bond and the ω_{22} axis.

In an isotropic sample (a "powder"), segments with all possible values of θ and φ are present, giving rise to an inhomogeneously broadened powder spectrum with a characteristic lineshape *(20,1)*. It extends from ω_{11} to ω_{33}, with steep edges at ω_{11} and ω_{33} from which the intensity increases monotonically towards the maximum at ω_{22}. If reorientational motions set in with rates on the scale of the width ω_{11}-ω_{33} of the powder spectrum, the frequencies change during the acquisition of the NMR signal. This results in lineshape changes which contain some information on the rate and geometry of the motion.

Conformation-Dependent NMR Frequencies. The angle dependence of the NMR frequency, though useful for many studies, results in an inhomogeneous broadening of the spectrum, which is undesirable in complex spectra. Magic-angle spinning (MAS) of the sample is a well-established technique for removing this broadening, leaving only lines at the isotropic-chemical-shift position. While very narrow lines may be observed in crystalline materials, amorphous materials show inhomogeneous linebroadenings of up to 20 ppm. This is due to the variation of the chemical shift with conformation-induced changes in the local electronic structure *(21)*. Recently, the spectra of a few aliphatic polymers have been simulated by ab-initio computer calculations of corresponding oligomers *(22)*. More traditionally, the conformational effect on the chemical shift has been analyzed semiempirically, most extensively for CH_2 units, in terms of the "γ-gauche effect" *(21)*. It is observed that for a given carbon site, the conformations that determine the positions of the carbons in the γ positions have the largest influence on the chemcial shift. This γ-gauche effect amounts to a 4-7 ppm upfield shift for each γ-gauche conformation. That means that the difference between t*.*t and g*.*g is twice that value. In solution, fast interconversion between the conformations removes the conformational linebroadening, and slightly different shifts (~ 1 ppm) remain only for different *configurations*, since they average the conformations with different weights.

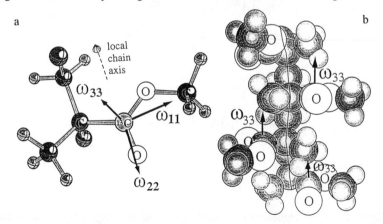

Figure 1. (a): Geometry of the PMMA repeat unit and the orientations of the carboxyl ^{13}C chemical-shift tensor, with ω_{11}, ω_{22}, ω_{33} axes. The ω_{33} axis is perpendicular to the OCO plane. (b): Typical chain segment in syndiotactic and commercial atactic PMMA. The local chain axis is approximately perpendicular to the OCO planes, since the bulky but relatively flat sidegroups have a tendency to stack perpendicular to the local chain axis. Therefore, the ω_{33} axes, indicated by the arrows, are nearly parallel to the (local) chain axis. Note also that the OCH_3 groups project out of the "core" of the chain and will thus be surrounded by atoms of other chains. Consequently, the environment of each sidegroup will be significantly asymmetric. (Adapted from ref. 15.)

2D Exchange NMR. The principle of two-dimensional exchange (2D) NMR *(23)* consists in the detection of slow frequency changes that occur during a mixing time t_m, by measuring the angular dependent NMR frequencies before and after t_m *(24,1)*. These frequency changes can be due to segmental reorientations, which alter the angle-dependent frequency of equation (1), or arise from conformational changes. The ^{13}C NMR pulse sequence for taking 2D time signals is shown Fig. 2(a). The 2D frequency spectrum $S(\omega_1,\omega_2;t_m)$ is generated by two successive Fourier transformations over t_1 and t_2 *(23,1)*. It represents the probability of finding a unit with a frequency ω_1 before t_m, and with a frequency ω_2 afterwards. If no frequency change (reorientation, conformational change) takes place during t_m, the spectral intensity is confined to $\omega_1 = \omega_2$, the diagonal of the frequency plane. Exchange between conformations corresponding to resolved peaks at ω_a and ω_b in 1D MAS spectra gives rise to *cross peaks* at the points (ω_a,ω_b) and (ω_b,ω_a), in addition to the diagonal peaks at (ω_a,ω_a) and (ω_b,ω_b). In 2D powder spectra, instead of cross-peaks, continuous intensity distributions with characteristic ridge patterns are observed for well-defined jumps *(24,1)*. In general, *large-angle* reorientations give rise to intensity in large parts of the frequency plane. Conversely, for anisotropic reorientations by *small angles*, the signal appears close to the diagonal, with a position-dependent broadening of the diagonal spectrum. Its most prominent effect is a change in the lineshape along the diagonal. The broadening is usually least pronounced at the edges of the powder spectrum, where the angular dependence of the frequency vanishes to first order *(20,1)*, so that these edges remain higher than the other, more strongly broadened regions along the diagonal.

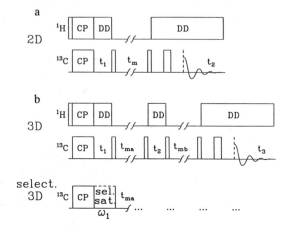

Figure 2. Pulse sequences of multidimensional exchange NMR as used for obtaining the spectra shown in this paper. ^{13}C magnetization is created by cross polarization (CP) from protons, which are then decoupled by high-power irradiation (dipolar decoupling, DD). Before detection, a Hahn spin echo is generated by means of a 180° pulse, in order to overcome spectrometer deadtime problems. (a): Pulse sequence for two-dimensional (2D) exchange NMR, with incremented evolution time t_1, mixing time t_m, and detection time t_2. (b): Three-dimensional (3D) exchange NMR with two mixing times, t_{ma} and t_{mb}. The angle-dependent frequency is measured three times by incrementing t_1, t_2, and detecting during t_3 in a full 3D experiment. In a "selective 3D" experiment, the frequency ω_1 is labelled by selective saturation of the rest of the spectral intensity.

For the understanding of the PMMA and PEMA ^{13}C spectra shown below, it is important to recognize how various motional processes can be distinguished due to their specific effects on the 2D intensity distribution (*1*). The analysis is particularly straightforward if the symmetries of the nuclear interaction tensor and the segmental reorientation are related. For instance, if the effective direction of the ω_{33} principal axis remains invariant in the motional process, no exchange occurs at the corresponding frequency ω_{33}, such that the spectral intensity remains on the diagonal at $(\omega_{33},\omega_{33})$ where it produces a notable extra peak compared to the rest of the spectrum with the intensity spread out into the 2D plane. Such invariance can result from a rotation around the corresponding principal axis. Due to the second-rank tensorial nature of the interaction, ω_{33} remains also unchanged under an inversion of the ω_{33} principal axis by a 180° flip of the segment.

Two-dimensional (2D) exchange NMR (*23,24*) studies of various polymers (*2-7*) have elucidated many details of the molecular reorientation geometry (*24,2,1*) of motions with correlation times in the range of 0.1 ms $< \tau < 10$ s. While these methods have been exploited to study the α relaxations of amorphous and semicrystalline polymers in great detail, only few 2D exchange NMR studies on localized (β) relaxations have been reported (*25,26*)· However, in 1D NMR certain localized relaxation processes, in particular phenylene ring flips, have been observed in many materials via their effects on ^{2}H quadrupolar as well as ^{13}C chemical-shift and dipolar line shapes (*27-29*).

3D Exchange NMR. The 2D exchange technique can be extended to 3D exchange NMR, where the orientation-dependent frequency is observed three times (*30,1*). The 3D pulse sequence, Fig. 2(b), contains two mixing time t_{ma} and t_{mb}, two incremented evolution times t_1 and t_2, and the detection period t_3. The 3D NMR spectrum resulting from Fourier transformation over t_1, t_2, and t_3 provides useful information on the sequence of jumps. In particular, it makes it possible to detect distinct signals from jumps back to the start position, regardless of the geometry of the reorientational motion. The full potential of 3D exchange NMR is discussed in reference (*1*).

3D exchange spectra are most conveniently analyzed in terms of 2D cross-sections at a constant ω_1. These 2D slices can also be measured individually by replacing the first evolution time t_1 by selective excitation of a frequency ω_1 (*15,1*). Thus, an ensemble of molecules with frequency ω_1 is labelled by this selection, which is followed by a mixing time t_{ma}, during which exchange to other frequencies can occur. This exchange intensity is monitored in the evolution period of an ensuing 2D experiment. If during its mixing time t_{mb}, a second frequency change occurs, it will be detected as off-diagonal intensity in the (ω_2,ω_3) 2D spectrum that is acquired. In particular, if the segments return to the orientations they had taken before t_{ma}, their detection frequency ω_3 will be equal to the initial frequency ω_1, so they produce contributions on the straight line $\omega_3 = \omega_1$ in the (ω_2,ω_3) plane (a *return-jump ridge*). Signals that do not exchange during t_{mb} appear on the diagonal $\omega_3 = \omega_2$ of the (ω_2,ω_3) plane. For exchange between two equivalent sites with a correlation time $\tau \ll t_{mb}$, the diagonal ridge and the return-jump ridge will have equal integral intensities, and their spectral lineshape will be the same. In contrast, for diffusive motions, involving a continuous distribution of sites, the exchange in the (ω_2,ω_3) plane would exhibit a clearly diffusive character and no return-jump ridge at $\omega_3 = \omega_1$

would be observed. It should be noted that, unlike the exchange patterns in the corresponding 2D spectrum, the return-jump ridge does not broaden for a system of segments undergoing exact N-site jumps with various reorientation angles. The return-jump ridge broadens only if the return jumps are imprecise, not taking the segments back to their original orientations.

The β Relaxation in PMMA

The β relaxation observed in poly(methyl methacrylate), PMMA, by means of both dielectric and dynamic-mechanical measurements is often considered as the archetype of a localized (β) relaxation in polymers. With the maximum of the mechanical loss near 10 Hz at ambient temperature, (*31*) it provides a mechanism for energy dissipation which can be linked to PMMA's favorable mechanical properties at and above room temperature (up to the softening setting in near the glass-transition temperature $T_g \approx 373$ K). Of a very similar nature as in PMMA are the β processes observed in other methacrylates, e.g. in poly(ethyl methacrylate) PEMA. Since these β relaxations are very prominent in dielectric relaxation, they have traditionally been attributed to the carboxyl sidegroup (see Fig. 1) with its large dipole moment, possibly involving a 180° rotational jump in spite of the asymmetry of the sidegroup, but their precise nature could not be determined.

Geometry of Large-Angle Ester-Group Motions. Figure 3 shows the 2D NMR spectrum of PMMA ^{13}C-labeled at 20% of the carboxyl carbons, measured at T = 333 K and t_m = 50 ms. An exchange pattern is seen clearly, exhibiting intensity far off the diagonal. This indicates large frequency changes, which correspond to large reorientation angles. A large portion of the frequency plane exhibits intensity, but not all of it, which proves that the motion is not isotropic.

As noted above, the simplest picture of the ester-group motion considers a rotation by 180° around the C-COO bond, which corresponds to a two-site jump. However, the diffuse intensity distribution in the experimental spectrum of Figure 3(a) might rather suggest a more or less isotropic diffusive motion of 25% of the sidegroups, while the rest would be quite rigid and give rise to the relatively sharp diagonal portion of the 2D spectrum. On the other hand, the ω_{33} frequency is invariant in the motional process. This is deduced from the 2D spectrum of Figure 3(a), where the cross-section through the spectrum at $\omega_1 = \omega_{33}$ has only intensity for $\omega_2 = \omega_{33}$. With the ω_{33} axis perpendicular to the backbone-sidegroup C-C bond, the invariance of ω_{33} requires an anisotropic rotation around this axis, and/or a two-site 180° rotational jump around the C-COO bond, which only inverts the ω_{33} axis without changing the NMR frequency of eq.(1).

That the dynamics indeed involves such two-site jumps can be corroborated by 3D exchange NMR. Figure 4 displays a spectrum equivalent to a two-dimensional slice at fixed ω_1 through a 3D spectrum, obtained in a selective-excitation 3D experiment. The two well-resolved straight ridges in the (ω_2,ω_3) plane, along the diagonal $\omega_3 = \omega_2$ and along the $\omega_3 = \omega_1$ line, prove that in spite of the diffusive appearance of the 2D exchange intensity in Figure 3(a), the reorientation of each segment involves only two, relatively well-defined, potential-energy minima. Thus, the featureless exchange intensity in the 2D NMR spectrum must have resulted from a broad distribution in the relative orientations of these pairs of potential energy

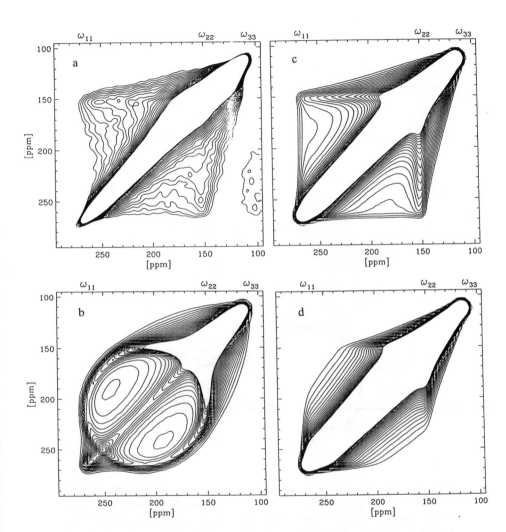

Figure 3.   (a): 2D exchange ^{13}C NMR spectrum of PMMA with 20% of ^{13}COO-labeled sidegroups. T = 333 K, t_m = 50 ms. (b): Attempted simulation of the 2D spectrum in (a) assuming a Gaussian distribution of side-group flip angles of ± 25° rms amplitude (i.e., 60° full width at half maximum) centered on 180°. The pronounced elliptical ridge is not observed experimentally. (c): Best simulation of the spectrum in (a), with a 180° ± 10° flip angle and a concomitant rotation around the ω_{33} direction (local chain axis) with a rms amplitude of ± 20°. (d): Unsatisfactory simulation which results for a rocking motion without 180° flips, i.e., when only rotations around the ω_{33} direction (local chain axis) with a rms amplitude of ± 25° are assumed. (Reproduced with permission from ref. 15. Copyright 1994 ACS.)

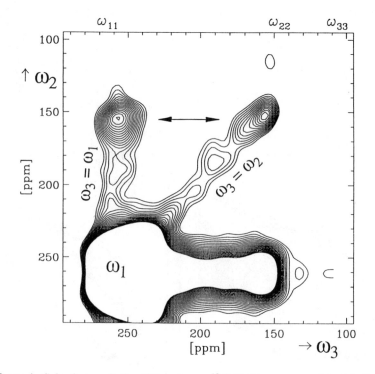

Figure 4. Selective-excitation 3D exchange ^{13}C NMR spectrum of the labelled PMMA carbonyl groups of PMMA at $T = 333$ K and mixing times $t_{ma} = t_{mb} = 50$ ms. Only two straight ridges, along $\omega_3 = \omega_2$ and $\omega_3 = \omega_1$, are observed in the relevant region of ω_2 unequal ω_1 , proving that the reorientation of each segment involves only two relevant potential-energy minima, in spite of the diffusive appearance of the corresponding 2D spectrum in Fig. 3(a). (Reproduced with permission from ref. 15. Copyright 1994 ACS.)

minima. This arises from the fact that different sidegroups have different environments in the glassy state, requiring varying degrees of rearrangement of the main chain in the energy-minimization after the jump.

In order to analyze the geometries of the two-site jumps in more detail, we compare the experimental 2D NMR spectrum of Figure 3(a) to simulated spectra for various motional models [Fig. 3(b)-(d)]. The simulated 2D exchange NMR spectrum resulting from 180° sidegroup flips without main-chain motion is plotted in Fig. 3(b). The simulation shows a pronounced elliptical, nearly circular ridge, which is, however, absent in the experimental 2D spectrum. There, except for straight ridges tapering at the ω_{33} end of the spectrum, the exchange intensity is rather featureless. In the simulation, the circular ridge cannot be broadened appreciably by assuming a distribution of the rotation angle around 180°. In fact, such a distribution with a root-mean-square (rms) amplitude of $\sigma = 25°$ (corresponding to a full width at half maximum of nearly 60°) was assumed in simulating the spectrum shown in Fig. 3(b).

Coupling of the Sidegroup Flips to Rotations around the Local Chain Axis. Because the ω_{33} and ω_{22} values are quite similar for the carboxyl chemical-shift anisotropy in PMMA, and since the ω_{33} axis is not involved in the exchange process, the exchange intensity mainly reflects the reorientation-angle distribution of the ω_{11} axis of the ^{13}C chemical-shift tensor. The featureless experimentally observed 2D intensity distribution in Fig. 3(a) indicates a wide reorientation-angle distribution for the ω_{11} axes. As indicated above, simulated 2D spectra such as that of Fig. 3(b) show that some limited deviations of the flip angle from a central value of 180° do not cause such a broadening in the spectral region between ω_z and ω_y. On the other hand, for these deviations an upper limit of $\sigma = 40°$ is established by detailed lineshape analysis, in particular by considering the approximate invariance of the ω_{33} frequency in the 2D spectrum. Distributions in this angle therefore cannot be invoked as the major source of the smearing-out of the exchange pattern.

However, a distribution of rotation angles around the ω_{33} axis, which is perpendicular to the OCO plane, is very effective in broadening the exchange features in the region between ω_{11} and ω_{22}, while producing no exchange at ω_{33}. Structural studies (*32-34*) and simulations (*35*) have shown that the normal of the OCO plane is parallel to the local chain axis within about 20°, as indicated in Figure 1. The spectral analysis of Fig. 3 shows that the amplitude of these restricted rotations around the local chain axis that accompany the 180° sidegroup flips is $\sigma = 20° \pm 7°$. The corresponding simulated spectrum is shown in Figure 3(c). If only such main-chain motions are assumed, without sidegroup flips, the spectral intensity is confined closer to the diagonal, Figure 3(d), than is found in the experimental spectrum, Figure 3(a).

As schematically indicated in Figure 5(a) - (c), due to the asymmetry of the sidegroup it is indeed expected that a rotation of limited amplitude around the normal of the OCO plane accompanies a sidegroup 180° flip. As shown in Figure 5(b), a simple 180° flip without main-chain readjustment will in general lead to steric clashes of the OCH_3 group with surrounding units. Figure 5(c) indicates how these can be avoided by a ~20° rotation around the normal of the OCO plane, which is approximately parallel to the local chain axis. Then, only slight changes in the environment are necessary to accomodate the asymmetric sidegroup after the 180° flip.

Significant lineshape changes on the diagonal are observed in the ^{13}C NMR 2D spectra. Figure 6 compares the stacked plots of ^{13}C 2D spectra at 233 K and 333 K.

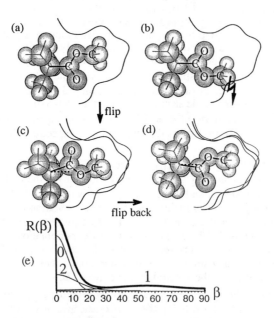

Figure 5. Schematic sketches of the dynamics of the asymmetric sidegroup in its correspondingly asymmetric environment [cf. also Fig. 1(b)]. (a): Initial sidegroup orientation. (b): Steric clash with the environment if an exact 180° flip without main-chain motion is assumed. (c): To fit the asymmetric sidegroup into the volume that it had occupied before the flip, a twist around the local chain axis is required. This in turn slightly deforms the environment. (d): A second jump takes the group back close to its original orientation in (a), but not exactly, due to the previous change in the environment in (c), which is enhanced by rotation of other sidegroups that make up that environment. Rearrangements of the main chain by neighboring flipping sidegroups will also induce such an effect of an effective small-angle rotation around the local chain axis. (e) Reorientation-angle distribution $R(\beta)$ for the ω_{11} axis of the carbonyl chemical-shift tensor that was used for the simulation of the ^{13}C 2D NMR spectrum of Figure 3(c). The component labelled "1" represents the 25% of the sidegroups that have flipped once (or 2N+1 times) and has a rms amplitude of 20°. Another 25% of the sidegroups have flipped back to the original orientation with an imprecision of 12° (two or 2N flips, labelled "2"), and the remaining 50% are trapped but rock with an amplitude of ca. 7° (no flips, labelled "0"). (Adapted from ref. 15.)

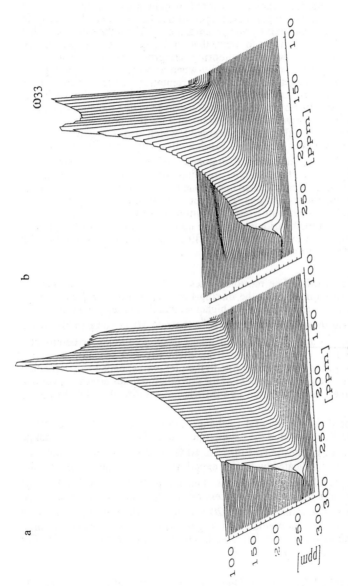

Figure 6. Stacked plots of ^{13}C exchange NMR spectra taken with $t_m = 50$ ms (a): Diagonal powder pattern for $T = 233$ K (b): At $T = 333$ K, significant changes of the intensity distribution on the diagonal are observed. The ω_{33} end of the spectrum is enhanced in part because the 180° inversion of the sidegroup leaves this frequency invariant, but mainly due to small-angle motions around the normal of the OCO plane. They occur naturally when the sidegroups have flipped twice [see Fig. 5(d)]. (Adapted from ref. 15.)

While a perfect diagonal powder pattern is observed at the lower temperature, at 333 K the intensity at the edges of the spectrum, in particular at the ω_{33} end, is enhanced relative to other parts of the spectrum. As explained in the theoretical section above, this is characteristic of the frequency-dependent broadening caused by anisotropic small-angle reorientations. The experimental spectrum is best matched by including rotations around the local chain axis by an rms amplitude of 7° for the sidegroups that do not flip, and by 12° rms amplitude for those that do. These reorientations arise due to the twisting of the chains when neighboring sidegroups reorient. The larger amplitude of the latter process results naturally when after two 180° flips, a given sidegroup does not return exactly to its original orientation due to a change in its environment. This is schematically displayed in Figure 5(d). The change in the environment could be due to the rearrangement of the surrounding moieties caused by the slight misfit of the given asymmetric sidegroup after the flip, or it might be caused by flips of other sidegroups within this environment, which also change the structure slightly. Figure 5(e) displays the overall reorientation-angle distributions of the carboxyl chemical-shift ω_{11} direction. Component "0" represents the 50% of the sidegroups that only rock, with an average amplitude of 7°. Component "1" results from the 25% of the sidegroups that have undergone one 180° flip, or an odd number of flips, and concomitant reorientations around the normal of the OCO plane with a rms amplitude of 20°, which result from the asymmetry of the sidegroup [see Fig. 5(c)]. Component "2" in the distributions, made up by another 25% of the groups, corresponds to two, or an even number, of flips with an effective rotation by typically 12° around the local chain axis.

Motional Rates in PMMA. The motional rates of the large-angle reorientation process have been determined from the mixing-time dependence of the exchange intensity found at different temperatures. At higher temperatures they can be estimated from the onset of 1D lineshape changes, where the rate $1/\tau$ is comparable to the powder linewidth $|\omega_{11} - \omega_{33}|$. Figure 7(a) displays these data in an Arrhenius plot together with literature values obtained from dielectric (30,36) and dynamic-mechanical (37) relaxation. The straight line in the plot corresponds to an activation energy of 65 kJ/mol. The good agreement of the corrrelation times around and above 300 K shows that the motion detected by NMR is indeed directly related to the β relaxation.

Integration of the experimental exchange intensity of the ^{13}C 2D NMR spectrum at $T = 333$ K, shows that the asymptotic value of the integrated exchange intensity for long mixing times makes up 25 ± 10 % of the total. In a symmetric two-site jump process as indicated by the 3D experiment, at least half of the intensity is found on the diagonal for any mixing time. This means that $2 \cdot (25 \pm 10)$% of the sidegroups participate in the exchange process on the time-scale of the correlation time of the β process. The remaining 50 ± 20% which are slow on that time scale must be trapped in environments with higher activation barriers. Such a heterogeneous distribution of correlation times has been confirmed by the MESSAGE experiment (15).

Relation to Mechanical Relaxation. Due to the asymmetry of the sidegroup, its motion occurs between energetically inequivalent sites; therefore, it is mechanically active. Our spectra show that the sidegroup dynamics are accompanied by main-chain motion and we can quantify the corresponding motional amplitudes to be as large as $(\pm)20°$. The ω_{33}-features of the ^{13}C 2D NMR spectra show that rearrangements of the

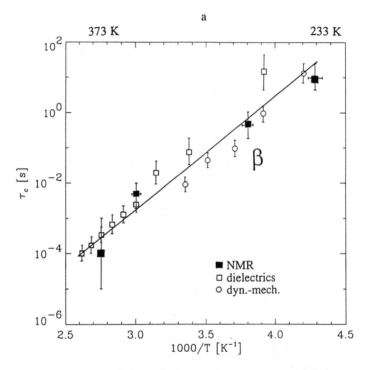

Figure 7. (a) Arrhenius plot of correlation times of the β-relaxation dynamics in PMMA, estimated from 1D and 2D NMR spectra, compared to both dielectric data obtained from the β-relaxation loss maxima (*30,36*) and relaxation times from dynamic-mechanical studies (*37*). (b) Arrhenius plot of the α and β-relaxation dynamics in PEMA, including the region where the two relaxations merge. Light-scattering (*38*), dielectric (*39*), and 1D/2D NMR (*16*) data are shown. *(Continued on next page.)*

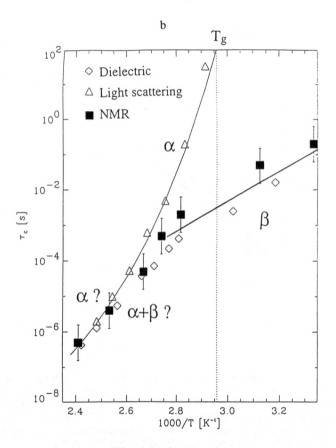

Figure 7. Continued.

main chain, which result in an effective rotation around the local chain axis, are at least as important in the energy minimization as are deviations from the 180° angle of the sidegroup motion. This indicates the dominance of steric interactions between different segments, compared to intramolecular bond-rotation potentials.

Merging of α and β Processes

Dielectric and dynamic-mechanical relaxation show similar activation energies of ca. 75 kJ/mol for the β processes in various poly(methacrylates) with different alkyl residues of the sidegroups. The α relaxation process, or glass-transition dynamics, has much larger (apparent) activation energies close to T_g. Therefore, the α and β processes merge within a relatively narrow temperature interval, as shown by the Arrhenius plot for PEMA in Figure 7(b), combining photon-correlation-spectroscopy (*38*), dielectric (*39*), and 1D/2D NMR (*16*) data. So far, only speculations have existed on the mechanism of the merging. PEMA is particularly suitable for investigating this α-β coalescence, since it occurs closer to Tg, and thus over a narrower range, than in PMMA, and within the range of correlation times accessible by 2D exchange NMR.

Figure 8(a) displays contour and stacked plots of a 2D exchange spectrum of ^{13}COO-labeled PEMA at T = 298 K and t_m = 500 ms. The exchange pattern closely resembles that of PMMA in Figures 3(a) and 6(b), which shows that the β relaxation processes in PMMA and PEMA have the same molecular origin.

The spectrum of Figure 8(b) was aquired at 17 K above the glass-transition temperature T_g with a mixing time of 500 ms, in the time-temperature region where the β process merges with the α process. The spectrum exhibits stronger exchange intensity but the pattern resembles that of the spectrum below T_g. The peak at ω_{33} is again the overall spectral maximum, which proves directly that the motion is pronouncedly anisotropic. The quantitative analysis shows that nearly all sidegroups are exchanging, undergoing 180° flips with similar imprecision as below T_g, but increased amplitude of rotation around the local chain axis.. This indicates that above T_g, the volume accessible to the sidegroups between chains is increased and trapped sidegroups are freed by slow fluctuations of their environment. 1D and 2D spectra at higher temperatures show the amplitude of the rotations around the main-chain axis increases strongly, while the reorientations of the chain axes themselves are restricted up to more than 60 K above T_g and the 180° flips remain relatively precise (*16,40*). In other words, the relaxation process at and above the region of α-β merging exhibits the features of both processes: The 180° sidegroup flip characteristic of the β process remains a distinctive mode of motion, while the rotation around the local chain axis, restricted for the original β process, becomes activated with very high apparent activation energy as typical for an α process. The spectra indicate that not only the correlation time, but also the amplitude of this rotation is activated with temperature.

The anisotropic glass-transition dynamics of PEMA is in contrast to the predominantly isotropic behavior of most other amorphous polymers investigated so far. The 2D spectrum of the COO group of PVAc (*3*) which has a spectral asymmetry parameter (η=0.27) comparable to that of PEMA (η=0.41), provides a good example, see Figure 9. In the PVAc spectrum of Figure 9(b) at T_g+20 K and t_m

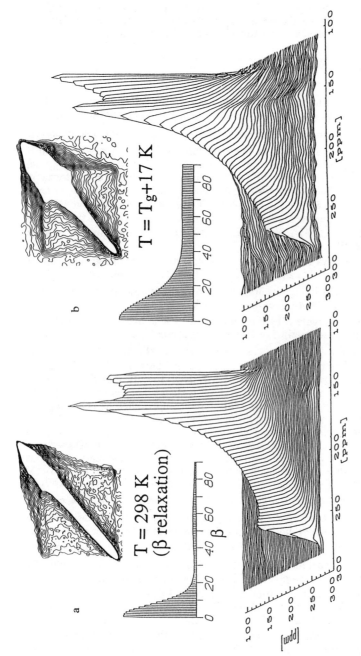

Figure 8. 2D exchange ^{13}C NMR spectra. (a) PEMA with 20% of ^{13}COO-labeled sidegroups with $t_m = 500$ ms, below T_g, at $T = 298$ K. Note the similarity to the spectrum of PMMA in Fig. 3(a) and 6(b). (b) Similar to (a), at 17 K above T_g, $T = 365$ K. Note the increase of the exchange intensity compared to (a) and the similarity of the exchange patterns. Reorientation-angle distributions for the ω_{11} axis of the carbonyl chemical-shift tensor are shown as inserts. (Adapted from ref. 16.)

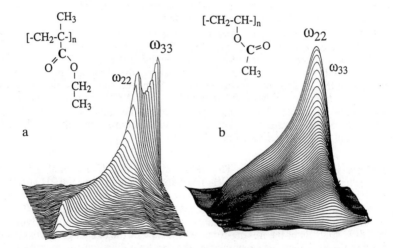

Figure 9. Dynamics above T_g, probed by carbonyl chemical-shift tensors. (a) PEMA at $T = T_g+17$ K, $t_m = 500$ ms. (b) Unlabeled PVAc at $T = T_g+20$ K, $t_m = 100$ ms. The 2D lineshapes show that the motion in PEMA is anisotropic compared to the reorientations in PVAc. In PEMA, the ω_{33} axis of the carbonyl tensor remains nearly invariant, which means that the local chain axis does not reorient significantly, and also proves that the sidegroup flip angles do not deviate far from 180°. (Adapted from ref. 16.)

= 100 ms, the diagonal ridge has nearly vanished, and the peak at ω_{22} is the overall spectral maximum, as is expected for any isotropic process.

Conformational Information from Structured CH$_2$ MAS Lines

The top of Figure 10 shows MAS spectra of three amorphous aliphatic polymers near T_g (17). The resonances of several peaks, in particular of the CH$_2$ bands, exhibit significant broadening and some fine structure. It is due to conformational effects on the chemical shifts, the γ-gauche effect, as discussed in the Theory section. The assignment of the chemical shifts to the conformations is particularly clear in armophous poly(propylene) from comparisons with isotactic and syndiotactic poly(propylene), where the conformations in the crystalline regions are known to be trans*.*gauche and t*.*t/g*.*g, respectively (41).

Below the 1D spectra in Figure 10 are shown corresponding 2D "exchange" spectra. They are exploited here not to obtain dynamical information, but instead to determine the homogeneous linewidth, perpendicular to the spectral diagonal, of the various resonances. The elongated patterns along the diagonal prove that the broadening of the 1D spectra is predominantly inhomogeneous. Thus, the 2D spectra show that the intensity between the maxima of the 1D CH$_2$ patterns is not due to homogeneous linebroadening. It strongly suggests that there are conformations intermediate between trans and gauche. This is in accordance with the diffusive reorientational dynamics discussed below.

Conformational Dynamics. The relatively small off-diagonal and large inhomogeneous broadening of the CH$_2$ resonances obvious from the 2D spectra of Figure 10 and Figure 11(a) show that no major conformational motions occur below T_g. Above T_g, strong conformational exchange is observed, see Figure 11(b). It is interesting to compare this isomerization dynamics with the reorientational dynamics. Quantitative analysis of the mixing-time dependence of the exchange intensities has shown that the exchange between gauche-like and trans-like conformations exhibits similar non-exponentiality as the reorientational motions. In the Arrhenius plot of Figure 12, the correlation times for conformational exchange (full symbols) and rotational dynamics (open symbols) are compared and found to be equal within the error margins. The data, inculding results from a ^{13}C relaxation study (42), follow a WLF (William-Landel-Ferry) relation shown as a solid line in Figure 12. By means of slow magic-angle spinning 2D exchange NMR (2), reorientational and conformational dynamics can actually be detected separately in one and the same spectrum. For a sample subject to slow MAS, the NMR powder spectrum splits up into a spinning sidebands spaced by the rotation frequency. In a suitable designed MAS 2D exchange experiment, off-diagonal sidebands indicate molecular reorientations. At the same time, conformational (i.e., isotropic-shift) changes can be detected in terms of exchange within each MAS band. Both features are observed in the experimental aPP spectrum of Figure 11(c).

In the interpretation of this equality of correlation times, it will be noted that the correlation times for the conformational exchange cannot be shorter that those of the reorientations, since the former must always be accompanied by the latter. On the other hand, it would have been well conceivable that the conformational exchange is slower than the rotational motion: The ill-defined, diffusive reorientational motions could have been produced by many small, possibly random, changes in successive torsion angles, with only a small percentage of large conformational changes. Only on a longer time-scale would all segment have undergone a conformational change.

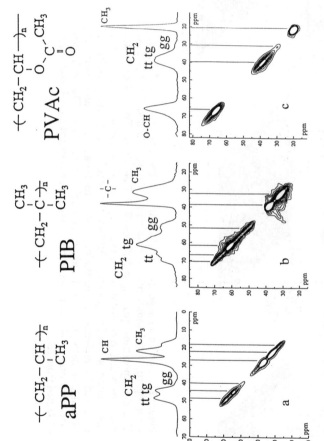

Figure 10. 13C CP MAS spectra of aliphatic polymers near Tg. Top: 1D spectra, bottom: 2D exchange spectra with short mixing time (no significant motion), providing a measure for the homogeneous linebroadening (width perpendicular to the spectral diagonal). These 2D spectra prove that the broadening of the 1D spectra is predominantly inhomogeneous. (a) Atactic poly(propylene). (b) Poly(isobutylene) with short (50 μs) cross polarization time to suppress the signal of the quaternary carbon. (c) Poly(vinyl acetate). (Adapted from ref. 17.)

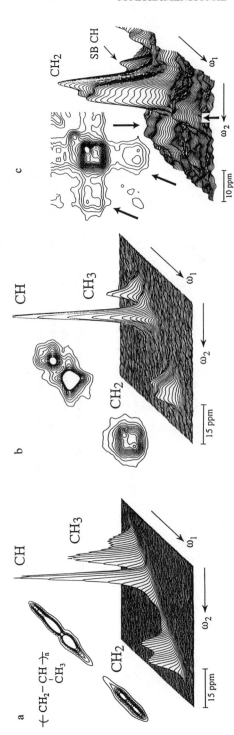

Figure 11. MAS exchange spectra of atactic poly(propylene). (a) and (b) fast spinning (no sidebands), (c) slow spinning. (a) $t_m = 5$ ms, $T = T_g - 5$ K, diagonal spectrum. (b) $t_m = 500$ ms, $T = T_g + 12$ K. The off-diagonal intensity is due to conformational exchange. (c) Slow spinning, zoom on the CH_2 region of the 2D spectrum. Arrows mark exchange spinning sidebands produced by molecular reorientation. Within each band, full conformational exchange is observed while the exchange sidebands are small. This indicates that conformational exchange is not slower than reorientational exchange. (Adapted from ref. 17.)

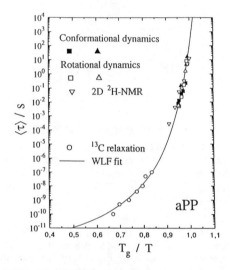

Figure 12. Reduced-temperature Arrhenius plot of correlation times τ_0 for various amorphous poly(propylene) samples. Filled symbols: conformational dynamics. Open symbols: reorientational dynamics. Solid line: WLF fit. (Adapted from ref. 17.)

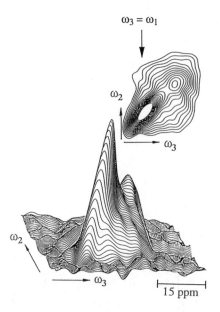

Figure 13. Selective-excitation 3D spectrum of aPP at $T = 252$ K $=$ Tg $+$ 11K, with $t_{ma} = t_{mb} = 50$ ms. The arrow marks the line $\omega_3 = \omega_1$ where the return-jump ridge would have been expected. Its absence proves the lack of significant orientational memory in the reorientational dynamics above the glass transition. Compare also the 3D spectrum of PMMA in Figure 4, where the return-jump ridge is prominent.

The fact that the two correlation times are not very different might suggest a description of the dynamics in terms of jumps on a more or less ill-defined diamond lattice. However, a selective 3D spectrum, Fig. 13, of the reorientational motions of aPP at 11 K above T_g does not show the orientational memory that would be expected according to such a model. On a diamond lattice, there are only four distinguishable bond orientations, and even for full exchange among them would the return-jump ridge remain a prominent feature. No return jump ridge at $\omega_3 = \omega_1$ is observed in the 3D spectrum, even though the ridge would occur even for quite imprecise return jumps, due to the location of ω_1 near the edge of the methyl powder pattern. This result is corroborated by the absence of return-jumps signal in a (full) 3D spectrum of PVAc at 10 K above T_g.

Summary

2D and selective-excitation 3D exchange ^{13}C NMR techniques have shown that in the β relaxation of PMMA, ca. 50% of the sidegroups undergo 180° flips, with less than 20° rms deviation in the flip angle, which are accompanied by rotational readjustments with an amplitude of ca. ±20° around the local chain axis. In a consecutive jump, the sidegroup returns to its original orientation with a precision of ca. 12°. At the merging of the β and α relaxation processes, nearly all sidegroups are flipping. The α process mainly activates the amplitude and correlation time of the rotation around the chain axis, but the sidegroup flip and chain motion remain distinctly anisotropic. In various other aliphatic polymers, conformational exchange above T_g has been studied by 2D MAS exchange NMR. The correlation times are very similar to those of the rotational motions in these polymers. Nevertheless, the absence of orientational memory in 3D NMR shows that dynamics on a slightly distorted diamond lattice is no suitable model for polymer dynamics above T_g.

Acknowledgments

It is a pleasure to thank Prof. H. W. Spiess for his enthusiastic support and supervision of the research presented in this article, and to acknowledge the contributions of our coworkers K. Zemke, C. Boeffel, A. Kulik, H. W. Beckham, U. Pawelzik, A. Ohlemacher, and D. Radloff. Financial support was provided by the Deutsche Forschungsgemeinschaft (SFB 262).

Literature Cited

(1) Schmidt-Rohr, K.; Spiess, H. W. *Multidimensional Solid-State NMR and Polymers*; Academic Press: New York, 1994.

(2) Hagemeyer, A.; Schmidt-Rohr, K.; Spiess, H. W. *Advances in Magn. Reson.* **1989**, *13*, 85.

(3) Schmidt-Rohr, K.; Spiess, H. W. *Phys. Rev. Lett* **1991**, 66, 3020.

(4) Schaefer, D.; Spiess, H. W.; Suter, U. W.; Fleming, W. W. *Macromolecules* **1990**, *23*, 3431.

(5) Zhang, C.; Wang, P.; Jones, A. A.; Inglefield, P. T.; Kambour, R. P. *Macromolecules* **1991**, *24*, 338.

(6) Lee, Y. K.; Emsley, L.; Larsen, R. G.; Schmidt-Rohr, K.; Hong, M.; Frydman, L.; Chingas, G. C.; Pines, A. *J. Chem. Phys.* **1994**, *101*, 1852.

(7) Chung, G. C.; Kornfield, J.A.; Smith, S. D. *Macromolecules* **1994**, 27, 964.

(8) Caravatti, P.; Neuenschwander, P.; Ernst, R. R. *Macromolecules* **1985**, *18*, 119.

(9) Clauss, J.; Schmidt-Rohr, K.; Spiess, H. W. *Acta Polymer.* **1993**, *44*, 1.

(10) Schmidt-Rohr, K.; Clauss, J.; Spiess, H. W. *Macromolecules* **1992**, *25*, 3273.

(11) Harbison, G. S.; Vogt, V.-D.; Spiess, H. W. *J. Chem. Phys.* **1987**, *86*, 1206.

(12) Henrichs, P. M. *Macromolecules* **1987**, *20*, 2099.

(13) Schmidt-Rohr, K.; Schaefer, D.; Hehn, M.; Spiess, H. W. *J. Chem. Phys.* **1992**, *97*, 2247.

(14) Chmelka, B. F.; Schmidt-Rohr, K.; Spiess, H. W. *Macromolecules* **1993**, *26*, 2282.

(15) Schmidt-Rohr, K.; Kulik, A. S.; Beckham, H. W.; Ohlemacher, A.; Pawelzik, U.; Boeffel, C.; Spiess, H. W. *Macromolecules* **1994**, *27*, 4733.

(16) Kulik, A. S.; Beckham, H. W.; Schmidt-Rohr, K.; Radloff, D.; Pawelzik, U.; Boeffel, C.; Spiess, H. W. *Macromolecules* **1994**, *27*, 4746.

(17) Zemke, K.; Schmidt-Rohr, K.; Spiess, H. W. *Acta Polymer.* **1994**, *45*, 148.

(18) Pines, A.; Gibby, M. G.; Waugh, J. S. *J. Chem. Phys.* **1973**, *59*, 569.

(19) Veeman, W. S. *Prog. NMR Spec.* **1984**, *16*, 193.

(20) Haeberlen, U. *Advances in Magnetic Resonance. Supplement 1*; Academic Press: New York, 1976.

(21) Tonelli, A. E. *NMR Spectroscopy and Polymer Microstructure*; VCH: New York, 1989.

(22) Born, R.; Spiess, H. W.; Kutzelnigg, W.; Fleischer, U.; Schindler, M. *Macromolecules* **1994**, *27*, 1500.

(23) Jeener, J.; Meier, B. H.; Bachmann, P.; Ernst, R. R. *J. Chem. Phys.* **1979**, *71*, 4546.

(24) Schmidt, C.; Blümich, B.; Spiess, H. W. *J. Magn. Reson.* **1988**, *79*, 269.

(25) Hansen, M. T.; Blümich, B.; Boeffel, C.; Spiess, H. W.; Morbitzer, L.; Zembrod, A. *Macromolecules* **1992**, *25*, 5542. Hansen, M. T.; Boeffel, C.; Spiess, H. W. *Coll. Polym. Sci.* **1993**, *271*, 446.

(26) Leisen, J.; Boeffel, C.; Dong, R. Y.; Spiess, H. W. *Liquid Crystals* **1993**, *14*, 215.

(27) Spiess, H. W. *Adv. Polym. Sci.* **1985**, 66, 23.

(28) Inglefield, P. T.; Amici, R. M.; O'Gara, J. F.; Hung, C.-C.; Jones, A. A. *Macromolecules* **1983**, *16*, 1552; O'Gara, J. F.; Jones, A. A.; Hung, C.-C.; Inglefield, P. T. *Macromolecules* **1985**, *18*, 1117.

(29) Schaefer, J.; Stejskal, E. O.; McKay, R. A.; Dixon, W. T. *Macromolecules* **1984**, *17*, 1749.

(30) Spiess, H. W.; Schmidt-Rohr, K. *Polym. Prepr. (ACS)* **1992**, *33(1)*, 68.

(31) McCrum, N. G.; Read, B. E.; Williams, G. *Anelastic and Dielectric Effects in Polymeric Solids*; Wiley: New York, 1967.

(32) Lovell, R.; Windle, A. *Polymer* **1981**, *22*, 175.

(33) Coiro, V. M.; De Santis, P.; Liquori, A. M.; Mazzarella, L. *J. Polym. Sci. C* **1969**, *16*, 4591.

(34) Kulik, A.; Spiess, H. W., *Macromol. Chem. Phys.* **1994**, *195*, 1755.

(35) Vacatello, M.; Flory, P. J. *Macromolecules* **1986**, *19*, 415.

(36) Gomez Ribelles, J. L.; Diaz Calleja, R. *J. Polym. Sci. Polym. Phys. Ed.* **1985**, *23*, 1297.

(37) Muzeau, E.; Perez, J.; Johari, G. P. *Macromolecules* **1991**, *24*, 4713.

(38) Patterson, G. D.; Stevens, J. R.; Lindsey. C. P. *J. Macromol. Sci.-Phys.* **1980**, *18*, 641.

(39) Ishida, Y.; Yamafuji, K. *Kolloid Z* **1961**, *177*, 97.

(40) Kulik, A. S.; Radloff, D.; Spiess, H. W. *Macromolecules* **1994**, *27*, 3111.

(41) Sozzani, P.; Simonutti, R.; Galimberti, M. *Macromolecules* **1993**, *26*, 5782.

(42) Dekmezian, A.; Axelson, D. E.; Dechter, J. J.; Bohra, B; Mandelkern, L. *J. Polym. Sci. Polym. Phys. Ed.* **1985**, *23*, 367.

RECEIVED February 2, 1995

Chapter 13

Applications of ^{1}H–^{19}F–^{13}C Triple-Resonance NMR Methods to the Characterization of Fluoropolymers

Peter L. Rinaldi[1], Lan Li[1], Dale G. Ray III[1], Gerard S. Hatvany[1], Hsin-Ta Wang[2], and H. James Harwood[2]

[1]Department of Chemistry, Knight Chemical Laboratory, University of Akron, Akron, OH 44325–3601
[2]Department of Polymer Science, Maurice Morton Institute of Polymer Science and Engineering, University of Akron, Akron, OH 44325

In this paper we illustrate some applications of 1D- and 2D-^{1}H/^{19}F/^{13}C triple resonance NMR techniques for characterizing fluoropolymers. These methods can be used to achieve spectral simplification, to disperse resonances permitting resolution of clearly spaced peaks due to nuclei in various stereo- and monomer sequences, and to establish one-bond and multiple-bond connectivities in order to identify structure fragments. Various permutations of INEPT, HMQC, and HMBC NMR experiments are used to obtain illustrative data from poly(1-chloro-1-fluoro-ethylene-co-isobutylene) (PCFE-IB).

The NMR analysis of polymers is very often difficult as a consequence of the numerous structures which result from the presence of variations in stereosequence and monomer distribution. The variety of structures present leads to complex spectra with numerous overlapping resonances. Very often, a particular structure or defect fragment in the molecule is of interest, such as the unique structures formed at the chain ends by initiation or termination reactions, the unique repeat units which occur at the junctions of two dissimilar blocks, or the structures which result from chain branching or grafting of functional groups onto the polymer chain. Our efforts to characterize fluoropolymers led us to exploit multi-dimensional (1) and triple resonance (2) NMR techniques which have been used effectively by biochemists over the past 5-10 years. In employing these techniques, a third NMR active nucleus (other than ^{1}H or ^{13}C), is incorporated in a substance, either through introduction of functionality or by artificial enrichment with an isotope that is normally present at low abundance levels, and then rf pulses are applied at the resonance frequencies of these nuclei. This permits filtering of most of the resonances from the normal NMR spectrum and enables the selective detection of resonances near the site of label incorporation.

Recently developed 2D-NMR methods have provided a plethora of NMR experiments for establishing the identity of various structure fragments (3). For example,

0097–6156/95/0598–0215$14.00/0

a one-bond heteronuclear shift correlation experiment (4) provides a means of identifying all of the ^{13}C-H fragments in a molecule. Similar experiments such as long range HETCOR (5), COLOC (6), and XCORFE (7) exploit two- and three- bond J_{CH} couplings permitting the detection of ^{13}C-C-H and ^{13}C-C-C-H fragments. These components of a molecule's structure, which are obtained from one or more 2D-NMR experiments, can be put together like the pieces of a puzzle to obtain a complete picture of a molecule's structure. Additionally, 2D-NMR experiments provide a means of dispersing the resonances from the 1D-NMR spectrum into a second dimension, thus achieving dispersion that is not possible in 1D-spectra even at the highest resonance frequency currently available with any NMR instruments. Triple resonance 2D-NMR techniques combined with isotopic labeling provide an infinite number of experimental methods for structure identification with an even higher level of spectral dispersion and simplification.

In studies on fluoropolymers, ^{19}F, which is naturally present in 100% abundance can be employed as the third nucleus in the triple resonance methods discussed above. This provides the opportunity to use a countless variety of pulse sequences to selectively detect different structural features of polymers. In this paper we will provide a few examples of ^{1}H/^{19}F/^{13}C triple resonance techniques which can be used to achieve spectral simplification and structure identification. These methods are not necessarily the ones which provide optimum sensitivity or spectral dispersion; they do serve as good examples of how these 2D-NMR techniques can be used in polymer chemistry.

1D-NMR Experiments

The 1D-NMR spectra of a relatively simple polymer, poly(1-chloro-1-fluoroethylene-co-isobutylene) (PCFE-IB) are shown in Figure 1. These spectra illustrate the difficulties usually encountered in basic 1D-NMR analyses. The ^{1}H NMR spectrum (Figure 1a) consists of two groups of resonances from the methylene (2.0-3.5 ppm) and methyl (1.0-1.5 ppm) groups of this polymer. Both resonance areas are complex due to the presence of resonances of a large number of monomer sequences and associated stereosequences present in the copolymer.

The ^{13}C-NMR spectrum also contains a large number of resonances. The shifts of ^{13}C atoms are generally more sensitive to their local environments and their resonances usually occur as singlets, making it easier to resolve distinct peaks for carbons in different environments. However, some of these occur as multiplets in the spectrum due to C-F coupling. Three groups of C-F doublets are observed in the 105-120 ppm region. The methylene, quaternary, and methyl carbon resonances are observed in the 50-60, 34-38, and 30 ppm regions of the spectrum, respectively.

Although there is only a single ^{19}F atoms per CFE repeat unit, the ^{19}F shift is most sensitive to changes in local environment. Consequently, the ^{19}F spectrum (Figure 1c) is the most complex of the three; a large number of resonances are detected and the spectrum is too complicated to be easily interpreted. We now show how these spectra can be simplified by employing new NMR experiments to study polymers.

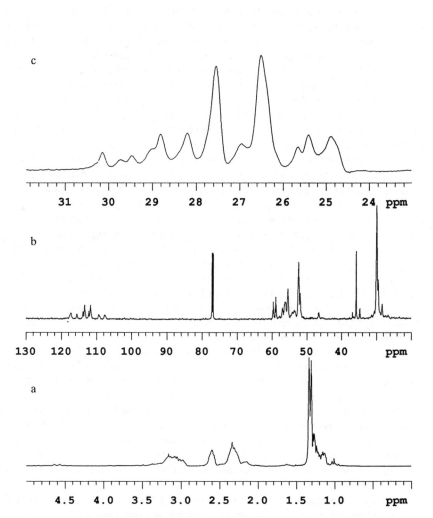

Figure 1. NMR spectra of poly(1-chloro-1-fluoroethylene-co-iso-butylene): a) ¹H spectrum, b) ¹³C spectrum, and c) ¹⁹F NMR spectrum.

$\{^{19}F\}^{13}C$ **Polarization Transfer.** Polarization transfer experiments such as INEPT (Insensitive Nuclei Enhanced by Polarization Transfer) (8) and DEPT (Distortionless Enhancement by Polarization Transfer) (9) were originally devised to improve the sensitivity for detection of nuclei such as ^{13}C, which have small magnetic moments, by transferring magnetization from ^1H which has a much higher magnetic moment. Polarization transfer occurs from ^1H (which has a much larger proportion of its spins in the ground state) to ^{13}C, providing a factor of 4 (proportional to n_H/n_C) increase in signal-to-noise for detection of ^{13}C. Furthermore, it is the ground state population of ^1H atoms, which have shorter T_1's than ^{13}C, that must recover before the pulse sequence can be repeated in signal averaging. Consequently, more transients can be averaged in a given period of time, providing an additional increase in sensitivity. One drawback of this method of signal detection is that only proton-bearing carbons are detected, since polarization transfer is accomplished using one-bond J-couplings between the observed and decoupled nuclei.

An outline of the INEPT sequence is shown in Figure 2. A combination of pulses is applied to the observed ^{13}C, and decoupled $\{X\}$ nuclei (note that by convention, the decoupled nucleus is generally surrounded by brackets when the two nuclei are listed in the name of an experiment). In this way X nucleus magnetization is transferred to ^{13}C. Figure 3 shows plots of how the final ^{13}C signal intensity varies as a function of the number of attached protons, and the lengths of the d2 and d3 delays. Usually, the experiments are performed with d2 = $1/(2 \ast J_{CX})$. The behavior of CH, CH_2, and CH_3 signal intensities as a function of d3 are shown by the plot in Figure 3b. Three different spectra are usually collected with d3 values of $1/(2 \ast J_{CH})$, $1/(3 \ast J_{CH})$ and $3/(4 \ast J_{CH})$ in order to provide three spectra containing only CH resonances, $CH/CH_2/CH_3$ resonances all positive, and CH/CH_3 resonances positive and CH_2 resonances inverted, respectively.

Shortly after the report of the INEPT experiment, it was realized that its only apparent shortcoming (i.e. the inability to observe quaternary carbon resonances) could be used to advantage in filtering undesired signals from complex NMR spectra by strategically placing labels such as ^2H in a structure and performing polarization transfer from the label to the observed nucleus (10). The ^{13}C spectra would then contain only ^{13}C resonances from carbons directly bound to ^2H. This strategy is ideally suited for selective detection of ^{13}C resonances in fluoropolymers because ^{19}F has nuclear properties and natural abundance similar to those of ^1H. Consequently, all of the sensitivity advantages achieved in $\{^1H\}^{13}C$-INEPT are also obtained in the $\{^{19}F\}^{13}C$-INEPT experiment.

Figure 4 contains the ^{13}C NMR spectra of 1-fluorohexane. Figure 4a is the normal ^{13}C spectrum with ^1H decoupling; doublets are observed for C-1 (83.9 ppm, $^1J_{CF}$ = 164 Hz), C-2 (30.4 ppm, $^2J_{CF}$ = 19.3 Hz), and C-3 (22.5 ppm, $^3J_{CF}$ = 5.9 Hz). When both ^1H and ^{19}F decoupling are performed (Figure 4b) the coupling to fluorine is eliminated and the spectrum is greatly simplified. Figure 4c is a $\{^{19}F\}^{13}C$-INEPT spectrum with both ^1H and ^{19}F broadband decoupling during the acquisition time. Delays were optimized for $^1J_{CF}$ = 164 Hz and only C-1 is detected as a singlet. The spectrum is greatly simplified by filtering all of the ^{13}C resonances of carbons not directly bound to ^{19}F. Only a trace of the C-2 resonance is detected in Figure 4c because although the delays are optimized for $^1J_{CF}$, they are set to only about

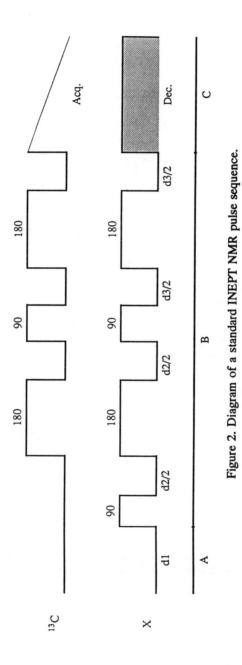

Figure 2. Diagram of a standard INEPT NMR pulse sequence.

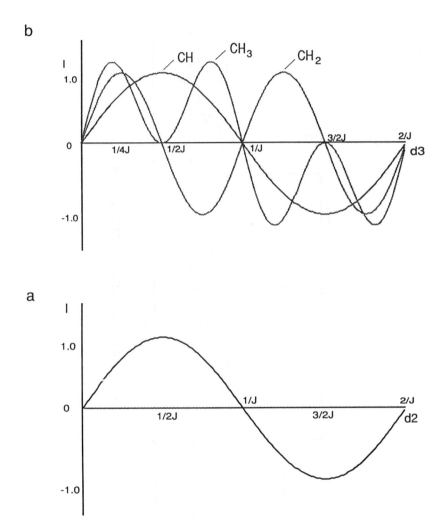

Figure 3. Plot of CH_n signal intensity variations in the INEPT experiment: a) as a function of the polarization transfer, d2, delay; and b) as a function of the refocussing, d3, delay.

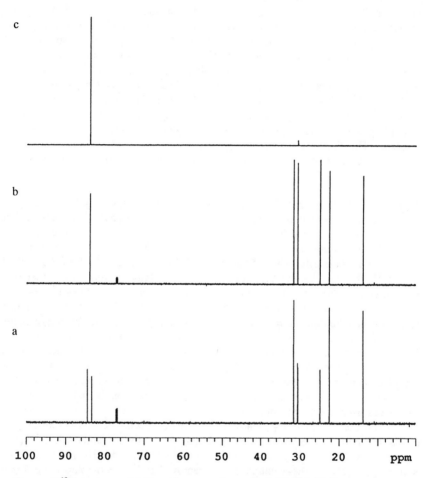

Figure 4. ^{13}C spectra of 1-fluorohexane: a) standard 1D-^{13}C NMR spectrum with ^{1}H decoupling during the acquisition time, b) 1D-^{13}C NMR spectrum with both ^{1}H and ^{19}F decoupling during the acquisition time, and c) {^{19}F}-^{13}C INEPT NMR spectrum with both ^{1}H and ^{19}F decoupling during the acquisition time.

$1/(16*^2J_{CF})$ and significant time for buildup of ^{13}C magnetization has not been achieved for carbons with couplings smaller than 100 Hz.

Figure 5 shows the results from similar experiments which were performed on PCFE-IB. In Figure 5a, the normal ^{13}C 1D-spectrum displays resonances from C-F, $CDCl_3$ solvent, methylene, quaternary and methyl carbons. The ^{13}C 1D-spectrum with both ^1H and ^{19}F decoupling during data acquisition (Figure 5b) is greatly simplified; note the three groups of C-F doublets between 105 and 120 ppm have been collapsed to groups of single lines. The methylene region of the spectrum (50-60 ppm) is also greatly simplified. The $\{^{19}F\}^{13}C$-INEPT spectrum in Figure 5c was also acquired with both ^1H and ^{19}F decoupling during data acquisition and contains only the three groups of resonances from the three nonequivalent permutations of triads. The $\{^{19}F\}^{13}C$-INEPT spectrum of PCFE-IB in Figure 5d was acquired with delays optimized for multiple-bond C-F couplings and with both ^1H and ^{19}F decoupling during data acquisition. Only methylene carbons a to C-F carbons are observed. The C-F resonances are greatly attenuated, and the resonances from aliphatic carbons which are not with two- or three-bonds from fluorine are completely eliminated.

$\{^{19}F\}^1H$ Polarization Transfer. INEPT can be use advantageously for the detection of ^1H resonances as well as ^{13}C resonances if polarization transfer is performed from ^{19}F to ^1H. However, because the ratio of the ^{19}F to ^1H resonance frequencies is close to one, the sensitivity gains achieved in detection of ^{13}C via INEPT experiments are not achieved. Ordinarily, ^1H-^{19}F J-couplings fall in the range of 20-50 Hz, 5-20 Hz and 0-5 Hz for $^2J_{HF}$, $^3J_{HF}$ and $^4J_{HF}$, respectively. The relatively well defined ranges for two-, three-, and four-bond couplings in acyclic hydrocarbons makes it possible to perform polarization transfer experiments which are edited based on the number of bonds between the fluorine and proton nuclei. Although the sensitivity gains achieved in detection of ^{13}C via INEPT experiments are not achieved, sensitivity approaching that obtained for direct detection of ^1H via a standard one-pulse experiment can usually be achieved. Fortunately, sensitivity in the detection of ^1H resonances is not as critical as in the detection of ^{13}C resonances. The spectral simplification achieved can be very useful since the ^1H chemical shift dispersion in not large, and ^1H resonances usually occur as multiplets.

The normal one-pulse and $\{^{19}F\}^1H$-INEPT spectra of 1-fluorohexane are shown in Figure 6. The ^1H spectrum in Figure 6a exhibits doublet splittings (from $^nJ_{HF}$) of $^2J_{HF} = 47.3$ and $^3J_{HF} = 24.6$ Hz in the multiplets at 4.42 (H-1) and 1.67 (H-2) ppm, respectively. These couplings are removed if ^{19}F decoupling is performed during the acquisition time, as shown in Figure 6b. Although it likely exists, four-bond H-F coupling could not be discerned in the complex multiplet from H-3. The spectrum in Figure 6c was obtained with $\{^{19}F\}^1H$-INEPT using delays which were optimized for $^2J_{HF} = 50$ Hz; only the multiplet from H-1 is seen in the spectrum. If the INEPT delays are optimized for $^3J_{HF} = 25$ Hz, then the resonances of both H-1 and H-2 become apparent (Figure 6d).

When using INEPT in this manner, the multiplicities of the resonances must be kept in mind. In $\{^1H\}^{13}C$-INEPT experiments, polarization transfer is always from n equivalent H atoms to a single ^{13}C atom in a ^{13}C resonance from a CH_n group.

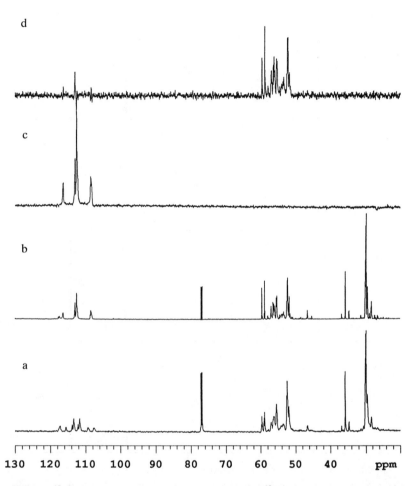

Figure 5. ^{13}C spectra of PCFE-IB: a) standard 1D-^{13}C NMR spectrum with ^{1}H decoupling during the acquisition time, b) 1D-^{13}C NMR spectrum with both ^{1}H and ^{19}F decoupling during the acquisition time, c) {^{19}F}^{13}C-INEPT NMR spectrum obtained with both ^{1}H and ^{19}F decoupling during the acquisition time and with delays optimized for detection of peaks from one-bond C-F couplings, and d) {^{19}F}^{13}C- INEPT NMR spectrum obtained with both ^{1}H and ^{19}F decoupling during the acquisition time and with delays optimized for detection of peaks from multiple-bond C-F couplings.

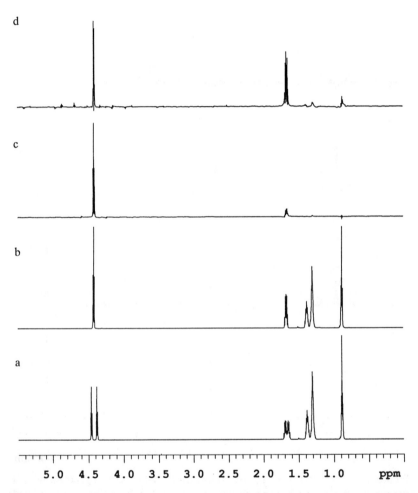

Figure 6. ^{1}H NMR spectra of 1-fluorohexane: a) normal 1D-spectrum, b) 1D-spectrum with broadband ^{19}F decoupling during the acquisition time, c) $\{^{19}F\}^1$H-INEPT spectrum with ^{19}F decoupling during the acquisition time and with the polarization transfer delay tuned for $^2J_{HF} = 50$ Hz, and d) $\{^{19}F\}^1$H-INEPT spectrum with ^{19}F decoupling during the acquisition time and with the polarization transfer delay tuned for $^3J_{HF} = 20$ Hz.

However, in the $\{^{19}F\}^1H$-INEPT experiment such as the one performed on 1-fluorohexane, polarization is from an isolated ^{19}F to n equivalent ^1H atoms (i.e. from F to CH_2). One important consequence of this difference is that the signal intensity variation as a function of d2 in $\{^{19}F\}^1H$-INEPT follows the behavior shown for d3 in the $\{^1H\}^{13}C$-INEPT experiment. The line for CH_2 signals from the plot in Figure 3b must be used to determine the optimum polarization transfer (d2) delay rather than the plot shown in Figure 3a; the optimum d2 is $1/(4*J_{HF})$ rather than the expected $1/(2*J_{HF})$. The signal variation as a function of d3 in the $\{^{19}F\}^1H$-INEPT experiment follows the behavior shown in Figure 3a. If polarization transfer is from two chemically equivalent ^{19}F atoms to two ^1H atoms, as would be the case in INEPT transfer from the geminal fluorines to the H-2 methylene protons in 1,1-difluorohexane, then curves analogous to the CH_2 line in Figure 3b must be used to determine the optimum values for both d2 and d3.

An illustration of the utility of the $\{^{19}F\}^1H$-INEPT experiment in analyzing polymers can be seen in Figures 7a, 7b, and 7c, which depict ^1H NMR spectra of PCFE-IB obtained from a one-pulse ^1H experiment, a one-pulse ^1H experiment with ^{19}F decoupling, and an $\{^{19}F\}^1H$-INEPT experiment with ^{19}F decoupling during data acquisition, respectively. ^{19}F decoupling alone does not provide significant new information; nor does it provide substantial simplification of the ^1H spectrum (compare Figures 7a and 7b). However, the $\{^{19}F\}^1H$-INEPT spectrum obtained with delays optimized for $^3J_{HF}$ (Figure 7c) is considerably simplified. The resonances from methyl protons (1.3 ppm) and methylenes protons flanked by two IB repeat units (2.6 ppm) which are present in Figures 7a and 7b, have been filtered from the spectrum in Figure 7c. The detection of the resonances at ca. 4.5 ppm are especially facilitated by INEPT ^{19}F decoupling as seen in the spectrum in Figure 7c.

By performing "filtered" experiments in this way, an additional advantage is realized. Since only the resonances of interest (i.e. those which are in the vicinity of the labeled site or third NMR active nucleus) are detected, dynamic range problems are minimized. This can be especially advantageous if the fluorine containing sites represent a small fraction of the repeat units in the polymer, as would be the case if a fluorine-containing initiator species were used.

2D-NMR Experiments

2D-NMR experiments provide a tremendously useful means of dispersing resonances of complex molecules. $\{^{19}F\}^{13}C$ HETCOR experiments are particularly useful since they provide shift correlation maps relating the shifts of ^{19}F to the shifts of directly bound ^{13}C; the shifts of both nuclei are extremely sensitive to their chemical environments and therefore provide phenomenal dispersion of resonances when these parameters are used as the basis for correlation in a 2D-NMR experiment.

$\{^{19}F\}^{13}C$ Heteronuclear Shift Correlation Spectroscopy (HETCOR). Figure 8 shows the $\{^{19}F\}^{13}C$ HETCOR spectrum of PCFE-IB obtained with both ^1H and ^{19}F decoupling during the acquisition time. Resonances from approximately one dozen different stereosequences are clearly resolved despite the fact that the 1D- ^{19}F and ^{13}C spectra reveal very little information. This experiment involved direct detection

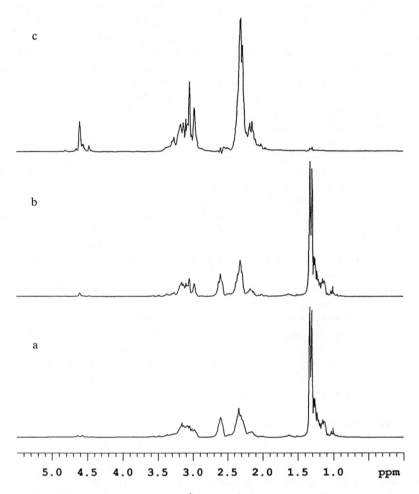

Figure 7. ^{1}H NMR spectra of poly(1-chloro-1-fluoroethylene-co-iso-butylene: a) normal 1D-spectrum, b) 1D-spectrum with broadband ^{19}F decoupling during the acquisition time, and c) {^{19}F}^{1}H-INEPT spectrum with ^{19}F decoupling during the acquisition time and with the polarization transfer delay tuned for $^2J_{HF}$ = 50 Hz.

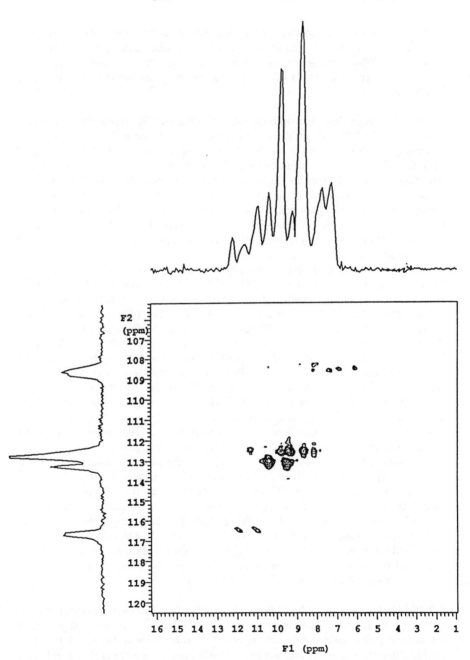

Figure 8. {¹⁹F}¹³C HETCOR spectrum of PCFE-IB with the 1D-¹⁹F spectrum plotted along the F1 axis and the 1D-¹³C spectrum of the C-F resonances plotted along the F2 axis.

of ^{13}C, and indirect measurement of the ^{19}F chemical shifts through the influence of the ^{19}F spins on the ^{13}C intensities during the evolution time (t_1) in a 2D-NMR experiment. While the information content is high in this type of 2D-NMR experiment, there is a better method of obtaining the same information with much higher sensitivity.

{^{13}C}^{19}F Heteronuclear Multiple Quantum Coherence Spectroscopy (HMQC). The {^{13}C}^{1}H-HMQC indirect detection experiment, depicted in Figure 9a, was first described in 1983 (11) as a means of providing the same information obtained from the HETCOR experiments, but with the sensitivity advantages associated with detection of the ^{1}H nucleus. In theory, sensitivity gains proportional to $(\upsilon^1_H/\upsilon^{13}_C)^3 =$ 64 are achieved compared to an experiment where ^{13}C resonances are measured directly in a one-pulse experiment. This corresponds to a 64-fold lower detection limit for small concentration species, or a 4000-fold reduction in the time to achieve the same signal-to-noise level (since signal-to-noise is proportional to the square root of the number of transient acquired).

Similar ideas work just as well when ^{19}F is substituted for ^{1}H; the resonance frequency of ^{19}F is almost as high as that of ^{1}H and its natural abundance is 100% (12). Little is lost in sensitivity and a tremendous increase in spectral dispersion is gained because the ^{19}F chemical shift range is so much larger than that of ^{1}H. Figure 10 shows a plot of the {^{13}C}^{19}F HMQC spectrum of PCFE-IB obtained with ^{1}H decoupling throughout the experiment and ^{13}C decoupling during the acquisition time; ^{13}C decoupling in the F1 dimension was accomplished with the appropriate use of 180^0 refocussing pulses during the evolution times. Note that only the downfield region of the ^{13}C (F1) dimension was contained within the spectral window as this region alone contains C-F resonances which will exhibit crosspeaks. Resonances from at least 40 separate C-F species are resolved. Many of these resonances were not observed in the ^{13}C-detected HETCOR spectrum in Figure 8 because they are present in low concentrations, and are below the ^{13}C detection limit.

Additionally, this experiment provides useful structure information about direct attachments of C-F atoms. Three groups of resonances associated with the three groups of peaks in the ^{13}C spectrum (arising from IEI, EEI/IEE, and EEE monomer sequences) are resolved (13). If we consider the group of resonances from EEE sequences, these can be further divided into three subgroups arising from EEE-centered pentad sequences with two E end-units, one I and one E end-unit, or two I end-units as illustrated in the Figure 10. Within each of these subgroups, additional fine structure is resolved which arises from the different permutations of stereosequences which are possible.

{^{13}C}^{19}F Heteronuclear Multiple Bond Correlation Spectroscopy (HMBC). HMBC (14) is an ^{1}H detected experiment related to HMQC, which was originally devised for obtaining correlations between ^{1}H and ^{13}C when there is two- or three-bond coupling between nuclei. A diagram of the HMBC pulse sequence is shown in Figure 9b; the experiment is performed with $\Delta = 1/(2*^1J_{CF})$ in order to suppress HMQC-type crosspeaks, and $\tau = 1/(2*^nJ_{CF})$ to optimize crosspeak intensities resulting from n-bond couplings. When {^{13}C}^{1}H-HMBC experiments are used, the ranges of

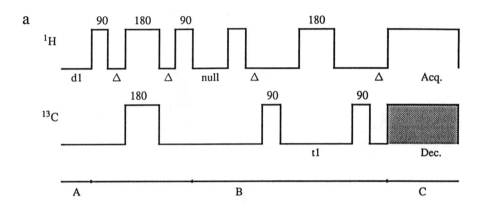

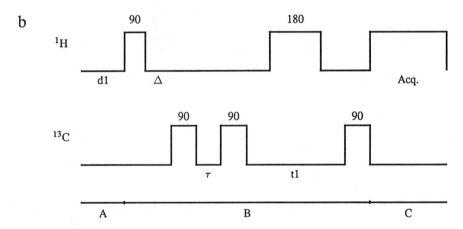

Figure 9. Pulse sequence diagrams for a) HMQC and b) HMBC 2D-NMR experiments; Δ is set to $1/(2*^1J_{CH})$ and τ is normally set to $1/(2*^nJ_{CH})$ where $^nJ_{CH}$ is the coupling range for which cross-peaks are desired.

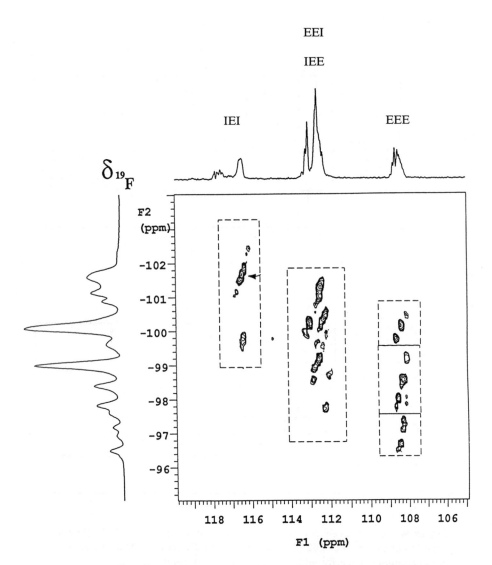

Figure 10. ^{19}F-$\{^{13}$C$\}$ HMQC spectrum of poly(1-chloro-1-fluoroethyl-ene-co-isobutylene with both ^{1}H and ^{13}C decoupling during data acquisition; tentative resonance assignments for some of the pentad sequences are labeled.

geminal and vicinal couplings overlap substantially, and are relatively small. It is therefore difficult in many instances to distinguish between cross peaks which arise from two-bond couplings and those which arise from three-bond couplings. Furthermore, when the experimental linewidths are comparable or larger than the magnitude of the coupling, rapid relaxation eliminates crosspeak intensities. Due to the nature of multiple bond couplings between ^{19}F and ^{13}C these issues are less of a problem in $\{^{13}C\}^{19}F$-HMBC: $^2J^{CF}$ is usually in the range of 30-50 Hz; $^3J_{CF}$ is usually in the range of 5-20 Hz; and $^4J_{CF}$ is usually in the range of 0-5 Hz (usually, $^4J_{CH}$ is not resolvable). This provides some interesting possibilities in terms of experiments which can be performed to gain additional structural information.

The first consequence of the larger couplings between ^{19}F and ^{13}C is that HMBC can be useful for studying the structures of high molecular weight polymers, even when the $\{^{13}C\}^{1}H$ HMBC are not effective. Ordinarily, the utility of $\{^{13}C\}^{1}H$ HMBC experiments for studying high molecular weight species is limited because the resonance linewidths are larger than the multiple bond J-couplings. Under these circumstances, relaxation occurs before information can transfer between spins. The much larger J_{CF} couplings make it possible to study larger molecules with broader resonance linewidths.

The relatively well defined ranges of two-, three-, and four-bond ^{19}F-^{13}C couplings also means that it is possible to perform several HMBC experiments, each with delays optimized to produce spectra with crosspeaks arising from a specific range of couplings. In this manner, a higher level of spectral simplification can be achieved, and better discrimination between correlations from two- and three-bond connectivities can be achieved. A schematic illustration of the information attainable from HMQC and HMBC spectra is depicted in Scheme 1. When the individual bits of information from these experiments are fit together like the pieces of a puzzle, a total picture of a fluoropolymer's structure can be obtained.

It is important not to have the mistaken notion that the peak intensity can be directly related to the magnitude of J-coupling in any given spectrum. The overall behavior of the peak intensity over a series of spectra provides a better indication of the approximate range of coupling. The HMBC crosspeak intensities follow a sinusoidal dependence on (τ/J_{CF}); a maximum signal is achieved when $\tau = 1/(2*J_{CF})$, however when longer t delays are used to detect crosspeaks from weaker couplings, crosspeaks arising from larger couplings may also reappear as t achieves values of $n/(2*J_{CF})$ where n is an odd integer.

The $\{^{13}C\}^{19}F$-HMBC spectra obtained from 1-fluorohexane obtained with delays optimized for 20, 5, and 1 Hz J_{CF} values are shown in Figures 11a, 11b, and 11c, respectively. In Figure 11a, an intense crosspeak is observed at F1 = 31 ppm from C-2 which is two bonds from ^{19}F. A weaker crosspeak at F1 = 25 ppm is also observed from C-3, but in this spectrum, the delays are clearly optimized for detection of correlations which arise from two-bond couplings.

The spectrum in Figure 11b was obtained with delays optimized for detection of correlations arising from three-bond (5 Hz) couplings. The crosspeak at F1 = 25 ppm which arises from three-bond C-F coupling to C-3 is considerably more intense than it is in Figure 11a. The crosspeak at F1 = 35 ppm from two-bond coupling is much weeker than it is in Figure 11a. In this molecule, the four-bond C-F couplings

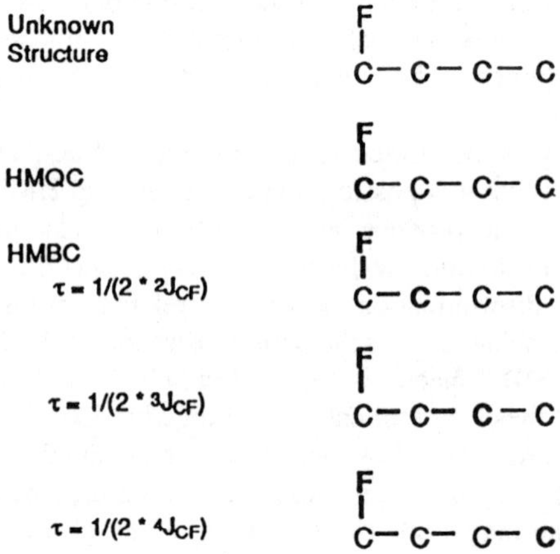

Unknown
Structure

$$F$$
$$|$$
$$C-C-C-C$$

HMQC

$$F$$
$$|$$
$$C-C-C-C$$

HMBC
$\tau = 1/(2 * {}^2J_{CF})$

$$F$$
$$|$$
$$C-C-C-C$$

$\tau = 1/(2 * {}^3J_{CF})$

$$F$$
$$|$$
$$C-C-C-C$$

$\tau = 1/(2 * {}^4J_{CF})$

$$F$$
$$|$$
$$C-C-C-C$$

Scheme 1.   Structure fragments identifiable from HMQC and HMBC experiments.

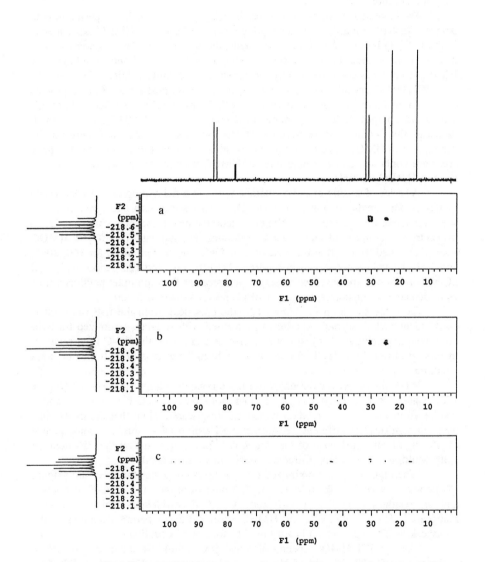

Figure 11. ^{19}F-$\{^{13}$C$\}$ HMBC spectrum of 1-fluorohexane with the τ delay optimized for a) $^2J_{CF}$ = 20 Hz, b) $^3J_{CH}$ = 5 Hz, and c) $^4J_{CF}$ = 1 Hz.

are too small to produce detectable crosspeaks when the delay is adjusted for a 1 Hz coupling (Figure 11c).

We have had considerable success in using $\{^{13}C\}^{19}F$-HMQC experiment with delays adjusted for multiple-bond coupling constants instead of HMBC experiments. In the $\{^{13}C\}^{1}H$-indirect detection experiments, the spectra are relatively complex and it is extremely desirable to suppress correlations from one-bond couplings; the HMBC technique was specifically designed to accomplish this. Because the $\{^{13}C\}^{19}F$-HMBC spectra are relatively simple and crosspeaks from C-F fragments occur in a distinct region of the spectrum, HMQC spectra usually suffice and provide several advantages. It is not possible to decouple ^{13}C in HMBC spectra, which increases their complexity relative the HMQC spectra. The ability to perform ^{13}C decoupling in HMQC spectra also provides improved signal-to-noise since the peak areas are contained under single lines in HMQC spectra rather than under several lines in HMBC multiplets.

A series of spectra must be obtained with the Δ delay progressively increased to accomodate smaller coupling constant. Once the resonances are assigned from the spectrum tuned for the largest coupling constant, the new resonances can be assigned in spectrum obtained with delays tuned for the smaller coupling. The $\{^{13}C\}^{19}F$-HMQC spectra obtained from 1-fluorohexane obtained with delays optimized for 160, 20, 5, and 1 Hz J_{CF} values are shown in Figures 12a, 12b, 12c, and 12d, respectively. All of the correlations are observed as single peaks rather than multiplet patterns which were detected in the corresponding HMBC spectra shown in Figure 11.

From the spectrum in Figure 12a, the crosspeak from the C-F fragment is easily identified. A second experiment performed with Δ delays optimized for two-bond couplings (Figure 12b) shows a clear crosspeak from the ^{13}C-C-F correlation; in this spectrum the delay is fortuitously set to null the crosspeak from one-bond coupling.

When the delays are optimized for three-bond couplings the spectrum (Figure 12c) contains a new crosspeak at 25 ppm from the ^{13}C-C-C-F correlation and a residue from the ^{13}C-F correlation which has reappeared. Note that the crosspeaks occur at significantly different ^{19}F shift in the F2 dimension; this is a consequence of the significant influence of the one-bond ^{13}C isotope shift on the ^{19}F chemical shift. Multiple bond isotope shifts are considerably attenuated.

When the delays are optimized for four-bond couplings the spectrum (Figure 12d) contains a very weak crosspeak at 22.5 ppm from the ^{13}C-C-C-C-F correlation and the residues from the ^{13}C-C-F and ^{13}C-C-C-F correlations which have both reappeared. The sensitivity in the HMBC spectra did not permit detection of the crosspeak at 22.5 ppm which is attributed to four-bond coupling.

The $\{^{13}C\}^{19}F$-HMQC spectra obtained from PCFE-IB obtained with delays optimized for 50, 33, 26, and 15 Hz J_{CF} values are shown in Figures 13a, 13b, 13c, and 13d, respectively. While these spectra are too complex to interpret here, it is evident that crosspeaks which have significant intensity in Figure 13a arise from larger two-bond coupling as well as residual peaks from one-bond coupling. Many of these signals have greatly diminished intensity in Figures 13b-d, while the crosspeaks arising from three-bond couplings increase in intensity. The crosspeaks at F1 = 105-120 ppm are not reappearing peaks from one-bond coupling, they arise

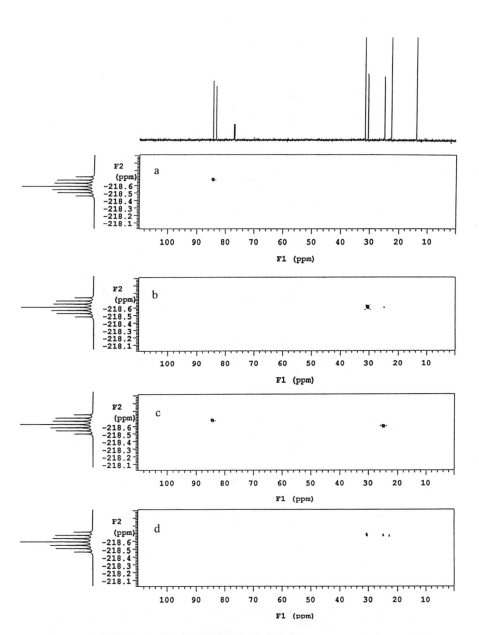

Figure 12. ^{19}F-$\{^{13}C\}$ HMQC spectra of 1-fluorohexane obtained with both ^1H and ^{13}C decoupling during data acquisition, and with Δ delays optimized for; a) $^1J_{CF} = 160$ Hz, b) $^2J_{CF} = 20$ Hz, c) $^3J_{CH} = 5$ Hz, and d) $^4J_{CF} = 1$ Hz.

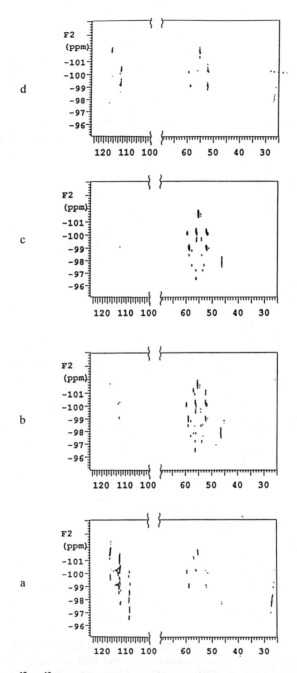

Figure 13. ^{19}F-$\{^{13}C\}$ HMQC spectra of poly(1-chloro-1-fluoroethyl-ene-co-isobutylene obtained with both ^1H and ^{13}C decoupling during data acquisition, and with Δ delays optimized for; a) $^2J_{CF} = 50$ Hz, b) $^3J_{CF} = 33$ Hz, c) ^3J_C, and d) $^3J_{CH} = 15$ Hz.

from three-bond coupling between ^{19}F and ^{13}C-F on a neighboring CFE repeat unit (i.e. F-C-C-^{13}C-F fragment).

Combined Use of 1D-Filtering and 2D-NMR Experiments

The 1D-NMR methods described above provide a useful method of obtaining spectral simplification; and the 2D-techniques provide spectral simplification as well as connectivity information. The combination of the two techniques can be an extremely powerful tool for obtaining spectral simplification while providing additional information about the fragments present in a polymeric structure. While there are many permutations of 1D- and 2D-NMR experiments which can provide spectral simplification and structural information, we have found combination of $\{^{19}F\}^1H$-INEPT and HMQC or HMBC to be extremely useful.

$\{^{19}F\}^1H$-INEPT/$\{^{13}C\}^1H$-HMQC and $\{^{19}F\}^1H$-INEPT/$\{^{13}C\}^1H$-HMBC. The $\{^{19}F\}^1H$-INEPT/$\{^{13}C\}^1H$-HMQC experiment (pulse sequence shown in Figure 14a) is similar to the $\{^1H\}^{29}Si$-INEPT/$\{^{13}C\}^{29}Si$-HMQC experiment reported by Berger (15). The $\{^{19}F\}^1H$-INEPT part of the experiment provides spectral simplification as well as some information about atomic connectivities. The spectral simplification can be extremely useful if the fluorine content of the polymer is low, since much more intense interfering resonances from the non fluorine-containing parts of the molecule are removed from the spectrum. The latter advantage can provide supplementary structural information about fluorine-proton proximities. The HMQC/HMBC part of the experiment provides valuable information regarding proton-carbon connectivities.

The structural fragments which can be identified from each series of experiments are outlined in Scheme 2. The $\{^{19}F\}^1H$-INEPT/$\{^{13}C\}^1H$-HMQC experiment can be performed in several different ways. In all cases, the HMQC Δ_3 delay is optimized for ca. 125-150 Hz one-bond C-H coupling; if the Δ_1 delay is optimized for a 50 Hz two-bond H-F coupling, then correlations are observed which identify all of the H-C-F fragments in the structure. If the Δ_1 delay is optimized for a 5-20 Hz three-bond H-F coupling, then cross peaks are observed in the 2D-spectrum which identify all of the C-H fragments which are separated by one bond from the C-F fragments. Similarly, if four-bond H-F coupling is resolved, the experiment could be repeated with a longer Δ_1 delay to identify the C-H fragments which are separated from the C-F fragment by two bonds.

A sample of the data obtained is shown in Figure 15. The standard $\{^{13}C\}^1H$-HMQC of 1-fluorohexane is shown in Figure 15a. Crosspeaks are observed which correlate the resonances from six nonequivalent carbon atoms with the resonances of directly bound protons. The $\{^{19}F\}^1H$-INEPT/$\{^{13}C\}^1H$-HMQC spectrum, obtained with Δ_1 delays set to perform INEPT polarization transfer from F to H separated by two bonds, is shown in Figure 15b. Only correlations between resonances of H and C atoms on the terminal methylene carbon are observed; the remaining resonances which were detected in Figure 15a are filtered from the spectrum in Figure 15b.

In a similar manner, $\{^{19}F\}^1H$-INEPT/$\{^{13}C\}^1H$-HMBC can be performed with a number of permutations of delays. Some of these permutations provide redundant information which is available from the $\{^{19}F\}^1H$-INEPT/$\{^{13}C\}^1H$-HMQC experiments

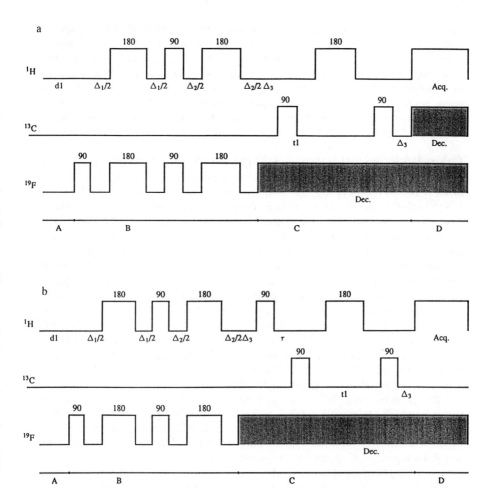

Figure 14. Pulse sequence diagrams for a) $\{^{19}F\}^1H$-INEPT/$\{^{13}C\}^1H$-HMQC and b) $\{^{19}F\}^1H$-INEPT/$\{^{13}C\}^1H$-HMBC; Δ_1 and Δ_2 are set based on $^nJ_{HF}$ as discussed in the section on $\{^{19}F\}^1H$-INEPT above; $\Delta_3 = 1/(2*^1J_{CH})$, and $\tau = 1/(2*^nJ_{CH})$ as described in the text.

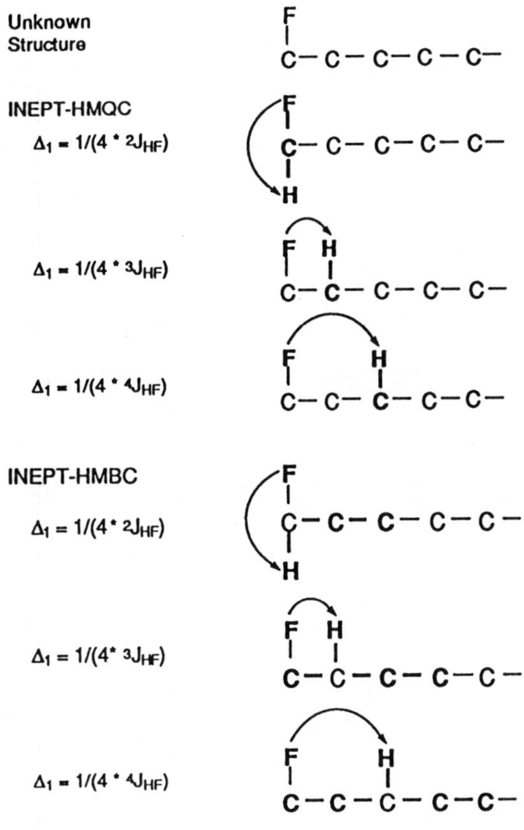

Scheme 2. Structure fragments identifiable from INEPT-HMQC and INEPT-HMBC experiments.

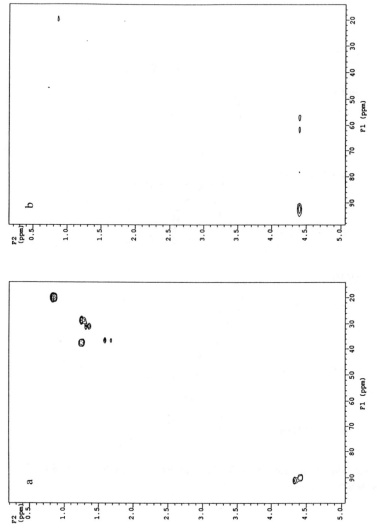

Figure 15. 2D-NMR spectra of 1-fluorohexane a) standard $\{^{13}C\}^1H$-HMQC spectrum obtained with $\Delta_3 = 1/(2*^1J_{CH})$ and ^{13}C decoupling during data acquisition; b) $\{^{19}F\}^1H$-INEPT/$\{^{13}C\}^1H$-HMQC spectrum obtained with $\Delta_1 = 1/(4*^2J_{HF})$, $\Delta_2 = 1/(2*^2J_{HF})$, $\Delta_3 = 1/(2*^1J_{CH})$ and ^{19}F decoupling during data acquisition.

described above. However, it is often useful to have this redundant information as reassurance that spectral artifacts are not being interpreted as real peaks. The information available from $\{^{19}F\}^1H$-INEPT/$\{^{13}C\}^1H$-HMBC experiments is illustrated in Scheme 2. In all cases, the spectra are obtained with the delay $\tau = 1/(2*^nJ_{CH})$ optimized for two- and three-bond C-H couplings (usually assuming ca. $^nJ_{CH} = 5$ Hz). As with $\{^{19}F\}^1H$-INEPT/$\{^{13}C\}^1H$-HMQC, the Δ_1 delay can be optimized for two-, three-, and in some cases four-bond H-F coupling. The HMBC part of the experiment then provides crosspeaks which identify carbons which are two or three bonds from the proton to which the fluorine magnetization has been transferred. As with the $\{^{19}F\}^1H$-INEPT/$\{^{13}C\}^1H$-HMQC the spectra obtained with longer delays (which are optimized for longer range couplings) will be contaminated with crosspeaks arising from larger, short range couplings. It is necessary to make assignments in the simpler spectra which are obtained with Δ_1 optimized for $^2J_{HF}$ before attempting to interpret spectra which were obtained with delays optimized for longer range couplings.

Summary

In this manuscript several examples of the spectral simplification and structural information which can be obtained using $^1H/^{19}F/^{13}C$ triple resonance and multidimensional NMR techniques have been provided. In the examples described, the ^{19}F chemical shift range is relatively narrow. Other fluoropolymers may contain ^{19}F resonances which are dispersed over more than 50 kHz. In those instances the pulse power on commercial instruments will be inadequate to flip all of the ^{19}F spins, and several spectra must be obtained which contain small portions of the ^{19}F chemical shift range.

While $^1H/^{19}F/^{13}C$ triple resonance studies of fluoropolymers have been described here, these techniques could just as easily have been used with a large number of other permutations of nuclei. For example, $^1H/^{31}P/^{13}C$ triple resonance NMR experiments could be useful for studying polymer additives and/or phosphorus-containing initiators. Elimination of the polymer background resonances might provide a means of studying the structures of byproducts from these materials in situ. Incorporation of ^2H at specific sites in a polymer (either by using deuterated initiator species or by deuteration of one monomer in a copolymer) would provide a means of studying polymer structures formed in unique chemical events which occur during a polymerization process (16). The selectivity achieved can also reduce dynamic range problems which make it difficult to detect resonances from low-concentration species in the presence of components which are present in much higher concentrations.

There are many different permutations of experiments which can be used to obtain information of the sort which has been extractable from the spectra shown here. The experiments described here are not necessarily the easiest or the best to perform, they are the ones which we have chosen to implement first in our laboratory. As additional experiments are implemented, it may be found that they provide better dispersion, filtering, and/or sensitivity. Clearly, there is much work to be done in this area. Hopefully, we have shown the potential of triple resonance NMR techniques for studying problems in polymer chemistry.

Acknowledgements

We wish to thank the National Science Foundation (DMR-9310642) for financial support of this work. We also wish to thank the Kresge Foundation and the donors to the Kresge Challenge Grant Program for funds used to purchase the 600 MHz NMR instrument used in this work.

Literature Cited

1.   a) Croasmun, W. R.; Carlson, R. M. K., Two-Dimensional Nuclear Magnetic Resonance Spectroscopy: Applications for Chemists and Biochemists, VCH Publishers, NY, 1994, b) Roberts, G. C. K., NMR of Macromolecules, A Practical Approach, IRL Press, NY, 1993, c) Bertini, J.; Molinari, H.; Niccolai, N., NMR and Biomolecular Structure, VCH Publishers, NY, 1991, d) Wagner, G., *Progress in NMR Spectroscopy*, **1990**, *22*, 101, e) Clore, G. M.; Gronenborn, A. M., *Crit. Rev. Biochem. Mol. Biol.*, **1989**, *24*, 479, Martin, G. E.; Zektzer, A. S., Two-Dimensional NMR Methods for Establishing Molecular Connectivity, VCH Publishers, N.Y., **1988**.
2.   a) Griffey, R. H.; Redfield, A. G., *Quart. Rev. Biochem.*, **1987**, *19*, 51, b) Clore, G. M.; Gronenborn, A. M., *Progress in NMR Spectroscopy*, **1991**, *23*, 43, c) Lee, J.; Feizo, J.; Wagner, G., *J. Magn. Reson. A*, **1993**, *102*, 322.
3.   Koenig, J., Spectroscopy of Polymers, Ch. 7, American Chemical Society, Washingtion, D.C., **1992**.
4.   Bax, A., J. Magn. Resonance, **1983**, *53*, 517.
5.   Freeman, R.; Morris, G. A., *J. Chem. Soc., Chem. Commun.*, **1978**, 684.
6.   Kessler, H.; Griesinger, C.; Zarbock, J.; Loosli, H. R., *J. Magn. Resonance*, **1984**, *57*, 331.
7.   Reynolds, W. F.; Hughes, D. W.; Perpick-Dumont, M.; Enriquez, R. G., *J. Magn. Resonance*, **1985**, *63*, 413.
8.   Morris, G. A., *J. Am. Chem. Soc.*, **1980**, *102*, 428.
9.   Doddrell, D. M.; Pegg, D. T.; Bendall, M. R., *J. Magn. Resonance*, **1982**, *48*, 323.
10.   a) Rinaldi, P. L.; Baldwin, N. J. *J. Am. Chem. Soc.*, **1982**, *104*, 5791, b) Ibid, *J. Am. Chem. Soc.*, **1983**, *105*, 7523.
11.   Bax, A.; Griffey, R. H.; Hawkins, B. L., *J. Magn. Resonance*, **1983**, *55*, 301.
12.   Ray, D. G.; Li, L.; Wang, H. T.; Harwood, H. J.; Rinaldi, P. L., *Makromol. Chem, Macromol. Symp.*, **1994**, *15*, 86.
13.   Ray, D. G.; Li, L.; Hatvany, G. S.; Wang, H. T.; Harwood, H. J.; Rinaldi, P. L., *Macromolecules*, submitted.
14.   Bax, A; Summers, M. F., *J. Am. Chem. Soc.*, **1986**, *108*, 2093.
15.   Berger, S., *J. Magn. Reson. A*, **1993**, *101*, 329.
16.   Johnston, J. A.; Rinaldi, P. L.; Farona, M. F., *J. Mol. Cat.*, **1992**, *76*, 209.

RECEIVED February 23, 1995

Chapter 14

Conformational Disorder and Its Dynamics Within the Crystalline Phase of the Form II Polymorph of Isotactic Poly(1-butene)

Haskell W. Beckham[1], Klaus Schmidt-Rohr[2,3], and H. W. Spiess[2]

[1]School of Textile and Fiber Engineering, Georgia Institute of Technology, Atlanta, GA 30332−0295
[2]Max-Planck-Institut für Polymerforschung, Postfach 3148, D−55021 Mainz, Germany

^{13}C 2D exchange NMR under conditions of magic-angle-spinning has revealed that the metastable form II polymorph of isotactic poly(1-butene) exhibits significant slow conformational exchange (in the milliseconds to seconds time regime) above the amorphous-phase glass transition. Comparison with spectra of the form I polymorph proves that this motion occurs within the crystalline regions. The observed dynamic conformational disorder within the crystalline regions of this semicrystalline polymer is characteristic of what some call a macromolecular *condis* crystal. The presence of motion within the crystalline regions is consistent with previous calculations of the conformational flexibility of this polymer.

Dynamic conformational disorder has recently been shown to accurately describe the chain motions within amorphous polymers above their glass transitions (T_g) (*1,2*). The disordered nature of these materials provides a wide distribution of intermolecular environments, and therefore potentials, which manifests itself by a distribution of correlation times for the molecular motions activated above T_g. The motions themselves have been analyzed experimentally in terms of both conformational transitions (e.g., *trans* ↔ *gauche*), and rotational motions of individual segments resulting from various changes in the torsional angles. While amorphous-phase motions are relatively nondescript, the ordered nature of crystalline arrays typically restricts the motions to more discrete processes. Due to bonding and symmetry arrangements, the energetically allowed conformational movements of segments within such an environment are fewer than those allowed in a disordered amorphous region. Note, nevertheless, that large-angle reorientations of chains about their long axes are possible if the atomic positions are occupied by equivalent units before and after the reorientation.

The presence of molecular motions within crystalline regions of semicrystalline polymers has been documented for some time (*3*). They have been attributed to the presence of dynamic mechanical α relaxations in semicrystalline polymers (*4*) and correlated with such bulk physical properties as solid-state

[3]Current address: Department of Polymer Science and Engineering, University of Massachusetts, Amherst, MA 01003

0097−6156/95/0598−0243$12.00/0

extrudability and temperature-utility range (5). However, the detailed nature of some of these motions has only been revealed recently through the application of solid-state NMR techniques. A variety of molecular motions are being fully characterized and catalogued in detail in the scientific literature (6,7).

Within the crystalline lamellae of semicrystalline polymers, one such motion proposed for some time is the helical jump (8,9) This involves the rotation of the chain about the helix axis by $360°·n/m$ where m is the number of repeat units per n turns of the helix, accompanied by translation of the chain by one monomer unit. Thus, for a single monomer unit, the relative position to the surrounding polymer chains within the crystallite is the same before and after the jump. For the following semicrystalline polymers exhibiting α relaxations, helical jump motions have been shown to occur at temperatures between the glass transition and melting transitions: (6) polyoxymethylene (10), isotactic polypropylene (11), polyethylene (12), and polyoxyethylene (13). Molecular model calculations indicate the mechanism involves a "defect" diffusion of the helix rotation through the crystal lattice, as opposed to a rigid-rod rotation of the crystal stem (although this latter process becomes more important as crystal thickness decreases) (14,15). Other types of crystalline-phase motions include conformational changes (16), functional group jumps, phenylene ring flips, and librations (17).

For each of the polymers shown to exhibit helical jump motions (POM, iPP, PE, PEO), the crystalline structure provides a relatively well-defined conformational energy landscape through which the chain stems jump between successive energy minima (12). None of these polymers have flexible side groups. The presence of side-group mobility might "smear out" a conformational energy map thereby making any main-chain motion less discrete. For longer side groups, a type of liquid-crystal-like translation/rotation might be expected, similar to those shown for discotic liquid crystals with flexible side chains (18,19). Another alternative is that the side chains would interdigitate in such a fashion that any main-chain motion at all would be prevented. These possibilities are being explored by the investigation of the chain dynamics of isotactic poly(1-butene) (PB), which has an ethyl side group, and isotactic poly(4-methyl-1-pentene), which consists of isobutyl side groups. The results for PB are presented here.

Polymorphism is quite common among the semicrystalline polyolefins. When processed from the melt, isotactic poly(1-butene) forms a metastable crystalline structure (form II, 11_3 helix) (20) that transforms over the course of days to a more stable structure (form I, 3_1 helix) (21). This conversion is greatly accelerated by the application of external stresses (22,23). Solid-state CP/MAS NMR studies on PB have been conducted (24) and revealed the isotropic chemical shifts to be dependent on the chain conformation of the respective helical microstructure: forms I and II, along with a form III (4_1 helix). While form III is prepared by crystallization from solution and therefore is not as commercially significant (25), forms I and II are well-known by melt processers of PB. Wide-angle X-ray diffractograms are shown in Figure 1 and confirm the existence of the different respective crystalline structures and high crystalline content for both forms: form I is hexagonal, form II is tetragonal. Dynamic mechanical spectra show no indication of motional processes above the glass transition (T_g = -20 °C) for either form (26). However, the crystalline transformation can only occur through some type of chain movement.

Experimental

Isotactic poly(1-butene) of high molecular weight was obtained from Aldrich. Differential scanning calorimetry (DSC) with a Mettler DSC 30 (10 °C/min) indicated T_m = 130 °C for PB (form I). After heating this sample to 200 °C and

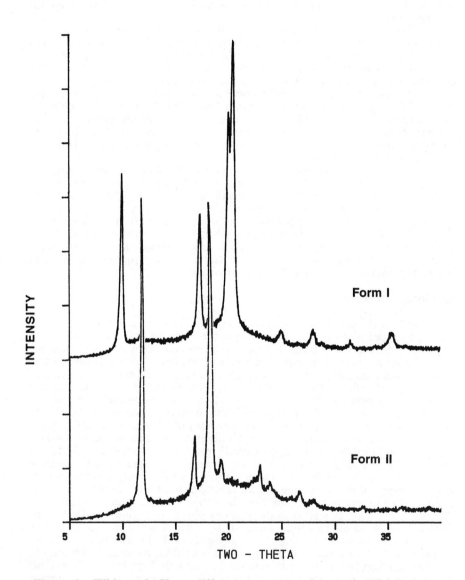

Figure 1. Wide-angle X-ray diffraction patterns of isotactic PB. Bottom diffractogram was taken one hour after cooling from the melt; top diffractogram is the same sample after annealing at room temperature for two weeks.

holding for five minutes before cooling, the subsequent DSC scan revealed a melting peak at 115 °C indicating production of the form II (27). For both forms the glass transition was detected at -23 °C.

Wide-angle X-ray diffractograms of PB melt-cast films were obtained with a Siemens D-500 diffractometer operated in reflection mode. The initial production of form II from the melt was confirmed. At ambient temperature with no applied stress, the transformation to form I was nearly complete after about 10 days. The crystallinity was similar for both forms: 40% in form II and 50% in form I.

The as-received polymer pellets were prepared for NMR studies by melting at 150 °C in a cylinder-shaped mold (5.5 mm diameter) followed by slow cooling to room temperature. For the PB, form II was generated by simply melting the solid polymer plug inside the rotor. The transformation from form II to form I is definitely accelerated by the applied stress of magic-angle spinning (MAS), the consequences of which are discussed below.

NMR measurements were conducted on a Bruker MSL 300 spectrometer (7 Tesla). Proton and carbon 90° pulses of 3.8 – 4.2 μs were used. The ^{13}C 2D exchange NMR experiment has been described previously (6,28,29,30). With magic-angle spinning (31), changes in isotropic chemical shifts, reflecting conformational changes of the polymer, can be detected as off-diagonal exchange intensity. Without magic-angle spinning, exchange due to isotropic chemical shift changes is superposed with exchange intensity due to reorientation of anisotropic chemical shift tensors. All experimental parameters were optimized and kept constant for the same polymer at different temperatures.

Results and Discussion

In Figure 2 are displayed the solid-state ^{13}C CP/MAS spectra of PB forms I and II. The well-defined peaks of form I can be identified with the chemical structure; the splitting of the methylene peaks indicates some nonequivalent solid-state packing due to the arrangement of the chains within the crystallites. The crystalline structure has hexagonal symmetry, with neighboring left- and right-handed helices which may exist either as isoclined or anticlined pairs in equal probability (32). The resulting two local environments affect only the methylene groups to produce two peaks of equal intensity. As opposed to the α modification of isotactic polypropylene (33,34), this coexistence is not disorder, but rather the ability of either chain modification to fit into the crystal lattice. Most significant for the present study is the poorly resolved methylene and methine region of the form II spectrum as compared to the form I spectrum. The broadening may be related to dynamics, structure, or both. It cannot be structure-related alone because the broad peaks would signify a distribution of isotropic chemical shifts so large that an amorphous material might be expected. The WAXD diffractogram of Figure 1 confirms the crystalline nature of PB form II. That the broadening is related to dynamics is proven in Figure 3, which is the CP/MAS spectrum of the same sample of PB form II taken at 273 K. Some resolution is revealed upon cooling of the sample. Thus, the broadening of the room-temperature spectrum is due to some type of chain motions present in the PB form II. However, the breadths of the peaks compared to the form I spectrum indicates that the room-temperature broadening is related to both structure and dynamics. Instead of well-defined chemical shifts, the form II crystalline structure at 273 K appears to be characterized by some distribution of isotropic chemical shifts for each chemical site. The chemical shifts for both forms are listed in Table I, along with the chemical shifts reported for form III (24). For form II, the chemical shifts are reported as ranges. Because these ranges encompass the chemical-shift values of both the form I and form III, the molecular conformations of form II must also include those of form I and form III.

Form II

$$\text{---}\underset{\underset{CH_3}{\overset{\displaystyle |}{CH_2}}}{\overset{\displaystyle |}{CH}}CH_2\text{---}_n$$

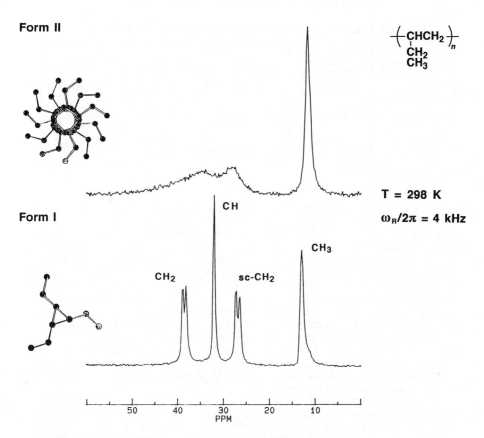

Form I

CH

CH₂   sc-CH₂   CH₃

T = 298 K

$\omega_R/2\pi = 4$ kHz

Figure 2. ^{13}C CP/MAS spectra of isotactic PB forms I and II at room temperature. The spinning speed is 4 kHz. Peaks are labeled for the form I ("sc" means side-chain). WAXD-determined helical structures are shown as projections along the helix axis.

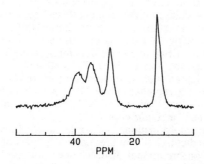

Figure 3. ^{13}C CP/MAS spectrum of isotactic PB form II at 273 K. Spinning speed is 4 kHz.

Table I. Isotropic Chemical Shifts of Isotactic Poly(1-butene)

polymorph	CH_2	CH	sc-CH_2[b]	CH_3
form I	38.9, 38.2	32.1	27.3, 26.5	13
form II	44 – 35.4	38.5 – 30.7	30.7 – 24.8	14.8 – 8.8
form III[a]	41.2	36.6	28.4	14.5, 13.7

[a]values adapted from those reported in ref. 24.
[b]"sc" means side-chain.

Since the form II obviously contains a variety of isotropic chemical shifts, as well as some degree of chain mobility, exchange among these chemical shifts is foreshadowed. This is proven in the ^{13}C 2D exchange NMR spectra of Figure 4 measured under conditions of magic-angle spinning. The methylene and methine regions of 2D exchange spectra for form II just above and below T_g are shown. Above T_g, the off-diagonal exchange intensity is evidence for conformational exchange with timescales in the milliseconds to seconds regime (the methyl region also shows the off-diagonal exchange signal). Upon lowering the temperature below T_g, these motions cease as indicated by all signal intensity confined to the diagonal. This diagonal spectrum contains not only the conformationally broadened peaks of form II, but also narrower peaks characteristic of form I. It is known that stress accelerates the transition from form II to form I in PB (35); this effect due to MAS has also been observed before. The ratio of the form I narrow peaks to form II broad peaks continuously increases under constant MAS and after about three days, the form I peaks dominate the spectrum.

The origin of the conformational exchange can be the crystalline phase, amorphous phase, or both. For the form II of PB, it is straightforward to establish the origin of the exchange by comparing the spectra with those of form I. The ^{13}C 2D exchange spectra of static samples of both forms are shown in Figure 5. By not doing the exchange experiment under MAS, any type of exchange, due to changes in isotropic chemical shifts or torsional reorientations, can be detected. The spectra of Figure 5 were measured at 258 K, the same temperature at which the conformational exchange was observed for the form II (Figure 4). Besides a small low-intensity signal under the maximum, the form I spectrum consists of a diagonal. Especially compared to the form II spectrum, no significant exchange intensity is observed. Thus, no significant molecular motions in the millisecond to second time regime are present, neither from the crystalline nor from the amorphous regions. Since the amorphous regions are practically identical in both forms with respect to content (see the X-ray diffractograms of Figure 1) and dynamics (same T_g), then the observed conformational exchange in form II above T_g is attributed to motions within the crystalline regions. Although such dynamic conformational disorder is well-known in amorphous polymers (or regions) above T_g, it has never before been observed within crystalline regions by 2D exchange NMR, and may in fact be the first such evidence of a macromolecular *condis* crystal (36).

For a mobile and conformationally disordered structure to exhibit a sharp X-ray pattern is not contradictory. There are many examples of such materials. The presence of dynamic conformational disorder does not preclude the existence of long-range positional order. For plastic crystals (37), long-range positional order is present despite orientational dynamic disorder. These materials certainly have relatively sharp X-ray reflections. Even with dynamic conformational disorder,

T = 258 K
t_mix = 500 msec

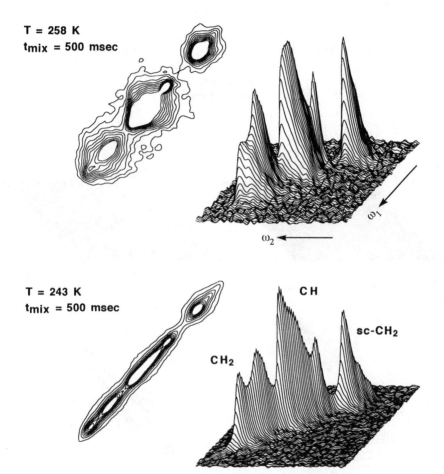

T = 243 K
t_mix = 500 msec

CH

sc-CH$_2$

CH$_2$

ω_1

ω_2

Figure 4. ^{13}C 2D MAS exchange spectra for isotactic PB form II (T$_g$ = 250 K): top, T$_g$ + 8 K; bottom, T$_g$ − 7 K. Only the methylene and methine regions of the spectra are shown. Partial conversion of form II to form I is observed in the spectra.

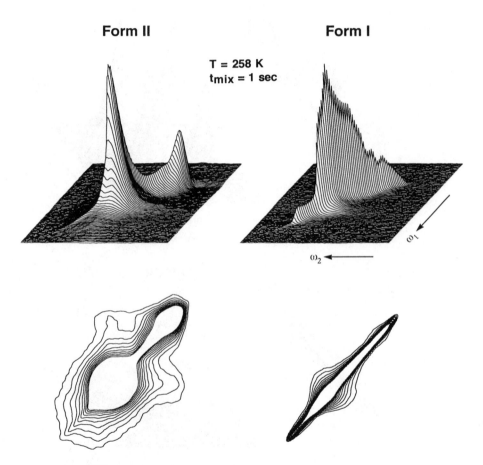

Figure 5. ^{13}C 2D exchange spectra of isotactic PB at T_g + 8 K: left, form II; right, form I. Contour plots are shown below their respective stacked plots.

long-range orientational as well as positional order may be preserved. The spatial autocorrelation function can retain a periodic part even in the presence of local disorder, and therefore its Fourier transform, the scattering pattern, can exhibit sharp peaks. The effects of molecular dynamics in X-ray diffraction studies are handled by a thermal-motion analysis (*38*), which provides structures with nuclei characterized by ellipsoids whose size represents the uncertainty in atomic positioning. A usual consequence of mobility-induced disorder is that the intensities of higher-order reflections are reduced. Comparison of the diffractograms of Figure 1 reveals that this is the case for form II in relation to form I.

Conformational energy maps have been computed for both forms of PB (*39*). For the form II, it has been shown that the potential energy minima describing the preferred conformations of the backbone are very flat and broad (*40,41*). Thus, the conformational flexibility exists. The frozen-in conformations of form II may be deduced from Figure 3 to cover isotropic chemical shift ranges of 8 – 9 ppm for the backbone carbons, and 5 – 6 ppm for the side-group carbons. These can be understood in terms of the semiempirical γ gauche effect (*42*), which has recently been quantitatively reproduced by ab initio calculations (*43*). Carbons whose γ substituents are in gauche conformations will experience an additional nuclear shielding of 4 – 5 ppm per γ gauche substituent. For a polymethylene chain, an 8 – 9 ppm spread would indicate that the two γ gauche substituents exist across the range from both trans to both gauche. Due to the presence of the ethyl side groups in PB, however, each backbone carbon has four γ substituents. The combination of shieldings from the four γ substituents easily provides the observed range in isotropic chemical shifts. The backbone γ substituents of a selected backbone carbon need not be both trans or both gauche. At the same time, the two side-chain carbons have only two γ gauche substituents each, which explains the smaller 4 – 5 ppm range of observed isotropic chemical shifts. The foregoing discussion also excludes side-group motion alone as the origin of the exchange, and side-group conformational disorder as the origin of the chemical-shift dispersion. The reason is that the chemical shifts of the backbone methine carbon would not be affected since there is no γ substituent for this carbon in the side group. The data of Figures 3 and 4 clearly indicate that the methine chemical shift is affected by the motion.

The transformation from the metastable form II to the stable form I can be followed spectroscopically as the form I peaks grow at the expense of the form II signal. It can also be followed with X-ray diffraction in which peaks attributed to both forms can be observed at times intermediate between the initial melting and the complete conversion. From the data presented here, some generalizations may be made regarding this transformation. The form II helix is in a constant state of motion in which the nuclei oscillate via rotations about torsion angles described by very flat potentials. As conformational space is explored via these motions, the torsional angles corresponding to the stable form I can be discovered. The energy barrier separating the form I from the form II is very small. It is perhaps instructive to point out that for screws of opposite handedness (representing helices with side chains), the closest packing density is achieved for tetragonal symmetry (*44*). Local preferred chain interactions and packing of side chains are not important for screws. When the PB solidifies from the melt, the production of form II (tetragonal) is obviously kinetically favored.

Another semicrystalline polymer shown to exhibit similar motional behavior is trans-1,4-polybutadiene (*45*). Using solid-state ^{13}C and ^{2}H NMR, the dynamics of the high-temperature polymorph of this polymer was characterized as conformational interconversions superposed with diffusive rotations. The presence of such motions in PB could result in a helix reversal, which would convert an isoclined helix into an anticlined helix. For the form I helix, such an arrangement is isoenergetic. For this form I helix, only the position of the methylene groups would be changed after the motion. As discussed above, the statistical placement of such helices within crystallites results in the observed splitting of the methylene peaks into two peaks of equal height. For the form II helix, however, a helix reversal would not leave all atoms in the same position thus providing a source of exchange signal as observed in Figure 5. The motion would proceed along the helix direction as a defect diffusion, analogous to the helical jump motions observed for other semicrystalline polymers without side groups, except that the motion in PB would be superposed on diffusive rotations. In fact, the presence of helical jump motions cannot be discounted; the disordered nature of the form II helix coupled with the presence of diffusive rotations, could very well lead to an exchange pattern as that shown in Figure 5 for helical jumps (6). Simulations and isotopic labeling studies will further resolve the motional mechanism.

Conclusions

Dynamic conformational disorder within the crystalline regions of the form II polymorph of isotactic poly(1-butene) has been proven to occur at temperatures above the glass transition ($T_g = 250$ °C). The crystalline-phase chain motions were directly revealed as off-diagonal exchange intensity in a 2D solid-state NMR exchange experiment under conditions of MAS. Comparison of form II spectra with those of form I containing equivalent amorphous regions (same T_g and WAXD-determined content) confirm the crystalline origin of the off-diagonal exchange signal. The existence of such crystalline chain mobility is undoubtedly related to the transformation of the metastable form II to stable form I of this semicrystalline polyolefin. The exact nature of the motion was postulated and is under further investigation.

Acknowledgment.   Support for HWB was provided by an NSF-NATO Postdoctoral Fellowship.

Literature Cited

1. Zemke, K.; Chmelka, B. F.; Schmidt-Rohr, K.; Spiess, H. W. *Macromolecules* **1991**, *24*, 6874.
2. Zemke, K.; Schmidt-Rohr, K.; Spiess, H. W. *Acta Polymer* **1994**, *45*, 148.
3. McCrum, N. G.; Read, B. E.; Williams, G. *Anelastic and Dielectric Effects in Polymeric Solids*; Dover: New York, 1991.
4. Boyd, R. H. *Polymer* **1985**, *26*, 323.
5. Aharoni, S. M.; Sibilia, J. P. *J. Appl. Polym. Sci.* **1979**, *23*, 133.
6. Schmidt-Rohr, K.; Spiess, H. W. *Multidimensional Solid-State NMR and Polymers*; Academic Press: London, 1994.
7. Schmidt-Rohr, K.; Kulik, A. S.; Beckham, H. W.; Ohlemacher, A.; Pawelzik, U.; Boeffel, C.; Spiess, H. W. *Macromolecules* **1994**, accepted.
8. Fröhlich, H. *Proc. Phys. Soc. Lond.* **1942**, *54*, 422.
9. Boyd, R. H. *Polymer* **1985**, *26*, 1123.
10. Kentgens, A. P. M.; de Boer, E.; Veeman, W. S. *J. Chem. Phys.* **1987**, *87(12)*, 6859.
11. Schaefer, D.; Spiess, H. W.; Suter, U. W.; Fleming, W. W. *Macromolecules* **1990**, *23*, 3431.
12. Schmidt-Rohr, K.; Spiess, H. W. *Macromolecules*, **1991**, *24*, 5288.
13. Spiess, H. W.; Schmidt-Rohr, K. *Polym. Prepr. (Am. Chem. Soc., Div. Polym. Chem.)* **1992**, *33*(1), 68.
14. Syi, J.-L.; Mansfield, M. L. *Polymer* **1988**, *29*, 987.
15. Rutledge, G. C.; Suter, U. W. *Macromolecules* **1992**, *25*, 1546.

16. Hirschinger, J.; Schaefer, D.; Spiess, H. W.; Lovinger, A. J. *Macromolecules* **1991**, *24*, 2428.
17. Hirschinger, J.; Miura, H.; Gardner, K. H.; English, A. D. *Macromolecules* **1990**, *23*, 2153.
18. Werth, M.; Vallerien, S. U.; Spiess, H. W. *Liq. Cryst.* **1991**, *10*, 759.
19. Leisen, J.; Werth, M.; Boeffel, C.; Spiess, H. W. *J. Chem. Phys.* **1992**, *97*, 3749.
20. Jones, A. T. *Polym. Lett.* **1963**, *1*, 455.
21. Jones, A. T. *Polymer* **1966**, *7*, 23.
22. Weynant, E.; Haudin, J. M.; G'Sell, C. *J. Mat. Sci.* **1982**, *17*, 1017.
23. Hong, K.-B.; Spruiell, J. E. *J. Appl. Polym. Sci.* **1985**, *30*, 3163.
24. Belfiore, L. A.; Schilling, F. C.; Tonelli, A. E.; Lovinger, A. J.; Bovey, F. A. *Macromolecules* **1984**, *17*, 2561.
25. Holland, V. F.; Miller, R. L. *J. Appl. Phys.* **1964**, *35*, 3241.
26. Goldbach, G. *Die Angew. Makromol. Chem.* **1973**, *29/30*, 213.
27. Kishore, K.; Vasanthakumari, R. *J. Macromol. Sci. – Chem.* **1987**, *A24(1)*, 33.
28. Szeverenyi, N.; Sullivan, M. J.; Maciel, G. E. *J. Magn. Reson.* **1982**, *47*, 462.
29. Hagemeyer, A.; Schmidt-Rohr, K.; Spiess, H. W. *Adv. Magn. Reson.* **1989**, *13*, 85.
30. Beckham, H. W.; Spiess, H. W. In *NMR - Basic Principles and Progress 32*; Blümich, B.; Kosfield, R., Eds.; Springer: Berlin, 1994.
31. Szeverenyi, N. M.; Bax, A.; Maciel, G. E. *J. Am. Chem. Soc.* **1983**, *105*, 2579.
32. Natta, G.; Corradini, P.; Bassi, I. W. *Del Nuova Cimento* **1960**, *15*, 52.
33. Corradini, P.; Giunchi, G.; Petraconne, V.; Pirozzi, B.; Vidal, H. M. *Gazz. Chim. Ital.* **1980**, *110*, 413.
34. Corradini, P.; Guerra, G. *Advances in Polymer Science 100*; Springer: Berlin, 1992; p 183.
35. Gohil, R.; Miles, M.; Petermann, J. *J. Macromol. Sci.–Phys.* **1982**, *B21(2)*, 189.
36. Wunderlich, B.; Möller, M.; Grebowicz, J.; Baur, H. *Advances in Polymer Science 87*; Springer: Berlin, 1988.
37. Sherwood, N., Ed. *The Plastically Crystalline State*; J. Wiley & Sons: Chichester, 1979.
38. Dunitz, J. D. *X-ray Analysis and the Structure of Organic Molecules*; Cornell University Press: Ithaca, 1979; p 244.
39. Corradini, P.; Napolitano, R.; Petraconne, V.; Pirozzi, B. *Eur. Polym. J.* **1984**, *20*, 931.
40. Petraccone, V.; Pirozzi, B.; Frasci, A.; Corradini, P. *Eur. Polym. J.* **1976**, *12*, 323.
41. Ajó, D.; Granozzi, G.; Zannetti, R. *Makromol.Chem.* **1977**, *178*, 2471.
42. Tonelli, A. E. *NMR Spectroscopy and Polymer Microstructure*; VCH: New York, 1989.
43. Born, R.; Spiess, H. W.; Kutzelnigg, W.; Fleischer, U.; Schindler, M. *Macromolecules* **1994**, *27*, 1500.
44. Wunderlich, B. *Macromolecular Physics, Vol. 1*; Academic: New York, 1973; p 86.
45. Möller, M. *Polym. Prepr. (Am. Chem. Soc., Div. Polym. Chem.)* **1987**, *28(2)*, 395.

RECEIVED February 2, 1995

Chapter 15

NMR Study of Penetrant Diffusion and Polymer Segmental Motion in Toluene–Polyisobutylene Solutions

Athinodoros Bandis[1], Paul T. Inglefield[2,3], Alan A. Jones[2], and Wen-Yang Wen[2]

[1]Department of Physics and [2]Carlson School of Chemistry, Clark University, Worcester, MA 01610

The self - diffusion coefficients of toluene in polyisobutylene (PIB) solutions were determined using the Pulsed Field Gradient Nuclear Magnetic Resonance technique. The volume fraction of toluene in the polymer was varied from 0.045 to 0.712 and the temperature was varied from 225 K to 368 K. The concentration dependence of the data was interpreted using both the Fujita and the Ventras-Duda free volume theories and the temperature dependence was interpreted with the WLF equation. These models describe separately the concentration and temperature dependencies of the toluene self - diffusion coefficients very well and the resulting free volume parameters are in a good agreement with the ones extracted from the analysis of viscosity data on the same system. Comparisons were made between the two free volume theories and the Fujita free volume parameters could be extracted from the Vrentas - Duda free volume parameters.
The spin - lattice relaxation times are determined for the methylene carbon of polyisobutylene (PIB), as well as for the ortho carbon of toluene in toluene - polyisobutylene solutions. The Hall - Helfand correlation function combined with restricted anisotropic rotational diffusion was used to treat the T_1 data of the methylene carbon of PIB to characterize the polymer segmental motion. A simple exponential correlation function was used to describe the local motion of toluene in the solutions which falls in the extreme narrowing limit for the solutions studied. Both models described satisfactorily the temperature and field dependence of the spin - lattice relation times. From the temperature dependence of the correlation times for the polymer segmental motion, the free volume of the solution at each concentration is extracted and compared with the values obtained from the diffusion of the toluene penetrant. The free volume values extracted from the T_1 data for the methylene carbon of PIB and the self - diffusion data for the toluene were found to be in substantial agreement. The inter-relationship of the timescale of segmental motion of the polymer and the translational diffusion of the toluene was also

[3]Corresponding author

examined and it was found that the two types of motion seem to be correlated in high polymer concentration solutions. The toluene reorientational motion was found to be much faster than both the polymer segmental motion and the toluene translational diffusion.

The diffusion of small molecules into polymers has been the subject of many investigations, because of its importance in industrial, medical and biological applications. Some examples are the membrane separation of gases, controlled drug delivery, control of polymerization, production of barrier materials for packaging and chemical defense. Diffusion of penetrants in polyisobutylene is of special interest since this material displays low permeability for a polymeric rubber well above the glass transition. This unusual property of polyisobutylene has been considered in computer simulations[1] of the rubber and in a detailed characterization of segmental motion[2] in the pure rubber by Nuclear Magnetic Resonance (NMR).

There are a variety of techniques for measuring the self - diffusion coefficients of a penetrant in a polymer matrix. These include sorption, permeation, light scattering, radioactive tracing, forced Rayleigh scattering and NMR. The Pulsed Field Gradient NMR Technique is valuable because it does not perturb the system. With this technique a direct measure of the self - diffusion coefficient of the penetrant is achieved by observing the molecules microscopically, while other methods (sorption method for example) indirectly determine the self - diffusion coefficient from macroscopic measurements. Also the experiment can give information about the geometry of the surroundings of the penetrant molecules in the case of heterogeneous systems; the dimension of the boundaries that the molecules may encounter and the mean square distance that the penetrant has travelled during the observation of diffusion. Other advantages of the technique are the ability to use small sample volumes, the precision of the measurements and the ability to be applied to both liquid and solid samples. The only limitation of the technique is that the lowest diffusion coefficient that can be commonly determined is around 10^{-9} cm^2 / sec to 10^{-10} cm^2 / sec.

In addition to the advantages of the Pulsed Gradient technique itself, with NMR the reorientational dynamics of the both the polymeric and penetrant components of the solution can be measured independently on the same sample. Thus both translational and rotational motion of the penetrant, and segmental motion of the polymer can be determined in the same samples. With modern NMR instrumentation, these measurements can be made over a wide range of temperatures and concentrations which lead to a fairly complete description of local dynamics. It is the goal of this work to present the results of such a complete set of NMR measurements on a single system beginning with translational diffusion of the penetrant.

Diffusion in polyisobutylene (PIB), has been studied before using different techniques[3 - 7]. In references (3) and (4) the study of the diffusion coefficients of six hydrocarbons (propane, n - butane, isobutane, n - pentane, isopentane and neopentane) in PIB was reported using the sorption method. The self - diffusion of benzene in PIB has been studied in reference (5) using the pulsed field gradient technique over a limited range of temperatures and concentrations. In reference (6) the diffusion of radioactively tagged cetane in PIB has been studied and finally in reference (7) the diffusion of toluene in butyl rubber has been studied using the sorption method. In all cases the data were analyzed using the Fujita free volume theory. Viscosity data on polyisobutylene solutions are also available[8 - 11]. In references (8 - 10) the intrinsic viscosities of PIB solutions in various solvents have been studied and their relations to polymer chain structure and to thermodynamic parameters governing the interaction between the polymer and the solvent have been made.

Free volume theories have been used widely for the interpretation of the concentration dependence of the self - diffusion of many small molecules into polymer matrices[12 - 15]. One frequently used theory, developed almost thirty years ago[12], is the Fujita theory and a more recent free volume theory is that of Vrentas and Duda[13 - 15]. These two free volume theories differ in several respects and it will be of value to compare them in the context of a rather complete data set, not only including information of translational motion of the penetrant but also reorientational motion of both components: polymer and penetrant. The molecular level information available from NMR will provide insight into some of the concepts introduced into these theories which are at least at first glance based on very local descriptions of polymer and solvent dynamics.

The Fujita theory begins with the empirical result of Doolittle[16 - 17] who showed that the viscosity η of a liquid of low molecular weight is given by:

$$\ln(\eta) = \ln A + B / f \tag{1}$$

where A and B are empirical constants for a given liquid and f is the fractional free volume, defined to be equal to the ratio of the volume of the space not occupied by the constituent molecules, v_f, over the total volume of the solution, v:

$$f = \frac{v_f}{v} \tag{2}$$

$$v = v_0 + v_f \tag{3}$$

where v_0 is the volume occupied by the molecules in accordance with Van der Waals radii and vibrational motions. Equation (1) measures the volume of a molecule relative to the volume of a void next to it. When the volume of the void is larger than the volume of the molecule the viscosity is low. When the volume of the void is smaller than the volume of the molecule the viscosity is high. Another assumption that is made is that the Doolittle relation applies to each jumping unit in polymer - solvent systems[18]. Jumping units are imagined to be of the size of small penetrant molecules or repeat units in the case of the polymer. Also the theory relates the mobility of the polymer and the solvent jumping units to the free volume per unit volume of solution.

The Vrentas - Duda theory is based on the Cohen - Turnbull result[19] showing that the mobility of a molecule in a pure liquid is related to the average free volume, v_f, per molecule through the following equation:

$$\ln (m) = E - \gamma v^* / v_f \tag{4}$$

where γ is a numerical factor between 0.5 and 1, also known as overlap factor, that is introduced because a given free volume may be available to more than one molecule and v^* is the minimum volume of a hole that a molecule must find in its vicinity in order for it to move. According to Cohen and Turnbull molecules are resident in cages bound by their neighbors for the majority of the time. When a density fluctuation occurs and a hole of enough size opens within the molecular cage, the molecule jumps into the hole. Diffusional motion is considered to be translation of a molecule across the space within its cage and bulk diffusion then occurs as a result of redistribution of the free volume within the liquid. The above equation was derived under the assumption that redistribution of free volume (holes) at constant volume requires no change in energy. Equation (4) demonstrates that when the minimum volume of a hole that a molecule must find for a jump to occur is larger than the average free volume per molecule then the mobility of the penetrant is small. Conversely, when v^* is smaller than v_f then the molecule can jump to a neighbouring position and the mobility is high.

Vrentas - Duda[13] formulated their theory based on the following assumptions as stated:

(i) V_f' is the free volume equal to the space not occupied by the actual molecules.

(ii) The free volume available to each jumping unit: solvent molecule or polymer segment, is equal to V_f' / N where N is the total number of solvent molecules plus polymer segments and V_f' is the total free volume of the solution.

(iii) The critical hole volume, v^*, of a jumping unit is equal to its intrinsic volume at zero Kelvin.

(iv) The partial specific volumes of the solvent and polymer are independent of composition i.e. there is no volume change on mixing and thermal expansion coefficients are given by their average values over the temperature range considered.

(v) As the temperature is increased, the increase in volume comes from the homogeneous expansion of the material due to the increasing amplitude of the anharmonic vibrations with temperature and from the formation of holes or vacancies which are distributed discontinuously throughout the material at any instant. The free space that is not occupied by the constituent molecules and is not associated with any holes or vacancies is called interstitial free volume. The energy for redistribution of the interstitial free volume is very large and this type of free volume therefore must be distributed uniformly among the molecules. The free volume associated with the discontinuous distribution of holes in the liquid is called the hole free volume. This free volume can be redistributed with no increase in energy.

In this study the temperature dependence will be analyzed using the WLF equation for diffusion[20], which is also commonly placed in the context of free volume theories. The well known form of the WLF equation is

$$\log_{10} \frac{\zeta}{\zeta_0} = \frac{-c_1^D (T - T_0)}{c_2^D + T - T_0} \tag{5}$$

where ζ is the translational friction coefficient of a penetrant molecule into the polymer matrix, ζ_0 is the value of ζ at T_0 and c_1^D and c_2^D are the WLF coefficients at the reference temperature T_0. The diffusion coefficient, D, can be introduced into this equation by using the well known expression :

$$D = \frac{kT}{\zeta} \tag{6}$$

where k is the Boltzmann's constant. The WLF equation can thus be applied to diffusion.

The local segmental motion of the polymer and the reorientational motion of the penetrant/solvent are determined using ^{13}C T_1 NMR measurements. Previous studies[2,21] have shown that segmental motion and diffusion of a penetrant in PIB may be correlated, which is not generally the case for other polymers. This was attributed to the very efficient molecular packing of the polymer[1]. This high density of PIB relative to many other polymers implies that chain reorientation must accompany translational motion of the penetrant. The objective of this work is to compare the time scale of segmental motion of the polymer with the self - diffusion coefficient of the toluene in the PIB - toluene solutions as a function of concentration to see if they are indeed correlated. Segmental motion in pure PIB has been examined in detail in a recent report[2] and that information will serve as a guide in the interpretation of the concentration dependent results to be presented here.

Experimental

The polyisobutylene sample used for the study was obtained from Cellomer Associates, Inc. of Webster, NY in the form of thick sheets of high molecular weight (Cat. #40E, Lot #02, mol. wt 1,000,000). HPLC Grade toluene was obtained from FisherChemical Fisher Scientific. The diffusion measurements were performed on a Bruker MSL - 300 spectrometer equipped with a High Resolution Diffusion Unit and gradient probe, with a proton frequency of 300.1 MHz. For the diffusion measurements the phenyl proton peak of toluene was observed.

The Stejskal - Tanner pulse sequence[22] π / 2 - G(δ) - π - G(δ) - (echo) was used to measure the diffusion coefficient of toluene in PIB solutions. The stimulated echo[23] pulse sequence π / 2 - G(δ) - π / 2 - τ - π / 2 - G(δ) - (echo) was also used for measuring the diffusion coefficient, employing longer diffusion time observations. The strength of the magnetic field gradients was varied from 1.4 T/m up to 14.0 T/m; the associated length δ, from 500 μsec up to 7 msec; and the distance between the two gradient pulses Δ, from 2 msec up to 80 msec. Those values of Δ correspond to diffusion distances from approximately 1.1 μm up to 9.1 μm. The error on the measured diffusion coefficients was estimated at 10%.

NMR spin - lattice relaxation (T_1) experiments were carried out on Bruker AC-200, MSL-300 and Varian Unity 500 spectrometers, which correspond to 50.33 MHz, 75.5 MHz and 125.7 MHz ^{13}C Larmor frequencies respectively.

Interpretation

The temperature dependence of toluene in PIB was analyzed with the following form of the WLF equation[20]:

$$\log_{10}\left(\frac{DT_0}{D_{T_0}T}\right) = \frac{c_1^D(T-T_0)}{c_2^D+T-T_0} \tag{7}$$

where $c_1^D = B_d$ / 2.303 $f(\phi_s)$, $c_2^D = f(\phi_s)$ / $\alpha(\phi_s)$, $f(\phi_s)$ is the fractional free volume of the solution at the reference temperature T_0, ϕ_s is the volume fraction of toluene in PIB and $\alpha(\phi_s)$ is the fractional free volume expansion factor and D_{T_0} is the self diffusion coefficient at T_0. Equation (7) can be rewritten as:

$$\ln\left(\frac{DT_0}{D_{T_0}T}\right) = \frac{\frac{\alpha(\phi_s)}{B_d}(T-T_0)}{\frac{f(\phi_s)}{B_d}\left(\frac{f(\phi_s)}{B_d}+\frac{\alpha(\phi_s)}{B_d}(T-T_0)\right)} \tag{7a}$$

The parameters that were floated in order to fit the data were $\alpha(\phi_s)$ / B_d and $f(\phi_s)$ / B_d. B_d is analogous to the B in equation (1) and will be defined precisely later. The fits along with the experimental self - diffusion coefficients at different concentrations and temperatures, are shown in Figure 1, and the reference temperature T_0 was chosen to be 298 K. The fitting was carried out using standard least square methods. The parameters obtained are shown in Table I. After $\alpha(\phi_s)/B_d$ and $f(\phi_s)/B_d$ were obtained, the WLF coefficients c_1^D and c_2^D were calculated (from $c_1^D = B_d$ / 2.303 $f(\phi_s)$ and $c_2^D = f(\phi_s)$ / $\alpha(\phi_s)$) and they are also shown in Table I.

To fit the data at a constant temperature, as a function of concentration, the Fujita free volume theory[12] was first applied. As mentioned above, the theory begins

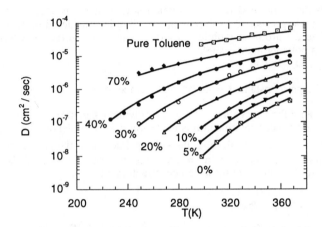

Figure 1: Self - diffusion coefficient of toluene in PIB solutions as a function of temperature at different concentrations. The solid lines represent fits to the WLF diffusion equation presented in the text. The zero penetrant (0%) concentration points are from Table II.

Table I. Parameters obtained from the temperature dependence fit of the toluene self - diffusion coefficient, using equation (6)

ϕ_s	$f(\phi_s) / B_d$	$\alpha(\phi_s) / B_d$ (deg^{-1})	c_1^D	c_2^D (K)
0.045	0.0904	5.83×10^{-4}	4.803	155.060
0.104	0.1050	6.84×10^{-4}	4.135	153.509
0.205	0.1270	7.25×10^{-4}	3.419	175.172
0.327	0.1550	8.44×10^{-4}	2.801	183.649
0.409	0.1900	9.60×10^{-4}	2.285	197.917
0.712	0.2700	1.10×10^{-3}	1.608	245.455
1.000	0.3700	1.70×10^{-3}	1.174	217.647

with the result of Doolittle[16 - 18] who showed that the viscosity η of ordinary liquids of low molecular weight is given by equation (1).

To relate this to diffusion, the mobility, m, of a molecule is considered to be the velocity with which a molecule translates under the action of unit force and is inversely proportional to the molecular friction coefficient and the viscosity η (Stokes law). Thus equation (1) can be written as:

$$\ln(m) = A - B_d / f \tag{8}$$

Where B_d is the value of B for the minimum hole which is needed to allow a given molecule to undergo such a displacement[12,19] and is considered to be a measure of hole size. The definition of mobility in terms of the self - diffusion coefficient of a penetrant in a polymer matrix is[24]:

$$D = R T m \tag{9}$$

where R is the gas constant, T is the absolute temperature and D is the self - diffusion coefficient.

In general the free volume of a given polymer - penetrant solution depends on temperature and concentration and it is more appropriate to denote the fractional free volume as $f(T, \phi_s)$ where again ϕ_s is the volume fraction of the penetrant. It has been shown[25] that if the increase in free volume by the addition of the diluent is proportional to the volume of the added penetrant, then:

$$v_f = (v_f)_2 + \gamma(T) v_1 \tag{10}$$
$$v = v_1 + v_2 \tag{11}$$

where v_f is the average free volume of the solution, $(v_f)_2$ is the average free volume of the pure polymer, $\gamma(T)$ is a proportionality factor that may be compared with the fractional free volume of the diluent, v is the total volume of the solution, v_1 is the volume of the penetrant and v_2 is the volume of the amorphous polymer. By dividing equation (10) with equation (11) the fractional free volume of the solution $f(T, \phi_s)$ is found to be a linear function of the volume fraction of the penetrant ϕ_s:

$$f(T, \phi_s) = f_p(T, 0) + (f_s(T) - f_p(T, 0)) \phi_s \tag{12}$$

where $f_p(T, 0)$ is the fractional free volume of pure polymer and $f_s(T)$ is a parameter that represents the fractional free volume of the solvent trapped in the polymer matrix. $f_s(T) - f_p(T, 0)$ represents the effectiveness of the penetrant for increasing the free volume when it is dispersed in the given polymer. Equation (12) can be rewritten as:

$$\frac{f(T,\phi_s)}{B_d} = \frac{f_p(T)}{B_d} + \left(\frac{f_s(T)}{B_d} - \frac{f_p(T)}{B_d}\right)\phi_s \tag{13}$$

If equations (12) and (8) are substituted into equation (9) considering that as $\phi_s \to 0$, D $\to D_0$ we obtain:

$$\ln\frac{D}{D_0} = \frac{B_d(f_s - f_p)\phi_s}{f_p^2 + (f_s - f_p)f_p\phi_s} = \frac{[(\frac{f_s}{B_d}) - (\frac{f_p}{B_d})]\phi_s}{(\frac{f_p}{B_d})^2 + [(\frac{f_s}{B_d}) - (\frac{f_p}{B_d})](\frac{f_p}{B_d})\phi_s} \tag{14}$$

where ϕ_s is the penetrant volume fraction, D_0 is the diffusion coefficient of toluene at the limit of zero penetrant concentration, f_p is the average fractional free volume of pure polymer and f_s is the fractional free volume of toluene in the solution.

For each temperature, the diffusion coefficient of toluene as a function of concentration was fitted using equation (14). The parameters varied are D_0, (f_p / B_d)

and (f_s / B_d). The reason for using (f_p / B_d) and (f_s / B_d) as fitting parameters and not f_p and f_s is to avoid making any assumptions about the parameter B_d. The parameters (f_p / B_d) and (f_s / B_d) do not have physical significance whereas f_p, f_s and B_d do. The parameters produced by the fit are shown in Table II and the lines corresponding to the fit in Figures 2 and 3.

From the diffusion coefficient of toluene at the limit of zero penetrant concentration in Table II and the WLF equation (7), WLF parameters can be obtained similar to those shown in Table I. Those are shown in the Table III .

The self - diffusion of toluene in PIB was also analysed via the Vrentas - Duda theory[13-15,26-28]. According to this theory the self diffusion coefficient of a penetrant into a rubbery polymer is given by:

$$D_1 = D_0 \exp(\frac{-E}{RT}) \exp[-\frac{-(\omega_1 V_1^* + \omega_2 \xi V_2^*)}{\omega_1(\frac{K_{11}}{\gamma})(K_{21} - T_{g1} + T) + \omega_2(\frac{K_{12}}{\gamma})(K_{22} - T_{g2} + T)}] \quad (15)$$

where V_i^* is the specific critical hole free volume of component i required for a jump (i can be equal to 1 or 2; 1 for penetrant or solvent and 2 for polymer), ω_i is the mass fraction of component i, T_{gi} is the glass transition temperature of component i, D_0 is a constant pre-exponential factor, E is the energy per mole that a molecule needs to overcome attractive forces which constrain it to its neighbors and γ is an overlap factor (between 1/2 and 1) which is introduced because the same free volume is available to more than one molecule. K_{11} and K_{21} are free-volume parameters for the solvent, while K_{12} and K_{22} are free volume parameters for the polymer.

The free volume parameters K_{11}, K_{21}, K_{12}, K_{22} are defined as follows[13]:

$$K_{11} = V_1^0(T_{g1})[\alpha_1 - (1 - f_{H1}^G)\alpha_{c1}] \quad (16)$$

$$K_{21} = \frac{f_{H1}^G}{\alpha_1 - (1 - f_{H1}^G)\alpha_{c1}} \quad (17)$$

$$K_{12} = V_2^0(T_{g2})[\alpha_2 - (1 - f_{H2}^G)\alpha_{c2}] \quad (18)$$

$$K_{22} = \frac{f_{H2}^G}{\alpha_2 - (1 - f_{H2}^G)\alpha_{c2}} \quad (19)$$

$$f_{H1}^G = \frac{K_{11}K_{21}}{V_1^0(T_{g1})} \quad (20)$$

$$f_{H2}^G = \frac{K_{12}K_{22}}{V_2^0(T_{g2})} \quad (21)$$

where α_i is the thermal expansion coefficient for the equilibrium liquid component i, α_{ci} is the thermal expansion coefficient for the component i for the sum of the specific occupied volume and the specific interstitial volume, V_i^0 is the specific volume of the pure equilibrium component i and f_{Hi}^G is the fractional hole free volume of the component i at T_{gi}. Finally, ξ is the ratio of molar volume of a solvent jumping unit to the molar volume of a polymer jumping unit. The parameter ξ can be a measure of the molar volume of the polymer jumping unit when the entire solvent molecule performs a jump during the process of the self - diffusion.

There are nine independent parameters in equation (15), D_0, E, ξ, K_{11}/γ, K_{21} - T_{g1}, K_{12}/γ, K_{22} - T_{g2}, V_1^*, V_2^* that need to be determined. However, most of them can be determined from viscosity data or sources other than diffusion data and in favorable cases all of them could conceivably be set from other data. In order to fit the diffusion data of toluene in PIB as a function of concentration, the parameters shown in Table IV were used. This parameter set was chosen from a number of literature sources and

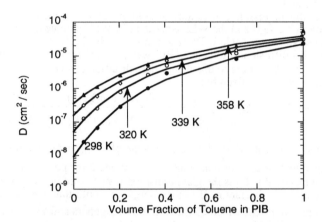

Figure 2: Self - diffusion coefficient of toluene in PIB solutions as a function of concentration at different temperatures. Solid lines represent fits to Fujita free volume theory presented in the text.

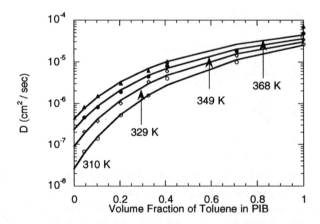

Figure 3: Self - diffusion coefficient of toluene in PIB solutions as a function of concentration at different temperatures. Solid lines represent fits to Fujita free volume theory presented in the text.

Table II. Parameters obtained from the concentration dependence fit of the toluene self - diffusion coefficient, using equation (14)

T(K)	f_s / B_d	f_p / B_d	D_0 (cm^2 / sec)
298	0.273	0.08735	$9.08 \ 10^{-9}$
310	0.300	0.09780	$2.55 \ 10^{-8}$
320	0.333	0.10700	$5.15 \ 10^{-8}$
329	0.364	0.11630	$8.70 \ 10^{-8}$
339	0.381	0.12500	$1.55 \ 10^{-7}$
349	0.398	0.13200	$2.30 \ 10^{-7}$
358	0.423	0.14100	$3.40 \ 10^{-7}$
368	0.449	0.14450	$4.10 \ 10^{-7}$

Table III. Parameters obtained from the temperature dependence fit of the toluene self - diffusion coefficient at zero toluene concentration (D_0, shown in Table 2), using equation (7)

ϕ_s	$f(\phi_s) / B_d$	$\alpha(\phi_s) / B_d$ (deg^{-1})	c_1^D	c_2^D (K)
0.000	0.0818	$5.34 \ 10^{-4}$	5.308	153.184

Table IV. Values of parameters for PIB / Toluene systems

V_1^* (cm^3 / gr)	0.917
V_2^* (cm^3 / gr)	1.004
K_{11} / γ (cm^3 / gr K)	$1.45 \ 10^{-3}$
K_{12} / γ (cm^3 / gr K)	$4.351 \ 10^{-4}$
$K_{21} - T_{g1}$ (K)	-86.32
$K_{22} - T_{g2}$ (K)	-97.6
ξ	0.892 ± 0.08
D_0 (cm^2 / sec)	$4.82 \ 10^{-4}$
E (kcal / mole)	0
M_1	92.13
V_1^0 (cm^3 / gr mole)	84.48
T_{g1} (K)	117
M_{2mono}	56.04
C_{12}^g	9.33
C_{22}^g (K)	107.4
T_{g2} (K)	205

is discussed in some detail elsewhere[29]. Since all the toluene and polymer free volume parameters mentioned above are close to each other, the actual choice of a parameter set is somewhat arbitrary and doesn't significantly change the quality of the fit. As a result of the fit, ξ, the only parameter floated, was found to be equal to 0.892 $\pm$ 0.08. The experimental self - diffusion coefficients of toluene in PIB along with the fitting curves (solid lines) are shown as a function of temperature and concentration in Figures 4(a) and 4(b).

The polymer segmental motion was characterized using the ^{13}C T_1 data from the methylene carbon of PIB and applying the model based on the Hall-Helfand description[30] of segmental motion combined with restricted anisotropic rotational diffusion[2,31]. The equations that relate the spectral density function corresponding to this motional model and the spin - lattice relaxation time are given by:

$$\frac{1}{T_1} = W_0 + 2W_{1C} + W_2 \qquad (22)$$

where

$$W_0 = \sum_j \gamma_C^2 \gamma_H^2 \hbar^2 \frac{J_0(\omega_H - \omega_C)}{20 r_j^6}$$

$$W_{1C} = \sum_j 3\gamma_C^2 \gamma_H^2 \hbar^2 \frac{J_1(\omega_C)}{40 r_j^6}$$

$$W_2 = \sum_j 3\gamma_C^2 \gamma_H^2 \hbar^2 \frac{J_2(\omega_H + \omega_C)}{10 r_j^6}$$

The methylene C - H distance is set at 1.09 Å. The expression for the spectral density function is described in detail elsewhere[2,29]. According to this model two correlation times describe the segmental motion of the polymer. One of them is τ_0 which is the correlation time for single conformational transitions and the other is τ_1 the correlation time for cooperative or correlated transitions. Each of these two correlation times was given an Arrhenius temperature dependence with activation energy E_{a0} and E_{a1} respectively and corresponding prefactors τ_∞. Also ℓ is the amplitude over which restricted rotational diffusion occurs and D_{ir} is the rotational diffusion constant. The form of the temperature dependence of ℓ and D_{ir} is shown below:

$$\ell = A\ T^{0.5} \text{ deg} \qquad (23)$$
$$D_{ir} = B\ T - C \text{ sec}^{-1} \qquad (24)$$

The results of fitting the T_1 field, temperature and concentration dependence is shown for a typical case in Figure 5 and the results are shown in Table V while the rest of the parameters have the values:

$$(\tau_\infty)_0 = 2.1\ 10^{-14} \text{ sec}; \quad (\tau_\infty)_1 = 1.0\ 10^{-14} \text{ sec} \qquad (25)$$
$$\ell = 5.15\ T^{0.5} \text{ deg}; \qquad D_{ir} = 3.26\ 10^7\ T - 3.5\ 10^9 \text{ sec}^{-1} \qquad (26)$$

When a WLF temperature dependence was given to the correlation times, there are six parameters to be established: $(\tau_0)_0$, $(\tau_1)_0$, ℓ, D_{ir}, c_1^0 and c_2^0 (c_1^0 and c_2^0 are the WLF coefficients). The reference temperature was chosen to be $T_0 = 298$ K and accordingly:

$$\ln \frac{\tau_0}{(\tau_0)_0} = \frac{-2.303 c_1^0 (T - T_0)}{c_2^0 + T - T_0} \qquad (27)$$

$$\ln \frac{\tau_1}{(\tau_1)_0} = \frac{-2.303 c_1^0 (T - T_0)}{c_2^0 + T - T_0} \qquad (28)$$

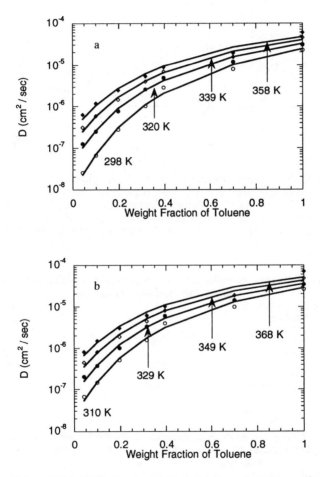

Figure 4(a) and 4(b): Diffusion coefficient of toluene in PIB as a function of concentration at different temperatures. Solid lines represent diffusion coefficients calculated from equation (15) of Ventras-Duda.

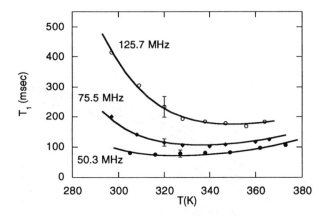

Figure 5: T_1 of the methylene carbon of PIB in toluene - PIB solutions as a function of temperature and Larmor frequency. The volume fraction of toluene in PIB is 0.045. An Arrhenius temperature dependence was given to the correlation times. The solid line corresponds to the fit for the Hall - Helfand function combined with anisotropic restricted rotational diffusion.

Table V. Activation energies for conformational transitions as a function of concentration of toluene in PIB

Volume Fraction of toluene in PIB.	E_{a0} (kJ / mole)	E_{a1} (kJ / mole)
0.045	34.0	34.5
0.104	33.4	33.7
0.205	32.7	32.0
0.327	31.9	30.0
0.409	31.7	28.4
0.712	30.4	25.7

Table VI. Fitting parameters of the spin - lattice relaxation data when a WLF temperature dependence was given to the correlation times

Volume Fraction of toluene in PIB	$(\tau_0)_0$ (sec)	$(\tau_1)_0$ (sec)	c_1^0	c_2^0 (K)
0.045	$3.00 \ 10^{-8}$	$1.20 \ 10^{-8}$	4.30	155.06
0.104	$1.80 \ 10^{-8}$	$8.40 \ 10^{-9}$	3.80	153.50
0.205	$8.85 \ 10^{-9}$	$4.25 \ 10^{-9}$	3.40	175.17
0.327	$6.23 \ 10^{-9}$	$1.88 \ 10^{-9}$	2.75	183.65
0.409	$3.40 \ 10^{-9}$	$9.70 \ 10^{-10}$	2.65	197.92
0.712	$3.00 \ 10^{-9}$	$4.00 \ 10^{-10}$	1.85	245.45

$$c_1^0 = \frac{B_0}{2.303f(\phi_s)} \tag{29}$$

$$c_2^0 = \frac{f(\phi_s)}{\alpha} \tag{30}$$

where $f(\phi_s)$ is the fractional free volume of the solution and α is the thermal expansion coefficient of the fractional free volume. B_0 represents the minimum void size necessary for a conformational transition. Comparable fittings to those obtained using the Arrhenius depedence are obtained and the fitting parameters are given in Table VI. Good agreement is evident for the fractional free volume of the solution derived from equation 29 and those values obtained from penetrant self diffusion analysis shown in Table I.

There is no field strength dependence in the T_1 data for the ortho carbon of toluene. Therefore, the extreme narrowing limit equation is used for the interpretation:

$$\frac{1}{T_1} = \frac{4}{3}\gamma_H^2\gamma_C^2\hbar^2 S(S+1)\tau\frac{1}{r_{CH}^6} \tag{31}$$

where $S = 1/2$, τ is the correlation time for molecular tumbling and $r_{CH} = 1.08$ Å is the distance between the proton and the carbon on the toluene phenyl ring.

Again, at first an Arrhenius temperature dependence was given to the correlation time, where E is the activation energy for rotation and τ_∞ is a prefactor. Two parameters were floated E and τ_∞. During the fitting procedure it was found that τ_∞ is around 1.0×10^{-14} sec and changes little as a function of concentration. Thus the value of τ_∞ was set equal to $1.0 \; 10^{-14}$ sec and E was the only parameter that was floated as concentration changed. The parameters obtained are shown in Table VII. Next a WLF temperature dependence was assigned to the correlation time

$$\ln\frac{\tau}{\tau_0} = \frac{-2.303c_1^0(T-T_0)}{c_2^0 + T - T_0} \tag{32}$$

where τ_0 is the value of the correlation time τ at the reference temperature T_0 which again was chosen to be 298 K. c_1^0 and c_2^0 are the WLF coefficients at the reference temperature:

$$c_1^0 = \frac{B}{2.303f(\phi_s)} \tag{33}$$

$$c_2^0 = \frac{f(\phi_s)}{\alpha} \tag{34}$$

where $f(\phi_s)$ is the fractional free volume of the solution and α is the thermal expansion coefficient of the fractional free volume of the solution. B represents the minimal void size necessary for rotation. For the fitting process of the T_1 data c_2^0 was set equal to c_2^D the WLF coefficient obtained from the temperature dependence of the self - diffusion of toluene in PIB - toluene solutions (Table I). The parameters that were floated in order to fit the data are c_1^0 and τ_0. the values obtained are shown in Table VIII. The fittings obtained are comparable to those obtained for the Arrhenius temperature dependence.

Discussion

It is clear that the self-diffusion of toluene in PIB is well described by free volume theories. Some comparisons between the different free volume treatments and the potential correlation between penetrant diffusion and polymer segmental motion can now be explored with the data in hand.

As a first step a comparison can be made between the fractional free volume determined from the temperature dependence of the translational diffusion of the toluene with that determined from the concentration dependence of the same quantity. Such a comparison is made in Figure 6 where the fractional free volume of the solution, $f(T, \phi_s)$, is presented as a function of the volume fraction of toluene in PIB. The points come from Table 1 and they were obtained from the WLF equation (7) where the temperature dependence of the toluene self - diffusion was analyzed, except the filled circle that represents the fractional free volume of pure toluene (at 293 K) obtained from viscosity data[32]. The solid line was obtained from equation (13) using the values of Table II for $f_p(T)$ and $f_s(T)$ (obtained from equation (14) where the concentration dependence of the toluene self - diffusion was analyzed). The error bars represent an uncertainty of approximately 20% for $f(T, \phi_s)$ in Table I, which comes from quality of the fitting (namely the fitting of the temperature dependence of the self - diffusion of toluene in PIB). As can be seen from Figure 6, the values of the fractional free volume of the solution that are obtained from the analysis of the temperature and concentration dependencies of the self - diffusion of toluene in PIB, are, in fact, close to each other.

A free volume analysis of the viscosity of the toluene - PIB solution was done by Fujita[32]. In that study the fractional free volume of the solution, $f(293 \text{ K}, \phi_s)$, and the fractional free volume of toluene trapped in the polymer matrix, $f_s(293 \text{ K})$, are reported as a function of concentration at 293 K. The fractional free volumes of the solutions at 293 K obtained from the self - diffusion data are in reasonable agreement with the values obtained from the viscosity data[32]. This establishes the utility of the free volume approach in relation to molecular mobility of a penetrant in a polymer. Viscosity is macroscopic measure, while the diffusion measurements presented here are microscopic measurements looking specifically at the penetrant molecule. Two different experiments probing two different levels of dynamics give approximately the same fractional free volume for the solutions, when the appropriate value for the minimum void size necessary for diffusion relative to viscosity was chosen.

Another result that has been confirmed, is the decrease of the fractional free volume of toluene when the latter is trapped in entangled PIB. This is seen by direct comparison between the free volume of toluene trapped in PIB (Table II) and the data for neat toluene from viscosity data[12,32] The same conclusion has been reached from the viscosity data[32], with a completely different approach. Also the fractional free volume parameters for pure PIB obtained from this work are similar to the values reported in the literature and extracted from other diffusion or viscoelastic measurements. This confirms the correctness of the magnitude of the free volume of the particular polymer.From the preceding it is obvious that B_d, the parameter that represents the minimum void size necessary for a penetrant molecule to diffuse, is important if results obtained from different methods that observe different kind of motions are to be compared. This is to be expected since the size of a moving unit is not the same for different kind of motions. The unfortunate fact is that it is not possible to obtain the dimension of the void when the parameter B_d is known.

The stated advantage of the Ventras-Duda theory is that the parameters can be independently determined and the theory has potential predictive capability. In our use of Ventras-Duda the only fitting parameter is ξ, the ratio of the molar volume of solvent jumping unit to the molar volume of polymer jumping unit, given by:

$$\xi = \frac{V_1^0(0)}{V_{2j}} = \frac{M_1 V_1^*}{M_{2j} V_2^*} \tag{35}$$

The value determined for ξ of 0.892 ± 0.08. corresponds to a polymer jumping unit of molecular weight 94.3 as derived from equation (35) when the toluene moves as a

Table VII. Fitting parameters of the ortho carbon T_1 data when an Arrhenius temperature dependence was given to the correlation time

Volume fraction of toluene in PIB ϕ_S	τ_{∞} (sec)	E (kJ / mole)
0.045	$1.0 \ 10^{-14}$	22.73
0.104	$1.0 \ 10^{-14}$	21.84
0.205	$1.0 \ 10^{-14}$	20.76
0.327	$1.0 \ 10^{-14}$	18.50
0.409	$1.0 \ 10^{-14}$	16.77
0.712	$1.0 \ 10^{-14}$	14.00

Table VIII. Fitting parameters of the ortho carbon T_1 data when a WLF temperature dependence was given to the correlation time

Volume fraction of toluene in PIB ϕ_S	τ_0 (sec) at T_0 298K	c_1^0 at 298 K	c_2^0 (K) at 298 K
0.045	$9.72 \ 10^{-11}$	2.45	155.06
0.104	$6.80 \ 10^{-11}$	2.35	153.50
0.205	$4.10 \ 10^{-11}$	2.25	175.17
0.327	$1.67 \ 10^{-11}$	2.10	183.65
0.409	$8.40 \ 10^{-12}$	2.00	197.92
0.712	$2.90 \ 10^{-12}$	1.81	245.45

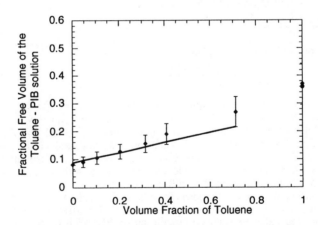

Figure 6: Comparison of the fractional free volume of the toluene - PIB solution, obtained from the temperature (points) and concentration (line) dependence of the diffusion coefficient. Solid line represents values obtained from equation (13). Filled circle represents the fractional free volume of pure toluene at 293K[32].

single unit. This value is close to the molecular weight of toluene (M_1 = 92.13). It is thus seen that the size of the polymer jumping unit is approximately the same as the size of the penetrant jumping unit. In view of this result, it would be appropriate to expect that both polymer and penetrant jumping units would require approximately the same minimum hole size necessary for a diffusion jump. It has been shown[13] previously that both Fujita and Vrentas - Duda free volume theories can be used when the molecular weights of the two jumping units are the same. As indicated in reference (33), the two theories are identical when the following condition is true:

$$\frac{\xi V_2^*}{V_2^0} = \frac{V_1^*}{V_1^0} \tag{36}$$

and when B_d (minimum hole size required for a given molecule to permit a displacement, according to Fujita theory[12]) is defined as:

$$B_d = \frac{\gamma V_1^*}{V_1^0} \tag{37}$$

This can be shown by rewriting equation (15) as:

$$\ln\left(\frac{D_1}{D_1(0)}\right) = \frac{\phi_1\left(\frac{\xi V_2^* f_1}{V_2^0 \gamma} - \frac{V_1^* f_2}{V_1^0 \gamma}\right)}{\left[\frac{f_2^2}{\gamma^2} + \phi_1 \frac{f_2}{\gamma}\left(\frac{f_1}{\gamma} - \frac{f_2}{\gamma}\right)\right]} \tag{38}$$

where $D_1(0)$ is the self - diffusion coefficient of the penetrant at the limit $\omega_1 = 0$, V_i^0 is the specific volume of pure component i at the temperature of interest, V_i^* is the specific critical hole free volume of component i required for a jump, ϕ_1 is the volume fraction of the solvent and f_i is the fractional hole - free volume of pure component i at the temperature of interest and the f_i's are given by:

$$f_1 = \frac{(K_{21} + T - T_{g1})K_{11}}{V_1^0} \qquad f_2 = \frac{(K_{22} + T - T_{g2})K_{12}}{V_2^0} \tag{39}$$

Using equations (36) and (37), equation (38) can be written as:

$$\ln\left(\frac{D_1}{D_1(0)}\right) = \frac{B_d(f_1 - f_2)\phi_1}{f_2^2 + f_2(f_1 - f_2)\phi_1} \tag{40}$$

Equation (40) is identical to the equation derived by Fujita[12]. For ξ = 0.892, V_2^* = 1.004 cm^3 / gr, V_1^* = 0.917 cm^3 / gr, V_2^0 = 1.091 cm^3 / gr, and V_1^0 = 1.154 cm^3 / gr we find that the left hand side of equation (36) is equal to 0.821 and the right hand side of equation (36) is equal to 0.794. Therefore we see that since the two ratios are close, either of the two theories can be used for the toluene - polyisobutylene system. Thus it is no surprise that both approaches accurately describe the temperature and concentration dependence of the self - diffusion constant.

It would be desirable if the ideas from these free volume concepts were able to shed some light on the rather unique behaviour of PIB relative to other polymers. In particular the unusual slow diffusion exhibited by PIB with respect to a variety of penetrants might be expected to be reflected in the value of certain free volume parameters. This is seen in the fact that the WLF parameters: c_1 and c_2, are considerably different from most other polymers, which can be adequately represented by the so-called universal values. The parameter α, the fractional free volume expansion factor, is considerably smaller than is typical for other polymers and does give some physical insight into the origin of the lower free volume at temperatures above the glass transition. This parameter is derived directly from the Fujita analysis. The Vrentas-Duda parameters, though useful in their potential for predictability, do not appear to yield any physical insight into uniqueness of PIB.

If we now include consideration of the relaxation data, three different kind of motions were considered: the self - diffusion of toluene in high molecular weight PIB, the molecular rotation of toluene and the segmental motion of the polymer. In general the c_1^0 WLF coefficient obtained from these three types of motion should be different, because of the different hole sizes required for each type of motion to occur, i.e., B's (in general $c_1^0 = B/2.303f(\phi_s)$ and $c_2^0 = f(\phi_s) / \alpha$). For the case of diffusion we have B_d, for the toluene rotation B and for the segmental motion of the polymer B_0.

In Table IX, $(f(\phi_s) / B_0)$ obtained from segmental motion and $(f(\phi_s) / B_d)$ obtained from translational diffusion are compared. From Table IX, it can be seen that the two parameters are close to each other, considering their uncertainties: of the order of ± 0.015 at low concentration, up to ± 0.05 at high toluene concentrations. The agreement is very good noting that the values are extracted by different methods from two completely different sets of data that describe different components of the solution. The toluene self - diffusion data were collected by observing directly the toluene molecule and the T_1 data were collected by observing the methylene carbon of PIB in the toluene - PIB solutions. From Table IX, since the fractional free volume of the solution, $f(\phi_s)$ and the thermal expansion coefficient of the free volume, α, are unique and independent of the type of motion that is examined, it can be concluded that $B_d \approx B_0$. Also the value of ξ, obtained from the Vrentas - Duda theory is 0.892 which corresponds to a polymer jumping unit with molecular weight 94.3. This is very nearly equal to the molecular weight of toluene which is 92. The fact that the two nearly coincide with each other means that the two jumping units will need the same minimal free volume for translation (for toluene) and a conformational transition (for the polymer) implying that B_d should equal B_0.

Another way to check to see if the two motions are correlated, is to directly compare the correlation time for translational diffusion τ_D and the correlation time for segmental motion (correlation time for single conformational transition). The first can be obtained from[34]:

$$\tau_D = \frac{b^2}{D} \tag{41}$$

where D is the self - diffusion coefficient of toluene, b is the distance of closest approach which is between 2.2 Å - 3.3 Å. Using 2.7 Å as the distance of closest approach and the self - diffusion of toluene in PIB - toluene solution, the correlation time for translational diffusion is calculated and shown in Table X The correlation time for a conformational transition, τ_0, is calculated from the Arrhenius equation and also shown in Table X It can be seen that the two correlation times are in agreement at the 0.045 toluene volume fraction which means at this low penetrant concentration the two motions (segmental and translational diffusion) are correlated. As we go to higher concentrations the difference between the τ_0 correlation time and the τ_D becomes larger up to two orders of magnitude at 0.712 toluene volume fraction. This might be expected since as we go towards higher concentrations a toluene molecule has fewer polymer molecules in its neighbourhood and it is not necessary for a polymer segment to move in order a toluene molecule to translate.

Thus, it can be seen that the two processes (diffusion and conformational transition) require similar minimum free volumes in order to occur and the time scale of the two motions are correlated at low concentrations of penetrant.

The spin - lattice relaxation data on the ortho carbon of toluene do not provide as much information about the motion of toluene as the corresponding data on PIB. However, at all concentrations the correlation time for toluene reorientation is two or more orders of magnitude faster than segmental motion and translational diffusion. Thus the reorientational motion of the toluene is uncoupled from the other two motions. The WLF analysis of the temperature dependence also leads to different values of c_1^0 and thus B further supporting the separation of these motions. Toluene reorientation would appear to be a much more rapid motion involving a smaller hole size.

Table IX. Fractional free volume comparison of the PIB - toluene solutions obtained from the diffusion data analysis[17] and from the T_1 data presented in this work

Volume fraction of toluene in PIB ϕ_s	$f(\phi_s) / B_d$ From Diffusion data	$f(\phi_s) / B_0$ From T_1 data
0.045	0.090	0.101
0.104	0.105	0.114
0.205	0.127	0.127
0.327	0.155	0.158
0.409	0.190	0.164
0.712	0.270	0.235

Table X. Comparison of the correlation times for diffusion, τ_D, and for single conformational transitions τ_0 for 0.045 toluene volume fraction

D (cm^2 / sec) Toluene Volume Fraction = 0.045	T (K)	τ_0 (sec)	τ_D (sec)
2.55×10^{-8}	298	1.915×10^{-8}	2.859×10^{-8}
6.66×10^{-8}	310	1.126×10^{-8}	1.095×10^{-8}
1.26×10^{-7}	320	7.453×10^{-9}	5.786×10^{-9}
2.04×10^{-7}	329	5.254×10^{-9}	3.574×10^{-9}
3.11×10^{-7}	339	3.642×10^{-9}	2.344×10^{-9}
4.55×10^{-7}	348	2.666×10^{-9}	1.602×10^{-9}
6.31×10^{-7}	357	1.982×10^{-9}	1.155×10^{-9}
8.14×10^{-7}	367	1.451×10^{-9}	8.956×10^{-10}

Acknowledgement

Financial support from the Army Research Office (Grant # DAAL03-91-G-0207) is gratefully acknowledged.

Literature Cited.

(1) R. H. Boyd, P. V. Krishna Pant, *Macromolecules*, **24**, 6325, (1991).
(2) A. Bandis, W. - Y. Wen, E. B. Jones, P. Kaskan, Y. Zhu, A. A. Jones, P. T. Inglefield, J. T. Bendler, *J. Polym. Sci.: Part B: Polym. Phys.*, **32**, 1707, (1994).
(3) S. Prager, F. A. Long, *J. Am. Chem. Soc.*, **73**, 4072, (1952).
(4) S. Prager, E. Bagley, F. A. Long, *J. Am. Chem. Soc.*, **75**, 1255, (1953).
(5) B. D. Boss, E. O. Stejskal, J. D. Ferry, *J. Phys. Chem.*, **71**, 1501, (1967).
(6) R. S. Moore, J. D. Ferry, *J. Phys. Chem.*, **66**, 2699, (1962).
(7) N. S. Schneider, J. A. Moseman, N. - H. Sung, *J. Polym. Sci. Part B, Polym. Phys.* **32**, 491, (1994).
(8) T. G. Fox, Jr., P. J. Flory, *J. Am. Chem. Soc.*, **73**, 1909, (1951)
(9) T. G. Fox, Jr., P. J. Flory, *J. Phys. Colloid. Chem.*, **53**, 197, (1949).
(10) P. J. Flory, *Principles of Polymer Chemistry*, Cornell University Press, Ithaca and London, (1990).
(11) A. A. Tager, V. Ye. Dreval, F. A. Khasina, *Polym. Sci., USSR*, **4**, 1097, (1963).
(12) H. Fujita, *Fortschr. Hochpolym. - Forsch., Bd.* 3, S. 1, (1961).
(13) J. S. Vrentas, J. L. Duda, *J. Polym. Sci.: Polym. Phys. Ed.*, **15**, 403, (1977).
(14) J. S. Vrentas, J. L. Duda, *J. Polym. Sci.: Polym. Phys. Ed.*, **15**, 417, (1977).
(15) J. S. Vrentas, J. L. Duda, *J. Polym. Sci.: Polym. Phys. Ed.*, **15**, 441 (1977).
(16) A. K. Doolittle, *J. Appl. Phys.*, **22**, 1471, (1951).
(17) A. K. Doolittle, *J. Appl. Phys.*, **23**, 236, (1952).
(18) H. Fujita, *Chemical Engineering Science*, **48** # 17, 3037 , (1993).
(19) M. H. Cohen, D. Turnbull, *J. Chem. Phys.*, **31**, 1164, (1959).
(20) J. D. Ferry, *"Viscoelastic Properties of Polymers"*, John Wiley & Sons, 3rd edition, (1980).
(21) Z. P. Dong, B. J. Cauley, A. Bandis, C. W. Mou, C. E. Inglefield, A. A. Jones, P. T. Inglefield, W. - Y. Wen, *J. Polym. Sci.: Part B: Poly. Phys.*, **31**, 1213, (1993).
(22) E. O. Stejskal, J. E. Tanner, *J. Chem. Phys.*, **42**, 288, (1965).
(23) J. E. Tanner, *J. Chem. Phys.* **52**, 2523, (1970).
(24) R. M. Barrer, R. R. Fergusson, *Trans. Faraday Soc.*, **54**, 989, (1958).
(25) H. Fujita, A. Kishimoto, *J. Chem. Phys.*, **34**, 393, (1961).
(26) J. S. Vrentas, J. L. Duda, H. - C. Ling, *J. Polym. Sci.: Polym. Phys. Ed.*, **23**, 275, (1985).
(27) J. S. Vrentas, J. L. Duda, H. - C. Ling, A. - C. Hou, *J. Polym. Sci.: Polym. Phys. Ed.*, **23**, 289, (1985).
(28) J. S. Vrentas, C. - H. Chu, M. C. Drake, E. von Meerwall, *J. Polym. Sci.: Part B:, Polym. Phys.*, **27**, 1179, (1989).
(29) A. Bandis, P. T. Inglefield, A. A. Jones, W. - Y. Wen, *J. Polym. Sci.: Part B:, Polym. Phys.* (submitted)
(30) C. K. Hall, E. Helfand, *J. Chem. Phys.*, **77**, 3275, (1982).
(31) W. Gronski, N. Murayama, *Makromol. Chem.*, **179**, 1521, (1978).
(32) H. Fujita, Y. Einaga, *Polymer*, **31**, 1486, (1990).
(33) J. S. Vrentas, C. M. Vrentas, *J. Polym. Sci.: Part B: Polym. Phys.*, **31**, 69, (1993).
(34) A. Abragam, *Principles of Nuclear Magnetism*, Oxford University Press, (1989).

RECEIVED February 2, 1995

Chapter 16

Miscibility, Phase Separation, and Interdiffusion in the Poly(methyl methacrylate)–Poly(vinylidene fluoride) System

Solid-State NMR Study

Werner E. Maas

Bruker Instruments, Inc., 19 Fortune Drive, Billerica, MA 01821

The miscibility of PMMA and PVF_2 is investigated using Solid State Nuclear Magnetic Resonance. Miscibility at the molecular level is detected using fluorine to carbon cross polarization experiments and fluorine to proton to carbon double cross polarization experiments. The proton to fluorine cross depolarization technique is used to determine the amounts of PMMA and PVF_2 that are intimately mixed. These NMR techniques probe miscibility in a range from 3 to 50 Å. The proton to fluorine cross depolarization technique is applied to the studies of phase separation and interdiffusion in the PMMA/PVF_2 system.

The occurrence of a single glass transition temperature (T_g) is the generally accepted criterion for distinguishing between miscible and immiscible polymer-polymer systems. While this criterion is satisfactory for many applications of polymer blends, the observation of a single T_g does not provide insight on miscibility on a molecular scale. It is generally believed that a single T_g is observed if the dimensions of the domains in which the separate constituents of a polymer blend occur, are smaller than 150 Å. Distance-sensitive techniques such as X-ray scattering, neutron scattering and Nuclear Magnetic Resonance (NMR) provide information on polymer miscibility on a molecular scale.

We have developed Solid State NMR techniques to characterize the molecular miscibility in blends of poly (methyl methacrylate) (PMMA) and poly (vinylidene fluoride) (PVF_2). These techniques detect the magnetic dipole interaction between nuclear spins of PMMA and PVF_2 and allow the determination of the distance between PMMA and PVF_2 fragments, as well as quantify the degree

0097–6156/95/0598–0274$12.00/0

of mixing in these blends. The presented Solid State NMR techniques probe molecular miscibility in a range from 3 to 50 Å.

In this paper we will briefly review the results of these experiments on blends of PMMA and PVF_2 and discuss the application of one of these techniques, the proton to fluorine cross depolarization technique, to the study of dynamic processes in $PMMA/PVF_2$ systems.

Experimental

Experiments are performed on a modified Bruker CXP-300 spectrometer. In order to study $PMMA/PVF_2$ blends an additional fluorine RF channel is added to the spectrometer. In addition the standard double-tuned Cross Polarization Magic Angle Spinning (CPMAS) probe is triple-tuned to enable simultaneous ^{13}C, ^{19}F and ^{1}H high-power excitation and decoupling. Figure 1 shows the ^{13}C CPMAS spectra of pure PMMA, obtained with high power proton decoupling, and of pure PVF_2, obtained with simultaneous high power proton and fluorine decoupling, together with the respective repeat units and the carbon resonance assignments. The improvement in resolution in the PVF_2 carbon spectrum by simultaneous decoupling of the fluorine and proton spins may be appreciated by comparison with the spectra shown in Figure 1C and D, which are obtained under ^{1}H-only decoupling and ^{19}F-only decoupling, respectively. For a detailed experimental description the reader is referred to reference (1).

NMR Techniques

Detection of Molecular Miscibility. The magnetic dipole interaction between two nuclear spins is proportional to the product of the gyromagnetic ratios γ of the nuclei and inversely proportional to the cube of the distance r between the nuclei. This r^{-3} dependence causes the magnitude of the dipolar interaction to fall off rapidly with increasing distance, thus providing the NMR spectroscopist with a tool to obtain short range information (several Å). The existence of a dipolar interaction between spins can be detected for instance through magnetization transfer experiments. A frequently used experiment is the Hartmann-Hahn cross polarization experiment in which spin polarization of one type of nuclei is transferred to another type of spins with which they share a dipolar coupling. The magnetization transfer is enabled through the use of radio frequent magnetic fields with frequencies equal to the Larmor frequencies of the spins involved and amplitudes that are matched to the Hartmann-Hahn condition (2). In a ^{19}F-^{13}C cross polarization (CP) experiment magnetization is transferred from fluorines of PVF_2 to carbon spins via the dipolar coupling. If subsequently the carbon spectrum is recorded, then the occurrence of PMMA ^{13}C resonances proves that PMMA carbons have a dipolar coupling with PVF_2 ^{19}F spins and thus are in close proximity to PVF_2 fluorines. An example of a ^{19}F-^{13}C CP spectrum of a $PMMA/PVF_2$ blend is shown in Figure 2A, where in addition to the carbon res-

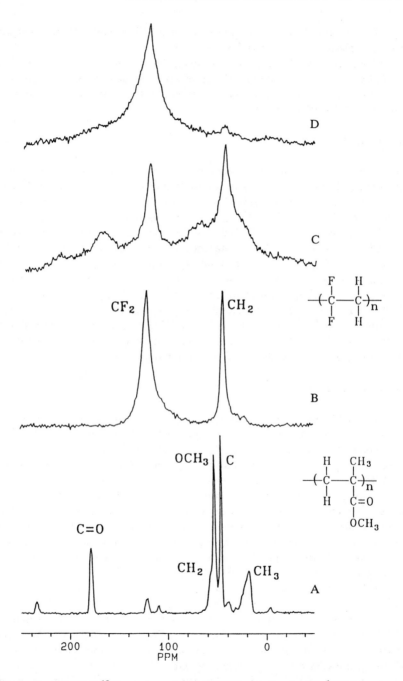

Figure 1. CPMAS ^{13}C spectra of (A) PMMA acquired with ^{1}H high power decoupling, (B) PVF_2 acquired with simultaneous ^{1}H and ^{19}F high power decoupling, together with the repeat units of PMMA and PVF_2, and PVF_2 spectra obtained with ^{19}F decoupling only (C) and ^{1}H decoupling only (D).

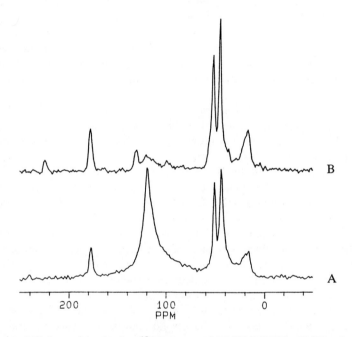

Figure 2. Magic angle spinning ^{13}C spectra of PMMA/PVF$_2$ 60/40, obtained with simultaneous ^{1}H and ^{19}F high power decoupling: A. cross polarized from ^{19}F and B. doubly cross polarized from ^{19}F to ^{1}H to ^{13}C.

onances of PVF_2 resonances of PMMA carbons are observed. The pulse sequence of the ^{19}F-^{13}C CP experiment is diagrammed in Figure 3A.

In addition to the detection of molecular miscibility, an average PVF_2 fluorine to a nearby PMMA carbon distance can be determined from the rate of magnetization transfer, which can be obtained by measuring the carbon intensities as a function of the cross polarization time. The cross polarization rate is proportional to the square of the dipolar coupling and therefore to r^{-6}, which limits the detection of the proximity of nuclear spins with the ^{19}F-^{13}C CP experiment to approximately 10 Å. For the various weight ratios of the PMMA/PVF_2 blend studied (1,3,4), the distance between a fluorine of PVF_2 and a carbon on a neighboring PMMA segment is found to be approximately 3 Å.

Since a single ^{19}F-^{13}C distance is not a good description for an amorphous blend, the carbon intensities as a function of the CP time were modeled with a distribution of distances(4). It was found that only a narrow distribution of ^{19}F-^{13}C distances around a mean value of 3 Å could account for the experimental data.

The intensities of the PMMA resonances observed in the ^{19}F-^{13}C CP experiment (Figure 2A) are small, tempting one to conclude that only a small portion of the PMMA chains is in close proximity to PVF_2 chains. However, one must take into account the relaxation processes which compete with the magnetization transfer. In particular for the PMMA/PVF_2 blends, the build-up of carbon magnetization, when cross polarizing from PVF_2 fluorines, is severely restrained due to the short fluorine rotating frame relaxation time $T_{1\rho}$ (≈ 1 ms). In unfavorable cases where the relaxation rate is larger than the cross polarization rate, the detection of molecular miscibility may be hindered or even prevented. An alternative technique which overcomes this problem is to transfer magnetization from fluorines to protons. Since the γ of protons is four times larger than that of carbons, the ^{19}F-^{1}H CP rate is sixteen times larger than the ^{19}F-^{13}C CP rate. This results in a more efficient build-up of proton magnetization in the ^{19}F-^{1}H CP experiment, as compared to the ^{19}F-^{13}C CP experiment. However, due to strong homonuclear dipolar couplings, solid state proton spectra often exhibit poor resolution and in order to distinguish between PVF_2 and PMMA proton magnetization, the proton magnetization is subsequently transferred to carbon via ^{1}H-^{13}C CP. Figure 2B shows a spectrum of the PMMA/PVF_2 60/40 blend, obtained via the ^{19}F-^{1}H-^{13}C double cross polarization experiment (Figure 3B). In comparing this spectrum to the ^{19}F-^{13}C CP spectrum of Figure 2A it is seen that the double cross polarization technique indeed results in much larger PMMA carbon intensities. Apart from the higher efficiency of the fluorine to proton to carbon magnetization transfer as compared to the fluorine to carbon transfer, this increase in PMMA carbon intensities is also due to the occurrence of proton spin diffusion, which causes proton magnetization from protons close to PVF_2 fluorines to be transferred to protons farther removed from fluorines, thus enabling the detection of PMMA segments at larger distances from PVF_2 fluorines than those detected in the direct fluorine to carbon magnetization transfer ex-

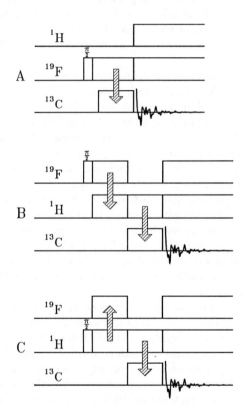

Figure 3. Pulse schemes of the NMR experiments: A. ^{19}F-^{13}C cross polarization; B. ^{19}F-^{1}H-^{13}C double cross polarization; C. ^{1}H-^{19}F cross depolarization followed by ^{1}H-^{13}C cross polarization.

periments. While the additional occurrence of proton spin diffusion hampers the determination of PVF$_2$ to PMMA distances, the ^{19}F-^{1}H-^{13}C double CP experiment provides qualitative proof of molecular miscibility up to approximately 15-20 Å.

Quantification of the degree of miscibility. Although the ^{19}F-^{13}C CP experiment and the ^{19}F-^{1}H-^{13}C double CP experiments are suitable to detect molecular miscibility in PMMA/PVF$_2$ blends, it is difficult to derive information from those experiments on the amounts of PVF$_2$ and PMMA that are intimately mixed. In order to derive quantitative information we developed the ^{1}H-^{19}F cross depolarization technique (3).

Instead of using the fluorines as a source of magnetization, we will now use the fluorine reservoir as a sink, through which proton magnetization can disappear. If, after creation of spin-locked proton magnetization in the whole sample, a fluorine RF field is turned on, its strength adjusted to the proton-fluorine Hartmann-Hahn condition, then all the protons that are dipolar coupled to fluorines will lose their magnetization to the lattice, via the fluorines. In a second step the remaining proton magnetization is transferred to carbons via ^{1}H- ^{13}C cross polarization, and the carbon magnetization is detected (see Figure 3C). The observed PMMA carbon magnetization is then from parts of PMMA molecules, whose protons do *not* have a dipolar interaction with fluorine. In other words, only those parts of PMMA molecules that are remote from PVF$_2$ molecules are observed. Through the additional occurrence of proton spin diffusion, PMMA proton magnetization from regions at larger distances from fluorines will be transported to the fluorine sink, which enlarges the detection limit for miscibility in the PMMA/PVF$_2$ blends to approximately 50 Å.

Apart from losing magnetization to fluorines, proton magnetization also disappears to the lattice (via $T_{1\rho}$ relaxation). This is corrected for by dividing the ^{13}C intensities obtained as a function of the ^{19}F-^{1}H transfer time by the intensities obtained from a blank experiment in which no fluorine RF field is used and where the loss of proton magnetization is thus entirely due to relaxation. The PMMA ($S^M(t)$) and PVF$_2$ ($S^V(t)$) carbon intensities as a function of the ^{1}H-^{19}F cross depolarization time t and corrected for ^{1}H-$T_{1\rho}$ relaxation, can thus be equated as:

$$S^M(t) = f^M_{mix} D(t) + (1 - f^M_{mix}) \tag{1}$$

$$S^V(t) = f^V_{mix} D(t) + (1 - f^V_{mix}) \exp\left(\frac{-t}{T_{HF}}\right) \tag{2}$$

where f^M_{mix} and f^V_{mix} are the mixed PMMA and PVF$_2$ fractions, respectively, $(1-f^M_{mix})$ and $(1-f^V_{mix})$ are the isolated fractions of PMMA and PVF$_2$, respectively, and $D(t)$ is a function which decays to zero and which describes the loss of proton magnetization to the sink through ^{1}H-^{19}F cross depolarization and proton spin diffusion. The equations for PMMA and PVF$_2$ differ in an extra term $\exp(-t/T_{HF})$ which arises from the fact that protons from isolated PVF$_2$

still lose their magnetization to the fluorine sink since all PVF_2 protons are close to fluorines.

Figure 4 shows the result for the PMMA proton magnetization detected via the C=O resonance, which levels off to a nearly constant value at 3.5 ms. This apparent constant level is attributed to the proton magnetization of the fraction $1-f_{mix}$ of PMMA that is not closely mixed with PVF_2. In the absence of proton spin diffusion PMMA units close to and PMMA units remote from PVF_2 can be distinguished by the distance aspect of ^1H-^{19}F cross depolarization. Proton spin diffusion complicates this distinction but clearly, the ability of proton spin diffusion to transfer magnetization from PMMA protons to PVF_2 protons can also be used to distinguish well-mixed PMMA from non-mixed PMMA.

The proton to fluorine cross depolarization experiments with subsequent carbon detection, directly provide information on the fractions of PMMA and PVF_2 that are intimately mixed. The isolated PMMA fractions are determined from the PMMA carbon intensities at longer cross depolarization times, whereas the isolated PVF_2 fractions are obtained from the PVF_2 carbon intensities at short depolarization times (the initial decay is mainly determined by the proton to fluorine transfer rate T_{HF}^{-1}). In addition, based on a spin diffusion model, information is obtained on the dimensions of the miscible domains in the $PMMA/PVF_2$ blends (see ref.(*3*) for details).

The cross depolarization technique has been applied to quantify the miscibility of $PMMA/PVF_2$ blends as a function of composition (*3*), and also to examine the effect of PMMA microstructure on mixing with PVF_2 by studying blends of isotactic, atactic and syndiotactic PMMA with PVF_2(*4*). In the remainder of this paper we report on the detection of dynamic processes in the $PMMA/PVF_2$ system.

Phase separation in PMMA/PVF₂ blends

The proton to fluorine cross depolarization technique, described above, can be employed to study the result of dynamic processes in $PMMA/PVF_2$ blends. In this section we will summarize the results of a study on the phase separation that occurs upon annealing of a $PMMA/PVF_2$ blend.

Samples of a $PMMA/PVF_2$ 60/40 blend (by weight ratio) are annealed at either 120°C or 140°C for a time t_{ann} and subsequently quenched in liquid nitrogen. A cross depolarization experiment then yields the fractions of not-mixed PMMA and PVF_2, which are diagrammed in Figures 5A and 5B, as a function of the annealing time. Both the fractions of isolated PMMA and PVF_2 increase sharply at short annealing times and the changes are slower at longer annealing times.

In addition to the NMR experiments Differential Scanning Calorimetry (DSC) experiments are performed. The DSC experiments show a melting peak from PVF_2 crystallites, indicating crystallization of PVF_2 to be the origin of the phase separation in the blend upon annealing. The DSC measurements after annealing

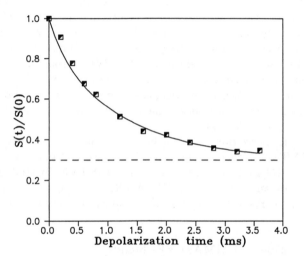

Figure 4. ^{13}C intensities as a function of the proton to fluorine cross depolarization time, scaled with respect to the intensity at zero depolarization time, for the C=O resonance of PMMA. The dashed line indicates the isolated PMMA fraction.

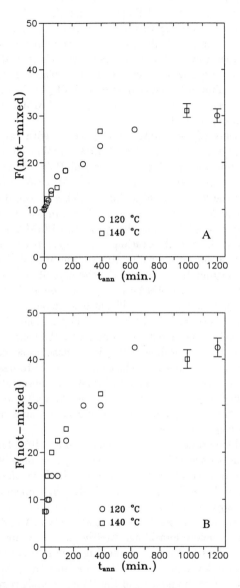

Figure 5. Fractions (in %) of isolated PMMA (A) and PVF$_2$ (B), as a function of the annealing time at 120°C and 140°C.

at 120°C or 140°C allow the crystalline fractions to be determined as a function of the annealing time; these crystalline fractions are depicted in Figure 6. Similar to the isolated fractions of PMMA and PVF$_2$ obtained from the NMR experiments, the crystalline fractions, observed with DSC, increase rapidly at short annealing times and slower at longer annealing times. A comparison between the data from DSC measurements and NMR experiments reveals that at least part of the phase separation, observed as an increase in the isolated PVF$_2$ fraction upon annealing, can be attributed to crystallization of PVF$_2$ in the blend. In addition, however, also an increase in the not-mixed fraction PMMA is revealed by the cross depolarization experiments.

In a previous study (3) on PMMA/PVF$_2$ blends with varying weight composition, it was found that the isolated PMMA fraction increases with higher PMMA content, while the fraction of not-mixed PVF$_2$ decreases with higher PMMA/PVF$_2$ ratio. The increase in the isolated PMMA fraction with annealing time thus indicates the formation of a PMMA-richer phase, due to the crystallization of PVF$_2$. This finding is also supported by a measured increase in proton T$_{1\rho}$ (see Papavoine et al.(5)). This change in the PMMA/PVF$_2$ ratio in the mixed phase also explains the observation that the growth of the isolated PVF$_2$ fractions slows down at longer annealing times. Due to depletion of PVF$_2$ in the amorphous regions surrounding the crystals, the crystal growth should continuously decrease, while leaving a PMMA-richer phase around the PVF$_2$ crystals (6). This, however, does not explain the significant discrepancy between the fractions of crystalline PVF$_2$, determined with DSC, and the fractions of not-mixed PVF$_2$, obtained from the NMR experiments (see Figure 7). We believe this difference to be caused by a crystalline-amorphous PVF$_2$-interphase, as suggested by Hahn et al.(7). Such an interphase is believed to be caused by head-to-head and tail-to-tail defects in PVF$_2$ and is expected to expel PMMA. The observation of a constant value of the difference between the NMR and DSC data after an initial annealing time also supports the existence of such an interphase, since this interphase is expected to have a constant thickness.

In an additional experiment the blend, annealed at 140°C for 16.5 hours, and containing according to the DSC measurements approximately 30% crystalline PVF$_2$, is heated in an oven at 190°C, which is well above the melting temperature of the blend and the PVF$_2$ crystals. NMR experiments performed after a certain time at 190°C revealed that after only 10 min. all the PMMA and PVF$_2$ is mixed again, i.e. the fractions of isolated PMMA and PVF$_2$ are again at the same values as before annealing. Melting at 190°C for longer times did not decrease these values further. This finding seems to suggest that no large PVF$_2$ crystallites have developed during the annealing time. Since the remixing of PVF$_2$ and PMMA after the melting of the PVF$_2$ crystals is governed by diffusion and the diffusion coefficient of PVF$_2$ chains is on the order of 10^{-10} cm^2/s at 190°C (see next section) the mean displacement L=$(2Dt)^{\frac{1}{2}}$ of the polymer chains during the 10 minutes at 190°C is approximately 3.5μm. This places therefore an

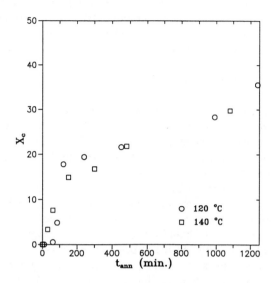

Figure 6. The crystalline fraction (in %) of PVF$_2$, determined with DSC, as a function of the annealing time at 120°C and 140°C.

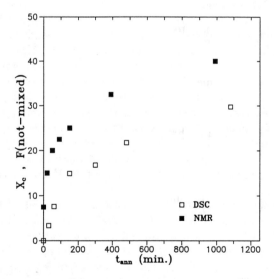

Figure 7. A comparison of the crystalline PVF$_2$ fraction determined by DSC and the isolated PVF$_2$ fraction determined by NMR, as a function of the annealing time at 140°C.

upper limit of a few micrometers on the size of the PVF_2 crystals. In the well-mixed $PMMA/PVF_2$ system, the depletion of PVF_2 in the regions surrounding the PVF_2 crystallites, may prevent the development of large crystals and lead to the formation of many small crystallites.

Interdiffusion of PMMA and PVF_2

The ability of the proton to fluorine cross depolarization experiment with subsequent carbon detection to distinguish between PMMA and PVF_2 chains that are in close proximity to each other, or mixed, and chains that are remote from each other, or isolated, is exploited to detect interdiffusion of PMMA and PVF_2 at temperatures above the T_g of both components.

In this study thin sheets of PMMA and PVF_2 (with an average thickness of approximately 90 μm) are stacked alternately in a ceramic magic angle spinner. The spinner with its content is heated in an oven at 190 °C for a certain time after which it is quenched in liquid nitrogen. During the time at 190 °C chains of PMMA and PVF_2 diffuse across the boundaries between the sheets. PMMA and PVF_2 form a miscible system and the negative free energy of mixing (8–10) will be the main driving force for mixing.

Before heating, at zero diffusion time, PMMA and PVF_2 are separated on a macroscopic scale. The distances between PMMA chains and PVF_2 chains are too large to be bridged by proton spin diffusion and no dipolar interactions exist between PMMA protons and PVF_2 fluorines. A ^{1}H-^{19}F cross depolarization experiment before annealing reveals 100 % isolated PMMA and 100 % isolated PVF_2. As the diffusion progresses PMMA and PVF_2 segments intermix, and near the boundaries between the sheets the distances between PMMA and PVF_2 will be small enough to be bridged by proton spin diffusion. This reveals itself in a decrease in the fractions of isolated PMMA and PVF_2 in the cross depolarization experiments with time at 190 °C. Some examples are shown in Figure 8 in which the PMMA OCH_3 carbon intensities are plotted as a function of the cross depolarization time for different diffusion times at 190 °C. The decrease in intensities at longer depolarization times for curves obtained at increasing diffusion times, indicates that as diffusion progresses more PMMA segments are close to PVF_2 segments, resulting in a decrease in the fraction of PMMA segments that are not close to PVF_2.

Figure 9 shows the data for both the isolated PMMA and PVF_2 fractions as a function of the diffusion time at 190 °C. These data can be interpreted based on a diffusion model which involves three diffusion fluxes: one of PMMA chains, one of PVF_2 chains and one vacancy flux due to the differences in mass and size between PMMA and PVF_2 chains. This net vacancy flux causes a mass flow of both PMMA and PVF_2 in the direction of the faster moving component. For details of this model the reader is referred to Wu et al. (11) and Maas et al. (12).

The solid lines in Figure 9 are fits calculated with a diffusion equation based on this model and yield the interdiffusion coefficients of PMMA ((7 ± 5)$\cdot10^{-11}$ cm^2/s)

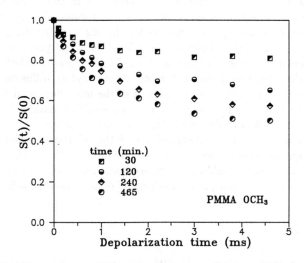

Figure 8. PMMA OCH_3 carbon intensities as a function of the ^1H-^{19}F cross depolarization time, obtained after different times at 190°C during which chain diffusion takes place.

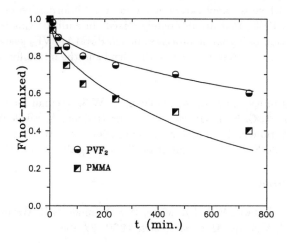

Figure 9. Fractions of isolated PMMA and PVF_2 as a function of the inter-diffusion time. The solid lines are fits calculated with a diffusion equation, from which the interdiffusion coefficients are obtained.

and PVF$_2$ $((15 \pm 5) \cdot 10^{-11}$ cm^2/s). Although the calculated curves approximate the experimental data reasonably well, we note that in the diffusion model the intrinsic diffusion coefficients are assumed concentration-independent. This assumption cannot, however, be exact since the driving force for the interdiffusion, the free energy of mixing, is known to be a function of composition (10,13).

In previous methods of detecting polymer diffusion, inert marker particles are inserted at the boundary between the components before diffusion takes place (11,13). The particles will move as a result of the mass flow, and from the displacement as a function of the diffusion time one can then obtain the ratio of the interdiffusion coefficients. The individual diffusion coefficients can then only be estimated based on prior knowledge of one of the diffusion coefficients. In the NMR technique presented here, data are acquired from both components involved in the diffusion, thereby enabling the determination of both diffusion coefficients simultaneously and without prior knowledge of either a single coefficient or their ratio.

Conclusions

The foregoing sections show that with the presented NMR techniques a detailed picture on the miscibility of polymer blends can be obtained. In the PMMA/PVF$_2$ blend, the fluorine to carbon cross polarization experiments and the fluorine to proton to carbon cross polarization experiments provide insight in miscibility on the molecular scale. The proton to fluorine cross depolarization experiments with subsequent carbon detection yield information on the amounts of PMMA and PVF$_2$ that are intimately mixed. In addition this technique provided information on dynamic processes in the PMMA/PVF$_2$ system, including phase separation and crystallization in the blend and the interdiffusion of PMMA and PVF$_2$.

Acknowledgments. The author thanks prof. W.S. Veeman for helpful discussions and support, C. Klein Douwel, P. van der Heijden, A. Eikelenboom, T. Papavoine, G. Werumeus Buning and J. Vankan for theoretical and experimental contributions and S. Pochapsky for critically reading the manuscript.

Literature Cited

(1) C.H. Klein Douwel, W.E.J.R. Maas, W.S. Veeman, G.H. Werumeus Buning, and J.M.J. Vankan. *Macromolecules*, **1990**, *23*, 406.

(2) S.R. Hartmann and E.L. Hahn. *Phys. Rev.*, **1962**, *128*, 2042.

(3) W.E.J.R. Maas, W.A.C. van der Heijden, W.S. Veeman, J.M.J. Vankan, and G.H. Werumeus Buning. *J. Chem. Phys.*, **1991**, *95*, 4698.

(*4*) A. Eijkelenboom, W.E.J.R. Maas, W.S. Veeman, J.M.J. Vankan, and G.H. Werumeus Buning. *Macromolecules*, **1992**, *25*, 18.

(*5*) C.H.M. Papavoine, W.E.J.R. Maas, W.S. Veeman, G.H. Werumeus Buning, and J.M.J. Vankan. *Macromolecules*, **1993**, *26*, 6611.

(*6*) B.S. Morra and R.S. Stein. *Polym. Eng. Sci.*, **1984**, *24*, 311.

(*7*) B.R. Hahn, O. Hermann-Schonherr, and J.H. Wendorff. *Polymer*, **1987**, *28*, 201.

(*8*) T. Nishi and T.T. Wang. *Macromolecules*, **1975**, *8*, 909.

(*9*) E. Roerdink and G. Challa. *Polymer*, **1978**, *19*, 173.

(*10*) J.H. Wendorff. *J. Polym. Sci., Polym. Lett. Ed.*, **1980**, *18*, 439.

(*11*) S Wu, H-K. Chuang, and C.D. Han. *J. Polym. Sci: Polym. Phys. Ed.*, **1986**, *24*, 143.

(*12*) W.E.J.R. Maas, C.H.M. Papavoine, W.S. Veeman, G.H. Werumeus Buning, and J.M.J. Vankan. *J.Polym Sci.,part B: Polym. Phys.*, **1994**, *32*, 785.

(*13*) E.J. Kramer, P.F. Green, and C.J. Palmstrom. *Polymer*, **1984**, *25*, 473.

RECEIVED February 2, 1995

Chapter 17

Oxygen Absorption on Aromatic Polymers
^{1}H NMR Relaxation Study

D. Capitani[1], A. L. Segre[1], and J. Blicharski[2]

[1]Istituto Strutturistica Chimica, M.B. 10, Monterotondo Stazione,
Roma 00016, Italy
[2]Institute of Physics, Jagellonian University, ul. Reymonta,
Kracow 30–059, Poland

In aromatic polymers, oxygen selectively adsorbed on aromatic rings acts as a strong relaxation contrast agent. The effect is maximal at rather low temperatures, in the 80-160K range, where a well defined minimum can be observed. The position of the minimum and the relative value of the spin-lattice relaxation time are modulated by the chemical nature of the polymer, by its packing (polymorphism), by the crystalline vs. amorphous ratio and by the maximal amount of adsorbed oxygen.
A full theoretical treatment of relaxation parameters has been done, leading to a best fit treatment of relaxation data. The proton-proton dipolar term at high temperature and the proton-oxygen scalar term at lower temperatures give the major contribution to spin-lattice relaxation. From the full theoretical equation, the best fit of experimental data on polymorphous polystyrenes leads to a large number of physico-chemical parameters. The activation energy for the phenyl ring libration was obtained, different for each polymorphous form. Moreover, for each aromatic polymer, the maximal number of adsorbable oxygen molecules can be obtained, giving an index for polymers suitable to act as oxygen scavengers.

Introduction. One major impact of NMR spectroscopy in polymer science comes from its capability to distinguish different configurations in homopolymers and different monomer sequences in copolymers (*1*) . These NMR studies in solution had as a main result the measure of the statistical distribution of configurations, which is extremely useful in chemistry for the study of catalytic systems and the mechanism of

0097–6156/95/0598–0290$12.25/0

polymerization reactions (*2*). Moreover some hint on the rheological properties can be obtained (*3*) . However, since polymers are mostly used in the solid state, a large effort was made in order to get spectra in the solid state as narrow as possible (*4*); this is today routinely performed on isotopically diluted nuclei by CP-MAS techniques (*5*). Relaxometric techniques have been known since the early beginning of NMR (*6*), while not so commonly used as spectroscopic measures, however NMR relaxometry is quite useful in polymer characterization and experimentally not very demanding (*7*). In this paper, the relaxometric behavior of aromatic polymers and the use of oxygen as a relaxation contrast agent will be discussed. Moreover a theory of proton relaxation in the presence of paramagnetic oxygen will be presented with some selected applications to oxygen-doped polymers.

Relaxation in degassed polymers. NMR relaxation in solid degassed polymers is a well known physico-chemical method (*5*). In polymers without side chains, the spin lattice relaxation as a function of temperature shows a single minimum at $\omega\tau \approx 1$; polymers with side chains may also show other secondary minima. Multiple relaxation times may be observed in semicrystalline polymers; the presence of crystalline domains in an amorphous matrix can usually be revealed since the T_2 for the crystalline domain often becomes significantly shorter at low temperature. However, in most cases only one T_1 value is observed due to spin diffusion. In well degassed polymers the dominating mechanism of relaxation is the dipole-dipole interaction , at high temperature in the weak collision case (*8*) and at low temperature in the strong collision case (*9*). On this regard aromatic polymers behave exactly like any other polymer ,(see Figure 1) for different polycrystalline polystyrenes . In this figure the relaxation behavior as a function of the temperature is shown for two different polymorphs of s-PS(syndiotactic polystyrene, namely the β and the γ forms) and a highly isotactic polystyrene. It is clear that NMR relaxometry is a technique not sensitive to polymorphism or to different tacticity.

Oxygen absorption in aromatic polymers. In the presence of oxygen, however, a dramatic shortening of the spin lattice relaxation time occurs (*10*),   (see the lower trace of Figure 2).   The same type of curve is shown for a fully degassed sample of syndiotactic polystyrene γ form to show the extent of the O_2 effect.   Since the absorption of oxygen is different in different aromatic polymers and since it must be modulated by the molecular packing, a full relaxation study was performed on s-PS having different polymorphism, (see Figure 3).   s-PS is in fact a semicrystalline polymer whose polymorphous   forms have been extensively studied and well characterized by X-Ray and CP-MAS (*11*-

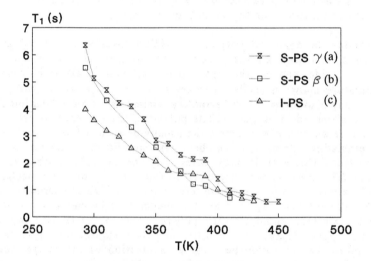

Figure 1. 30MHz. Proton T_1 relaxation times (in seconds) as a function of the temperature for three well degassed polystyrenes:
(a) γ syndiotactic polystyrene;
(b) β syndiotactic polystyrene;
(c) isotactic polystyrene.

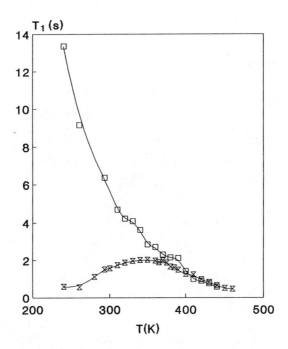

Figure 2. Semilogarithmic plot of proton T_1 relaxation times (in milliseconds), measured at 30 MHz, as a function of 1000/T (K^{-1}) for a well degassed sample of syndiotactic polystyrene in γ polymorphous form (upper trace) and for the same undegassed sample (lower trace); solid line through experimental points results from the best fit procedure.

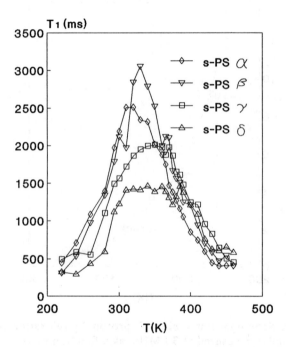

Figure 3. 30MHz Proton T_1 relaxation times as a function of the temperature for α, β, γ and δ s-PS polymorphous forms.

12). Even without any theoretical interpretation, the relaxometric data allow some preliminary conclusions for aromatic polymers in the presence of oxygen:

i)proton spin lattice relaxation is a technique able to discriminate among different polymorphs.

ii)differences in proton spin lattice relaxation might be used to gain improved CP-MAS ^{13}C spectra (*13*).

iii)using the observed differences in proton relaxation times, optimal temperatures may be found for obtaining CP-MAS spectra.

Multi-exponential decays: the spin diffusion process.

In s-PS, at temperatures higher than ≈220K, after a 180°−τ−90° sequence (or any equivalent sequence such as an aperiodic saturation recovery or a simple saturation recovery) a single exponential decay was observed; however by decreasing the temperature a bi-exponential decay appears (*14*) (see Figure 4). The two components correspond to two spin lattice relaxation times: one of the order of 1 s the second one of the order of 0.1 s. The relative intensity of the two components is ≈3:5. Since in polystyrenes there are three backbone protons (aliphatic) and five aromatic protons, the hypothesis was forwarded that different relaxation times might correspond to backbone and aromatic protons. To verify this hypothesis a backbone deuterated s-PS was synthesized and crystallized in all possible polymorphous forms. The lower trace of Fig. 4 corresponds to a backbone deuterated s-PS in the same polymorphous phase and the same crystalline content than the upper trace. Since the slow relaxing component is missing, it is clear that indeed this component is due to backbone protons and that a full separation in T_1 occurs between aliphatic and aromatic protons. A similar result was obtained for an amorphous polyphenylene oxide (*15*) (see Figure 5) ; in fact at temperatures lower than ≈250 K two spin lattice relaxation times can be measured, whose relative intensity is   3:1. These relaxation times have been attributed respectively to the methyls and to the aromatic protons. Similarly in an alternate copolymer between carbon monoxide and styrene, at 200 K , the phenyl rings relaxation time is different and much shorter than the T_1 relaxation time of the backbone, as confirmed by the corresponding data on a backbone deuterated copolymer (Capitani, D.; Segre, A.L.; Barsacchi, M.; Lilla, G. paper in preparation) . Thus it appears clear that the spin diffusion process, which equalises T_1 values, is not acting when the differences in relaxation values become sufficiently large. This behaviour is by no means confined to polymeric systems since an analogous behaviour was also found in benzoic acid (*16*).

In Table I, the temperature at which the spin diffusion process is no more active is reported.

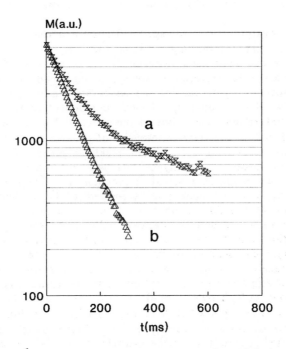

Figure 4. ¹H NMR , 30 MHz .   Semilogarithmic plot of
experimental points obtained from an IR experiment at   T=200 K
on a s-PS in γ polymorphous form:
a) fully protonated
b) deuterated on the backbone.

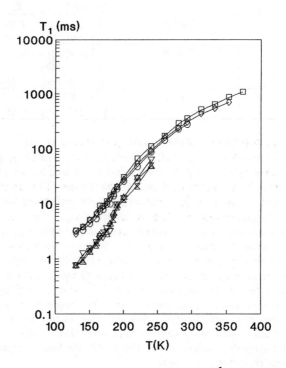

Figure 5. Totally amorphous silylated PPO. ^{1}H NMR , 30MHz. T_1 values as a function of the temperature.

Table I.   **Temperature at which the efficiency of the spin diffusion process is lowered and multiexponentiality is observed with full separation between spin-lattice relaxation of aromatic and aliphatic protons**

Polymer	Temperature	Relaxation mechanism
s-PS α	200 K	CH, CH_2 dipolar phenyl ring O_2
s-PS β	200 K	CH, CH_2 dipolar phenyl ring O_2
s-PS γ	220 K	CH, CH_2 dipolar phenyl ring O_2
PPO	240 K	CH_3 rotational+ dipolar+O_2 phenyl ring O_2
CO-Styrene	200 K	CH, CH_2 dipolar phenyl ring O_2

Relaxation in backbone deuterated polystyrenes. Due to the well known complexity of solving multi-exponential decays (*17*), and with the purpose of quantifying oxygen absorption on aromatic polymers , s-PS was synthesized fully deuterated on the backbone. The deuterated s-PS was crystallized to give α,β,γ semicrystalline polymorphous polymers, which were characterized by X-Ray diffraction.   On these polymers, proton T_1 relaxations were measured at 30 and 57MHz as a function of temperature,   (see Figures   6 and 7) . It must be noted that at low temperatures a multiple decay is again observed for T_1 relaxation times. The observed relaxation multiplicity in these samples can be attributed to the different amount of oxygen absorbed in the crystalline phase on respect to its amorphous counterpart. On this respect oxygen acts as a true relaxation contrast agent, allowing the observation of different domains. A comparison between X-Rays and NMR data is reported in Table II , showing the reliability of the NMR method.   Details of these data have been reported elsewhere (*14*).

Table II. Percentage of different phases in syndiotactic polymorphous polystyrenes

	α		β		γ	
	NMR	X-Ray	NMR	X-Ray	NMR	X-Ray
crystalline phase	35-40	40	60	60	40	40
amorphous phase	$\begin{Bmatrix}50\\10\end{Bmatrix}$	60	$\begin{Bmatrix}30\\10\end{Bmatrix}$	40	60	60

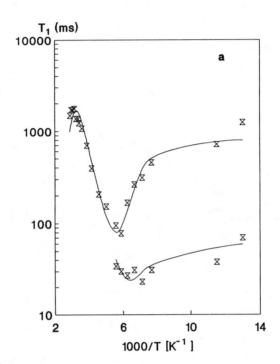

Figure 6. ^{1}H NMR, 30MHz . Semilogarithmic plot of T_1 relaxation times (in milliseconds) as a function of 1000/T (K^{-1}) respectively for: a) Backbone deuterated syndiotactic polystyrene in the α polymorphous form. b) Backbone deuterated syndiotactic polystyrene in the β polymorphous form. c) Backbone deuterated syndiotactic polystyrene in the γ polymorphous form.
Relaxation times were measured at 30 MHz within the temperature range 77-400 K; solid line through experimental points results from the best fit procedure. *(Continued on next page.)*

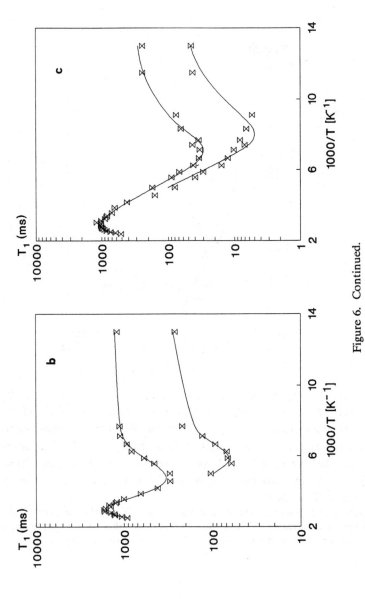

Figure 6. Continued.

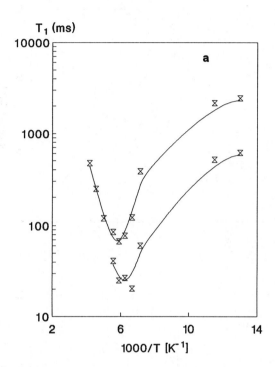

Figure 7. ^{1}H NMR, 57MHz   Semilogarithmic plot of T_1 relaxation times (in milliseconds) as a function of 1000/T (K^{-1}) respectively for: a) Backbone deuterated syndiotactic polystyrene in the α polymorphous form.   b) Backbone deuterated syndiotactic polystyrene in the β polymorphous form. c) Backbone deuterated syndiotactic polystyrene in γ polymorphous form.
Relaxation times   were measured at 57 MHz within   the temperature range 77-298 K;   solid line through experimental points results from the best fit procedure. *(Continued on next page.)*

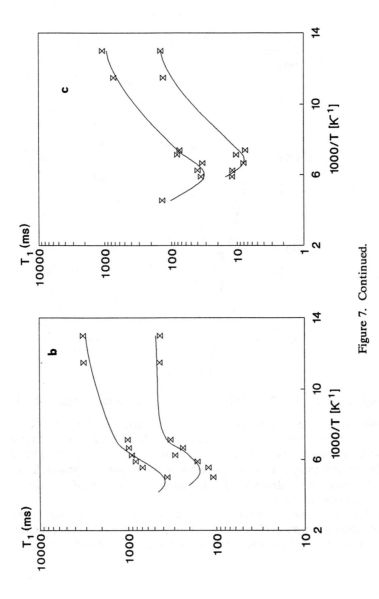

Figure 7. Continued.

Theory of spin lattice relaxation in oxygen-doped polymers.

Consider a system of interacting spins I_i and S_i with magnetic moments μ_{I_i} and μ_{S_i} the spin-Hamiltnian of this system can be considered as a sum of two terms: $\mathcal{H} = \mathcal{H}_0 + \mathcal{H}'(t)$ where $\mathcal{H}_0$ is the unperturbed Zeeman Hamiltonian and is $\mathcal{H}'(t) = \mathcal{H}_d^{II}(t) + \mathcal{H}_d^{IS}(t) + \mathcal{H}_s^{IS}(t)$ a perturbation taking into account the following time dependent interactions between spins:

1) the dipole-dipole interaction between nuclear spins I ;
2) the dipole-dipole interaction between nuclear spins I and electron spins S;
3) the scalar interaction between nuclear and electron spins.

The energy of interaction between the nuclear magnetic field and an electron spin S with magnetic moment μ_S is :

$$E = \mu_S B =$$

$$= g_e \mu_B g_N \mu_N \left\{ -\frac{I \cdot S}{r^3} + 3\frac{(I \cdot r)(S \cdot r)}{r^5} \right\} + g_e \mu_B g_N \mu_N \frac{8\pi}{3} \delta(r) I \cdot S$$

where $\mathcal{H}_d = -g_e \mu_B g_N \mu_N \left\{ -\frac{I \cdot S}{r^3} + 3\frac{(I \cdot r)(S \cdot r)}{r^5} \right\}$

is the Hamiltonian for the dipole-dipole interaction and

$\mathcal{H}_s = g_e \mu_B g_N \mu_N \dfrac{8\pi}{3} \delta(r) I \cdot S = A \dfrac{8\pi}{3} \delta(r) I \cdot S$ is the scalar Hamiltonian.

Let us introduce polar coordinates and the raising and lowering operators $I^{\pm}, S^{\pm}$, and time dependent random functions of the angles $\Theta(t), \Phi(t)$ defining the orientation of the inter-spin vector. These random functions are proportional to the spherical harmonics $F_l^{(m)}(\Theta(t), \Phi(t)) \propto Y_l^{(m)}(\Theta(t), \Phi(t))$, and in the simplest case of a two spin system, (l=2 ; m=0, ±1,±2), the following expression can be obtained (*18*):

$$\mathcal{H}_d^{IS} = \left(\frac{\mu_0}{4\pi}\right)\frac{h^2}{r^3}\gamma_I\gamma_S \left\{ \begin{array}{l} \left[I_z S_z - \frac{1}{4}(I_+ S_- + I_- S_+)\right]F_2^{(0)} + \\[2mm] \left[I_+ S_z + I_z S_+\right]F_2^{(1)} + \left[I_- S_z + I_z S_-\right]F_2^{(1)*} + \\[2mm] I_+ S_+ F_2^{(2)} + I_- S_- F_2^{(2)*} \end{array} \right\}$$

and $\mathcal{H}_s^{IS} = A\left[I_z S_z + \frac{1}{2}(I_+ S_- + I_- S_+)\right]$

It is possible to calculate the transition probability between an initial state i and a final state f:

$$W_{if} = \frac{1}{th^2}\left|\int_0^t \langle m_f |\mathcal{H}_d(t)| m_i\rangle e^{-i\omega_{if}t} dt\right|^2 \qquad \text{where } \omega_{if} = \frac{E_f - E_i}{h}.$$

Applying the master equation for spin population and following the Solomon treatment (19,20) it is possible to obtain the following equations for the longitudinal component of I and S:

$$\frac{dI_z}{dt} = -(W_0 + 2W_1 + W_2)(I_z - I_0) - (W_2 - W_0)(S_z - S_0)$$

$$\frac{dS_z}{dt} = -(W_2 - W_0)(I_z - I_0) - (W_0 + 2W_1' + W_2)(S_z - S_0)$$

where W_0, W_1, W_1', W_2 are the transition probabilities of a two spins system and I_0, S_0 are the equilibrium values respectively for I_z and S_z. In the case of a "spin-like" dipole-dipole interaction $W_1 = W'$; the only observable quantity is the sum $(I_z + S_z)$ and the following equation can be obtained:

$$\frac{d(I_z + S_z)}{dt} = -2(W_1 + W_2)(I_z + S_z - I_0 - S_0)$$

This equation represents a decay with relaxation rate:

$$\frac{1}{T_1} = 2(W_1 + W_2).$$

In the presence of "spin-unlike" dipole-dipole interaction between a nuclear spin I and an electron spin S it is possible to consider $S_z = S_0$. In this case in fact the dipolar interaction generates a relaxation process which is negligible for the electron spin on the NMR time of observation. The following equation is obtained:

$$\frac{dI_z}{dt} = -(W_0 + 2W_1 + W_2).$$

Introducing the calculated expressions for the transition probabilities W_0, W_1, W_2 the following equations can be respectively obtained in the case of "spin-like" and "spin-unlike" dipole-dipole interaction:

$$\frac{1}{T_{1d}^{II}} = G_{II}\left\{\frac{F}{F^2 + \omega_I^2} + \frac{4F}{F^2 + 4\omega_I^2}\right\} \qquad (1)$$

where G_{II} is the coupling constant for the dipole-dipole proton interaction.

$$\frac{1}{T_{1d}^{IS}} = \frac{4}{15}G_{IS}\left\{\frac{C}{C^2 + (\omega_I - \omega_S)^2} + \frac{3C}{C^2 + \omega_I^2} + \frac{6C}{C^2 + (\omega_I + \omega_S)^2}\right\}$$

and G_{IS} is the coupling constant for proton-oxygen interaction; since $\omega_S \gg \omega_I$ the following equation is obtained:

$$\frac{1}{T_{1d}^{IS}} = \left[A_{IS}\frac{C}{C^2 + \omega_S^2} + B_{IS}\frac{C}{C^2 + \omega_I^2}\right] \qquad (2)$$

where $A_{IS} = \dfrac{28}{15}G_{IS}$ and $B_{IS} = \dfrac{12}{15}G_{IS}$

In the presence of scalar interaction the only contribution to the relaxation rate is the transition probability W_0; the equation for the relaxation rate is:

$$\frac{1}{T_{1s}^{IS}} = \frac{2}{3}A^2S(S+1)\frac{E}{E^2 + (\omega_I - \omega_S)^2}$$

If $E_{IS} = \dfrac{4}{3}\left(\dfrac{A^2}{h^2}\right)$ and $\omega_S \gg \omega_I$ the above equation can be semplified

as follows: $\dfrac{1}{T_{1s}^{IS}} = E_{IS}\dfrac{E}{E^2 + \omega_S^2} \qquad (3)$

Correlation times. As previously described the following interactions between spins were taken into account:
1) the dipole-dipole proton interaction.
2) the dipole-dipole proton electron interaction: the interaction between nuclear spins (protons) and the unpaired electrons (S=1) belonging to O_2 molecules has been taken into account.
3) the scalar proton-oxygen interaction.
Average effective correlation times for molecular motions depend on the temperature; these are:
a) τ_R, correlation time for isotropic motion which obeys to the Arrhenius law:

$$\frac{1}{\tau_R} = R_0 \exp\left(\frac{-E_R}{RT}\right)$$ where E_R is the activation energy for motions of

the polymeric backbone.

b) τ_L, correlation time for the phenyl ring libration,

$$\frac{1}{\tau_L} = L_0 \exp\left(\frac{-E_L}{RT}\right)$$

where E_L is the activation energy for librational motion.

c) τ_{O_2}, the effective correlation time for the oxygen in the complex; it has been considered in turn, as a sum of two terms:

$$\frac{1}{\tau_{O_2}} = \frac{1}{\tau_E} + \frac{1}{\tau_S}.$$

The correlation time τ_E is an Arrhenius-like term accounting for the exchange in the complex aromatic ring·O_2: $\frac{1}{\tau_E} = E_0 \exp\left(\frac{-E_E}{RT}\right)$, where

E_E is the binding energy of the complex.

The correlation time τ_S is a collision-like term taking into account for the random motion due to the collisions in the oxygen gas; it has the

form: $S = \frac{1}{\tau_S} = pT^{1/2}$

If other processes occur which affect the scalar proton-oxygen interaction, their effect could be included in the collision term as

additional correlation rates s; in this case $S = \frac{1}{\tau_S} = pT^{1/2} + s$

Equation for the spin-lattice relaxation rate. Assuming:

$$F = R_0 \exp\left(\frac{-E_R}{RT}\right) + L_0 \exp\left(\frac{-E_L}{RT}\right)$$

$$C = F + S$$

$$E = E_0 \exp\left(\frac{-E_E}{RT}\right) + S$$

the spin-lattice relaxation rate is expressed by the following equation

$$\frac{1}{T_1} = G_{II}\left(\frac{F}{F^2 + \omega_I^2} + \frac{4F}{F^2 + 4\omega_I^2}\right) +$$

$$u\left(A_{IS}\frac{C}{C^2 + \omega_S^2} + B_{IS}\frac{C}{C^2 + \omega_I^2} + E_{IS}\frac{E}{E^2 + \omega_S^2}\right)$$

(4)

It must be outlined that dipolar and scalar proton-oxygen interactions strongly depend on the motion of oxygen molecules and on their paramagnetic relaxation (*20-22*).

Moreover, by lowering the temperature, the molar fraction of adsorbable oxygen increases. Thus while at higher temperatures, in the weak collision case, the proton dipole-dipole interaction is active, at low temperature the only effective term is that due to paramagnetic oxygen. Note that in the equation the coefficient u is the maximum molar fraction of absorbable oxygen: *a scale of u terms for different polymer gives an evaluation of the polymer as oxygen scavenger.*

Equation (4) is by no means exhaustive of all possible cases, in particular a low amount of oxygen such as in badly degassed polymers requires additional terms (*9*). Moreover, if polymers adsorb high quantities of oxygen it might be necessary to introduce for the oxygen a translational correlation time and the diffusion coefficient (*23*). The general theory, in a slightly different and more complete notation, has been reported elsewhere (Capitani, D.; Segre, A.L.; Blicharski, J.S. *Macromolecules*, in press).

Best Fit of experimental data. Equation (4) can be written in a parametric form and optimized with a fit program using a least squares procedure (Sykora, S. Program "Stefit", part of the software of spectrometer "Spinmaster"). Filled curves in Figures 6 and 7 have been obtained by the optimizing eq. 4 . Even if a full discussion of the parameters will be reported elsewhere , some of the chemico-physical information deserves a comment. In particular two kinds of parameters can be obtained, one is due to molecular motions of the polymeric system itself, see Table III; while the second set is due to the interactions between the polymeric system and oxygen , see Table IV.

The parameter E_R has been previously defined as the activation energy for motions of the polymeric backbone. These motions become relevant to the relaxation process at high temperature. As a consequence this energy must show little or no dependence on the polymorphism and on the presence or absence of oxygen. In fact, as shown in Table III, all E_R values are the same within the experimental error and are not sensitive to the presence of oxygen.

The parameter E_L has been defined as the activation energy for the phenyl ring libration.

Table III. Best fit parameters E_R, E_L are reported for α, β and γ polymorphous polystyrenes; last column in the table refers to a well degassed γ (deg-γ) polymorphous form. Data collected at 30 MHz.

30 MHz	α	β	γ	deg-γ
E_R	31.3±0.2	31.3±0.2	31.1±0.2	31.1±0.2
E_L	5.5±0.2	6.7±0.1	4.6±0.2	4.6±0.2

Table IV. Mean values $\bar{u}$ (molar fraction %) and $\overline{E_E}$ (KJ/mol) of best fit values of parameter u and E_E, obtained from experimental data collected at 30 and 57 MHz, are reported for crystalline ("c") and amorphous ("a") fractions of α, β and γ polymorphous polystyrenes

	$\bar{u}$	$\overline{E_E}$
α_c	5±2	15.0±1
β_c	1.1±0.2	12.0±0.6
γ_c	12±3	10.4±0.5
α_a	15±5	13.5±0.5
β_a	3.6±0.3	11.7±0.3
γ_a	59±5	11.2±.7

From Table III it can be seen that different values for E_L are obtained in different polymorphous forms with the following order: $E_L(\beta) > E_L(\alpha) > E_L(\gamma)$. In agreement with this observation the β polymorphous form is characterized by a tighter molecular packing than the α form (crystallographic density of the α form is 1.033 g/cc while it is 1.076 g/cc for the β form ; De Rosa, C.; Guerra, G.; Corradini, P. , personal communication).

In Table IV, $\bar{u}$ has been defined as the maximal molar fraction, $°/_{\infty}$, of adsorbable oxygen in each phase. The evaluation of the term $\bar{u}$ constitutes an important chemical application of the outlined theory.

In fact a scale of the term $\bar{u}$ gives an evaluation of the polymer as an oxygen scavenger. This evaluation is fundamental in the study of membranes used for gas separation. The evaluation of $\bar{u}$ in different polymorphs is shown in Table IV. The following index for oxygen adsorption can be obtained: $\beta_c < \beta_a \leq \alpha_c < \gamma_c \leq \alpha_a << \gamma_a$ where the subscript "c" means crystalline phase and the subscript "a" means amorphous phase. This scale relates well to the permeability scale (24,25) for organic solvents, thereby showing that: $\beta < \alpha < \gamma$.

The parameter E_E has been defined as the binding energy for the complex aromatic ring·O_2.

The evaluation of E_E in different polymorphous is also reported in Table IV. The mean value of E_E for all polymorphous forms is about 12Kj/mol.

Conclusions
The good match between the previously outlined theory and the experimental relaxation values allows the accurate determination of physico-chemical parameters related to oxygen adsorbed on aromatic polymers. Oxygen acts as a powerful relaxation contrast agent. As a consequence those motional parameters which are active in the Larmor

frequency range can be evaluated. In aromatic polymers and, in particular for the case of polymorphous polystyrenes, these parameters are of two types, i.e. relative to motions of the polymer, such as backbone overall motions or phenyl ring librations, and physico-chemical parameters relative to the binding of the oxygen in the complex with the aromatic polymer. The differentiation of libration energies in different polymorphs of syndiotactic polystyrene demonstrates the strength of the method.
The possibility of measuring the actual amount of adsorbed oxygen as a function of temperature is the most important application of the theory. In this regard, other aromatic polymers that are able to act as oxygen scavengers will be investigated further.

Acknowledgements. This research was partly supported by the CNR special ad hoc program "Chimica Fine II".

Literature cited

1)Tonelli, A.E. "*NMR Spectroscopy and Polymer Microstructure: The Conformational Connection*" VCH NewYork USA,**1989**.

2)Corradini, P.; Busico, V.; Guerra ,G. "Monoalkene polymerization.Stereospecificity " in "*Comprehensive Polymer Sci.*" Pergamon Press Oxford pp.29-50, **1989**.

3)Avella, M.; Martuscelli, E.; DellaVolpe, G.; Segre, A.L.; Rossi, E. *Makrom.Chem.* **1986**, *187*, 1927

4) Andrew,E.R. "*High Resolution NMR in Solids*" Intnl.Rev.Sci. , Phys.Chem., Series II, vol.4, Magnetic Resonances pp.173-220, **1987**.

5)McBrierty, V.J.; Packer, K.J. *Nuclear Magnetic Resonances in Solid Polymers* Cambridge Univ.Press, Cambridge G.B., **1993**.

6)McCall, D.W. "Relaxation in solid polymers", in "*Molecular Dynamics and and Structure*" , Carter,R.S. and Rush, J.J.(eds.), National Bureau of Standards, Special Publn 301, Washington, pp.475-537, **1969**.

7)Fukushima, E.; Roeder, S.B.W. "*Experimental Pulse NMR*"
Addison-Wesley Pub.Co. Reading, Mass. USA, **1981**.

8)Bloembergen, N.; Purcell, E.M.; Pound, R.V. *Phys. Rev.* **1948**, *73*, 679.

9)Slichter, C.P. Ailion, D. *Phys. Rev.* **1964**, *135*A, 1099.

10)Capitani, D.; Segre, A.L.; Grassi, A.; Sykora, S. *Macromolecules* **1991**, *24*, 623.

11) Guerra, G.; Vitagliano, V.M. De Rosa, C,; Petraccone, V.; Corradini, P. *Macromolecules* **1990**, *23*, 1539.

12) Grassi, A.; Longo, P.; Guerra, G. *Makrmol. Chem. Rapid. Comm.* **1989**, *10*, 687.

13) Hochmann. J; Kellerhals , H. *J.Magn.Res.* **1980** , *38*, 23.

14)Capitani, D.; De Rosa, C.; Ferrando, A,; Grassi,A.; Segre, A.L. *Macromolecules* **1992**, *25*, 3874.

15)Capitani, D.; Clericuzio, M.; Fiordiponti, P.; Lillo, F.; Segre, A.L. *Eur. Polym. J.* **1993**, *29*, 1451

16) Stöckli, A.; Meier, B.H.; Kreis, R.; Meyer, R.; Ernst, R.R.
J. Chem. Phys. **1990**, *93*, 1502.
17)Clayden,N.J.; Hesler,B.D.   *J.Magn.Res.* **1992**, *98*,271
(18) Weissbluth, M. *Atoms and Molecules*, Academic Press **1978**.
(19) Solomon, I. *Phys. Rev.* **1955**, *99*, 559.
(20) Solomon, I.; Bloembergen, N. *J. Chem. Phys.* **1956**, *25* ,261.
(21) Bloembergen, N. *J. Chem. Phys.* **1957**, *27*, 572.
(22) Bloembergen, N.; Morgan, L.O. *J. Chem. Phys.* **1961**, *34*, 842.\
(23) A. Abragam *The Principles of Nuclear Magnetism* , Oxford,
Clarendon Press, 1961
(24) Immirzi, A.; De Candia, F.; Iannelli, P.; Vittoria, V.; Zambelli, A.
Makromol. Chem. Rapid Comm. **1988**, *9*,761.
(25) Rapacciuolo, M.; De Rosa, C.; Guerra, G.; Menfichieri, G.; Apicella,
A.;   Del   Nobile,   M.   *J.   Mat.   Sci.*   **1991**,   *10*,184.

RECEIVED February 2, 1995

Chapter 18

Chain Packing and Chain Dynamics of Polymers with Layer Structures

Hans R. Kricheldorf, Christoph Wutz, Nicolas Probst, Angelika Domschke, and Mihai Gurau

Institut für Technische und Makromolekulare Chemie, D–20146 Hamburg, Germany

Two types of layer structures were studied, so called sanidic and normal smectic layer structures. Based on substituted terephthalic acid rigid rod polyesters and polyamides with flexible side chains were synthesized. They form sanidic layer structures consisting of parallel packed main chains and partially ordered paraffin domains between them. Comparison of measured and theoretical long spacing suggests that the alkyl side-chains interdigitate. The alkyl domains exhibit reversible melting and crystallization, whereas the backbone structure persists. The influence of number and length of the side chains as well as the influence of the temperature on the mobility of the main chain was studied by ^{2}H NMR spectroscopy of deuterated hydroquinone or phenylenediamine units. Only the tetra-substituted polyester shows isotropization at 60°C, whereas the corresponding polyamides do not melt due to the strong H-bonds between neighboring main chains.

Furthermore, poly(ester-imide)s with different alkane spacers were synthesized forming smectic LC-phases as well as solid smectic phases. In the layers of long spacers (>16 CH_2) the chain segments partially adopt trans-conformation leading to ordered domains. ^{2}H NMR measurements reveal that the spacers are highly mobile between T_g and T_m, though the poly(ester-imide)s are hard solids in this temperature range. Surprisingly, a sizable fraction of the spacers can undergo an isotropic motion in the solid smectic state.

Layer structures are the most simple, and thus, one of the most frequently occuring supermolecular structures of both low molecular weight compounds and polymers. Layer structures are formed by ions in minerals such as brucite ($Mg(OH)_2$) and sili-

0097–6156/95/0598–0311$12.50/0
© 1995 American Chemical Society

cates with $[Si_2O_5]^{2-}$ anions or in synthetic salts (e.g. $CdCl_2$, CdI_2). A layer structure where all atoms forming one layer are covalently interconnected is represented by graphite. However, most layer structures formed by organic molecules are based on weak interactions such as dipole-dipole interactions "hydrophobic bond" or hydrogen bonds. Such layer structures are characteristic for the walls of micels and vesicles formed in aqueous environment by most tensides, soaps or natural amphiphiles such as phosphatides and lecithins. Typical for layer structures of organic amphiphiles is the exposure of a hydrophilic layer (frequently containing ions) to the aqueous phase, whereas a hydrophobic layer contacts other hydrophobic components. Double layers of amphiphile molecules are for example vesicles, membranes in living cells and the walls of nerves. In these cases a hydrophobic layer is covered on both sides with layers of hydrophilic groups.

Another important class of biological layer structures are the β-sheets of polypeptides and proteins (*1,2*). The layers are formed by a parallel or (mostly) antiparallel array of peptide chains interconnected in one plane by H-bonds. This combination of strong main chains with numerous strong H-bonds has the consequence of a high mechanical stability. Therefore, β-sheet structures typically occur where proteins with high elastic modulus and tensile strength are required, for instance in silk or spider webs. Synthetic aliphatic or aromatic polyamides form analogous layer structures. A planar array of antiparallel main chains is interconnected by H-bonds (*1,3-5*). The similarity between the β-sheets and sheet structures of polyamides becomes even more evident, when the polyamides bear substituents. Quite recently several research groups have started to study poly(phenylene terephthalamide)s derived from substituted terephthalic acids and/or substituted p-phenylene diamines. Taking into account that the polyamide chain cannot adopt a zig-zag conformation the so called sanidic layer structure (Figure 1) is the analog of the β-sheet structure. Quite similar sanidic layer structures may be formed by polyesters derived from substituted terephthalic acids and hydroquinones (*6-8*), by polyimides of substituted diamines (*9*) and by polybenzobisoxazoles (*10*) derived from substituted terephthalic acids.

Another kind of layer structures are the smectic systems, which are characterized by layer planes more or less perpendicular to the chain axis (Figure 2). Depending on the tilting of the layer planes and on the molecular order inside the layers several subclasses of smectic layer structures were defined for low molecular compounds (*11*). Most of them are liquid crystalline phases. In the case of main chain polymers only smectic-A and smectic-C are LC-phases. Here the interaction between the mesogens is weak enough to allow motion of chains or chain segments relative to each other. All higher ordered smectic systems are solid mesophases because the mesogens form "two dimensional crystals" tied to each other by numerous spacers.

A third kind of layer structures are the discotic systems (*12,13*) (Figure 3) which can be formed by low molecular weight compounds or by main chain polymers. Characteristic for this kind of layer structures is a hexagonal array of the columns formed by disk-type molecules (mesogens). Because of the diversity and complexity of layer structures the resarch in this field is still in its beginning. The present work has the purpose to contribute to the knowledge about chain packing, chain dynamics and phase transitions of polymers forming sanidic or smectic layer structures.

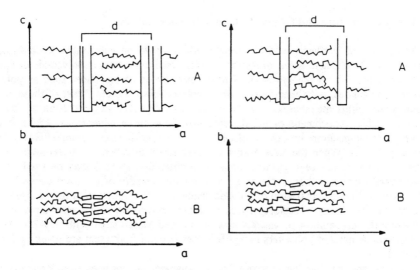

Figure 1: Schemes of sanidic layer strutures formed by rigid rod polymers of mono- or disubstituted building blocks

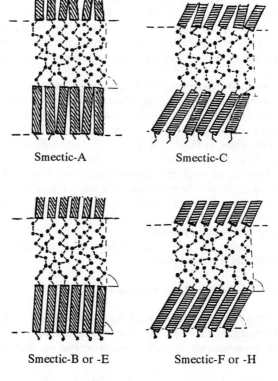

Smectic-A   Smectic-C

Smectic-B or -E   Smectic-F or -H

Figure 2: Schemes of different types of smectic layer strutures

Chemical Structure and Chain Packing of Sanidic Layers

The concept of synthesis in the present work is based on substituted terephthalic acids. Terephthalic acid is one of the most widely used monomers in polycondensation chemistry and allows the synthesis of numerous classes of polymers, such as polyesters, polyamides, poly(benzobisoxazole)s, poly(oxadiazole)s etc. Most of these polymers are neither meltable nor soluble in common solvents (with exception of conc. sulfuric acid). Attachment of substituents may have three advantages: Meltability without decomposition, improved solubility and - in special cases - compatibility with other polymers. When the substituents are exclusively attached to the terephthaloyl unit the comonomer (e.g. hydroquinone or p-phenylenediamine) may be used in a deuterated form, and thus, serve as a probe for the mobility of the main chain. Furthermore, selective deuteration of the side chains will give information on the dynamics of the side chains. Therefore, it was important to develop new synthetic methods allowing the preparation of various new substituted terephthalic acids (14-23). The selectively deuterated polyesters and polyamides derived from them are given by formulas 1 - 8.

At this time the chain packing of all these polymers has not been completely elucidated yet. However, the X-ray diffraction patterns in combination with IR- and ^{13}C NMR CP/MAS spectra and DSC measurements suggest the following scheme. Polymers derived from monosubstituted terephthalic acids form sanidic layers with a back-to-back array of the main chains (Figure 1A) (10,21,23). Figure 4 shows the X-ray diffraction pattern of polyester 6b (n=15) The sharp reflection at 2θ = 2.8° corresponds to a d-spacing of the sanidic layers of 31.5Å. Whereas the weak and broad wide angle reflections indicate small amounts of crystallinity within the paraffin domains. The side chains are "interdigitated" but details, such as a tilting of the side chains relative to the layers of the main chains, may vary with the nature of the substituent.

The polymers of disubstituted terephthalic acids form "monolayers" of main chains, again with full interdigitating of the side chains. Surprisingly, even in the case of tetraalkyl-substituted terephthalic acids (8a, b) interdigitating takes place. However, the analytical data available so far do not allow a differentiation between the normal interdigitating as illustrated in Figure 1 and a pairwise interdigitating (Figure 5). Anyway, it is a result of particular interest that the properties of the polyaramides 8a and 9 are completely different, although the number (and lengths) of the alkyl side-chains per repeating unit is identical. The side chains of the polyaramides 9 do seemingly not interdigitate at room temperature. Furthermore, these polyaramides melt around 200 °C and are soluble in chloroform and other organic solvents. In contrast, the polyaramides 8a do not melt below 400 °C and they are insoluble in all common neat solvents. In other words, the substitution pattern plays an important role for the properties of substituted "rigid rod polymers".

Chain Dynamics of Substituted Polyaramides

The influence of the substitution pattern on the mobility of the aromatic main chain can be studied by means of ^{2}H NMR spectroscopy, performing line shape analysis as well as relaxation measurements. Since the phenylene rings are deuterated four typical

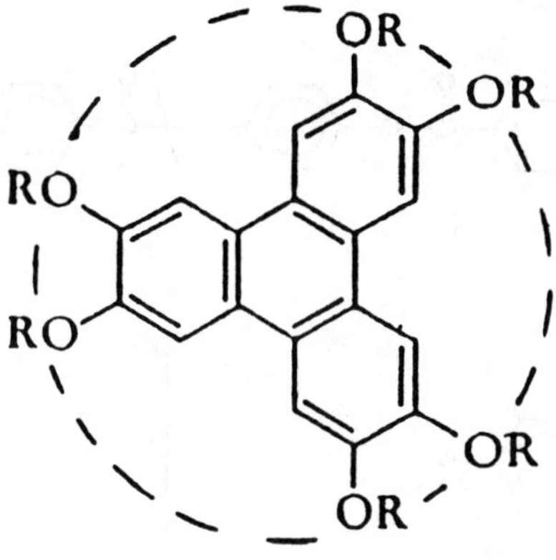

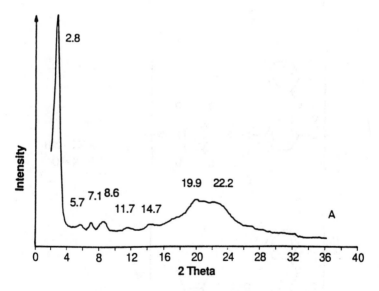

Figure 3: Scheme of a discotic layer structure

Figure 4: WAXD pattern of polyester <u>6b (n=15)</u>

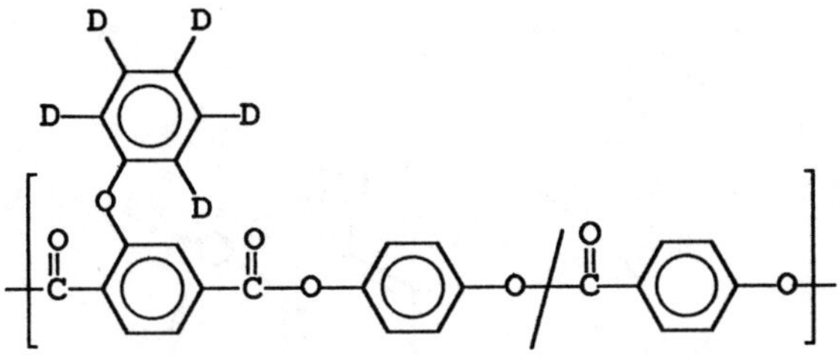

1

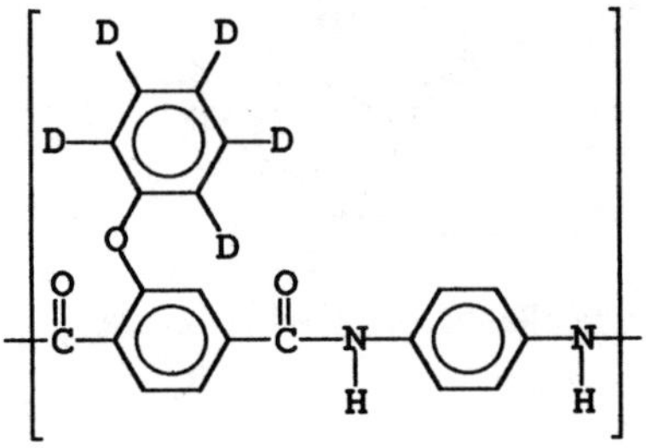

2

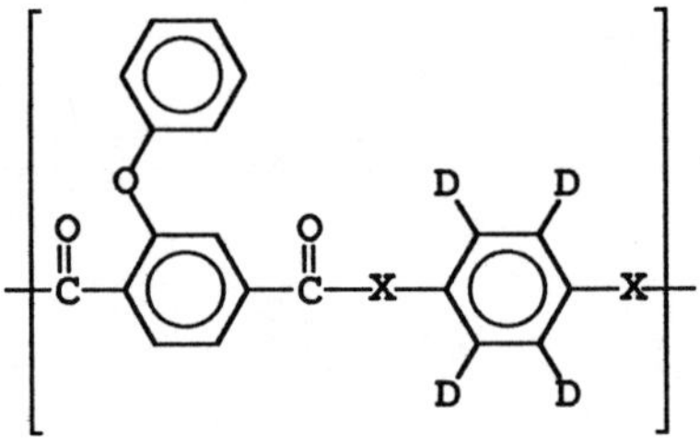

3

a : X = NH
b : X = O

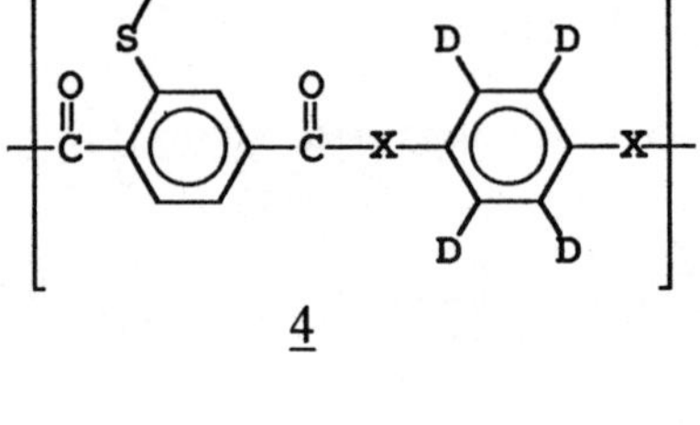

4

a : X = NH
b : X = O

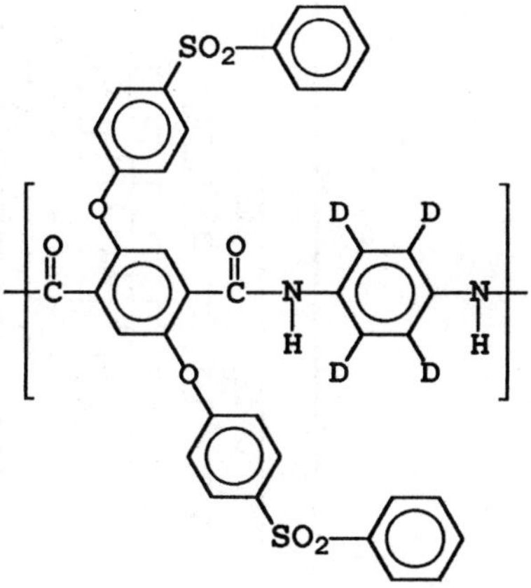

5

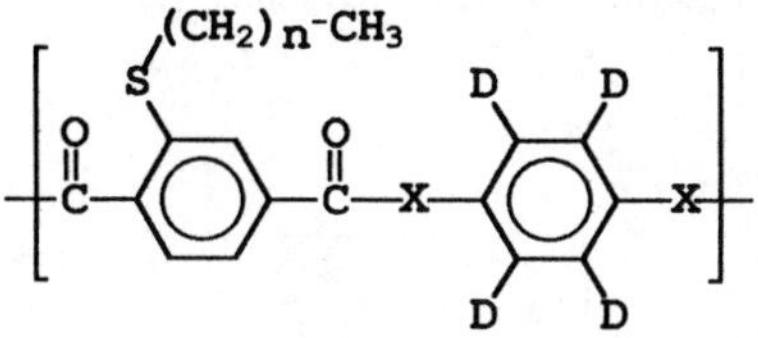

a : X = NH
b : X = O
n = 7, 11, 15

6

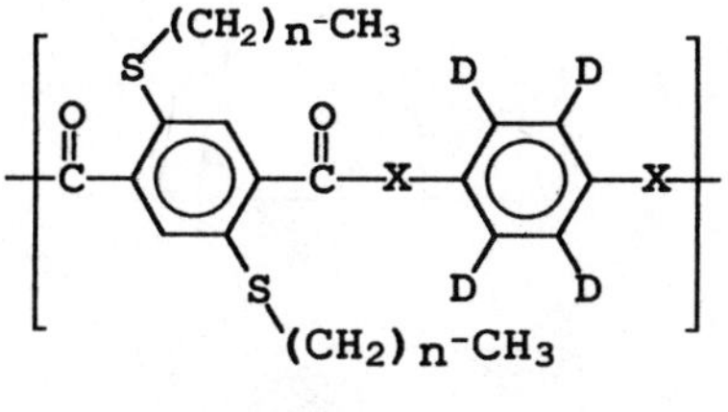

a : X = NH
b : X = O
n = 7, 11, 15

7

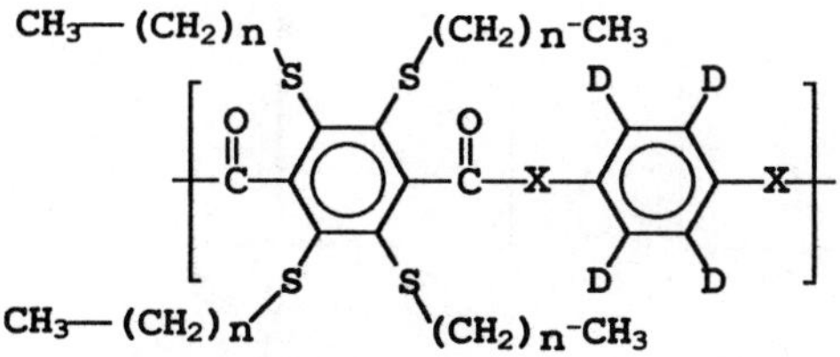

a : X = NH
b : X = O
n = 7, 11, 15

8

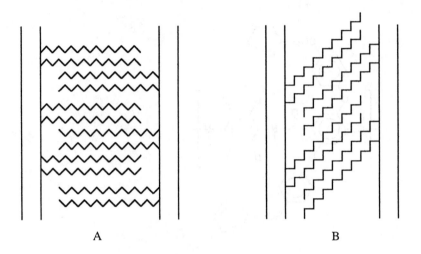

Figure 5: Scheme of pairwise interdigitating side chains of tetrasubstituted terephthalic acids. (A) side chain rectangular to the main chains, (B) side chains tilted

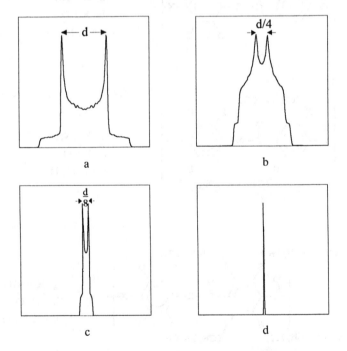

Figure 6: Calculated ^{2}H NMR spectra for different motions of a deuterated phenylene ring: rigid (a), flip (fast limit) (b), free rotation (c) and isotropic motion (d)

molecular motions and corresponding line shapes are possible (Figure 6): rigid powder pattern (Pake diagram) (a), ring flips (fast limit) (b), free rotation (c) and isotropic motion (d).

The fully relaxed spectra of the monosubstituted polyamid 6a (n=15) at different temperatures are shown in Figure 7. At 193 K the line shape corresponds to the rigid powder pattern, at 253 K inner singularities occur at $\Delta\varpi/\delta=\pm\frac{1}{4}$, which indicate flips of the phenylene rings. At higher temperatures the ring flips are dominating the line shape, but no free rotation (singularities at $\Delta\varpi/\delta=\pm1/8$) or isotropic motion is observed in the polyamides due to the strong main chain lattice with H-bond forces. The rise of the central line at 543 K turn out to be irreversible upon cooling and is thus due to decompostion of the polymer.

Comparing the spectra of the samples 6a, 7a and 8a (Figure 8) reveals the influence of the number of side chains. While the attachment of the second substituent increases the mobility compared to the mono-substituted analog significantly, the introduction of four substituents cause no further increase in mobility.

Figure 9 shows exemplary the relaxation of longitudinal magnetization of sample 6a at 293 K. The relaxation curve was fitted by one, two and three exponentials, resp. , each corresponding to a discret relaxation time, or by combinations of distributions of relaxation times.

For all polyamides and polyesters 6-8 a+b a sum of two exponential functions gives the best fit (according to standard deviation and correlation matrix), whereas the fit with three exponnetials is overdefined. The corresponding fractions and relaxation times of the components are listed in Table I.

Table I. Relaxation Times T_1 and their Amounts in Aromatic Polyamides and Polyesters with 1,2 and 4 Thio-hexadecyl Side Chains at Different Temperatures

	Polyamides			Polyesters		
	Thio-hexadecyl side chains			Thio-hexadecyl side chains		
Temperature	1 (6a)	2 (7a)	4 (8a)	1 (6b)	2 (7b)	4 (8b)
20 °C	0.48 s 55%	0.25 s 34%	0.32 s 34%	0.93 s 82%	0.38 s 49%	0.55 s 52%
	15 ms 43%	16 ms 63%	13 ms 64%	27 ms 17%	11 ms 49%	13 ms 45%
60 °C	0.33 s 24%					
	10 ms 73%					
100 °C	0.07 s 19%					
	7 ms 79%					

When sample 6a (n=15) is annealed to 333 K both relaxation times become shorter and the fraction of the faster relaxing component increases. When heating to 373 K the shortening of the longer relaxation time becomes the dominating effect.

Comparison of the relaxation times of the samples 6a, 7a and 8a confirms the result of the line shape analysis: The disubstituted polyamid 7a has more short relaxing components than the mono-substituted 6a, whereas attachment of two more side chains does not increase the mobility further.

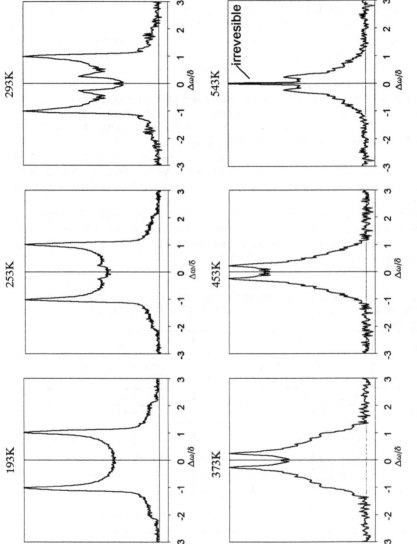

Figure 7: Fully relaxed ^{2}H NMR spectra of monosubstituted polyamid $\underline{6a}$ (n = 15) at different temperatures

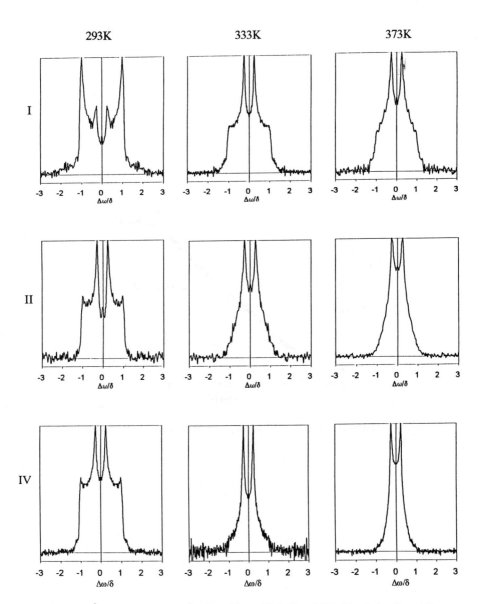

Figure 8: ^{2}H NMR spectra of polyamides with I, II and IV side chains (n = 15) at different temperatures

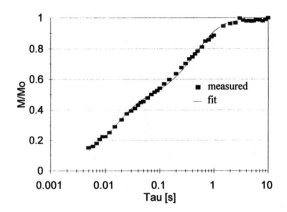

Figure 9: Relaxation of longitudinal magnetization in polyamid <u>6a</u> (n = 15) at 293 K

This result is in agreement with IR spectra which indicate the existance of perfectly H-bonded layer structures even in the case of tetrasubstituted polyaramids (23).

Chain Dynamics of Substituted Polyesters

The line shapes of the ^{2}H NMR spectra of the mono- and disubstituted polyesters 6b and 7b (Figure 10) are comparable to the analogous polyamides (Figure 8). But while the latter exhibit a splitting of the inner singularities of 32 kHz, which is the theoretical value for phenylene ring flips, the splitting of the polyesters is lower (24 kHz) and decreases during heating, indicating additional motion by librations. Attachment of four substituents 8b changes the properties absolutely: In this case a central lorentzian line appears in the ^{2}H NMR spectrum indicating isotropic motion already at 293 K. It becomes more pronounced with increasing temperature and dominates at 333 K, whereas again no free rotation of the rings is observed.

While the structure of the mono- and disubstituted polyesters is dominated by the rigid backbones, it is broken by four substituents in lack of the H-bonds present in the case of polyaramides.

Surprisingly, the isotropic motion in polyester 8b at 293 K has no remarkable influence on the relaxation of magnetization. The relaxation times T1 become shorter and the amount of short relaxing component becomes higher by attaching two instade of one side chain, but they do not change when proceeding to tetra-substitution (Table I).

Comparison of the ^{2}H NMR spectra of polyester 6b (n = 7 and 15) (Figure 11) indicates, that the length of the substituent has almost no influence on the mobilty of the main chain.

Smectic Poly(ester-imide)s

Recent studies (24) indicate that LC-main chain polymers tend to form smectic type layer structures when they possess a regular sequence of polar and non polar groups (or segments) along the main chain. In this connection it should be emphasized that the term smectic is not a synonym for liquid crystalline, but a label for a certain kind of layer structure which may be true-liquid-crystalline phases or solid mesophases. In contrast to low molecular weight LC-compounds so called LC-main chain polymers form a true smectic LC-phase only, when adopting the smectic-A or smectic-C phase. In both cases the interaction between the mesogens is relatively weak and the individual chains are mobile relative to each other. However, in the case of smectic-B and higher ordered smectic phases (e.g. F,G,E,H) the electronic forces between the mesogens are strong enough to cause the formation of "two dimensional crystals". Typical for the crystalline character of the highly ordered smectic-E and -H phases is the supercooling effect in DSC measurements (cooling curves) which results from slow nucleation steps. In contrast the formation of the typically liquid-crystalline smectic-A and -C phases is spontaneous and no supercooling effect occurs.

Poly(ester-imide)s based on alkane spacers and mesogens containing imide groups are good candidates for the formation of smectic layer structures, because the contrast between the polar imide groups and the nonpolar alkane spacers favors a registration (or molecular phase separation) of these groups. Two typical examples of such a

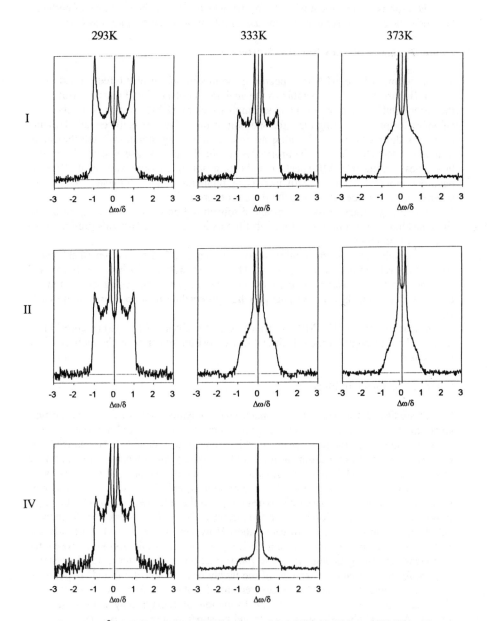

Figure 10: ^{2}H NMR spectra of polyesters with I (6b), II (7b) and IV (8b) side chains (n = 15) at different temperatures

chemical structure are the poly(ester-imide)s 10 and 11. Both classes of PEI's form smectic layer structures in the solid state, but most of their properties are quite different despite the isomeric character of their repeating units. For instance, the PEI's 10 form a nematic melt and only one kind of smectic layers in the solid state. In contrast, the PEI's 11 yield an isotropic melt upon heating and short living smectic-A phase upon rapid cooling (monotropic LC-character). Depending on the thermal history the solid state of the PEI's 11 may exist as a frozen smectic-A phase (smectic glass), as a smectic-B or as a smectic-E phase (*24,25*).

The problem studied in this work concerns the mobility and conformation of the alkane spacers in the solid smectic phases of 10 and 11. If these spacers are long enough (n > 11) they might be capable of forming small domains of ordered paraffin phases in between the layers of mesogens (Figure 12). ^{13}C NMR CP/MAS spectra should allow a distinction between the all-trans conformation of ordered paraffin domains (as illustrated for the side chains of sanidic systems) and gauche conformations of mobile chain segments. Therefore, the protonated poly(ester-imide)s 11 with n = 12, 16 and 22 were characterized by means of this ^{13}C NMR method. In all cases a strong and broad signal around 31 ppm was found representing the gauche conformations, analogous to the assignment of LDPE-signals (26) . However, in the case of n = 12 a weak shoulder is detectable at 33 ppm, in the case of n =16 a weak signal and in the case of n = 22 a strong signal shows up at 33-34 ppm (Figure 13A). These sharp signals at 33-34 ppm indicate the presence of ordered domains built up by chain segments in trans-conformation. In agreement with this interpretation the intensity of the "trans peaks" decrease with increasing temperature. But in the case of n = 22 the trans peak has not completely vanished yet when the sample is heated to 100 °C (Figure 13B).

The ^{2}H NMR spectra of 12 provide more information. The temperature dependence of the line shape indicates little motion at -100 °C and increasing mobility with higher temperatures. The most surprising and important result is the observation of isotropic motion at 100 °C (Figure14) still 50 °C below the melting point (as defined by DSC measurements). This means that considerable fraction of the long spacer can undergo an isotropic motion in the highly anisotropic solid smectic state. Such a mobility may be understood assuming that:

-   The cross-section of the mesogens is by a factor of 2 greater than the cross-section of an aliphatic spacer in trans conformation.
-   The layer distances are short enough to allow patial coiling of the spacers.

This unexpected high mobility contrasts sharply with the situation of the spacers in the copoly(ester-imide) 13. This copolyester forms a nematic melt above 100 °C and the ^{2}H NMR spectra (Figure 15) prove that the spacers undergo preferentially anisotropic motions up to 140 °C. At 160°C the isotrpoic motion leads to spontaneous ordering in the magnetic field. Obviously the spacers are more extended in the nematic phase, so that an isotropic motion is impossible.

These results demonstrate again that more information on the chain dynamics is required for a satisfactory understanding of solid layer structures.

Experimental

The NMR spectra were recorded with a BRUKER MSL-300 FT spectrometer.

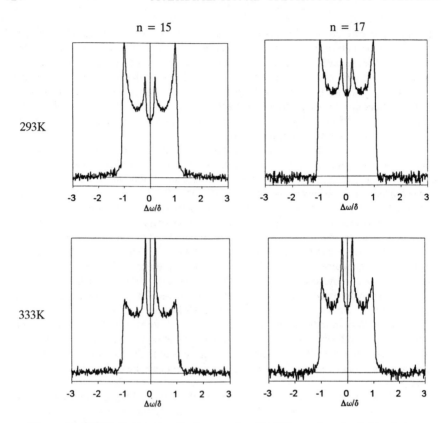

Figure 11: ^{2}H NMR spectra of polyester <u>6a</u> with different length of side chains (n = 7, 15) at different temperatures

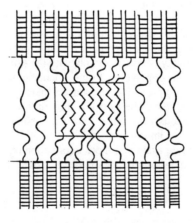

Figure 12: Schematic representation of an ordered paraffin phase in between layers of mesogens

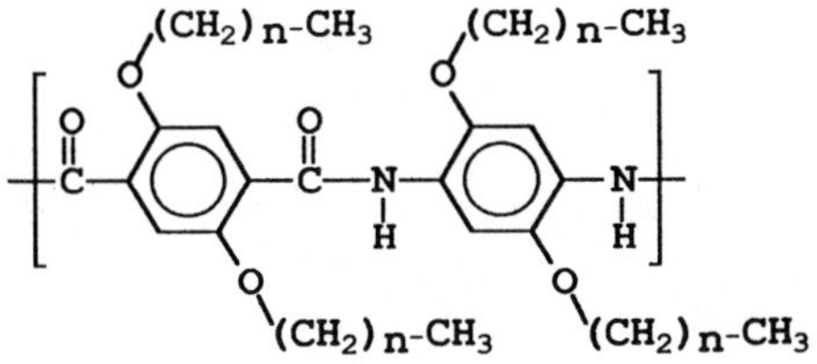

$$n = 3 - 5$$

9

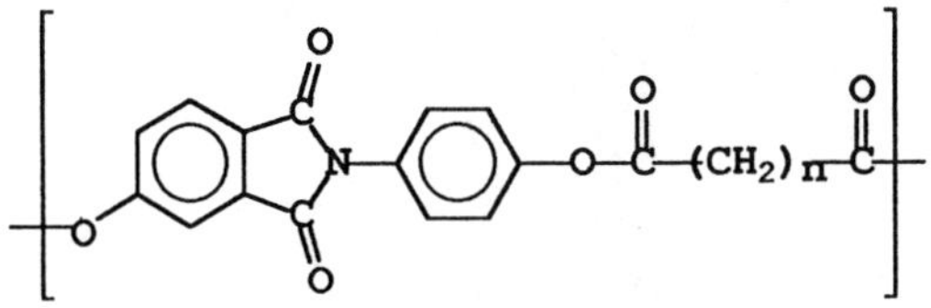

$$n = 4 - 12, 14, 20$$

10

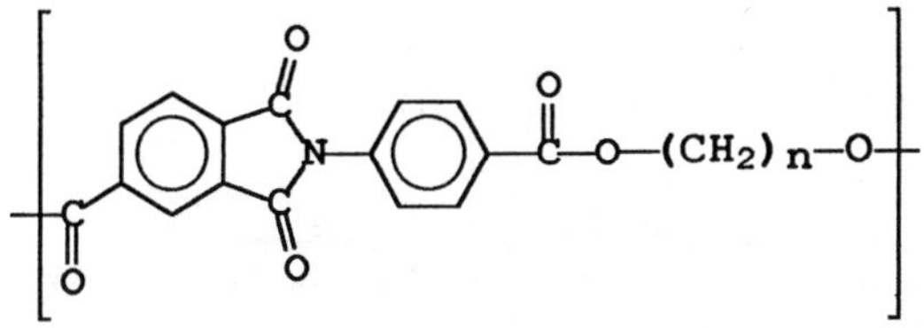

$$n = 4 - 10, 12, 16, 22$$

11

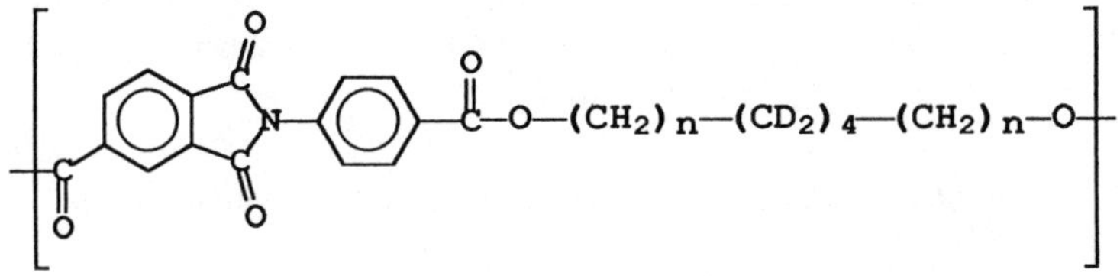

$$n = 9$$

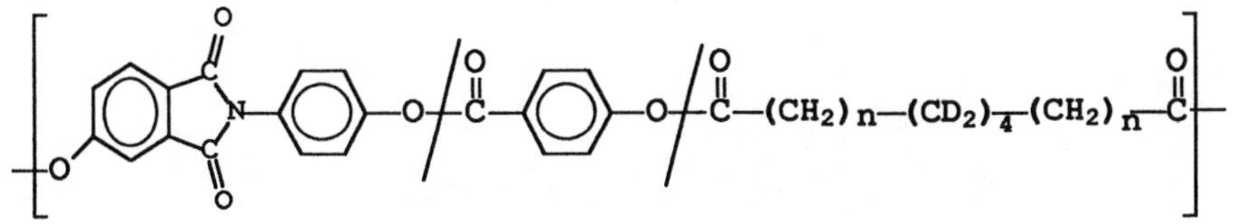

12

$$n = 8$$

13

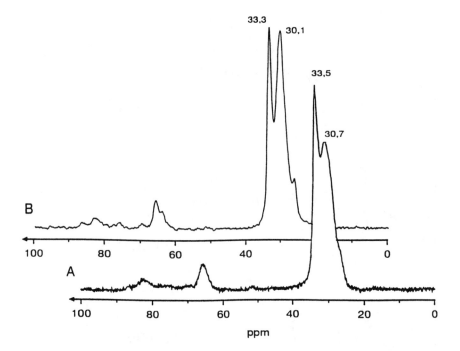

Figure 13: ^{13}C NMR CP/MAS spectra of poly(ester-imide)s 11 with n = 22 at 20°C (A) and 100°C (B)

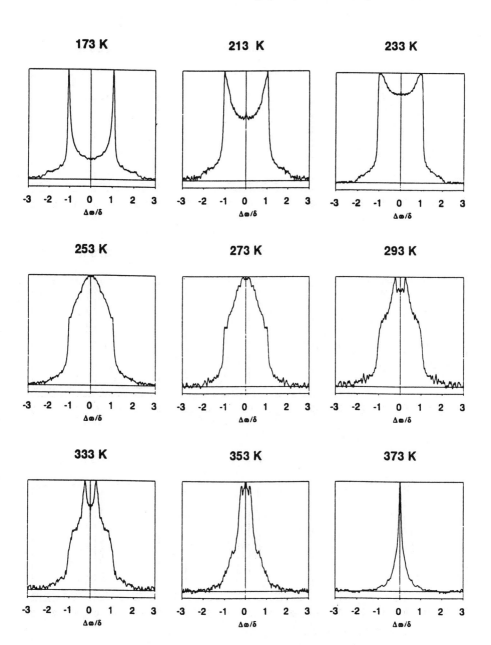

Figure 14: ^{2}H NMR spectra of poly(ester-imide) 12 at different temperatures

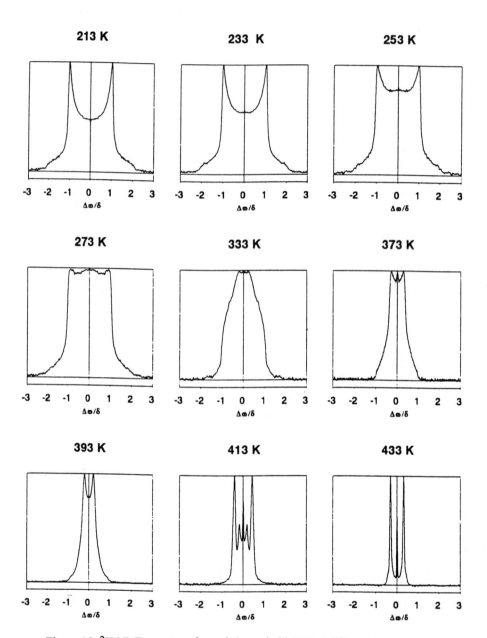

Figure 15: ^{2}H NMR spectra of copoly(ester-imide) $\underline{13}$ at different temperatures

For ^{13}C NMR CP/MAS ZrO$_2$ double bearing rotors at a spinning rate of 4 kHz were used. The contact time amounts to 1 ms, the recycle-delay 4 s. The ^{2}H NMR spectra were obtained by using the solid-echo pulse technique with a pulse delay τ_1 of 40 µs and quadrature phase detection. Fully relaxed spectra were obtained after letting the magnetization to build up after a saturation sequence for a time $\tau > 3$ $T_{1,max}$, were $T_{1,max}$ describes the long component of the spin-lattice relaxation. For the spin-lattice relaxation measurements the entire spin system is first saturated by a 90° pulse train. The spin-echo is then evaluated as a function of the delay τ, which is varied between 5 ms and 10 s.

Literature Cited

1.   Walton, A.G.; Blackwell, J. *Biopolymers;* Academic Press: New York, 1973
2.   Fraser, R.D.B.; McRae, T.P.G.; Rogers, E. *Keratins;* C.C. Thomas Inc.: Springfield Ill., 1972
3.   Müller, A.;. Pflinger; R. *Kunststoffe* **1960**, *50*, 203
4.   Northolt; M.G. *Eur. Polym. J.*, **1974**, *10*, 799
5.   Elias, H.G. *Macromolecules;* 5th ed. Hüthig & Wepf: Basel, New York, 1990; Chapter 19
6.   Ringsdorf, H.; Tschirner,P.; Hermann-Schönherr, O.; Wendorff, J. *Makromol. Chem.* **1987**, *188*, 1431
7.   Ballauff, M.; Schmidt, G.F. *Makromol. Chem. Rapid Commun.* **1987**, *8*, 93
8.   Frech, C.B.; Adam, A.; Falk, U.; Boeffel, C.; Spiess, H.W. *New Polymeric Mater.* **1990**, *2*, 267
9.   Helmer-Metzmann, T.; Rehahn, M.; Schmitz, L.; Ballauff, M.; Wegner, G. *Makromol. Chem.* **1992**, *193*, 1847
10.   Kricheldorf, H.R.; Domschke, A. *POLYMER* **1994, *35*, 199
11.   Gray, G.W.; Goodby, J.W. *Smectic Liquid Crystals;* Heyden & Son Inc.: Philadelphia, 1984
12.   Karthaus, O.; Ringsdorf, H.; Ebert, M.; Wendorff, H.J. *Makromol. Chem.* **1992**, *193*, 507
13.   Bauer, S.; Plesnivy, T.; Ringsdorf, H.; Schumacher, P. *Makromol. Chem. Macromol. Symp.* **1992**, *64*, 19
14.   Kricheldorf, H.R.; Döring, V. *Makromol. Chem.* **1988**, *189*, 1425
15.   Kricheldorf, H.R.; Schwarz, G. *Makromol. Chem. Rapid Commun.* **1989**, *10*, 243
16.   Kricheldorf, H.R.; Engelhardt, J. *J. Polym. Sci. Part A Polym. Chem.* **1990**, *28*, 2335
17.   Kricheldorf, H.R.; Weegen-Schulz, B.; Engelhardt, J. *Makromol. Chem.* **1991**, *192*, 631
18.   Kricheldorf, H.R.; Bürger, R. *Makromol. Chem.* **1993**, *194*, 1197
19.   Kricheldorf, H.R.; Bürger, R. *Makromol. Chem.* **1993**, *194*, 2183
20.   Kricheldorf, H.R.; Bürger, *J. Polym. Sci. Part A Polym. Chem.* **1994**, *32*, 355
21.   Kricheldorf, H.R.; Domschke, A. *Macromol. Chem. Phys.* **1994**, *195*, 943
22.   Kricheldorf, H.R.; Domschke, A. *Macromol. Chem. Phys.* **1994**, *195*, 957

23. Kricheldorf, H.R.; Domschke, A. *Macromolecules* **1994,** *27,* 1509
24. Kricheldorf, H.R.; Schwarz, G.; Berghahn, M.; de Abajo, J.; de la Campa, J. *Macromolecules* **1994,** *27,* 2540
25. Kricheldorf, H.R.; Schwarz, G.; de Abajo, J.; de la Campa, J. *POLYMER* **1991,** *32,* 942
26. Clauss, J.; Schmidt-Rohr, K.; Adam, A.; Boeffel, C., Spiess, H.W. *Macromolecules* **1992,** *25,* 5208

RECEIVED February 2, 1995

Chapter 19

Electron Paramagnetic Resonance and ^{1}H and ^{13}C NMR Study of Paper

D. Attanasio[1], D. Capitani[2], C. Federici[3], M. Paci[4], and A. L. Segre[2]

[1]Istituto Chimica dei Materiali and [2]Istituto Strutturistica Chimica,
Consiglio Nazionale delle Ricerche, C.P. 10, 00016 Monterotondo
Stazione, Roma, Italy
[3]Istituto Centrale per la Patologia del Libro, Via Milano 76, 00184 Roma,
Italy
[4]Dipartamento di Chimica, II Università di Roma, Tor Vergata,
Via Ricerca Scientifica, 00136 Roma, Italy

High quality antique sheets of paper have been characterized by ^{1}H NMR relaxation, ^{13}C CP MAS spectra, and Electron Paramagnetic Resonance spectroscopy. Paper can be regarded as a bi-component material made by cellulose and water plus a small amount of organic and inorganic additives and impurities.
Semi-crystalline fibrous cellulose, rich in water, is present as polymorphs I_α and I_β. Amorphous cellulose, with a lower water content, presents a higher amount of paramagnetic impurities and is characterized by quite short ^{1}H spin-lattice relaxation times and by ^{13}C resonances with noticeable chemical shifts. "Ad hoc" tailored sequences are able to produce ^{13}C CP MAS spectra in which the amorphous content of cellulose in paper is quite well observable. The nature of water as fully bound to the cellulose lattice has also been proved. Low-temperature EPR spectra have shown the presence of measurable amounts of different inorganic paramagnetic impurities, such as Fe^{3+}, Mn^{2+}, Cu^{2+} often found in different stereochemical environments. The spectra are all, qualitatively, closely similar. However, quantitative data have shown that in paper the state of conservation does not depend on the amount of pseudo-octahedral iron, but is strongly correlated to the concentration of this metal ion in a rhombic stereochemistry and to the presence of even very small amounts of copper.

Introduction. It is widely known that paper is the most commonly used writing material. While books in their current form have been produced for almost 18 centuries, paper has been used systematically for this purpose only during the last 6 centuries.

In Europe , the first kind of paper was made of rags. This was the rule until the first half of 19th century when wood pulp began to be commonly used as a raw material, causing the rapid degradation of paper which has occurred in the last 150 years.

Ancient paper, despite being always made with high quality material, can be very damaged and even fully destroyed. With the aim of characterizing paper of all possible types and its degradation, a spectroscopic characterization of paper was attempted, starting from high quality ancient paper.

Paper can be considered as a bi-component material made by cellulose and water containing small amounts of organic or inorganic additives or impurities (Paci M.; Federici C.; Capitani D.; Perenze N.; Segre A.L. *Carbohydrate Polymers* in press).

In the present work all the studied samples are unprinted sheets of well-documented historical origin, spanning between the 15th and 18th century.

This choice has many advantages, the main one being the high quality of paper made of rags, compared to modern paper made of wood- pulp.

Another advantage for the study of paper degradation is the possibility of obtaining paper in a good state of preservation, or in poor condition, or even completely destroyed, taken from a homogeneous source, i.e. an old book.

As far as this article is concerned, we will have to refer to paper conservation conditions as "destroyed" , "ruined" and "good".

We are aware that these adjective do not have scientific meaning, but they will be useful to classify the conservation state as it looks to an observer. Therefore a very brittle paper , breaking as soon as it is touched, usually brown-coloured will be labeled as "destroyed" ; a weakened paper, usually spotted and spoiled , but which keeps its own cohesive features will be labeled " ruined"; while " good" is the paper which shows no alteration marks.

Characterization of solid paper was performed as follows: pulsed ^{1}H low resolution NMR relaxation measurements at 57 MHz, high resolution ^{1}H NMR line width and T_1 measurements at 200 MHz, ^{13}C NMR spectroscopy with CP-MAS at 100 MHz also with the use of "ad hoc" tailored sequences, and EPR measurements at room and liquid nitrogen temperatures.

Results and Discussion. Cellulose, the major component of a sheet of paper, is always present as a semicrystalline polymer. The crystalline fraction is present as a mixture of different polymorphous forms always accompanied by a variable amount of amorphous material. While the complex polymorphism existing in cellulose has been demonstrated by ^{13}C CP-MAS techniques (*1*), the structure of two polymorphous forms of cellulose has been recently solved by electron diffraction (2). The existence of amorphous cellulose has been recently demonstrated by ^{13}C CP-MAS NMR on materials originating from algae (3). With the purpose of characterizing the material "paper" a full NMR study was performed using both ^{1}H and ^{13}C spectra. Finally, following the indications of NMR relaxometry, the presence of paramagnetic centers was studied by EPR spectroscopy. The information obtained by all these techniques is complementary and for the sake of clarity will be discussed separately.

Low Resolution Pulsed NMR. A proton free induction decay (FID) of the nuclear magnetization is shown in Figure 1a at 57MHz; in the time domain the fast decaying component is due to cellulose, while the barely observable slow decaying component is due to water; by Fourier transformation of the FID into the frequency domain, Figure 1b, the cellulose component appears as a broad and noisy hump, while the water component appears as a sharp peak at the top of the broad one. An appropriate choice of points within the time domain allows the measure of T_1 for both components (4). An aperiodic sequence of saturation-recovery was used with at least 128 different delays (5).

In all examined samples, both the water and the cellulose signals exhibit multiple T_1 relaxations corresponding to at least three different T_1 components, see Figure 2a and 2b. Thus for both signals a fast , an intermediate and a slow relaxing T_1 components can be observed. For all studied samples , the T_1 values of the long relaxing component of cellulose are shown in Figure 3a, plotted against the corresponding component of water. All these values , without any correction, are well correlated around the y=x line (correlation coefficient larger than 98% ,while the angular coefficient is 1.09 ± 0.06). This finding shows that a very efficient spin diffusion mechanism exists, able to transfer the magnetization from the water pool to the cellulose pool and vice versa. Thus, *water is strongly bound to the cellulose* and the amount of free water, if any, is indeed negligible.

By plotting the intermediate T_1 value of the cellulose vs. the corresponding value of water, see Figure 3b, it can be seen that all points lie in the lower part of the y=x line; the correlation coefficient is only 68% , and the slope of the best fit straight line is only 0.4±0.1. Thus the spin diffusion process is rather poor, possibly due to the presence of other relaxation mechanisms such as those due to interactions with paramagnetic impurities. Another possible explanation is that the total amount of water is not enough to be fully efficient in the spin diffusion process. The fastest T_1 component, both for cellulose and water is of the order of few milliseconds with a rather high error; thus, the operating relaxation mechanism should be dominated by paramagnetic impurities such as Cu and Fe . Indeed, as shown below, an EPR study reveals the presence of these and other paramagnetic impurities.

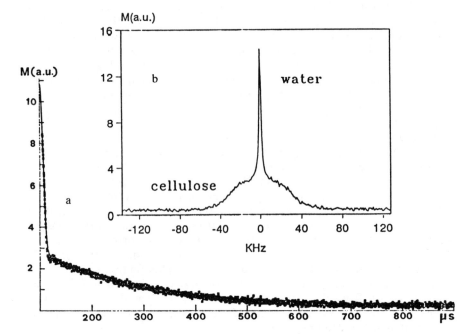

Figure 1. ^{1}H NMR at 57 MHz. a); Free Induction Decay in the time domain of a piece of antique paper. The fast decaying component is due to cellulose, while the slow decaying one is due to water.   b); Same experiment; the solid line through the experimental points is the best fit curve.

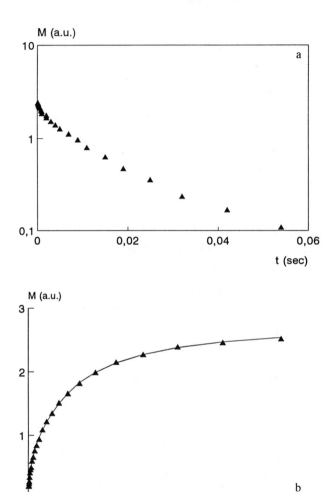

Figure 2.   ^{1}H NMR at 57 MHz. Aperiodic Saturation Recovery
experiment on the cellulose component. a)For the sake of clarity a
multi-exponential   decay   is   shown   in   a   semilogarithmic
representation. b)Same experiment in a linear scale; the full line
trough experimental points results from a best fit procedure.

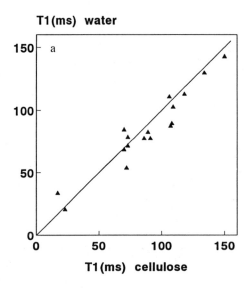

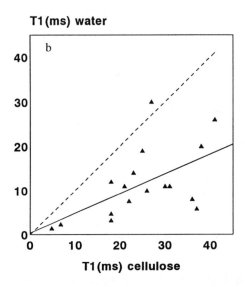

Figure 3 . a) T_1 values due to the long relaxing component of
water are reported vs. the corresponding T_1 values of cellulose.
b) T_1 values due to the intermediate relaxing component of water
are reported vs. the corresponding T_1 values of cellulose.
Solid lines through the experimental points result from a linear
regression analysis. All T_1 values were measured at 57 MHz.

High Resolution Proton NMR Spectra   The sharpness of the water resonance points to the possibility of performing (*4*) ^1H NMR measurements on a conventional high resolution NMR spectrometer, see Figure 4. Inversion-recovery experiments performed on this peak show again the presence of three spin-lattice values, whose values agree with the corresponding data at 57MHz, see Table I.

In this way, due to the efficiency of the spin diffusion process previously observed, simply by measuring the spin-lattice relaxation time of water on a conventional high resolution spectrometer, it is possible to attain the corresponding value for the slow relaxing component of cellulose.   The relative amount of water was measured on sheets of paper showing different degradation. In all sheets of paper which appear in a good state of conservation or only slightly deteriorated the amount of water is the same within 10% , while in all samples which appear in a very bad state of conservation (strong degradation) a net loss of water can be measured, see Figure 5. By comparing the proton line width of paper belonging to the same book, but showing a different state of degradation, a definite enlargement can be measured, as reported in Table I. This enlargement may be due to a higher (or different) paramagnetic contribution, as will be discussed in the EPR section. Moreover the net loss of water seems associated with a net degrade of the material; this in turn may be associated with an increase in the amorphous fraction.

In all semicrystalline polymers, the higher content of impurities can be found in the amorphous fraction, hence our attention was focused on two different samples, one as a source of cellulose rich in long fibers and highly crystalline (sample 1), the other one as a source of amorphous cellulose (sample 4), this one also with a very low water content.   On these two samples ^{13}C CP-MAS NMR spectra were performed with the aim of observing spectroscopic differences between sheets having different degradation.

^{13}C CP-MAS NMR (100MHz).

The ^{13}C CP-MAS spectra of two sheets of ancient paper either well preserved or almost completely destroyed do not show major differences. This is due to the fact that long fibers, well packed and corresponding to the crystalline fraction of the material, cross polarize much better than the amorphous material (*6,7*); this is well known and it is due to differences in relaxation times between the crystalline fraction and its amorphous counterpart. As a consequence to evidence the " amorphous" component of paper in presence of a "long fiber" component it is necessary either to enhance the amorphous component or to lower the crystalline one; all this can be actually done with " ad hoc" tailored sequences.

The pulse sequence proposed by Torchia (*8*) and taking advantage of differences in T_1 relaxation does not seem to give any significant result; likewise a preparatory sequence usually called DEFT (*9*) which realigns selectively the magnetization of the slow relaxing spins gives only partial

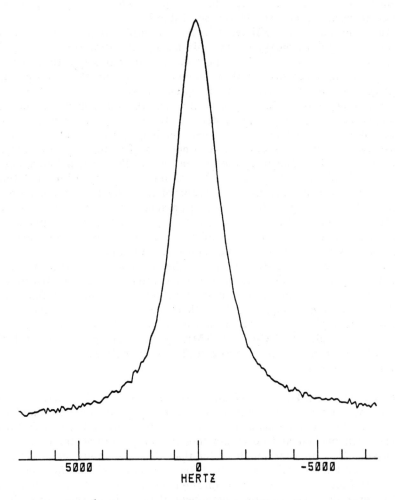

Figure 4. The ^{1}H NMR spectrum at 200.13 MHz is reported for a sample of ancient paper; only the resonance due to the water component can be observed.

Table I. Line width, Area and T_1 spin-lattice relaxation times of ancient paper samples

Sample	Line width (Hz)	Area	T_1 water 200 MHz	T_1 cellulose 57MHz	T_1 water 57MHz
1 15th century (north Italy) good	2014	96	130 40 3.48	149 41 3.5	143 26 2.0
2 15th century (north Italy) deteriorated	2014	100	124 28 3.00	73 18 2.0	80 12 1.9
3 15th century (north Italy) good	1307	101	220 53 3.7	120 30 2.0	114 11 2.2
4 15th century (Italy o Germany) destroyed	1816	80	108 40 2,1	70 23 1.4	69 14 1.7
5 15th century (north Italy) destroyed	1683	79	138 56 4.1	90 37 2.7	82 5.8 1.7
6 18th century (Sardegna) destroyed	3759	80	30 9.8 0.9	24 6.7 0.8	21 2.3 0.5
7 18th century (Sardegna) destroyed	3682	72	26 8.9 1.0	17.2 4.7 0.4	34 1.4 0.6
8 15th century (Italy) very ruined	1414	104	127 34 3.4	71 18 1.8	62 3.2 2.0
9 15th century (Italy) ruined	1403	101	98 9.3 2.4	71 18 1.4	81 4.7 1.1
10 15th century (Italy) good	1120	101	167 56 4.7	90 21 2.4	89 11.0 2.1
11 15th century (France) very ruined	1704	87	146 38.3 2.86	73 22 1.6	72 7.55 3.09
12 15th century (France) very good	1557	93	280 88 3.9	107 27 2.9	101 30 2.89
13 15th century (France) destroyed	1857	83	177 56 2.9	86 25 2.0	81 19 2.5
14 15th century (France) very good	1683	95	227 49 3.7	105 31 2.9	111 11.0 2.1
15 16th century (Germany) very good	1215	108	169 33 6.1	108 26 3.4	99 10 3.2
16 15th century (Germany) little ruined	1195	110	106 13 4.3	108 36 2.6	102 8 1.7
17 15th century (Germany) very good	1180	102	209 42 6.1	136 38 3.9	129 20 3.4

Errors on T_1 components are: <3% on the longest component; <6% on the intermediate component;<20% on the shortest components.Errors in the line width <5Hz.
Errors in the relative water content <5%.

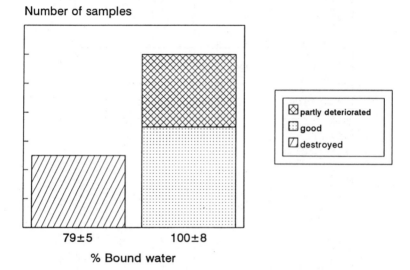

Figure 5. The relative percentage of bound water is reported respectively for samples of ancient paper in a good or only slightly deteriorated state of conservation and for samples showing full degradation. The amount of water is the same, within 10%, for samples in a good state of preservation and much lower in the destroyed samples.

results. The best results were obtained taking advantage of different proton transverse relaxation times with the following sequence:

90(1H)-tau-Spin lock(CP)

---------(CP)----------AQ(13C)

The results obtained on "good" or "destroyed " sheets of paper are shown in Figure 6a and 6b.
Here in the upper trace of Figure 6a the ^{13}C spectrum of amorphous cellulose is obtained, rather similar but with a better resolution than spectra obtained on "amorphous" cellulose from algae (3). In particular the resonances of C atoms in position 1,4,6 are split and none of the observed values is coincident with the values of the crystalline form. This strong splitting of the resonances of carbons 1,4,6 might be due to a twisted conformation of the ring around the glycosidic bond (10).

EPR Spectroscopy.   The results discussed so far led us to investigate in more detail the presence and role of paramagnetic impurities in paper. All the measured samples contain substantial amounts of paramagnetic transition ions and exhibit spectra which are all qualitatively similar. Signals ascribable to organic free radicals impurities, quite common in modern paper, are almost undetectable in these old samples. The quite general sample spectrum reported in Figure 7 is dominated by signals due to high-spin Fe^{3+}, present in two different chemical environments. The broad line centered at g $\cong$ 2.0 (peak to peak width W_{pp} = 45 $\div$ 55 mT) is typical of quasi octahedral Fe^{3+} with only minor components of a lower symmetry crystal field (11).   Such line represents the envelope of five unresolved spin transitions and does not allow precise determination of the zero field splitting parameters, which are, however, of the order of 0.1 cm^{-1} or smaller. The sharper, quasi isotropic signal found in the g $\cong$ 4.3 region (W_{pp} = 8 $\div$ 12 mT) is the well known rhombic iron spectrum very commonly found in biological materials (11,12). It represents a metal environment as far removed as possible from axial symmetry and corresponds to full inequivalence of the three magnetic principal axes. However, although the EPR spectrum for such a situation is unique, the symmetry is not and therefore the precise environment around the metal ion remains undefined.
Superimposed on the octahedral iron absorption is the typical spectrum of high-spin Mn^{2+}, i.e. the sextet due to the -1/2 $\rightarrow$ 1/2 spin transition split by hyperfine interaction of the electron magnetic moment with the manganese nuclear spin (I = 5/2).
About one third of the samples showed more or less intense additional lines, clearly due to the presence of the copper ion. In some other cases similar lines were too weak to be identified with certainty, underneath the intense iron spectrum. The copper spectra measured in different paper

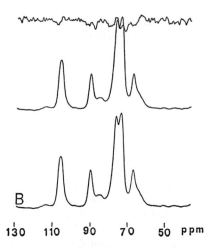

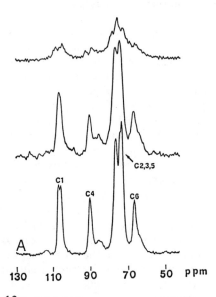

Figure 6. The ^{13}C CP MAS spectra, at 100MHz, are respectively reported for a sample of ancient paper showing full degradation (A) and for a sample of ancient paper in a good state of conservation (B). Spectra were collected with the sequence described in the text, i.e with a proton T_2 selection. The lower traces in A and B were obtained with a delay $\tau = 43\ \mu s$, the middle traces in A and B were obtained with 0 μs delay , while the upper traces are the difference spectra. In the case of the sample in a good state of conservation (B) the upper trace, showing the difference between spectra collected at different delays, is null. Carbon resonances assignment is also reported.

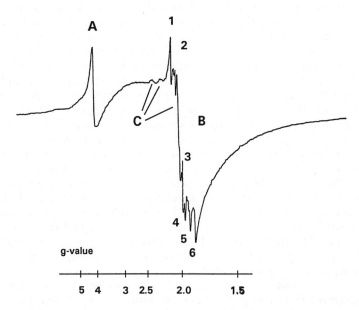

Figure 7. The low-temperature (110 K), X-band EPR spectrum of a 10 mg specimen of a 15th century, deteriorated sheet of paper coming from northern Italy (sample 2). The spectrum is a quite general example of the results obtainable measuring ancient paper. The most prominent features are assigned to rhombic and pseudo-octahedral Fe^{3+} (signals A and B, respectively). Superimposed on this latter are the parallel and perpendicular Cu^{2+} lines (signals C) and the characteristic Mn^{2+} sextet (lines 1-6).

samples belong to two different types (CuA and CuB, see Figures 8 and 9) with the following sets of magnetic parameters: CuA $g_{||}$ =2.252; $g_{\perp}$ = 2.065 ÷ 2.020; $A_{||}$= 192·10^{-4} cm^{-1}; $A_{\perp}$ unresolved. CuB $g_{||}$ = 2.307; $g_{\perp}$ = 2.11 ÷ 2.04; $A_{||}$= 129·10^{-4} cm^{-1}; $A_{\perp}$ unresolved.

They differ in that   CuA shows smaller g values coupled with a much larger nuclear hyperfine splitting constant $A_{||}$. Within the framework of the most common copper symmetry, i.e. tetragonal, these variations are known to depend on the relative strength of the axial and equatorial ligand field perturbations (13,14). Going stepwise from the square-planar to the octahedral limit, the total splitting of the 3d levels, and hence the energy difference between the ground state and the first excited state, decreases. The prime consequence of this behavior is to bring about exactly the increase of g   and the decrease of $A_{||}$. observed in CuB.

This latter spectrum closely resembles that of the free, hydrate copper ion suggesting that only loose interactions occur, in this case, with the paper polymeric matrix. As opposed to this,   the CuA signal reflects a coordination geometry closer to the square-planar, possibly due to a more defined coordination of the metal ion to the cellulose backbone.

As already pointed out all the experimental spectra are closely similar and do not show any qualitative dependence from the exact origin or conservation state of the paper samples. Therefore EPR data were used to measure the concentration of the spectroscopically active transition ions and to see whether such quantitative data might be more informative. The experimental procedure was based upon well established literature methods (15,16) and is reported elsewhere (Attanasio D.; Capitani D.; Federici C.; Segre A.L. *Archaeometry* in press). The essential experimental results are collected in Table II.

Both rhombic and pseudooctahedral iron show similar ranges of concentrations from 100 to 400 ppm, the total amount of this ion ranging from *ca.* 300 to 700 ppm. However, the relative amounts of iron in the two different chemical environments do not show any obvious correlation. The concentration of Mn^{2+} has not been measured systematically, however, its amount is estimated to be about 5% of the total iron content. Copper, when detectable, is usually in the 50 ÷ 100 ppm concentration range. In two cases extremely high amounts (2100 and 3400 ppm) of this ion were found. Additional information available on these two samples suggests that the high copper concentration is the result of mistaken restoration attempts carried out in the past.

The plots reported in Figures 10 and 11 illustrate the essential outcomes which can be drawn from the quantitative data reported above. The former shows that the state of conservation of paper nicely correlates with the amount of rhombic Fe^{3+} present. The number of available data is too low for a fully significant statistical analysis. However, the linear regressions reported in Figure 10 clearly indicate that degraded or well-preserved paper cluster in two rather well separated regions, according

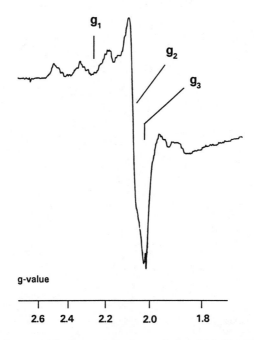

Figure 8. A magnified presentation of the 110 K, X-band EPR copper signal displayed by sample 4 and referred to in the text as the CuA spectrum.

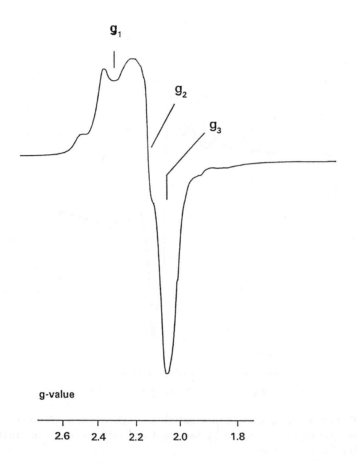

Figure 9. A magnified presentation of the 110 K, X-band EPR copper signal displayed by sample 6 and referred to in the text as the CuB spectrum.

Table II. Characteristics of Ancient Paper Samples and ESR Concentrations (a) (ppm) of Paramagnetic Transition Ions (b,c)

	SAMPLE	State of Conservation	Fe (oct.)	Fe (rh.)	CuA	CuB
1.	15th Century (northern Italy)	good	230	230		
2.	15th Century (northern Italy)	deteriorated	385	290	70	
3.	15th Century (northern Italy)	good	295	370		
4.	15th Century (Italy or Germany)	destroyed	155	340	130	
5.	15th Century (northern Italy)	destroyed	255	320		
6.	18th Century (Sardinia)	destroyed	masked	220		2100
7.	18th Century (Sardinia)	destroyed	masked	105		3400
8.	15th Century (Italy)	very ruined	385	265		
9.	15th Century (Italy)	ruined	295	260	70	
10.	15th Century (Italy)	good	335	150		
11.	15th Century (France)	very ruined	345	250		
12.	15th Century (France)	very good	250	65		
13.	15th Century (France)	destroyed	325	220		
14.	15th Century (France)	very good	415	150		
15.	16th Century (Germany)	very good	274	90		
16.	15th Century (Germany)	slightly ruined	295	190		
17.	15th Century (Germany)	very good	250	120		
18.	15th Century (northern Italy)	ruined		390	50	

a. Control measurements indicate that total errors approximately are within 10% of the measured values. b. Fe(oct.), Fe(rh.), CuA, and CuB refer to the different iron and copper signals present in the samples. For the corresponding magnetic parameters and for a full discussion of the ions chemical environments see text.c. All the samples show also the typical signal of Mn^{2+}. Its concentration, not measured directly, is estimated to be of the order of 50 ppm or lower.

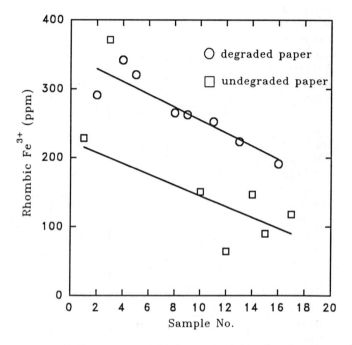

Figure 10. The EPR measured concentration of rhombic Fe^{3+} (ppm) vs. the sample number and its state of conservation. The solid lines represent a linear regression analysis of the concentrations measured in the two kinds of samples. The well-preserved sample 3, with an unexpectedly high concentration of rhombic Fe^{3+} , well outside the standard deviation , has not been included in the regression.

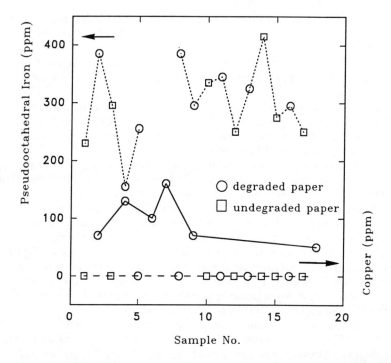

Figure 11. The EPR measured concentration of pseudo-octahedral Fe^{3+} and Cu^{2+} vs. the sample number and its state of conservation. Two points are illustrated by these plots: the state of conservation of the sample and the amount of octahedral iron (broken line) are totally unrelated. Secondly several degraded samples do not contain copper (full line at zero concentration) but all the samples which do contain this metal ion are badly deteriorated. For a clearer presentation, the anomalously high copper concentration of samples 6 and 7 has been scaled down of a factor 100.

to their high or low rhombic iron content. The limiting value is found to be around 250ppm.

Some additional, qualitative results are presented in Figure 11 and can be summarized as follows:

i) Samples of both degraded and undegraded paper contain randomly varying amounts of pseudo octahedral Fe^{3+} . Apparently no correlation exists between the concentration of this chemical species and the state of conservation of paper.

ii) The role of copper is crucial in that samples containing this ion, even in very small amounts, were always badly deteriorated or totally destroyed.

iii) About 50% of the samples which do not contain copper are degraded samples. These are precisely those which exhibit large amounts of rhombic iron.

Obviously, the presence of trace metals is only one out of many possible reasons for paper degradation. However we note that in all the samples analyzed degradation was invariably connected with the presence of copper or high amounts of rhombic Fe^{3+}.

In chemical terms the degradation of paper is essentially the conversion of fibrous and highly crystalline cellulose into a largely amorphous material. Such transformation is the result of different, complex processes, among which hydrolysis is, probably, the most important.

Oxygen induced, radical degradation is another possible mechanism, which , however, is more likely to be relevant in the degradation of modern paper. This latter, in fact, contains large amounts of lignin, a complex aromatic polymer, which is the source of organic free radicals, acting as chain initiators. Whatever the contribution of these and other mechanisms to the degradation may be, copper appears to be, by far, the most effective metal catalyst in promoting such reactions.

Of course the concentration of copper and other trace metals gives also a measure of the concentration of corresponding anions, such as sulphate and others. These latter are known to affect deeply the degradation processes essentially by determining the acidity of paper. Therefore effects both cation- and anion- dependent are present and difficult to discriminate and estimate separately.

Conclusions .   This study presents the preliminary results of an investigation of the physico-chemical properties of paper considered as a composite material made with cellulose, water and organic and inorganic impurities. The material is heterogeneous, due to the presence of different polymorphs of cellulose and to amorphous material, characterized by a different amount of water. In ancient paper, paramagnetic impurities are found clearly related to the degradation. Degraded ancient paper presents an increase in the amorphous content, a loss of water and the presence of characteristic EPR resonances .

These results taken together give an insight into the structure of amorphous cellulose. This structure should be poor in water content, rich in paramagnetic impurities and with a conformation twisted with respect to the long fibers common in cellulose Ia and Ib.

Aknowledgements. This work was supported by CNR, special Project "Beni Culturali" .

Literature cited
1) VanderHart, D.L.; Atalla, R.H. *Macromolecules,* **1984,** *17,* 1465.
2) Sugijama, J.; Vuong, R.; Chanzy, H.   *Macromolecules,* **1991,** *24,* 4168.
3) Torri, G.; Sozzani, P.; Focher, B. *From Molecular Materials to Solids;*   ed. F.Morazzoni ; Polo Ed.Chimico: Milano, **1993.**
4) McBrierty, V.J.; Packer, K.J. *Nuclear Magnetic Resonance in Solid Polymers*; Cambridge Univ.Press: Cambridge, **1993.**
5) Fukushima, E.; Roeder, S.B.W. *Experimental Pulse NMR. A Nuts and   Bolts approach*; Addison-Wesley C. Ed.: Reading , **1981.**
6) Chanzy, H.; Henrissat, B.; Vincendon, M.;Tanner, S.F.; Belton, P.S. *Carbohydrate Res.,* **1987,** *160,* 1.
7) Cael, K.B.; Patt, S. *Macromolecules,* **1985,** *18,* 821.
8) Torchia, D.A.   *J.Magn.Reson.,* **1978,** *30,* 613.
9) Hochmann, J.; Kellerhals, H. *J.Magn.Reson.,* **1980,** *30,* 613.
10) Tonelli, A.E. *NMR Spectroscopy and Polymer Microstructure; the Conformational Connection*; Deerfield Beach, FL, **1989.**
11) Gibson, J.F. in *ESR and NMR of Paramagnetic Species in Biological and Related Systems*; eds. Bertini, I.; Drago, R.S.; D. Reidel: Dordrecht, **1980.**
12) Griffith, J.S. *Mol.Phys.* **1964,** *8,* 213.
13) Ammeter, J.; Rist, G.; Günthard, Hs.H. *J.Chem.Phys.* **1972,** *57,* 3952.
14) Attanasio, D. *J.Magn.Reson.* **1977,** *26,* 81.
15) Poole, C.P.Jr. *Electron Spin Resonance;* Wiley - Interscience: New York, **1983.**
16) Chang, T.T. *Magn.Reson.Rev.* **1984,** *9,* 65.

RECEIVED February 2, 1995

FLUORESCENCE SPECTROSCOPY

Chapter 20

Applications of Luminescence Spectroscopy in Polymer Science

Section Overview

Ian Soutar

School of Physics and Chemistry, Lancaster University,
Lancaster LA1 4YA, United Kingdom

Luminescence spectroscopy is an extremely powerful and versatile means of interrogation of polymer systems. Photophysical techniques may be used to study energy transfer and trapping phenomena: in this case, the chromophores involved are *generally* intrinsic to the macromolecule and present in relatively high local concentrations. Alternatively, luminescence methods may be applied in explorations of the physical behavior and characteristics of macromolecular assemblies: in this instance the "reporter" is usually a luminescent species which is introduced to the system under investigation either as a molecular probe or as a covalently bound label. In such probe and label studies, the guest sensor is incorporated at extremely low levels of concentration (typically, 10^{-5} - 10^{-6}M) in the hope that minimal perturbation of the system will result. It is the high degrees of sensitivity afforded by luminescence spectroscopy which allows luminescent sensors to be added at such low concentrations. The sensitivity of the technique is such that, even in labelling experiments, polymer concentrations can often be accessed which are below the limits of resolution of most other spectroscopic and scattering approaches.

Luminescence techniques have enjoyed increasing application in polymer science in recent years. Several reviews of the topic are available (see, for example, references 1-5) and only the briefest of overviews is offered below.

Luminescence Processes

The term "luminescence" refers to the emission of radiation from an excited state and covers two phenomena; fluorescence and phosphorescence. The unimolecular processes leading to these radiative deactivation steps are conveniently introduced by reference to a "Jablonski Diagram" of the type shown in Figure 1.

The lowest energy state of the vast majority of organic molecules is a singlet state (i.e. one in which electrons with opposite spin angular momenta are paired in molecular orbitals). This state is depicted as S_0 in Figure 1. Absorption, occurring with retention of spin, promotes an electron from the highest occupied molecular orbital into one of higher energy, creating a molecule in a higher excited state S_n.

0097–6156/95/0598–0356$12.00/0

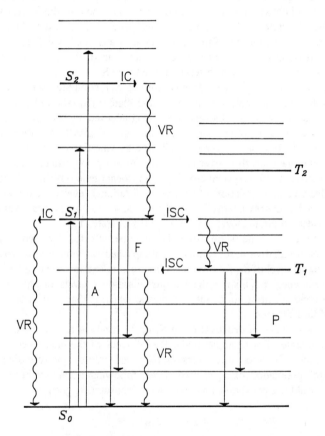

Figure 1   State diagrm depicting the unimolecular processes of
A - absorption; F - fluorescence; IC - internal conversion;
ISC - intersystem crossing; P - phosphorescence and
VR - vibrational relaxation.

Vibrational levels of the excited state are also populated and a range of wavelengths are absorbed in the excitation process as revealed in the absorption spectrum.

In condensed media, the processes of internal conversion (a radiationless, isoenergetic transition from a vibrational level of an upper excited state to a vibrational level of a lower state) and vibrational relaxation (the intermolecular exchange of energy whereby a vibrationally "hot" species achieves thermal equilibrium with its surroundings) rapidly ($<10^{-12}$s) render the molecule in the lowest vibrational level of the S_1 state. (Exceptions to this generalisation exist, but are rare). Consequently, the proportion of the energy initially absorbed which can be accessed by the spectroscopist, as luminescence, is determined (in part) by the relative efficiencies of the processes which serve to deactivate S_1.

Fluorescence (the radiative deactivation of S_1) competes with non-radiative routes of returning the molecule to its ground state. ($S_1 \rightarrow S_0$ radiative transitions occur to the various vibrational levels of S_1: in condensed media, fluorescence results in the appearance of a broad "band" spectrum at energies lower than that of the absorption band). Internal conversion, followed by vibrational cascade is one means whereby S_1 is radiationlessly converted into S_0. Another non-radiative route is that of intersystem crossing. Intersystem crossing is an isoenergetic transition, involving spin inversion, leading to population of an upper vibrational level of a triplet state. In condensed media, excess energy is rapidly lost and subsequent deactivation occurs from the lowest vibrational level of the lowest triplet state, T_1.

Return to the ground state from T_1 may be effected by two unimolecular processes. Non-radiative deactivation is accomplished *via* intersystem crossing to an upper vibrational level of S_0, followed by vibrational relaxation. The radiative process of phosphorescence, populating the various vibrational levels of the ground state produces a band spectrum, in condensed media, which is disposed at lower energies than that of fluorescence.

The various modes of deactivation of S_1 or T_1 combine to determine the lifetime of the excited state. For example, under the influence of the unimolecular processes of fluorescence (F), internal conversion (IC) and intersystem crossing (ISC), a "steady-state" population of S_1 excited states, following instantaneous cessation of excitation, would decay exponentially with a rate constant, k, given by

$$k = k_F + k_{IC} + k_{ISC} \tag{1}$$

where k_F, k_{IC} and k_{ISC} are the rate constants governing the first-order decays of S_1 through F, IC and ISC, respectively. The excited state lifetime, τ, is the reciprocal of this composite rate constant

$$\tau = k^{-1} \tag{2}$$

(and corresponds to the time that it would take for the concentration of excited states to decrease to $1/e$ of its initial value). Naturally, in the presence of other deactivation pathways such as chemical reaction from S_1 or bimolecular quenching, τ would be further reduced and additional terms would need to be incorporated into equation (1), in determination of k.

In condensed media, excited singlet states are generally characterized by lifetimes which lie in the nanosecond to microsecond range. Phosphorescent states on the other hand are longer lived (conversion to the ground state being spin "forbidden") and, dependent upon chromophore, environment, etc., exist upon timescales which might range from the microsecond domain to tens of seconds. As a consequence, by choosing appropriate combinations of fluorescent and phosphorescent sensors, the polymer chemist can access an extremely broad range of "test frequencies".

The intensity of luminescence which may be observed under a given set of conditions is governed, in part, by the quantum yield for the emission. The quantum yield is the rate of the process concerned, relative to that of photon absorption. In the case of fluorescence, for example, where the processes of F, IC and ISC are the sole processes leading to deactivation of S_1, the quantum yield of fluorescence, ϕ_F, would be given by

$$\phi_F = \frac{k_F}{k_F + k_{IC} + k_{ISC}} \tag{3}$$

In the presence of bimolecular quenching, ϕ_F (and the observed fluorescence intensity) would be reduced from this limit and a further term(s) would have to be added to the denominator of equation (3) to account for the quenching process(es).

Applications in Polymer Science

The versatility of the luminescence technique derives from the diversity of approaches which might be adopted in probing macromolecular properties. Dependent upon the nature of the system involved and the information required, a number of strategies may be employed, including use of the following:

Luminescence Quenching.   As noted above, both the quantum yields of luminescence and excited state lifetimes can be reduced by the action of quenchers. The quenching process involves transfer of energy from the molecular excited state, M^*, and the quencher, Q, as depicted in equation 4.

$$M^* + Q \rightarrow M + Q^* \tag{4}$$

The energy transfer process can involve either long-range or short-range interactions between M^* and Q. The kinetic expressions governing the quenching of emission from M^* vary with the operative energy transfer mechanism, the relative mobilities of M^* and Q and the importance of energy migration within the system.

In polymer science, a form of parlance has evolved in which the term "luminescence quenching" (especially when applied to fluorescence) is often used specifically to describe experiments designed to examine the quenching of excited states *via* short-range interactions. Within this terminology, the term "(non-radiative) energy transfer" is frequently applied in description of situations in which dipole-dipole, long-range interactions are assumed to be dominant. In either case, the majority of "quenching" experiments in polymers involve covalently bound energy donors (either in the form of labels or as chromophores intrinsic to the macromolecule under investigation).

Luminescence studies of labeled macromolecules employing diffusing, collisional quenchers have been widely applied in polymer science. The information obtained can be used to quantify the permeability of the medium to the quencher and may, in turn, reveal details regarding the morphology of the system. For example, triplet excited states dispersed in polymer matrices will exhibit degrees of quenching by oxygen which are dependent upon the relaxation characteristics of the host polymer (1,6,7). Phosphorescence quenching measurements can be used to study oxygen diffusion in polymer matrices (1). Quenching has also been used to good effect in studies of the morphology of colloidal particles in non-aqueous dispersions (see, for example, references 8 and 9 and references therein); conformational changes in polyelectrolytes (see, for example reference 10 and references therein) and energy migration in polymers (11-14).

Long-range energy transfer measurements have been used to study macromolecular interpenetration in polymer blends (15,16) as colloidal particles coalesce during film formation (17,18).

Excimer Formation. A particularly useful and interesting form of quenching involves the formation of an excited state dimer or *excimer*. Excimer formation is common in aromatic species and may be represented as

$$M^* + M \rightleftharpoons E^* \qquad (5)$$

The excimer, E^*, is characterized by a structureless emission band located at longer wavelengths than the "normal" fluorescence from the unassociated or "monomeric" M^*. In steady-state spectroscopy, the influence of excimer formation upon the overall photophysical behavior of the system is often expressed in terms of I_E/I_M, the ratio of the intensities of fluorescence observed in the spectral regions dominated by emission from E^* and M^*, respectively.

The phenomenon has featured prominently in studies of the intrinsic photophysical behavior of macromolecules (see, for example, references 5 and 19). In addition, intramolecular excimer formation has been used as a tool in studies of matrix microviscosities (20,21) and end-to-end cyclization dynamics of polymers (5,22).

Luminescence Anisotropy. The emission anisotropy approach allies the orientational dependence of luminescence intensity with its temporal characteristics to yield a powerful interrogative combination. This technique, the basic principles of which are outlined elsewhere in this volume (10) can be used to study the orientational distribution of chromophores within a matrix, energy transfer phenomena and rotational dynamics. In the latter context, dispersed probes can be used as sensors of the microviscosity of their environment whereas luminescent labels can furnish information regarding the dynamics and mechanisms of macromolecular relaxation processes which might prove difficult to acquire by other means. Applications of emission anisotropy techniques in polymer science have been reviewed (see, for example, references 5, 23 and 24).

Medium-sensitive Chromophores. The emission intensity and lifetime characteristics of many of the chromophores which are commonly used as luminescent labels (in anisotropy experiments, for example) of macromolecules, reflect the nature of their microenvironment. For instance, both the fluorescence intensity and excited state lifetime of naphthalene-based labels of poly(methacrylic acid) are sensitive to the conformational change which accompanies neutralization of the polyacid (*10*). In this case, the chromophore experiences a marked change in environment from that of the hydrophobic domains which exist within the hypercoiled conformations of the acidic form of the polyelectrolyte to the aqueous rich habitat afforded in the open-coil structure of the polysalt. However, there is a range of luminescent species whose spectral and/or photophysical properties are particularly affected by the nature of the medium in which they are dispersed. Such medium-sensitive molecules find application in polymer science (as probes or labels) as sensors of polarity or microviscosity.

Pyrene is a molecule which has found widespread use as a polarity probe: both the vibrational structure of the molecule's fluorescence spectrum (*26*) and excited state lifetime is dependent upon the polarity of its surroundings. Dansyl derivatives, on the other hand, can reflect changes in the polarity of the medium in which they are dispersed through the accompanying shifts in their spectra and changes in ϕ_F (*27*). Molecular rotors, molecules in which internal rotation serves as a means of radiationless deactivation of the S_1 state have been used (*28,29*) over a considerable number of years, as fluorescent probes of local viscosity in e.g. studies of polymerizing media, polymer melts and solids. Molecular rotors whose S_1 states possess reasonable degrees of charge transfer character exhibit sensitivity to medium polarity (through ϕ_F and spectral shifts) in addition to their ability to reflect changes in the free volume distribution within the system.

The triplet states are particularly sensitive to the influence of molecular motion within the surrounding matrix and phosphorescent probes and labels can be used as simple sensors of the onset of transitions in polymer solids (*6,7*) or of the conformational behavior of polyelectrolytes in aqueous media (*10*).

Conclusions and the Future

Photophysical techniques have matured into an accepted tool in polymer science for investigation of the physical behavior of macromolecular systems. The information that they afford forges valuable links to other forms of spectroscopy and other, more conventional techniques for the study of macromolecular properties. The scope of the luminescence approach is considerably enhanced, in this respect, by the use of time-resolved methodologies which are becoming increasingly common in the study of polymers. This is a trend which is likely to continue.

The articles which follow in this section of the current volume, bear testament to the diversity of situations in which luminescence spectroscopy is applied and demonstrate the variety of approaches available to the polymer scientist: they contain examples of the application of fluorescence probing of curing resins and thin films; energy transfer studies of Langmuir-Blodgett films and self-ordering amphiphilic polymers; the use of excimer formation in investigations of a variety of water-soluble polymers and applications of fluorescence anisotropy measurements in the study of energy transfer and polymer relaxation phenomena, both in solution and at interfaces.

Having served its apprenticeship in polymer science, luminescence spectroscopy is being applied increasingly in the study of complex systems, of technological importance. This trend is likely to continue into the future, with the power of time-resolved measurements, in particular, extending the scope of studies of macromolecular interactions in blends, adsorption at surfaces, hydrophobic domain formation in aqueous solutions of macromolecules, film formation and the study of relaxation processes in polymer solids (to cite but a few examples!).

Literature Cited

1. *Polymer Photophysics and Photochemistry*; Guillet, J.E., Cambridge University Press: Cambridge, 1985.
2. *Polymer Photophysics*; Phillips, D., Ed.; Chapman and Hall: London, 1985.
3. *Photophysical and Photochemical Tools in Polymer Science*; Winnik, M.A., Ed.; NATO ASI Ser. C; D. Reidel Publ. Co.: Dordrecht, 1986; Vol 182.
4. *Photophysics of Polymers*; Hoyle, C.E.;Torkelson, J.M., Eds.; ACS Symp. Ser.; ACS: Washington, DC, 1987; Vol 358.
5. Soutar, I.; *Polym. Int.* **1991**, *26*, 35.
6. Somersall, A.C.; Dan, E.; Guillet, J.E. *Macromolecules* **1974**, *7*, 233.
7. Rutherford, H.; Soutar, I. *J. Polym. Sci. Phys. Ed.* **1977**, *15*, 2213.
8. Winnik, M.A. **1986**, in ref. 3, pp 611-627.
9. Winnik, M.A. **1987**, in ref 4, pp 8-17.
10. Soutar, I.; Swanson, L., this volume.
11. Ishii, T.; Handa, T.; Matsunaga, S. *Macromolecules* **1978**, *11*, 40.
12. Ueno, A.; Osa, T.; Toda, F. *Macromolecules* **1977**, *10*, 130.
13. Hargreaves, J.S.; Webber, S.E. *Macromolecules* **1984**, *17*, 1741.
14. Holden, D.A.; Ng, D.; Guillet, J.E. *Br. Polym. J.* **1982**, 159.
15. Morawetz, H. *Science* **1988**, *240*, 172.
16. Zhao, Y.; Prud'homme, R E. *Macromolecules* **1990**, *23*, 713.
17. Pekcan, O.; Winnik, M.A.; Croucher, M.D. *Macromolecules* **1990**, *23*, 2673.
18. Zhao, C.-L.; Wang, Y.; Hruska, Z.; Winnik, M.A. *Macromolecules* **1990**, *23*, 4082.
19. Soutar, I.; Phillips, D. **1986**, in ref. 3, pp 97-127.
20. Bokobza, L.; Monnerie, L. **1986**, in ref. 3, pp 449-466.
21. Bokobza, L.; *Prog. Polym. Sci.* **1990**, *15*, 337.
22. Winnik, M.A. **1986**, in ref. 3, pp 293-324.
23. Nobbs, J.H.; Ward, I.M. **1985**, in ref. 2, pp 159-220.
24. Soutar, I. In *Developments in Polymer Photochemistry*, Allen, N.S., Ed; Applied Science Publishers; London, 1982, Vol 3; pp 125-163.
25. Monnerie, L. **1985**, in ref. 2, pp 279-339.
26. Kalyanasunderam, T.; Thomas, J.K. *J. Amer. Chem. Soc.* **1977**, *99*, 2039.
27. Morawetz, H. **1986**, in ref. 3, pp 85-95.
28. Oster, G.; Nishijima, Y. *Fortschr. Hochpolym.-Forsch.* **1964**, *3*, 313.
29. Loufty, R.O. **1986**, in ref. 3, pp 429-448.

RECEIVED February 2, 1995

Chapter 21

Fluorescence Studies of the Behavior of Poly(dimethylacrylamide) in Dilute Aqueous Solution and at the Solid–Liquid Interface

Ian Soutar[1], Linda Swanson[1], S. J. L. Wallace[1], K. P. Ghiggino[2], D. J. Haines[2], and T. A. Smith[2]

[1]School of Physics and Chemistry, Lancaster University, Lancaster LA1 4YA, United Kingdom
[2]Department of Chemistry, University of Melbourne, Parkville, Victoria 3052, Australia

Fluorescence techniques, including quenching and anisotropy measurements, have been used to study the dynamic behavior of poly(dimethylacrylamide), PDMAC, in dilute aqueous solutions and the interactions between PDMAC and poly(methacrylic acid), PMAA, and at a silica/water interface. These studies, involving PDMAC species bearing naphthyl and pyrenyl moieties as fluorescent labels, indicate that the polymer exists as a flexible and relatively open coil in aqueous media whose behavior is largely unaffected by changes in pH. The polymer is adsorbed on the surface of colloidal silica and appears to "lie flat" on the particles under high surface:polymer conditions. Complexation with PMAA produces relatively rigid species in which the segmental mobility of the PDMAC is dramatically reduced.

Fluorescence techniques have been shown to be very powerful, versatile and sensitive tools for the investigation of polymer behavior (see for example references 1-5 and references therein) and have proved popular, in recent years, in the study of water-soluble macromolecules. In this context, fluorescence anisotropy measurements have been used to gain information at the molecular level regarding the dynamic behavior of polymers in aqueous solution (6-13). In this paper we report upon preliminary investigations using fluorescence techniques, particularly time-resolved anisotropy measurements (TRAMS), to study the dynamic behavior of poly(dimethylacrylamide) in dilute solution and its interactions at a water/silica interface and with poly(methacrylic acid).

Experimental

Materials. 1-Vinylnaphthalene, VN, was synthesized from 1-acetonaphthone (Aldrich) by reduction (NaBH$_4$) to the carbinol, followed by dehydration (KHSO$_4$) (13). The monomer was purified by fractional distillation under reduced pressure immediately prior to use.

Acenaphthylene, ACE, (Aldrich) was triply recrystallized from ethanol and triply sublimed.

1-Vinylpyrene, VPy, was prepared using a modification of the procedure described by Webber (*14*).

Methacrylic acid, MAA, (Aldrich) and dimethylacrylamide, DMAC, (Aldrich) were prepolymerized (UV radiation) and fractionally distilled immediately prior to use.

Benzene (BDH) and diethylether (May and Baker) were purified by fractional distillation.

Methanol (Aldrich, spectroscopic grade) was used as supplied.

Ludox AM (Du Pont) was used as supplied.

Polymers of MAA or DMAC were prepared by free radical polymerization in benzene solution at 60°C under high vacuum using AIBN as initator. Fluorescently labeled samples were prepared by copolymerization of DMAC with 0.5 mole% of ACE, VN, or VPy, respectively. Polymers were purified by multiple reprecitations from methanol into ether.

Instrumentation. Fluorescence spectra and lifetimes were obtained using Perkin-Elmer LS50 and Edinburgh Instruments 199 spectrometers, respectively.

Time-resolved anisotropy measurements were made using radiation either from the SRS, Daresbury, or a picosecond synchronously-pumped dye laser as excitation. Details of the experimental configurations of the synchrotron and laser sources have been previously published (references 13 and 9, respectively).

Results and Discussion

Poly(dimethylacrylamide), PDMAC, in Dilute Solution.

Fluorescence Intensity and Lifetime Measurements. The intensity of radiation emitted by a fluorescently labeled water-soluble polymer and its mean lifetime are often relatively simple and convenient means of detecting the occurrence of pH and/or temperature induced conformational changes in polyelectrolytes, for example. Figure 1 shows the pH dependence of the fluorescence intensity, I_f, and average lifetime, τ_f, of the fluorescence from a VN labeled sample of PDMAC at 298K. At pH values of greater than *ca.* 2, both I_f and τ_f are, within experimental error, independent of pH. At high degrees of acidity (represented by nominal values of pH<2) both I_f and τ_f decrease. Clearly, in these highly acidic media, changes occur in the local environment of the VN label which affect the rate of deactivation of its excited state. Similar results are obtained when an ACE label (*15*) is used as an alternative to VN. Most probably the "transition" observed in the photophysical behavior of the label is caused by protonation of the PDMAC. It remains to be seen whether the changes induced in the fluorescence characteristics of the labels are indicative of a transition in the physical behavior of the polymer itself in dilute aqueous solutions, as discussed below.

Fluorescence Quenching Data. Fluorescence quenching of the emission from luminescent labels can provide useful insights into the microenvironment provided for the excited state by its polymeric host and thereby into the nature of the conformation adopted by the polymer under a given set of conditions (see, for example, references 7,16-19 and references therein). In the current studies of PDMAC in dilute aqueous solutions we have employed two neutral (nitromethane and acrylamide) and two charged species (iodide and thallous ions) as quenchers with a view to gaining information regarding the microviscosity within the polymer coil.

For truly dynamic, collisional processes, the Stern-Volmer equation [equation (1)] applies

$$I^o/I = \tau^o/\tau = 1 + k_q \, \tau^o \, [Q] \qquad (1)$$

where I^o, τ^o and I, τ are the fluorescence intensity and lifetime in the absence and presence of quencher, Q, respectively and k_q is the rate constant for the bimolecular quenching interaction. Typical data using nitromethane as quencher are shown in Figure 2. The data show clear evidence for the contribution from a static component to the quenching: the intensity and lifetime data are not equivalent and the intensity plot shows a slight curvature concave to the ordinate. Indeed, static components appear to be a general feature of the quenching, by nitromethane, of the fluorescence from aqueous solutions of naphthyl-labeled macromolecules (*19,20*).

The dynamic quenching (represented by lifetime data) of the fluorescence of PDMAC/VN by CH_3NO_2, Tl^+ or I^- seems to be unaffected by pH, over the accessible range. [Problems are to be anticipated in the attempted use of CH_3NO_2 as a quencher in basic media (*21*), particularly where naphthyl labels are employed (*19*)]. Values of k_q of the order of $2-3 \times 10^9 M^{-1}s^{-1}$ are obtained. These are close to that expected in the limit of diffusion control for the interaction, in aqueous solution, between a mobile "contact quencher" and a fluor, the translational mobility of which is hindered by binding to a polymer chain. In this respect, the quenching behavior of PDMAC/VN using nitromethane is similar in efficiency to that reported earlier in this volume (*22*) for poly(acrylic acid) at low pH but is much more efficient than that shown for the hypercoiled form of poly(methacrylic acid) (*22*). These observations are consistent with the view that PDMAC exists as a relatively open, hydrated random coil in dilute aqueous solution.

When acrylamide is used as a quencher in the PDMAC/VN system at pH values between 4 and 10, k_q values of *ca.* $4.7 \times 10^8 M^{-1}s^{-1}$ are obtained; almost an order of magnitude less than those observed for quenching by nitromethane or iodide. Similar effects have been observed by Pascal et al. (*23*) in the quenching, by acrylamide, of ACE-labeled polyacrylamide in water and it was concluded that the acrylamide diffusion must be significantly slower inside the polyacrylamide coil than it is in water. Clearly in the case of PDMAC there is also evidence for interactions between the polymer and the acrylamide (which either reduces the rate of diffusion of the quencher within the polymer coil or effectively lowers the local concentration of acrylamide).

Time-Resolved Anisotropy Measurements (TRAMS). Time-resolved anisotropy studies of the intramolecular segmental mobility of PDMAC, using either 1-VN or ACE as label, reveal that at pH values ≥ 2 the relaxation of the polymer is unaffected by pH and is characterized by a correlation time, τ_c, of 3 ± 0.4 ns in analyses

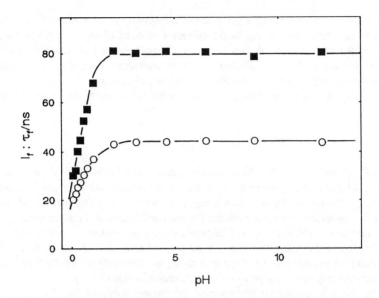

Figure 1.   pH dependence of the intensity (■) and lifetime (o) of the fluorescence from PDMAC/VN at 298K.

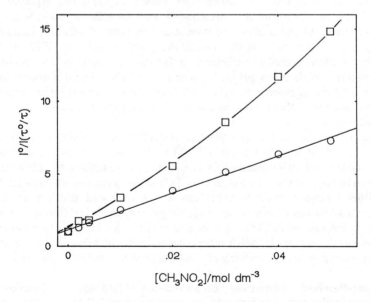

Figure 2.   Stern-Volmer plots for the quenching of the fluorescence from PDMAC/VN in dilute aqueous solution at pH=4. [lifetime data (o); intensity data (□)].

in which the true time-dependence of the fluorescence anisotropy, r(t), was modeled by a single exponential function of the form

$$r(t) = r_o \exp(- t/\tau_c) \tag{2}$$

At no pH was there any statistical justification in the use of a more complex function in modeling r(t). Decay curves, typical of the time-dependence of the observed orthogonal components, I_{11} and $I_{\perp}$, of fluorescence intensity, used to construct the "raw" anisotropy function, R(t), are shown in Figure 3 for PDMAC/ACE.

In highly acidic media (pH<2), there is an indication that the rate of segmental motion of the polymer increases but the reduction in τ_c is small and just outside the limits of experimental error. It appears that whatever effect is responsible for the changes in τ_f and I_f of the fluorescent labels in highly acidic media, has little significance in determination of the dynamic behavior of the PDMAC in aqueous solution.

The TRAMS experiments indicate that PDMAC exists as a reasonably flexible species in aqueous solution. The observed value of τ_c (*ca.* 3 ns) is intermediate between that of e.g. poly(methyl methacrylate) in toluene [*ca.* 2.1 ns (*24*)] or poly(methacrylic acid) in methanol [*ca.* 4 (*25*) ns] at 298K. In comparison with other polymers in aqueous solution, PDMAC exhibits a segmental relaxation rate which is comparable to that of poly(acrylic acid) in its fully neutralized form but less than that of the polyacid itself (*22*) and markedly less than that of hypercoiled poly(methacrylic acid) (*13,22*). To our knowledge, there are no published data concerning the segmental dynamics of PDMAC in aqueous solution against which the current results may be compared. Pascal et al. (*23*) have estimated τ_c for an ACE-labeled sample of poly(acrylamide) as *ca.* 10ns. It is to be supposed that the 3-fold difference in rates of segmental motion between poly(acrylamide) and its dimethyl derivative must reflect differences in the H-bonding (e.g. polymer-solvent and intramolecular polymer-polymer) interactions exhibited by the two polymers.

Pyrene Probe Studies. When the spectral charcateristics of pyrene dispersed (nominal concentration 10^{-5}M) in a 10^{-2}wt% solution of PDMAC in water were compared to those of pyrene in water alone no differences in I_3/I_1 ratio (*26*), fluorescence lifetime or excitation spectra were observed. Either the coils of the polymer do not act as a solubilizing medium for pyrene at all or the environment experienced by solubilized pyrenes can not be resolved from that of pyrene dispersed in the bulk aqueous phase.

The results of TRAMS, fluorescence quenching and pyrene probe investigations of PDMAC indicate that the polymer exists, in aqueous media, as a relatively flexible, solvent-swollen, random coil structure.

Adsorption of PDMAC at a Silica-Water Interface. Macromolecular behavior at interfaces is important to considerations of adhesion, colloid stabilization, lubrication and flocculation phenomena and has attracted a great deal of attention, both theoretical and experimental [see, for example, references 27-29 and references therein]. In this section, we describe preliminary experiments aimed at assessing the

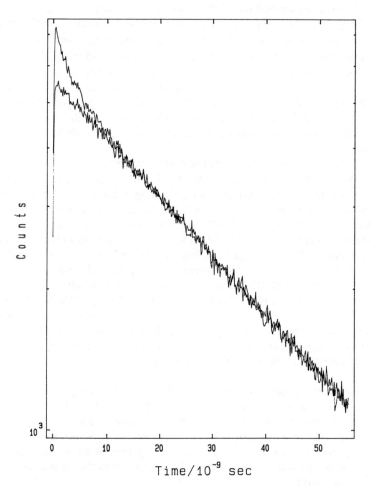

Figure 3. $I_{\Pi}(t)$ and $I\perp(t)$ for PDMAC/ACE following vertically polarized synchrotron excitation. [pH=3.3]

applicability of luminescence techniques to the study of adsorption of polymers at solid-water interfaces. In particular, we have used TRAMS to investigate the effects of adsorption at a silica surface upon the dynamic behavior of PDMAC in aqueous dipersions. In these experiments Ludox AM (Du Pont) silica particles were used. This is a colloidal silica dispersion in which some of the Si surface atoms have been replaced with Al in order to produce a dispersion which is relatively stable across a wide range of pH values.

The intensity of fluorescence emitted by a labeled (whether VPy or ACE) sample of PDMAC is a simple means whereby adsorption of the polymer on the silica surface may be detected: I_f increases when adsorption occurs (*30*). The adsorption of polymer can be monitored also through the associated increase of the average fluorescence lifetime of the VPy label (*30*). In either case, a plateau is attained in the fluorescence parameter at high silica/polymer concentration ratios as adsorption of the polymer becomes complete. Naturally, the silica content at which the maximum (in e.g. fluorescence intensity) is observed varies with the amount of PDMAC present in the system. For a polymer concentration of 10^{-3}wt%, the intensity of fluorescence (and its associated decay time) of PDMAC/VPy attain a constant value at Ludox contents of *ca.* 0.1 wt% and above.

The amount of labeled polymer adsorbed on the colloid may be readily quantified either through the absorbance of or, better, the fluorescence from, the label remaining in the supernatant liquid, isolated following centrifugation (at 77,000g) of the equilibrated polymer/colloid dispersion. In such experiments it was shown that 100% adsorption of the polymer is attained, from an initial polymer concentration of 10^{-3} wt%, at a Ludox content of *ca.* 0.1 wt% (*30*). As reported above, this is the concentration of colloid at which a constant intensity of fluorescence is obtained in equilibrated systems *containing* colloid (regardless of whether the label be ACE or VPy), reinforcing the fact that the changes in fluorescence of the label can be a simple, if crude, sensor of polymer adsorption.

TRAMS experiments have the potential to yield information regarding the segmental dynamics of macromolecules adsorbed on solid surfaces and thence about the conformation adopted by the polymer at the liquid/solid interface. Initial TRAMS investigations, using naphthyl-labeled polymers (*30*) have shown that the relaxation behavior of macromolecules adsorbed on to colloidal silica from aqueous solutions can be characterized by correlation times greatly in excess of the excited state lifetimes of the labels (which, in turn, essentially establish the "time-base" for the experiment). Consequently, in the current work we have employed a longer-lived pyrenyl label in attempts to characterize the interactions between PDMAC and Ludox AM in aqueous media. The fluorescence lifetime of this label varies from *ca.* 130ns for the polymer in fluid solution to *ca.* 160ns in the adsorbed state.

Figure 4 shows the changes in the intensities of fluorescence observed in planes oriented parallel, $I_{\parallel}(t)$, and perpendicular, $I_{\perp}(t)$, respectively to that of vertically polarized excitation, induced by varying amounts of Ludox AM. In the absence of colloid, $I_{\parallel}(t)$ and $I_{\perp}(t)$ are coincident at all but the shortest times following excitation [Figure 4(a)]. However, even in the prescence of a relatively small amount of silica $(6 \times 10^{-3}$wt%), anisotropy is clearly evident in the emission [Figure 4(b)]: the curves are only just beginning to converge at the longest times (*ca.* 700ns) shown. As shown in

Figure 4(c), the timescale over which the anisotropy persists, increases at higher silica contents.

The resultant observed anisotropy decay curves, R(t), are complex and problematic in analyses. Direct analyses of R(t) in terms of a single exponential decay function of the form

$$R(t) = A \exp(- t/\tau_c) + B \tag{3}$$

produce fits to the data which are reasonably justified in terms of statistical criteria of "goodness of fit". Typical correlation times, resultant upon such fits are listed in Table I.

Table I : Correlation times for PDMAC/VPy, at various silica contents, obtained from unconstrained fits to R(t) using equation 3

Silica content (wt%)	τ_c/ns
0	3
6×10^{-3}	124
8×10^{-3}	143
10^{-2}	72
10^{-1}	180
1	435
2	328
5	278

Reference to Table I reveals that the TRAMS data are responsive to adsorption of the PDMAC/VPy to the surface of the modified silica particles. Recovery of a value of τ_c of *ca.* 3ns for the relaxation of PDMAC in the absence of added colloid is particularly gratifying since the VPy label, with its long-lived singlet excited state, is not particularly well suited to the characterization of the segmental motion of such mobile systems. Nevertheless, this estimate of τ_c is in excellent agreement with those obtained using VN (see above) or ACE (*15*) labels. Furthermore, the increase in τ_c which accompanies the introduction of increasing amounts of colloidal silica into the system is in general accord with "expectation" for an adsorbing species. However, closer scrutiny of the data gives cause for concern:

(i) τ_c tends towards a maximum at 1wt% added colloid, rather than the plateau which might otherwise be expected. Furthermore, the data show a distinct lack of uniformity at concentrations less than 1wt%.

(ii) as silica is introduced to the PDMAC/VPy system, the "goodness of fit" to a single exponential function is disrupted. The effect is more evident at lower silica contents.

(iii) as the silica content of the system is increased, the increase in τ_c is accompanied by a concomitant increase in the fitting parameter B defined in equation 3.

(iv) relaxation parameters evolved in direct analyses of R(t) data sets do not agree well with those derived using other forms of analysis such as impulse reconvolution (*31*).

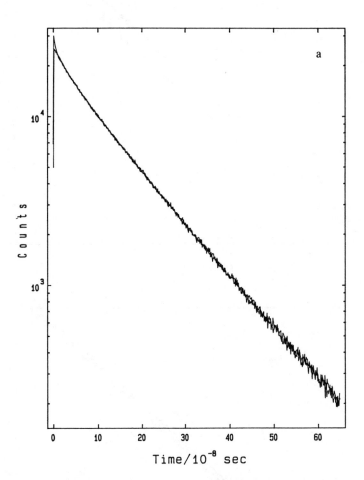

Figure 4. $I_{\Pi}(t)$ and $I_{\perp}(t)$ for PDMAC/VPy following vertically polarized laser excitation (a) in the absence of Ludox AM and (b), (c) at Ludox AM concentrations of 6×10^{-3} and 5 wt%, respectively. *(Continued on next page.)*

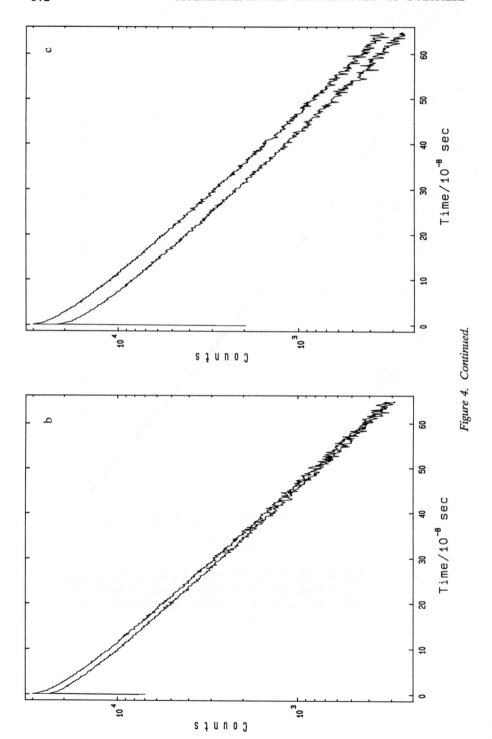

Figure 4. Continued.

Apparently, the TRAMS data contain information concerning the adsorption of the PDMAC at the surface of the Ludox AM colloid which is not evident from the superficial examination afforded by either fluorescence intensity or lifetime data from the VPy label: both I_f and τ_f indicate that, as adsorption becomes complete (at *ca.* 0.1wt% colloid) the *photophysical* behavior of the label becomes established. However, apart from the "incongruity" in τ_c which seems to arise at a colloid concentration of 10^{-2}wt%, the TRAMS data would seem to imply that the final equilibrium conformation of adsorbed PDMAC is *not* attained at the colloid concentration corresponding to complete adsorption of the polymer. In contrast, the anisotropy data *might be* interpreted as being indicative of the fact that the polymer which is initially adsorbed enjoys a greater degree of segmental freedom than is enjoyed by species "laid down" on the colloid surface at higher contents of the adsorbing particles (since τ_c continues to increase at colloid contents greater than that at which adsorption of polymer, from solution, becomes complete). It is our opinion that such is the case. However, prior to establishment of such a proposition, we should have to consider the limitations imposed by the form of data analysis that we have adopted.

The fact that the raw R(t) data are not amenable (directly) to reconvolution analysis can pose problems when τ_f and/or τ_c are comparable to the width of the excitation pulse and R(t) is analysed directly (i.e. without "deconvolution"). However, this is not the case in this instance since the instrumental response function of the picosecond laser system and associated detection electronics has negligible duration compared to either τ_f or τ_c of the labeled adsorbate. The problems associated with direct analysis of R(t) lie, in this case, in the very slow relaxation behavior exhibited by the adsorbed polymer. At low silica contents where polymer adsorption is incomplete, the fluorescence anisotropy decay curves will contain contributions from both adsorbed species and "free" polymer, in solution. A "forced fit" to a single exponential analysis will result in the generation of a mean value of τ_c which will reflect the presence of fluorescent components for which $\tau_c=3$ns and others for which τ_c may be a composite of correlation times of segments of adsorbed species existing in varying proportions as loops, tails and trains according to the amount of surface available. As is observed, the quality of fit would be expected to improve as the silica content of the system increases, since the overall anisotropy behavior will contain ever decreasing contributions from rapidly relaxing unadsorbed polymer molecules in the system. At higher silica concentrations the anisotropy relaxation curve will be dominated by slower polymer relaxation processes but problems in attaining a unique mathematical solution are to be envisaged since the fitting procedure will have difficulties in distinguishing between a long-lived anisotropy and the "background". This will result in a strong cross-correlation between τ_c and B (equation 3) when both are allowed to vary freely as fitting parameters. In turn, large values of B will result, as is observed and noted, in (iii), above.

Large values of B would imply that the relaxation behavior of the adsorbed polymer species results in a significant residual anisotropy contribution, r_∞, to the fluorescence anisotropy of the labeled system. Since there is no reason to suppose that the relaxation of the adsorbed PDMAC should produce a residual anisotropy in the fluorescence of its VPy label, we have analysed the TRAMS data, imposing the restriction that $B(= r_\infty) = 0$ to help the fitting procedure converge to a solution which produces a physically meaningful value of τ_c, representative of the dynamic behavior of the adsorbed polymer species. The resultant data are presented in Table II.

Consideration of the τ_c data presented in Table II prompts the following comments:

(i) Setting r_∞ to zero (a realistic value considering the nature of the polymer and interface under consideration) results in the evolution of estimates of τ_c, for the totally adsorbed polymer, of the order of 1.5-2.2μs. Such estimates are not unreasonable. [Given a nominal mean diameter for the colloid of 12nm, values of τ_c of the order of 0.2μs, calculated using equation 4,

$$\tau_c = \frac{\eta V}{RT} \qquad\qquad (4)$$

(where η=solvent viscosity and V=molar volume of the rotating entity) might be expected. Values for τ_c of the order of 1-2μs for the colloid (plus adsorbed polymer) might be anticipated considering the "drag" imposed by its associated solvent sheath and its interactions with the bulk aqueous medium].

(ii) The τ_c data show that the relaxation behavior of the VPy label (and, therefore, of the polymer with which it is associated) varies with the amount of silica present in the system. Furthermore, changes in τ_c are apparent even in systems in which the silica content exceeds that required (ca. 10^{-1}wt%) for complete polymer adsorption.

Table II : Correlation times for PDMAC/VPy, at various silica contents, obtained from impulse reconvolution fits to D(t) using a single exponential function to model r(t)

Silica content (wt%)	$\tau_c/\mu s$
0	0.004
6×10^{-3}	0.4
8×10^{-3}	0.4
10^{-2}	0.8
10^{-1}	1.0
1	1.5
2	1.6
5	2.2

Other observations relevant to considerations of the interaction of PDMAC with Ludox AM include the facts that

(a) analyses of anisotropy data using raw R(t) information yields values of τ_c which are comparable with those resultant upon impulse reconvolution (31) analyses, provided both fitting procedures are constrained to achieving r_∞=0.

(b) the quality of fit deteriorates from that observed in the absence of silica as the silica content of the system is initially increased. Once the polymer is completely adsorbed, single exponential analyses become increasingly appropriate in modeling the fluorescence anisotropy behavior of the system. At higher silica contents a single correlation time serves to characterize the relaxation behavior of the colloid and adsorbed polymer.

Recognizing these considerations, the anisotropy data were reanalyzed for data sets obtained at lower silica contents (<0.1wt%) in terms of dual exponential models of the form

$$r(t) = A_1 \exp(-t/\tau_{c1}) + A_2 \exp(-t/\tau_{c2}) \qquad (5)$$

These analyses resulted in recovery of correlation times of the order of 7-10ns for τ_{c1}, and 0.5-0.6μs for τ_{c2}, respectively for systems containing 6x10-3wt% and 8x10-3wt% Ludox AM. It would appear, on the basis of these preliminary investigations, that the correlations times τ_{c1} and τ_{c2} represent mean correlation times characteristic of the relaxation behavior of the PDMAC/Ludox AM system which are composite values dominated by contributions from unassociated polymer molecules in the bulk aqueous medium and adsorbed species, respectively. In addition, it is tempting to assume that the values of τ_{c2} observed under conditions of incomplete adsorption upon the colloid indicate that when the polymer has to compete for limited adsorption sites on the colloid, it will adopt configurations, (such as loops and tails) which afford greater degrees of segmental mobility than is evident in the trains of adsorbed polymer formed in the fully "plated out" conformations assumed at higher silica contents. This proposal requires testing using fluorescent labels (such as naphthyl species) which are more sensitive than VPy to the shorter relaxation components within the overall dynamic behavior of the polymer/adsorbant system. Such experiments are the subject of continued study in our laboratories.

Complexation of PDMAC with Poly(methacrylic acid). Complexation of PDMAC/VPy with PMAA, in acidic media, affects the photophysical characteristics of the VPy label in a manner similar to that observed when the polymer is adsorbed on a silica surface: complexation results in an increase in both fluorescence intensity and lifetime (*30*). The extent to which anisotropy persists to longer times in the decay curves, $I_\parallel(t)$ and $I_\perp(t)$, is also affected by the degree of complexation experienced by the PDMAC. Again, as with the relaxation behavior observed in adsorption studies, complexation introduces very long components into the anisotropy decays and, in a similar fashion to the data analysis procedures adopted earlier, r_∞ was fixed at zero to help the fitting program converge to a physically meaningful solution. Correlation times resultant upon single exponential fitting of the anisotropy data obtained at various ratios of PDMAC/VPy:PMAA (holding the PDMAC/VPy content of the system constant at $6x10^{-4}$wt%) are shown in Table III.

Clearly, the τ_c data, in a semi-quantitative fashion, show the effects of complexation with PMAA upon the segmental motion of the PDMAC. At low amounts of PMAA complexing agent the anisotropy data are dominated by the relaxation characteristics of the uncomplexed PDMAC species. As the PMAA concentration is progressively increased, the segmental mobility of the PDMAC is hindered as complexation occurs and appears to minimise at a 1:1 molar ratio (in terms of repeat units). In analysing the anisotropy decays directly at PMAA contents of

0.1wt% or greater, we have largely excluded the influence of the uncomplexed PDMAC/VPy species (characterized by a τ_c of *ca.* 3ns) and the listed data serve to describe the relaxational characterisitics of the complexed PDMAC. Within the limits imposed by attempting to characterize, using a single correlation time,what is likely to be an extremely broad and heterogeneous range of segment environments, the τ_c data would seem to imply that at a 1:1 base molar ratio, the PDMAC segments experience the greatest restraint in complexation. (This might be as expected since this ratio offers the greatest number of potential polymer-polymer "contact points" to each of the complexing species. At lower PMAA contents, PDMAC molecules and segments will compete for binding sites on the PMAA species and "the complex" is likely to consist of a PMAA molecule bound to segments of more than one PDMAC species. The PDMAC molecules present in the complex formed at a PMAA:PDMAC ratio of 0.1, for example, would be envisaged as having a greater number of loops and tails (which help maintain solubility of the complex) than those found in a 1:1 complex. Consequently the apparent segmental mobility of the PDMAC would be expected to be greater in the former case. In the presence of excess PMAA, the PDMAC appears to be more mobile than the 1:1 complex. In this respect the data might be indicative of the fact that under these conditions PDMAC molecules are complexed to more than one PMAA species. (Otherwise, if a stoichiometric complex were to be preferred, there would not be any change in the rotational freedom of the VPy label at higher PMAA:PDMAC ratios). Furthermore, the data would seem to suggest that the competition for binding sites on the PDMAC between PMAA segments produces a situation in which the PDMAC is less constrained than in the 1:1 complex. (If the value of τ_c simply characterized whole molecule, end-over-end tumbling of the complex, τ_c would be expected to increase as more PMAA species become associated with the complex).

Table III : Correlation times for PDMAC/VPy obtained at various contents of PMAA in single exponential analyses of R(t) in which r_∞ is constrained to a value of zero

PMAA : PDMAC/VPy[†]	τ_c/ns
0.01	5.6
0.1	120
1	915
10	660
100	670

[†] Total concentration of PDMAC/VPy = 6×10^{-4} wt%.

The quality of fit furnished by single exponential modeling of the anisotropy decay decreases as the PMAA:PDMAC ratio is varied from that of the 1:1 complex and fitting to a more complex function is justified on statistical grounds. The use of a dual exponential model results in the evolution of two correlation times of the order of 60ns and 1.4μs respectively, in systems containing relative proportions of PMAA:PDMAC of 100, 10, and 0.1. It is unlikely, however that these correlation times have physical significance beyond merely serving to better parameterize the decay of anisotropy of the fluorescence of the complexed species. It is likely that the PDMAC segments exist in a wide variety of heterogeneous environments.

Conclusions

1. Fluorescence intensity, lifetime, quenching and anisotropy measurements indicate that PDMAC exists in a flexible, open-coiled conformation, in dilute aqueous solutions.
2. PDMAC is adsorbed at the surface of the colloidal silica (Ludox AM) from dilute aqueous solution. At low silica contents, dispersions contain both adsorbed polymer and free species in solution. At higher silica contents, the polymer is adsorbed completely in conformations which are dependent upon the amount of particle surface available per unit mass of polymer.
3. PDMAC complexes with PMAA in acidic media. Complexation produces a dramatic reduction in the segmental mobility of the PDMAC, maximal effect occurring at equimolar ratios of PMAA:PDMAC repeat units.

Acknowledgments

The authors wish to thank the following organizations for sponsorship: The Leverhulme Trust (Research Fellowship to LS); EPSRC and Zeneca (CASE studentship to SJLW); The Australian Research Council (grants to KPG).

Donation of sample of Ludox AM by Du Pont UK Ltd. is gratefully acknowledged.

Literature Cited

1. Ghiggino, K.P.; Roberts, A.J.; Phillips, D. *Adv. Polym. Sci.* **1981**, *40*, 69.
2. *Polymer Photophysics*; Phillips, D. Ed.; Chapman and Hall: London, 1985.
3. *Photophysical and Photochemical Tools in Polymer Science*; Winnik, M.A. Ed.; NATO ASI Ser. C; **182**, D. Reidel Publ. Co.: Dordrecht, 1986, Vol 182.
4. *Photophysics of Polymers*; Hoyle, C.E.; Torkelson, J.M., Eds.; ACS Symp. Ser.; ACS: Washington, DC, 1987, *Vol 358.*
5. Soutar, I. *Polym. Int.* **1991**, *26*, 35.
6. Anufrieva, E.V.; Gotlib, Yu. Ya. *Adv. Polym. Sci.* **1981**, *40*, 1.
7. Ghiggino, K.P.; Tan, K.L., chapter 7 of reference 2.
8. Ghiggino, K.P.; Bigger, S.W.; Smith, T.A.; Skilton, P.F.; Tan, K.L., chapter 28 of reference 4.
9. Heyward, J.J.; Ghiggino, K.P. *Macromolecules* **1989**, *22*, 1159.
10. Soutar, I.; Swanson, L. *Macromolecules* **1990**, *23*, 5170.
11. Soutar, I.; Swanson, L. *Polymer* **1994**, *35*, 1942.
12. Ebdon, J.R.; Lucas, D.M.; Soutar, I.; Swanson, L. *Macromol. Symp.* **1994**, *79*, 167.
13. Soutar, I.; Swanson, L. *Macromolecules* **1994**, *27*, 4304.
14. Hargreaves, J.S.; Webber, S.E. *Macromolecules* **1982**, *15*, 424.
15. Soutar, I.; Swanson, L; Thorpe, F.G.; Zhu, C, *to be published.*

16. Chu, D.Y. and Thomas, J.K. *Macromolecules* **1987**, *25*, 259.
17. Delaire, J.A.; Rodgers, M.A.J.; Webber, S.E., *J. Phys. Chem.* **1984**, *88*, 6219.
18. Arora, K.S.; Turro, N.J. *J. Polym. Sci. Polym. Chem. Ed.* **1987**, *25*, 259.
19. Soutar, I.; Swanson, L. *Eur. Polym. J.* **1993**, *29*, 371.
20. Ebdon, J.R.; Lucas, D.M.; Soutar, I.; Swanson, L., *to be published*.
21. Olea, A.F.; Thomas, J.K. *Macromolecules* **1989**, *22*, 1165.
22. Soutar, I.; Swanson, L., this volume.
23. Pascal, P.; Duhamel, J.; Wang, Y., Winnik, M.A.; Zhu, X.X.; Macdonald, P.; Napper, D.H.; Gilbert, R.G. *Polymer* **1993**, *34*, 1134.
24. Soutar, I.; Swanson, L. *to be published.*
25. Soutar, I.; Swanson, L.; Imhof, R.E.; Rumbles, G. *Macromolecules* **1992**, *25*, 4399.
26. Kalyanasunderam, K. and Thomas, J.K. *J. Amer. Chem. Soc.* **1977**, *99*, 2039.
27. Robb, I.D. In *Comprehensive Polymer Science* Allen, G.; Bevington, J.C., Ser. Eds.; Booth, C.; Price, C., Vol. Eds.; Pergamon Press: Oxford, 1989, Vol. 2; ch 24; pp 733-754.
28. Takashi, A.; Kawaguchi, M. *Adv. Polym. Sci.* **1982**, *46*, 1.
29. Fleer, G.J.; Lyklema, J. In *Adsorption from Solution at the Solid/Liquid Interface* Parfitt, G.D.; Rochester, C.H. Eds.; Academic Press: London, 1983; pp153-220.
30. Lumber, D.C.J.; Robb, I.D.; Soutar, I.; Swanson, L., *unpublished data.*
31. Barkley, M.D.; Kowalczyk, A.A.; Brand, L. *J. Chem. Phys.* **1981**, *75*, 3581.

RECEIVED February 2, 1995

Chapter 22

Fluorescence Studies of Pyrene-Labeled, Water-Soluble Polymeric Surfactants

Michael C. Kramer, Jamie R. Steger, and Charles L. McCormick

Department of Polymer Science, University of Southern Mississippi, Hattiesburg, MS 39406-0076

The synthesis of a fluorescently labeled water-soluble terpolymer based on acrylamide (AM) and the surface-active monomer sodium 11-Acrylamidoundecanoate (SA) is reported. Incorporation of 2-(1-pyrenylsulfonamido) ethyl acrylamide (APS) into the monomer feed yields a terpolymer with solution properties that are different from previously synthesized AM/SA copolymers. APS fluorescent label acts as a model hydrophobe; changes in pH and ionic strength that drive the viscosity response in AM/SA copolymers also affect the fluorescence emission properties of the APS label. Steady-state fluorescence emission studies reveal significant constriction of the polymer chain as pH decreases or electrolyte concentration increases. Fluorescence quenching studies suggest that the salt-induced chain collapse results in enhanced structuring of mixed aggregates formed by SA and APS units. Highly sensitive photophysical techniques confirm the pronounced pH and salt-responsiveness of AM/SA-based polymers observed in viscosity studies.

Previous work in our research laboratories has focused on the synthesis and characterization of fluorescently labeled water-soluble copolymers. The use of a covalently bound fluorescent moiety allows microscopic evaluation of the solution properties via steady state and transient fluorescence studies. Incorporation of a bulky hydrophobic group such as pyrene yields a polymer with associative properties via ground state inter- and intrapolymer hydrophobic interaction. Currently underway is an investigation of the solution properties of amphiphilic polyelectrolytes based on the surface-active monomer sodium 11-acrylamidoundecanoate (SA). The primary objective is the elucidation of ionic strength and pH effects on polymer conformation in aqueous media. Changes in these parameters affect the ability of these systems to sequester hydrophobic molecules. Assessment of salt- and pH-driven polymer

0097–6156/95/0598–0379$12.00/0

solution behavior leads to a better understanding of the phase-transfer properties exhibited by such materials.

Experimental

Materials. All reagents and solvents were purchased from Aldrich Chemical Co.. Acrylamide was recrystallized twice from acetone. Other materials were used as received. Water for synthesis and solution preparation was deionized and possessed a conductance $< 10^{-7}$ mho/cm.

Instrumentation. A Bruker AC-200 NMR spectrometer was used to determine ^{1}H and ^{13}C NMR spectra. Molecular weights were obtained with a Chromatix KMX-6 low angle laser light scattering photometer equipped with a 633 nm HeNe laser. Refractive index increments (dn/dc) were measured with a Chromatix KMX-16 laser differential refractometer. UV-VIS spectra were recorded with a Hewlett Packard 8452A diode array spectrophotometer. HPLC was carried out on a Hewlett Packard 1050 Series chromatograph fitted with a photodiode array UV detector and an Alltech Versapak C_{18} reversed-phase column. Viscosities were measured with a Contraves LS-30 rheometer at 25 °C and a shear rate of 6 sec^{-1}. Steady-state fluorescence spectra were measured on a Spex Fluorolog-2 fluorescence spectrometer equipped with a DM3000F data system. Excitation and emission slit widths of 1 mm and right angle geometry were employed.

Sodium 11-Acrylamidoundecanoate (SA). The methods of Gan (*1*) were employed in the synthesis of SA monomer. Purity was confirmed via NMR and melting point determination.

2-(1-Pyrenylsulfonamido) Ethylacrylamide. The procedure employed by Ezzell and McCormick (*2,3*) was followed to give pyrene monomer in good yield and purity according to NMR and HPLC.

Pyrene-labeled AM/SA Copolymer (AMSA/Py) (*4*) (Figure 1). A solution of sodium 11-acrylamidoundecanoate (1.8 g, 6.5 mmol) and acrylamide (9.0 g, 127 mmol) in 340 ml H_2O was added to the reaction flask immersed in a 50 °C water bath. After 2-(1-pyrenylsulfonamido) ethyl acrylamide (0.51 g, 1.3 mmol) that had been finely ground with a mortar and pestle was added to the flask, the solution was degassed with nitrogen for one hour, then sodium dodecyl sulfate (21.7 g, 75.2 mmol) added. At this point direct bubbling of nitrogen through the solution was stopped to prevent excessive foaming. After stirring the solution under nitrogen for 3 hours, most of the fluorescent comonomer had dissolved into SDS micelles. A degassed solution of potassium persulfate (50 mg, 0.19 mmol) in 5 ml H_2O was then injected into the monomer/surfactant solution. After stirring under nitrogen for three hours, the viscous polymer solution was added to 1400 ml acetone to yield a white, rubbery precipitate. The polymer was washed with refluxing methanol for 16 hours in a Soxhlet extractor to remove residual monomer and surfactant. Good yield (> 70 %) is obtained.

Results and Discussion

Polymer Syntheses. Gan (5) first reported the copolymerization of acrylamide with sodium 11-acrylamidoundecanoate (SA) at a monomer feed level of $\geq$ 30 mole %. Reactivity ratio studies indicated a random incorporation of SA monomer, and the copolymer composition approximates that of the monomer feed. When a growing radical encounters a micelle containing polymerizable vinyl monomers, the localized monomer concentration is much higher than in the aqueous phase. Micellization also serves to orient the monomers. This is observed as a fast overall polymerization rate yielding high molecular weight polymer. When SA is homopolymerized, high conversion and molecular weight ($M_w > 10^6$ g/mol) is obtained in a short period of time (< 1 hr) (1). Even though a low feed level of SA is employed, high conversion within a short period of time is obtained.

Gan reported molecular weights in the 5×10^5 - 2×10^6 g/mol range. The molecular weight of AM/SA copolymers is observed to decrease to the lower limit as AM content is increased from 0 to 70 mole % (5). AMSA/Py (Figure 1) has a molecular weight of 1.8×10^6 g/mol. Good monomer and surfactant purity may account for the higher molecular weight obtained relative to other AM/SA copolymers.

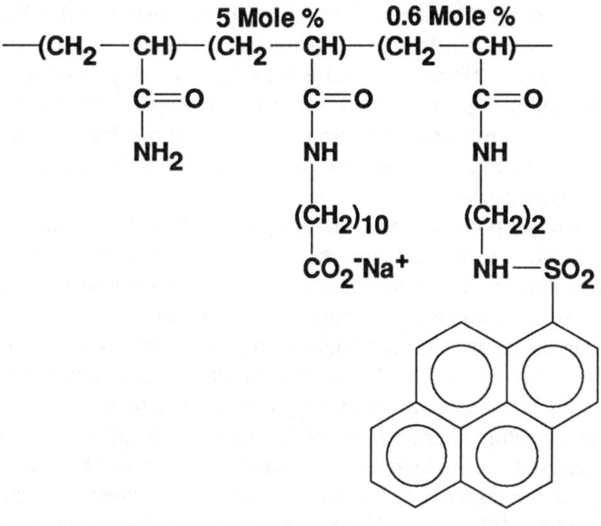

Figure 1. Pyrene-labeled AM/SA copolymer.

SA incorporation is assumed to equal monomer feed based on Gan's work and the high conversion obtained. When a copolymer of AM and SA with 90/10 AM/SA (mol/mol) in monomer feed was characterized by integration of ^{1}H NMR spectra, a copolymer composition of 10 ± 1 mole % was determined. If a similar weight percent incorporation of SA is assumed, 0.6 mole % incorporation for AMSA/Py is determined from UV-Vis studies ($\varepsilon = 24,120$ M^{-1} cm^{-1} at 352 nm (2,3,6)). The polymerization utilized 1 mole % fluorescent monomer in the monomer feed, but only 60 % of the monomer is incorporated. This is not an unusual result. Ezzell and McCormick (3) also reported low pyrene comonomer content in acrylamide copolymers synthesized both homogeneously and in the presence of added surfactant. Pyrene labels may be sufficiently hydrophobic such that the solubility in SDS micelles is limited, and incomplete dissolution into SDS micelles would result.

The use of an externally added surfactant to copolymerize a hydrophobic monomer with a hydrophilic monomer in aqueous solution can impart unique microstructural characteristics (7). A "blocky" microstructure may result from the inherent heterogeneity of the medium. When such a polymerization is carried out, the propagating radical in aqueous solution adds to the hydrophilic monomers in the aqueous phase. When the macroradical encounters a micelle, polymerization of all monomers within the micelle can occur if the micellar lifetime is sufficiently long. At high surfactant concentrations, this is a valid assumption, as k_p is quite high (10^3 - 10^4 M^{-1} sec^{-1}) for acrylamide and its related monomers (8). Since virtually all of the hydrophobic monomer, along with a fraction of the acrylamide (5,9) is partitioned into the micellar interior, a "block" of hydrophobic monomer is formed.

Viscosity and Fluorescence Studies. The block-like microstructure exhibited by these systems manifests itself in the rheological, as well as the photophysical response. Ezzell and McCormick (2,3,6) reported microstructural effects on the solution behavior of pyrene-labeled polyacrylamides synthesized by homogeneous and heterogeneous techniques. Pyrene labeled acrylamide copolymer synthesized in the presence of SDS exhibited a pronounced upwards curvature in the viscosity profile. This viscosity response is reflective of a low C*, or critical overlap concentration. At this concentration, interchain hydrophobic association drives the chain overlap that results in polymer aggregation.

Viscosity studies provide a reasonable assessment of the bulk, macroscopic solution behavior, but a detailed analysis requires the use of a more sensitive characterization technique. Fluorescence emission studies are inherently sensitive to low chromophore concentrations and angstrom-level motions.

Our research group has investigated pyrene label excimer emission to elucidate solution behavior. As the degree of hydrophobic association increases, so does the interaction between two isolated pyrenes bound to polymer to form the sandwich-like conformation characteristic of an excimer. Interpolymer hydrophobic association is observed as an increase in excimer relative to that of "monomer" emission. A typical fluorescence spectrum of AMSA/Py is shown in Figure 2. The peak maxima at 380, 400, and 420 nm arise from the fluorescence emission of isolated pyrenes (monomer emission). The broad, structureless band centered around 520 nm results from emission of excited dimeric pyrene (excimer). The intensity of excimer emission relative to that of the monomer emission (I_E/I_M) increases with polymer concentration. This response parallels the macroscopic viscosity behavior (*4,6*). I_E/I_M and apparent viscosity are plotted as a function of polymer concentration in Figure 3. I_E/I_M steadily increases with polymer concentration. The intermolecular hydrophobic associations that drive the viscosity profile also increase the population of excited pyrene dimers. Enhancements in intramolecular associations are also observed as an increase in I_E/I_M. As salt or acid is added to solution of SA copolymers, shielding or elimination of electrostatic repulsions between carboxylate groups of the SA unit leads to coil collapse that reduces the viscosity and I_E/I_M in these systems(*4,10*).

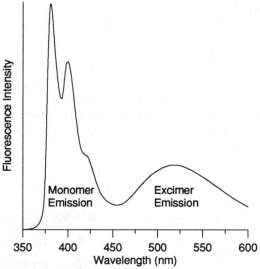

Figure 2. Fluorescence emission spectrum of pyrene-labeled polymer AMSA/Py. Polymer concentration: 0.052 g/dl in H_2O. Chromophore concentration: 3.4×10^{-5} M. Excitation wavelength: 340 nm.

Salt and pH Effects. The effects of salt and pH on the degree of excimer emission are shown in Figure 4. Viscosity studies have shown that pH decrease and salt addition lower the viscosity of AM/SA copolymers. Intrapolymer micellization of SA units is enhanced at the expense of charge-charge repulsions that interfere with association of the surfactant groups. Figure 5 illustrates this trend. As salt or acid are added, the shielding or elimination of electrostatic interactions collapses the coil. Pyrene groups within the polymer coil are closer to one another. The increase in localized chrompohore concentration is observed as an increase in excimer emission relative to monomer emission.

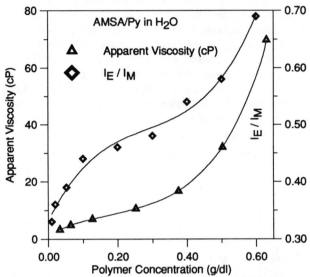

Figure 3. Apparent viscosity and excimer emission/monomer emission (I_E/I_M) as a function of AMSA/Py concentration in H_2O. (I_E/I_M = intensity at 519 nm / intensity at 400 nm).

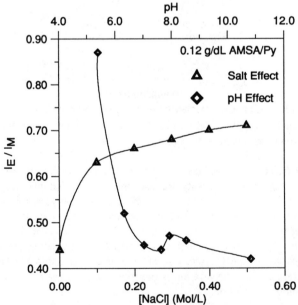

Figure 4. Salt and pH dependence of I_E/I_M. AMSA/Py polymer concentration: 0.12 g/dl.

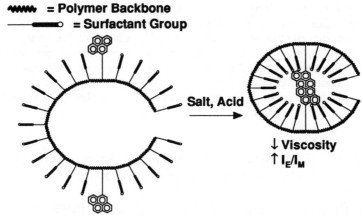

Figure 5. Proposed mechanism of salt/pH-triggered polymeric micelle structuring and photophysical enhancement.

Fluorescence Emission Quenching Studies. In order to probe the accessibility of the aqueous environment to the pyrenylsulfonamido chromophore, fluorescence quenching experiments were carried out using the amphiphilic quencher nitromethane ($MeNO_2$). First order quenching constants were calculated using the Stern-Volmer equation (*11*).

$$\frac{I_0}{I} = 1 + K_{SV}[Q] \tag{1}$$

Equation 1 describes dynamic Stern-Volmer fluorescence quenching, where I_0 = emission intensity in the absence of quencher Q, I = emission intensity in the presence of Q, [Q] = quencher concentration, and the Stern-Volmer quenching constant K_{SV} is expressed as:

$$K_{SV} = k_q \tau_0 \tag{2}$$

where k_q is the quenching rate constant usually expressed in M^{-1} sec^{-1}, and τ_0 is the lifetime of unquenched chromophore. If quenching occurs by a dynamic mechanism, a plot of I_0/I versus [Q] yields a straight line with slope = K_{SV}.

Shown in Table I are the results of Stern-Volmer plots for nitromethane quenching of 0.12 g/dL AMSA/Py in deionized water and aqueous sodium chloride solutions. The linearity of the I_0/I versus [Q] plots verifies the dynamic nature of fluorescence quenching. With the average APS fluorescence lifetime determined by Ezzell and McCormick (*6*) ($\tau_0 \cong 13 \times 10^{-9}$ sec), quenching rate constants can be calculated. The Stern-Volmer quenching constants and quenching rate constants approximate previously reported values (*6,12*). The magnitude of the quenching rate constants signifies a dynamic, diffusion-controlled process.

As electrolyte concentration increases, K_{SV} decreases. It should be noted that sodium chloride does quench APS fluorescence, but only slightly. Therefore, it is valid to assume that any change in the nitromethane Stern-Volmer quenching constant with added sodium chloride arises from electrolyte-induced conformational changes and not from sodium chloride quenching. As salt is added, the coil collapses. Any changes in K_{SV} would indicate a change in the accessibility of the amphiphilic quencher to the pyrene label. The results from this study suggest that salt-triggered collapse creates a rigid environment that restricts quencher-label diffusion. K_{SV} decrease would then reflect an increase in the viscosity of the mixed microdomain formed by SA and pyrene label.

Table I. Nitromethane Quenching of Pyrenesulfonamide Fluorescence in 0.12 g/dl Aqueous AMSA/Py

[NaCl] (M)	K_{SV} (M^{-1})	k_q (M^{-1} sec^{-1})
0	35	2.7×10^9
0.1	32	2.5×10^9
0.3	25	1.9×10^9
0.5	20	1.5×10^9

Summary

A description of the photophysical response of a fluorescently labeled water-soluble polymer containing surface-active functional groups is presented. The enhancement in AMSA/Py excimer emission with increasing polymer concentration follows the association-induced viscosity response. pH and salt-induced collapse of the polymer coil also enhances excimer emission due to intramolecular label aggregation. Results from fluorescence quenching studies suggest salt-induced microdomain structuring to give SA/pyrene aggregates with increased rigidity.

Acknowledgments

Funding from the U.S. Department of Energy, the U.S. Office of Naval Research, and Unilever is gratefully acknowledged.

References

1.   Yeoh, K.W.; Chew, C.H.; Gan, L.M.; Koh, L.L.; Teo, H.H. *J. Macromol. Sci.-Chem.* **1989**, *A26*, 663.
2.   Ezzell, S.A.; McCormick, C.L. In *Water-Soluble Polymers: Synthesis, Solution Properties, and Applications*; Shalaby, S.W.; McCormick, C.L.; Butler, G.B. Eds.; ACS Symposium Series No. 467; ACS: Washington, DC, 1991; Chapter 8.

3. Ezzel, S.A.; McCormick, C.L. *Macromolecules* **1992**, *25*, 1881.
4. Kramer, M.C.; Steger, J.R.; McCormick, C.L. *Proc. Am. Chem. Soc. Div. Polym. Mat.: Sci. Eng.* **1994**, *71*, 413.
5. Yeoh, K.W.; Chew, C.H.; Gan, L.M.; Koh, L.L.; Ng, S.C. *J. Macromol. Sci.-Chem.* **1990**, *A27*, 711.
6. Ezzell, S.A.; Hoyle, C.E.; Creed, D.; McCormick, C.L. *Macromolecules* **1992**, *25*, 1887.
7. Peer, W.J. *Proc. Am. Chem. Soc. Div. Polym. Mat.: Sci. Eng.* **1987**, *57*, 492.
8. *Polymer Handbook, Second Edition*; Brandrup, J.; Immergut, E.H.,- Eds.; Wiley: New York, 1975; Chapter II, p 47.
9. Biggs, S.; Hill, A.; Selb, J.; Candau, F. *J. Phys. Chem.* **1992**, *96*, 1505.
10. Kramer, M.C.; Welch, C.G.; McCormick, C.L. *Polym. Prepr. (Am. Chem. Soc. Div. Polym. Chem.)* **1993** *34 (1)*, 999.
11. Lacowicz, J.R. *Principles of Fluorescence Spectroscopy*; Plenum: New York, 1983. Chapter 9.
12. Morishima, Y; Ohgi, H.; Kamachi, M. *Macromolecules* **1993**, *26*, 4293.

RECEIVED February 2, 1995

Chapter 23

Luminescence Spectroscopic Studies of Water-Soluble Polymers

Ian Soutar and Linda Swanson

School of Physics and Chemistry, Lancaster University,
Lancaster LA1 4YA, United Kingdom

Luminescence spectroscopy is a versatile technique for studying polymer systems. In this article, we hope to illustrate the value of applying a range of luminescence approaches (such as anisotropy measurements, energy transfer and excimer formation) to investigations of macromolecular behavior, with particular reference to studies of the behavior of water-soluble polymers and their hydrophobically-modified forms.

Photophysical techniques continue to feature prominently in recent reports of investigations of the physical behavior of polymer systems (e.g. see references 1-3 and references therein). The popularity of such approaches to the study of macromolecular behavior is derived largely from the combined degrees of sensitivity, selectivity and specificity that luminescence measurements afford to the interrogation of a chemical system.

The intensity of luminescence emitted by a particular chromophore, within a given molecular assembly, is dependent upon a variety of factors, the most important of which are shown schematically in Figure 1. The opportunity to apply combinations of these parameters in characterization of the emission from a given chromophore, confers a distinct advantage to the use of luminescence spectroscopy over that afforded by its absorption counterpart in the characterization of molecular behavior. (See, for example, reference 3). Arguably, time-resolved anisotropy measurements (TRAMS), which combine the dual dependence of emission intensity upon molecular orientation and its time of sampling following excitation, represent the most powerful means whereby luminescence spectroscopic techniques can be applied to the study of polymer behavior and, in particular, the study of macromolecular dynamics. However, complementary luminescence approaches involving the use of e.g. luminescence quenching, fluorescence and phosphorescence spectra, excimer formation and energy transfer are extremely useful in providing additional information regarding the physical behavior of a given polymer system. In the current presentation, we hope to demonstrate the power of combining a variety of luminescence experiments to the

0097–6156/95/0598–0388$12.50/0

investigation of polymeric media. In particular, we seek to illustrate the potential of luminescence spectroscopy in studies of macromolecular phenomena, through a series of examples concerning the behavior of aqueous-borne polymer species which have been studied recently in our laboratories.

Emission Anisotropy Measurements

The anisotropy of the luminescence emitted by an electronically excited species is a function of the extent to which molecular reorientation occurs within the timescale imposed by the lifetime of the excited state. Consequently, anisotropy measurements provide a means of studying molecular dynamics. If the luminescent species is simply dispersed in the system as a *probe*, it can report directly upon the local viscosity e.g. in polymer solutions, curing resins, coalescing colloids during film formation, polymer solids etc. If the luminescent moiety is covalently bound as a *label* to a macromolecule, it can yield information, directly, upon polymer relaxation.

The principles of the technique are shown in Figure 2. In its most commonly adopted form, the chromophores to be examined are photoselected in absorption using polarized excitation. If the luminescent molecule remains stationary during its excited state lifetime, the resultant radiation will remain highly polarized, characteristic of the intrinsic anisotropy of that particular species. On the other hand, if molecular motion serves to randomize the photoselected species during the average lifetime of the ensemble of excited states, the emitted radiation will become depolarized to a degree dependent upon the relative rates of molecular reorientation and excited state deactivation.

Information regarding molecular motion is contained within the relative intensities, $I_\parallel$ and $I_\perp$, of luminescence observed in planes parallel and perpendicular, respectively, to that used for photoselection. "Decoding" of this information makes use of the anisotropy function, r, defined in equation (1):

$$r = \frac{I_\parallel - I_\perp}{I_\parallel + 2I_\perp} = \frac{D}{S} \tag{1}$$

Under photostationary steady-state conditions and assuming that molecular relaxation is such that it can be characterized by a single average correlation time, τ_c, kinetic information may be obtained using the Perrin equation:

$$r^{-1} = r_o^{-1} (1 + \tau/\tau_c) \tag{2}$$

where r_o is the intrinsic anisotropy and τ is the lifetime of the emissive state.

Steady state anisotropy experiments are relatively simple and convenient means of studying polymer relaxation processes both in fluid solution and in the solid phase (e.g. see references 3 and 4 and references therein). However, the steady-state approach is limited, in general, by the necessity to assume that the relaxing species behaves as a pseudo-spherical rotor, the motion of which *is* capable of being modeled by a single correlation time. Furthermore, problems can occur, especially in the case of aqueous

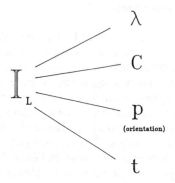

Figure 1.   Parameters characteristic of luminescence intensity.

DEPOLARIZATION RESULTANT UPON ROTATIONAL MOTION

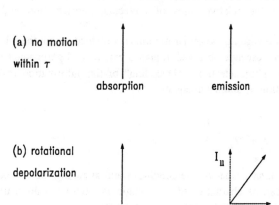

Figure 2.   Principle of the luminescence anisotropy approach to the study of rotational diffusion.

solutions of certain water-soluble polymers (5), in estimating r_0 in a reliable fashion. In principle, time-resolved anisotropy measurements (TRAMS) can overcome these shortcomings of the steady-state approach.

Figure 3 shows the time-dependent orthogonal components of intensity, $I_\parallel(t)$ and $I_\perp(t)$, of the fluorescence from an acenaphthylene-labeled sample of PMAA in dilute aqueous solution (10^{-3} wt %) at pH = 4.4, following pulsed, polarized excitation from a synchrotron (SRS, Daresbury, UK). $I_\parallel(t)$ and $I_\perp(t)$ contain information regarding the reorientation of the acenaphthyl label (and thence of the segmental relaxation of its PMAA "host") but are also distorted by the finite width of the excitation pulse from the synchrotron. This complicates analysis of the data to an extent which depends upon the magnitudes of both the fluorescence lifetime and the reorientational correlation time relative to the breadth of the excitation pulse profile. Analytical procedures for recovery of rotational relaxation parameters from TRAMS studies of labeled polymers have recently been discussed (6,7).

Figure 4 shows the pH-dependent segmental relaxation behavior of poly(methacrylic acid), PMAA, poly(acrylic acid), PAA, and hydrophobically-modified forms of these polyelectrolytes produced by copolymerization of the parent acid with styrene. [At high pH, modeling r(t) using a single exponential decay function proved statistically adequate in the impulse reconvolution (8) analytical procedure adopted. At low pH, minor deviations from purely exponential behavior are apparent in some data sets, especially those obtained from PMAA]. Clearly, neutralization of PMAA is accompanied by a dramatic and rather abrupt transition in the rate of intramolecular segmental motion of the polyelectrolyte which contrasts with the relatively smooth reduction in segmental mobility of PAA which is effected by increasing the pH of the medium. These observations agree broadly with those of other workers: for example, Katchalsky et al. using viscosity (9) and diffusion coefficient (10) data have observed marked transitions in their pH dependences for PMAA which are not apparent for aqueous solutions of PAA. These observations have been interpreted in terms of a dramatic increase in chain dimensions as the PMAA undergoes a conformational transition between the "hypercoiled" form of the polyacid and the expanded coil adopted by the polysalt. The TRAMS data show that in its hypercoiled form, inter-segmental interactions reduce the intramolecular mobility of the polyelectrolyte to such an extent that a correlation time of *ca.* 50 ns results. In contrast, the expanded coil form of PMAA is reasonably flexible; its segmental motion being characterized by a value of τ_c of less than 10 ns (11). At all values of pH, PAA is an altogether more flexible species: τ_c varies from *ca.* 6 ns in highly acid media to between 1 and 2 ns at high pH.

Reference to Figure 4 reveals that the conformational behavior of the polyelectrolyte can be modified through the introduction of hydrophobic species during copolymerization. For example, incorporation of styrene affects both the segmental mobility of the resultant polyelectrolyte and the pH range over which the conformational transition occurs (12). In the case of AA-based polymers, hydrophobic interactions between the styryl residues significantly reduce the segmental mobility relative to that of PAA. (In the absence of such intra-coil aggregations, steric effects, induced by the presence of styrene units within the polymer backbone, would

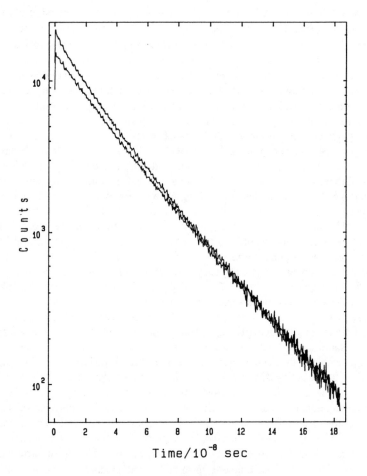

Figure 3.   Parallel and perpendicular components of fluorescence
              intensity of PMAA/ACE following excitation using
              vertically polarized synchrotron radiation.

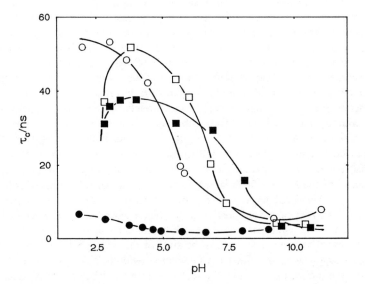

Figure 4.   pH dependence of correlation times for segmental motion in ACE labeled PAA (●), PMAA (O), a methacrylic acid-styrene copolymer containing 24 mole% styrene (□) and an acrylic acid-styrene copolymer containing 39 mole% styrene (■).

have little effect on the segmental mobility of the polymer, as indeed is observed in the completely neutralized species).

The hydrophobic microdomains created within the confines of the polymer coil determine the capacity of the macromolecule for solubilization of organic guests, as revealed by spectroscopy and discussed below. Anisotropy experiments (whether steady-state or time-resolved) upon solubilized fluors can yield information regarding the fluidity of the microenvironment afforded to the guest and the interactions between the guest and hydrophobic components of its "confining cavity". For example, when pyrene (10^{-6} M) is dispersed in a series of acidic aqueous solutions of AA and MAA polymers (10^{-2} wt % in polymer) its fluorescence is virtually completely depolarized in the presence of PAA itself but becomes increasingly anisotropic as the styrene content of the hydrophobically-modified styrene-AA copolymeric host is increased (12). These observations reinforce the indications of more conventional "spectroscopic" approaches, that dilute solutions of PAA have little (if any) solubilizing influence upon pyrene (and other guests) whereas incorporation of a hydrophobic modifier such as styrene confers such a capacity upon the resultant polymers. In contrast, the hydrophobic domains created with PMAA at low pH, as reflected both by fluorescence spectroscopy (see below) and anisotropy measurements, can accommodate organic guests without the need for hydrophobic modification of the polymer. The fluorescence from the solubilized pyrene guest molecules is highly anisotropic: both time-resolved and steady-state data lead to estimates of τ_c in the region of 60 ns for rotational reorientation of pyrene and its associated solubilizing sheath (12). Styrene modification of the MAA system increases τ_c for rotational relaxation of solubilized pyrene to ca. 85 and 250 ns respectively for styrene/MAA copolymers containing 17 and 24 mole percent of styrene, respectively. These correlation times exceed those characteristic of the segmental motions of the hydrophobically-modified polymers and it is likely that the nature (and perhaps the size) of the hydrophobic domains created within such copolymers, are affected by interactions with the solubilized guest (12).

Since the anisotropy of the fluorescence from a labeled macromolecule is dependent upon its segmental mobility, TRAMS may be used in the study of inter-polymer complexation and the adsorption of water-soluble polymers at interfaces.

Figure 5 shows the decays of the "parallel and perpendicular components", $I_\parallel(t)$, and $I_\perp(t)$, of the fluorescence from a vinylpyrene labeled sample of poly(dimethylacrylamide), PDMAC, complexed (1:1 mole ratio) with PMAA at a pH of 2.2. In aqueous media, the vinylpyrene label exhibits a fluorescence lifetime of the order of 130 ns. Since the correlation time characteristic of the segmental motion of uncomplexed PDMAC is of the order of 3 ns (13) under these conditions, $I_\parallel(t)$ and $I_\perp(t)$ are convergent at all but the very shortest of times following pulsed excitation in the absence of PMAA. The formation of an inter-polymer complex with PMAA has a dramatic effect upon the segmental mobility of the PDMAC, leading to a considerable anisotropy of fluorescence from the labeled complex which (as is shown in Figure 5) persists until long times [$I_\parallel(t)$ and $I_\perp(t)$ do not fully converge over the 700 ns time range shown]. Direct analysis of the anisotropy decay yields a correlation time of ca. 900 ns for the labeled segments of PDMAC in 1:1 complexes with PMAA (13). Fluorescence anisotropy measurements have previously been applied in studies of the

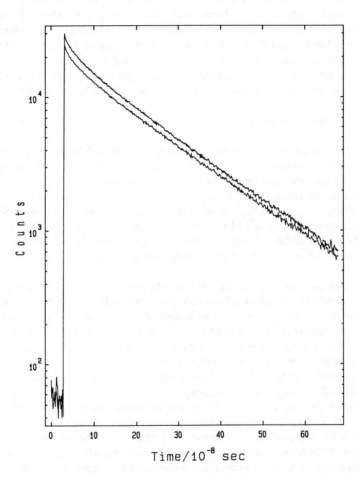

Figure 5.   Parallel and perpendicular components of fluorescence of PDMAC/VPy complexed with PMAA at a mole ratio of 1:1 and a pH of 2.2.

complexation, in aqueous media, between poly(ethylene oxide) and PAA (*14*) or PMAA (*15*).

PDMAC adsorbs, over a wide range of pH, on the modified surface of the colloidal silica, Ludox AM (Du Pont) (*13,16*). The fluorescence anisotropy decay rate is dependent upon the relative amounts of polymer and colloid present in the system. At high silica contents, the polymer is essentially completely adsorbed and direct analyses of the anisotropy decays resultant upon rotational reorientation of a vinylpyrene-labeled adsorbate yield τ_c values of the order of 1-2μs (*13*). This slow relaxation occurs on a timescale comparable to that expected for whole particle rotation in the aqueous colloidal dispersion and might be indicative that the polymer "lies flat" upon the colloid surface (rather than form trains or loops) in adsorption.

"Spectroscopic" Information

The study of water-soluble polymers can often be enhanced by simple spectroscopic and lifetime measurements. For example, the fluorescence spectrum of pyrene is sensitive to its local environment (*17,18*). In particular, the intensity ratio of two specific vibronic bands (the I_3/I_1 ratio) may be used as a sensor of the molecule's surroundings (*18*). Thomas et al. (*19,20*) have shown how the pH-induced conformational transition of PMAA is revealed in the change which occurs in the I_3/I_1 ratio of a pyrene probe (solubilized in the hypercoiled form of the polymer at low pH) as its macromolecular host expands during neutralization. The effect is shown in Figure 6.

The I_3/I_1 ratio can also be used to give information regarding the conformational behavior of hydrophobically-modified polymers of MAA (*cf*. Figure 6). The transition occurring in I_3/I_1 as the pyrene solute is ejected into the aqueous phase, appears at higher values of pH in the styrene/MAA copolymers than in PMAA itself. The effects of hydrophobic modification upon the capacity of the polymer to solubilize organic guests such as pyrene are clearly demonstrated. The data show shifts (in pH) of the conformational transition of the polyacids which are in broad agreement with those revealed in TRAMS (*cf*. previous section).

Interestingly, the I_3/I_1 ratio for pyrene solubilized in the acidic forms of the polyelectrolytes would appear to indicate that the polarity of the hydrophobic domains created in the styrene/MAA copolymers is greater than that of the corresponding domains in PMAA itself. Such differences in domain characteristic are not translated into differences in intramolecular chain mobilities (*cf*. Figure 4) and might appear surprizing in the light of anisotropy data from the solubilized probe itself (see earlier): the latter clearly indicate that the solubilizing "pockets" in the polymer are highly compacted, in the presence of the pyrene, and impose severe restrictions upon the rotational freedom of the guest.

The intensity of fluorescence emitted from probes and labels and excited state lifetimes are frequently sensitive to the conformation adopted by the polymeric hosts (*21,22*). Figure 7 shows the dependence of the fluorescence lifetime of pyrene upon pH in the presence of PMAA or styrene/MAA copolymers. The excited singlet state of pyrene solubilized in the hydrophobic domains of the polyacids is long-lived (≥ 300 ns) whereas that of pyrene dispersed in water is *ca*. 120 ns. The lifetime data clearly

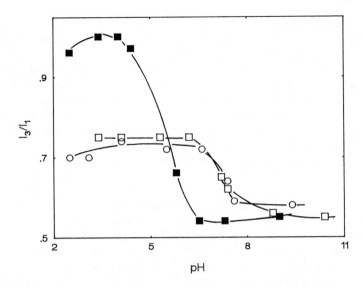

Figure 6.   pH dependence of the I_3/I_1 ratio for pyrene dispersed in PMAA (■) and methacrylic acid-styrene copolymers containing 17 (O) and 24 (□) mole% styrene, respectively.

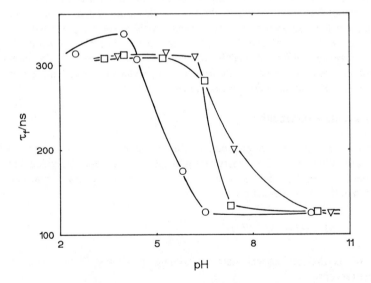

Figure 7.   pH dependence of the average fluorescence lifetime of pyrene (10^{-6}M) dispersed in PMAA (O) and methacrylic acid-styrene coploymers containing 17 (□) and 24 (▽) mole% styrene, respectively.

serve as a simple means of *observing* the conformational transition between the polyacid and polysalt forms of the MAA-based systems and of demonstrating the effects of hydrophobic modification upon the position of the transition in the pH range. In a similar fashion, both the fluorescence intensity and lifetime of the ACE and VN labels, used in anisotropy studies of the chain dynamics of these water-soluble polymers, may be used to simply demonstrate that a conformational transition occurs and to define its position upon the pH scale. The lifetime of the triplet state of these (21) and other (23) labels and the intensity of the resultant phosphorescence which can be sustained in these media at room temperature, can also be used to study the conformational behavior of water-soluble polymers (21,23).

The fact that changes in the microenvironment of a chromophore can affect excited state deactivation rates means that fluorescence intensity and lifetime measurements can be used as simple and convenient sensors of the fact that some form of interaction involving the polymer is indeed occurring in a system (prior, say, to embarking upon more time-consuming studies involving e.g. TRAMS). For example the sensitivity of the lifetime of the pyrene chromophore to its microenvironment provides a relatively straightforward means of detection of the complexation between e.g. vinylpyrene-labeled PDMAC and PMAA or PAA or the adsorption of vinylpyrene-labeled PDMAC on colloidal silica particles (13). In studies of the solubilization capacity of water-soluble polymers (and their hydrophobically-modified forms) for organic guests, perylene is an extremely useful fluorescent probe: its low solubility in water $[1.2 \times 10^{-9} \, M \, (24)]$ ensures a low "background" intensity of emission. Solubilization within the hydrophobic domains of suitable water-soluble polymers is marked by the observation of intense fluorescence from the system. As an example, the conformational transition between the expanded polysalt of a styrene/AA copolymer (10^{-2} wt%; containing 44 mole% styrene) and the compact form adopted by the polyacid may be readily detected by the concomitant increase in the intensity of fluorescence which can be detected from $10^{-5} \, M$ perylene dispersed in the system and solubilized at pH values lower than *ca.* 7 (12).

Energy Transfer (Quenching)

The luminescent excited state may be quenched through transfer of energy to another species. For convenience, these energy transfer interactions may be divided into short-range and long-range processes, respectively. The general process of quenching may be written in the form of equation 3

$$M^* + Q \rightarrow M + Q^* \tag{3}$$

where M* is the luminescent state undergoing quenching (donor) and Q is the quencher (acceptor).

Short-range processes generally involve production of a "collisional complex", (MQ*), and the observed quenching kinetics depend upon the relative rates of dissociation of the complex and, under appropriate circumstances, will be controlled by the mutual diffusion rates of M* and Q. Thus, fluorescence quenching is appropriate to the study of the diffusion of molecular species in fluid solutions and can furnish valuable information regarding the range of microenvironments which can be created within the coils of water-soluble polymers. The extent to which any given quencher can access a fluorescent label or guest is dependent upon polymer chain mobility, domain hydrophobicity and the nature of the quencher. The range of quenchers which have been used in probing the behavior of water-soluble polymers includes ionic species [such as Cu^{2+} (25,26), Tl^+ (21,22,25) and I^- (21,22,25,26)] as well as neutral molecules [notably, nitromethane; see, for example, references 21,22,25,26]. Ionic quenchers report upon their specific interactions with polyelectrolytes in addition to reflecting their preference for the aqueous-rich regions of the system over hydrophobic domains.

The data listed in Table 1 reflect the ease of access by quenchers to ACE labels incorporated into a range of polyacids. The effect of the hydrophobic methyl backbone substituent of PMAA upon the conformation adopted by the polymer in acidic media is apparent in comparison of the bimolecular rate constants, k_q, for the quenching of the fluorescences of labeled PMAA and PAA respectively. ACE labels within the hypercoiled form of PMAA are much less accessible to quenchers [whether neutral (CH_3NO_2) or cationic (Tl^+)] than those attached to the acidic form of PAA. In the latter instance, quenching efficiencies approaching to that expected for diffusion-controlled interactions are apparent: these observations support evidence from both anisotropy and spectroscopic measurements (see previous sections) that PAA exists in a relatively open-coiled conformation even in acidic media.

Table 1 : Fluorescence Quenching Data for ACE- and VN[†]-labeled Polymers in Aqueous Media at 298 K

System	Quencher	pH	$k_q{}^a/10^9\,M^{-1}\,s^{-1}$
PMAA	CH_3NO_2	4.6	0.6 (21)
PAA	CH_3NO_2	3.7	4.0 (12)
PMAA	Tl^+	3.1	0.3 (21)
PAA	Tl^+	2.3	4.2 (12)
PMAA	I^-	12.7	0.6 (21)
PMAA	Tl^+	11.4	700 (21[†])
AA/MMA (34 mole%)	CH_3NO_2	3.6	0.7 (12)
AA/STY (39 mole%)	CH_3NO_2	4.1	0.2 (12)
AA/MMA (34 mole%)	CH_3NO_2	6.9	4.2 (12)
AA/STY (39 mole%)	CH_3NO_2	7.0	0.3 (12)

[a] Estimated from fluorescence lifetime data using the Stern-Volmer equation viz. $\tau^\circ/\tau = 1 + k_q\,\tau^\circ[Q]$ where τ°, τ are the lifetimes of the ACE or VN label in the presence or absence of quencher, Q, respectively. No attempt has been made to compensate for the effects of differential partitioning of quenchers between the polymer domains and the bulk aqueous phase.

The k_q values shown in Table 1 for the CH_3NO_2 induced quenching of the ACE fluorescences from labeled PMAA and PAA are in good agreement with those reported by Chu and Thomas (22) for pyrenyl-labeled samples of the same polyelectrolytes and by Webber et al.(26) for PMAA bearing a diphenylanthryl label. Data presented by these authors (22,26) show that ionic species such as Tl^+, I^- and Cu^{2+} show low efficiencies of quenching of excited states located within hypercoiled PMAA relative to those observed (22) for PAA in acid media.

Fluorescence quenching data obtained for water-soluble polymers can reveal the presence of specific interactions with potential quenchers. For example, acrylamide-induced quenching of the fluorescences of ACE-labeled polyacrylamide (27) or VN-labeled PDMAC (13) is lower than is to be expected on the basis of unperturbed diffusion of acrylamide, as a result of polymer-quencher interactions. Similarly, the negatively charged ionic atmosphere of the polysalt of PMAA can reduce the efficiency of anionic quenchers such as I^-, and enhance that of cationic species, such as Tl^+. (cf. Table 1).

Hydrophobic modification of the acrylic acid polymer, using copolymerized methyl methacrylate (34 mole%) leads to formation of hydrophobic domains within its acidic form which appear to be similar to the PMAA hypercoil in their ability to inhibit access of the label by CH_3NO_2. Use of copolymerized styrene (39 mole%) produces hydrophobic pockets which appear even less permeable, in acid conditions (cf Table 1). As the pH is increased, the quenching data offer convincing evidence of the ability of styrene modification to extend the range of pH over which hydrophobic domains can be maintained in aqueous media. The absence of this characteristic in the methyl methacrylate copolymer is also evident. Once again, the quenching data provide information complementary to that afforded by TRAMS and spectroscopy.

The study of long-range energy transfer processes between labeled polymers can provide valuable information regarding macromolecular interactions and interpenetration phenomena (see, for example, references 28 and 29 and references therein). In aqueous media, non-radiative energy transfer has been used, for example, in the study of inter-polymer interactions (30) and of the behavior of poly(N-isopropylacrylamide) (31) and hydrophobically-modified forms of this polymer (32).

As we have described above, the introduction of hydrophobic species can modify the dilute solution behavior of water-soluble macromolecules, encouraging the formation of hydrophobic pockets which are, in turn, capable of solubilizing organic guests. If these hydrophobic modifiers are aromatic chromophores, the potential for energy transfer from the host hydrophobic cavity to sequestered guests may be realized. The first example (33) of this effect emerged from the pioneering work of Guillet and coworkers upon such "photozyme" systems. It was shown that a copolymer of acrylic acid and 1-naphthylmethyl methacrylate (6.9 mole%), terminated with one anthryl acceptor per chain, exhibited marked enhancement of the singlet energy transfer from the naphthyl donors to the terminal traps when the solvent was changed from dioxane to aqueous NaOH (pH=12). This observation, combined with the fact that aqueous solutions produced substantial fluorescence from naphthyl excimeric states (an emission absent from the fluorescence of dioxane solutions of the polymer), led to the suggestion that the hydrophobic naphthyl copolymerized units aggregated in aqueous media (33). This initial report was supported by a further publication (*34*) which emphasized the importance of the "antenna effect" (the subject of a series of publications by Guillet et al. initiated (*35*) in 1982) in the close confines of the hydrophobic domains created in aqueous solutions of appropriate water-soluble copolymers. The photocatalytic properties of the "photozyme polymers" has been amply demonstrated (see, for example, reference 36 and references therein). The general topic of "photon-harvesting" polymers has been the subject of an excellent review (*36*).

Energy transfer experiments can be used to reveal differences in the microenvironments available to guests occluded in the hydrophobic domains of water-soluble polymers. For example, VN/MAA copolymers will solubilize species such as pyrene and perylene over extended regions of pH compared to PMAA itself. Excitation of the VN components of the copolymer results in efficient energy transfer to a proportion of the solubilized guests. Alternatively, the entire guest population can be sampled by excitation directly, at wavelengths which are "invisible" to the polymer. When perylene is solubilized at a concentration of 10^{-5}M in a 10^{-3}wt% solution of MAA/VN{36} [i.e. a copolymer containing 36 mole% of VN], its fluorescence spectrum consists of emissions from both monomer and excimer. The spectral profile of the subset of perylene molecules which can be accessed by energy transfer from the naphthyl species shows negligible excimer fluorescence. Clearly the perylene molecules are distributed throughout a range of local environments within the MAA/VN coil. Those located within the more hydrophobic domains are revealed in energy transfer. The more polar regions of the host environment might contain aggregates of perylene which exhibit excimer formation or may simply be less viscous than the more compact domains, thus allowing excimers to be formed diffusionally. Similarly, differences in the vibronic structures of pyrene fluorescences which result from direct excitation and energy transfer from the VN units suggest that there are differences in polarity between the local environments of the various pyrene guests. Evidence of extremely efficient "photon harvesting" may be found in the phosphorescence excitation spectra of triplet pyrene species formed using heavy ion quenchers, such as Tl^{+} (*37*). These spectra indicate that this population of excited states, unlike its singlet counterparts, is created preferentially by energy transfer from the VN component of the solubilizing macromolecule (*37*).

As discussed above, energy transfer between aromatic moieties, themselves part of the hydrophobic pockets housing energy traps (whether guest species or integral to the polymer) can be extremely efficient, both as a result of the enforced proximity between donor and acceptor and through migration of energy between closely packed donors, prior to the final transfer step. The efficiency of a resonance transfer is critically dependent upon the relative distance between the donor excited state and the acceptor. Evidence of the compact nature of the hydrophobic domains formed in hydrophobically-modified polyelectrolytes such as AA/STY{44} and the contraction thereof that the inclusion of organic guests might effect is apparent in the restrictions imposed upon the rotational freedom of solubilized probes (see earlier). Further evidence is afforded by energy transfer experiments.

The compact form of an AA/STY{44} copolymer, containing 0.5 mole% VN as label, formed at low pH contains hydrophobic domains capable of solubilizing perylene (*cf.* see previous section). These domains are sufficiently compact to promote energy transfer between the VN label and the perylene guest, as shown in Figure 8. As the pH is increased the cavity expands and the average distance between the VN labels and the perylene guest molecules increases. The perylene is also progressively excluded from the expanding cavities until, at high pH, there is no evidence for perylene fluorescence from energy transfer (*nor* from direct excitation into the perylene lowest energy absorption band itself).

If the transfer of energy from the VN labels to the perylene guests were governed by Forster-type (*38*) interactions, it might be expected (*39*) that the time-resolved intensity of fluorescence of the naphthyl donor $i_D(t)$ would be described by a function of the form

$$i_D(t) = i_D(0) \, exp[- t/\tau_o - \beta(t/\tau_o)^{1/2}] \tag{4}$$

where $i_D(0)$ is the fluorescence intensity at t=0, τ_o is the fluorescence lifetime of the unquenched donor (assuming that the decay of the donor *can* be described by a single exponential function) and β is a term dependent upon, *inter alia*, the (unknown) local concentration of acceptor. Tenability of such a model function would necessitate, amongst other stringent criteria, that donor and acceptor are homogeneously dispersed within the system. In the AA/STY{44} aqueous dispersions, it is to be expected, for any given (overall) concentration of perylene, that hydrophobic cavities will exist in which VN donors are located but do not contain a solubilized perylene probe. In such cases of true kinetic isolation of a proportion of the VN donors it might be expected that the decay of the fluorescence of the total donor population might more properly be described by a function of the form

$$i_D(t) = A_1 \, exp[- t/\tau_o - \beta(t/\tau_o)^{1/2}] + A_2 \, exp(- t/\tau_o) \tag{5}$$

where the ratio of the terms A_1 and A_2 should reflect the relative proportions of donor species participating in energy transfer to solubilized acceptor species to those isolated in microdomains which do not contain a solubilized acceptor.

Equation 5 is identical to that proposed by Winnik et al. (*40*) to describe the developing degrees of energy transfer between donor and acceptor species localized within coalescing aqueous-borne colloids during film formation and which has enjoyed much success (See, for example references 41 and 42 and references therein) in the study of such systems. Figure 9 shows the fluorescence decay behaviors of a VN-labeled copolymer of AA/STY{44} at different (total) concentrations of occluded perylene guest, under acid conditions. The general shapes of the decay curves are similar to those observed by Winnik et al. (*40*) in coalescing colloidal media and of the form expected on the basis of equation 5.

Reconvolution analyses using the functional form represented by equation 5 to model the observed emission from the VN label of the AA/STY{44} system do not provide fits to the data that might be judged adequate using the stringent criteria normally adopted in photophysical investigations: for example, values of $\chi^2 \geq 1.6$ are generally obtained. However, high quality fits to the data should not be expected in such complex systems since,

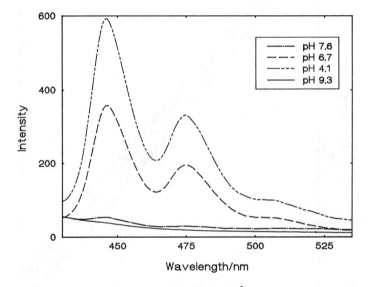

Figure 8.   Emission spectra of perylene (10^{-5}M) dispersed in solutions of a 1-VN labeled copolymer of acrylic acid and styrene (styrene content = 44 mole%) at different values of pH.

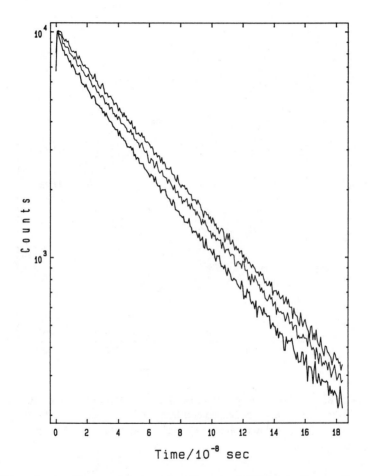

Figure 9.   Decays of fluorescence from the 1-VN label of an acrylic acid-styrene copolymer, containing 44 mole% styrene, in the presence of (top to bottom) 0, 10^{-5} and 10^{-4}M perylene at pH = 4.2.

(i) the fluorescence decays of labels such as VN, attached to water-soluble polymers, are generally not strictly single exponential in aqueous media.

(ii) the donor and acceptor might not remain stationary during the excited state lifetime of the donor, introducing transient effects at shorter times. Despite these shortcomings, data resultant upon analyses of this type are plausible The ratio $A_1:A_2$, for example, increases as the amount of solubilized perylene increases. Further investigations are in progress to determine how useful such an approach might be in the study of the microdomains formed in water-soluble polymers.

Excimer Formation

Excimer formation, a feature of the photophysical behavior of most aromatic species, occurs through the interaction between an electronically-excited molecule, M*, and a ground state molecule, M, of the same chemical type. The process may be represented as

$$M* + M \rightarrow D* \tag{6}$$

where D* is the excimer (or "excited dimer"). This self-quenching of M* and associated formation of D* is promoted at higher concentrations of chromophore. Dependent upon a number of factors, excimer formation can occur intramolecularly, in bi-or poly-chromophoric species. In general excimer formation is characterized by the appearance, in the fluorescence spectrum, of a structureless emission band of energies lower than that of the quenched emission of M*. The effect is illustrated in Figure 10 which shows the emission spectrum of a MAA/VN copolymer (containing 36 mole% VN) in MeOH (10^{-4}M in chromophore) exhibiting quenched "normal" fluorescence from the naphthyl chromophores with a maximum at *ca.* 340nm and excimer fluorescence centred around 400nm.

Low molar mass bichromophoric species have been widely used as sensors of the microviscosity of the environment in which they are dispersed. In polymer science such probes can be used to study the effects of segmental diffusive processes in polymer matrices (e.g. see references 43 and 44 and references therein) but have enjoyed little application in the study of water-soluble polymers. Relatively recently, however, Winnik et al. (*45*) have used the hydrophobic probe bis(1-pyrenylmethyl)ether to study the fluidity and micropolarity of the polymeric micelles formed in aqueous solutions of copolymers of N-isopropylacrylamide with various other N-alkylacrylamides.

The occurrence of excimer formation in water-soluble polymers which have been "tagged", at relatively low loading levels, with pyrenyl derivatives has been used to study their behavior in aqueous solution (e.g. references 46 and 47), in complexation with other polymers (e.g. references 48-51) and in interactions with colloidal substrates (e.g. references 52 and 53). The photophysical behavior of pyrene-labeled polymers in aqueous solution has formed part of a recent, excellent review (*54*).

At higher loading levels the probability of interactions between chromophores within the lifetime of the excited state is enhanced.   Furthermore, ground state interactions become important in determining the behavior of the polymer in aqueous solution (as described above) and the aromatic species become hydrophobic modifiers of the system.   Intramolecular excimer formation in macromolecules has been the

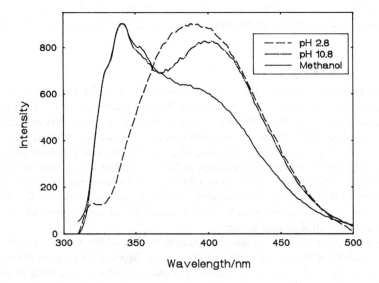

Figure 10.   Fluorescence spectra of a methacrylic acid-vinylnaphthalene
              copolymer (36 mole% VN).

subject of much study (see e.g. references 3 and 55 and references therein). The intensity ratio of fluorescence from excimer to monomer I_D/I_M depends *inter-alia* upon the thermodynamic quality of the solvent (*56*): the poorer the solvent, the more compact the polymer coil and the greater I_D/I_M. The increase in the local chromophore concentration encourages both formation of excimer traps and energy migration to such traps. Similar effects can be observed in aqueous media as shown in Figure 10. In highly acidic media the MAA/VN{36} copolymer exists in a highly compacted conformation promoted by hydrophobic interactions between both the backbone methyl groups of the MAA and the VN units, as hydrophobic modifiers, themselves. The resultant high local concentration of naphthyl chromophores favors excimer formation and the fluorescence spectrum of the copolymer is dominated by emission from excimeric traps. As the MAA species in the polymer are neutralized, however, fluorescence from unassociated or "monomeric" naphthalenes becomes apparent and at high pH, is of greater intensity than that of the excimer. Similar trends in I_D/I_M with pH have been observed in a random copolymer of 2-vinylnaphthalene (44 mole%) and acrylic acid (*57*). The fact that excimer emission makes a significant contribution to the copolymer fluorescence at high pH shows that close contacts between the naphthyl species are maintained in basic media. This might be indicative of the existence of intrachain hydrophobic domains promoted largely by interactions between the VN modifiers and further studies of the photophysical behavior of copolymer systems such as these are in progress: our understanding of the nature of hydrophobically-modified polymers of this type is far from complete.

It might be expected on the basis of luminescence and other experimental data, that the 2-VN units in Webber's acrylic acid copolymer would provide the major driving force for creation of compact hydrophobic domains at low pH. In our MAA copolymer on the other hand, it is likely that both VN and MAA units (through involvement of their backbone methyl substituents) would be incorporated into the hydrophobic structures. In this instance, VN-VN approaches do not seem to be unduly hindered and an excimer emission, enhanced relative to that observed in basic media, is evident. In contrast, methacrylic acid copolymers containing naphthyl species [copolymerized 2-(1-naphthylacetyl)ethylacrylate (*58*) or 2-(1-naphthylacetamido)-ethylacrylamide (*59*)] pendant to the polymer *via* (relatively) long, flexible chains, show considerable I_D/I_M ratios in basic media which are *reduced* as the system is made more acidic (*58,59*).

Conclusions

The photophysical behavior of luminescent probes and labels can provide remarkable insights into the behavior of their water-soluble, macromolecular hosts. The ability to apply combinations of approaches involving e.g TRAMS, energy transfer experiments and studies of excimer formation makes luminescence spectroscopy a powerful means of interrogating polymer systems.

Acknowledgments

The authors wish to thank the Leverhulme Trust and the EPSRC for financial support for much of the work described in this article.

Literature Cited

1. *Photophysical and Photochemical Tools in Polymer Science*; Winnik, M.A., Ed.; NATO ASI Ser. C; D. Reidel Publ. Co.: Dordrecht, 1986; Vol 182.
2. *Photophysics of Polymers*; Hoyle, C.E.;Torkelson, J.M., Eds.; ACS Symp. Ser.; ACS: Washington, DC, 1987; Vol 358.
3. Soutar, I.; *Polym. Int.* **1991**, *26*, 35.
4. Soutar, I. In *Developments in Polymer Photochemistry*; Allen, N.S., Ed.; Applied Science Publishers; London, 1982, Vol 3; pp 125-163.
5. Soutar, I.; Swanson, L. *Polymer* **1991**, *35*, 1942.
6. Soutar, I. *Makromol. Chem. Macromol. Symp.* **1992**, *53*, 393.
7. Soutar, I.; Swanson, L.; Imhof, R.E.; Rumbles, G. *Macromolecules* **1992**, *25*, 4399.
8. Barkley, M.D.; Kowalczyk, A.A.; Brand, L. *J. Chem. Phys.* **1981**, *75*, 3581
9. Katchalsky, A.; Eisenberg, H. *J. Polym. Sci.* **1951**, *6*, 436.
10. Kedeem, O.; Katchalsky, A. *J. Polym. Sci.* **1954**, *22*, 159.
11. Soutar, I.; Swanson, L. *Macromolecules* **1994**, *27*, 4304.
12. Ebdon, J.R.; Lucas, D.M.; Soutar, I.; Swanson, L., to be published.
13. Soutar, I.; Swanson L.; Wallace, S.J.L.; Ghiggino, K.P.; Haines, D.J.; Smith, this volume.
14. Heyward, J.J.; Ghiggino, K.P., *Macromolecules* **1989**, *22*, 1159.
15. Soutar, I.; Swanson, L., *Macromolecules* **1990**, *23*, 5170.
16. Lumber, D.C., Robb, I.D., Soutar, I. and Swanson, L., *unpublished data.*
17. Nakajima, A. *Bull. Chem. Soc. Jpn.* **1971**, *B44* , 3272.
18. Kalyanasunderam, K.; Thomas, J.K. *J. Amer. Chem. Soc.* **1977**, *99*, 2039.
19. Chen, T.S.; Thomas, J.K. *J. Polym. Sci. Polym. Chem. Ed.* **1979**, *17*, 1103.
20. Olea, A.F.; Thomas, J.K. *Macromolecules* **1989**, *22*, 1165.
21. Soutar, I.; Swanson, L. *Eur. Polym. J.* **1993**, *29*, 371.
22. Chu, D.Y.; Thomas, J.K. *Macromolecules* **1984**, *17*, 2142.
23. Turro, N.J., Caminati, G.; Kim, J. *Macromolecules* **1991**, *24*, 4050.
24. Pearlman, R. and Baunerjee, S.; Yalkowsky, S.H. *J. Phys. Chem. Ref. Data* **1984**, *13*, 555.
25. Arora, K.S.; Turro, J.J. *J. Polym. Sci. Polym. Chem. Ed.* **1987**, *25*, 259.
26. Delaire, J.A.; Rodgers, M.A.J. and Webber, S.E. *J. Phys. Chem.* **1984**, *88*, 6219.
27. Pascal, P.; Duhamel, J.; Wang, Y.; Winnik, M.A.; Zhu, X.X.; Macdonald, P.; Napper, D.H.; Gilbert, R.G. *Polymer* **1993**, *34*, 1134.
28. Morawetz, H. **1986**, in reference 1, p547.
29. Morawetz, H. *Science* **1988**, *240*, 172.
30. Nataga, I.; Morawetz, H. *Macromolecules* **1981**, *14*, 87.
31. Winnik, F.M. *Polymer* **1990**, *31*, 2125.
32. Ringsdorf, H.; Simon, J.; Winnik, F.M. *Macromolecules* **1992**, *25*, 5353.

33. Holden, D.A.; Rendall, W.A.; Guillet, J.E. *Ann. N.Y. Acad. Sci.* **1981**, *366*, 11.
34. Guillet, J.E. and Rendall, W.A. *Macromolecules* **1986**, *19*, 224.
35. Ng, D.; Guillet, J.E. *Macromolecules* **1982**, *15*, 724.
36. Webber, S.E. *Chem. Rev.* **1990**, *90*, 1469.
37. Soutar, I.; Swanson, L. *to be published.*
38. Förster, Th. *Disc. Faraday Soc.* **1959**, *27*, 7.
39. Birks, J.B. *J. Phys. B. Ser. 2* **1948**, *1*, 946.
40. Zhao, C.-L.; Wang, Y.; Hruska, Z.; Winnik, M.A. *Macromolecules* **1990**, *23*, 2673.
41. Wang, Y.; Winnik, M.A. *J. Phys. Chem.* **1993**, *97*, 2507.
42. Pekcan, O. *TRIP* **1994**, *2*, 236.
43. Bokobza, L.; Monnerie, L. **1986**, in ref. 1, p449.
44. Bokobza, L. *Prog. Polym. Sci.* **1990**, *15*, 337.
45. Winnik, F.M.; Winnik, M.A.; Ringsdorf, H.; Venzmer, J. *J. Phys. Chem.* **1991**, *95*, 2583.
46. Turro, N.J.; Arora, K.S. *Polymer* **1986**, *27*, 783.
47. Ringsdorf, H.; Venzmer J.; Winnik, F.M. *Macromolecules* **1991**, *24*, 1678.
48. Bednar, B.; Li, Z.; Huang, Y.; Chang, L.C.P.; Morawetz, H. *Macromolecules* **1985**, *18*, 1829.
49. Arora, K.S.; Turro, N.J. *J. Polym. Sci. Polym. Phys. Ed.* **1987**, *25*, 243.
50. Oyama, H.T.; Tang, W.T.; Frank, C.W. *Macromolecules* **1987**, *20*, 474.
51. Oyama, H.T.; Tang, W.T.; Frank, C.W. *Macromolecules* **1987**, *20*, 1839.
52. Char, K.; Gast, A.P.; Frank, C.W. *Langmuir* **1988**, *4*, 989.
53. Char, K.; Frank, C.W.; Gast, A.P. *Langmuir* **1990**, *6*, 767.
54. Winnik, F.M. *Chem. Rev.* **1993**, *93*, 587.
55. Soutar, I. and Phillips, D. **1986**, in ref. 1, p97.
56. Soutar, I. *Ann. N.Y. Acad. Sci.* **1981**, *366*, 24.
57. Morishima, Y.; Kobayashi, T.; Nozakura, S.; Webber, S.E. *Macromolecules* **1987**, *20*, 807.
58. McCormick, C.L.; Hoyle, C.E.; Clark, M.D. *Macromolecules* **1990**, *23*, 3124.
59. McCormick, C.L.; Hoyle, C.E.; Clark, M.D. *Macromolecules* **1991**, *24*, 2397.

RECEIVED February 23, 1995

Chapter 24

Photophysical Study of Thin Films of Polystyrene and Poly(methyl methacrylate)

E. H. Ellison and J. K. Thomas

Department of Chemistry and Biochemistry, University of Notre Dame,
Notre Dame, IN 46556

The extent of the coverage of planar quartz surfaces by ultrathin films of polystyrene (PS) and poly(methylmethacrylate) (PMMA) is examined by using time-resolved fluorescence spectroscopy. Cations of the fluorescent pyrene derivative (1-pyrenyl)butyltrimethylammonium bromide (PBN) are adsorbed on quartz plates from aqueous solutions. Polymer deposition is achieved by controlled withdrawal of PBN-derivatized plates from polymer solutions. An analysis of the rate of quenching and prompt quenching by O_2 of the PBN fluorescence in the presence of films prepared from dilute solutions reveals exposed regions of the quartz surface. The extent of the surface coverage by PMMA is much greater than PS when films are prepared from either 0.001M or 0.0025M solutions. The difference in coverage is attributed to the relative affinities of the two polymers for the quartz surface.

Thin polymer films are frequently prepared by evaporation of solvent from polymer solutions coated on solid substrates. For example, spin coating involves the removal of excess polymer solution on a planar substrate by rapid spinning followed by evaporation of the solvent from a liquid film adhered to the substrate. Dip coating involves the deposition of a liquid film onto a planar substrate by controlled withdrawal of the substrate from a polymer solution followed by evaporation of the solvent. Both methods can be used to prepare ultrathin films, or films less than ca. 1000Å thick. Certain properties of ultrathin films, such as glass transition temperature, free volume, or coverage, may differ from the usual bulk polymer properties associated with thicker films. Unfortunately, many of the usual physical or spectroscopic methods used to characterize thick films are not amenable to the analysis of ultrathin films, due to the low abundance of material. Therefore, novel techniques must be developed to study ultrathin films.

0097–6156/95/0598–0410$12.00/0

Recently, methods utilizing STM and AFM have been developed to describe the extent of the coverage of PS (MW = 900,000) spin-coated from dilute solutions onto surface-oxidized silicon wafers (*1*). AFM images indicate that pseudo-continuous films exist at concentrations between 0.05 and 0.075 wt% PS in toluene, and that continuous films exist at and above 0.1 wt% (0.009M). Lower concentrations than 0.05 wt% result in polymer aggregation with regions of exposed silica between the aggregates.

In this investigation, we develop an alternative method that utilizes fluorescence spectroscopy to assess the extent of the surface coverage on planar substrates. The fluorescent pyrene derivative 4-(1-pyrenyl)butyltrimethylammonium bromide is employed; PBN is adsorbed as the cation onto fused-quartz plates (or slides) from aqueous solutions. Ionic interactions with ionized surface silanols render PBN cations immobile on the surface. The PBN-derivatized slides are coated with polymer films by dip coating (*2*). The deposition of organic coatings onto planar substrates by dip coating occurs during withdrawal of the substrate from a polymer solution in which it was immersed. Prior to withdrawal, polymer adsorption is necessary to reduce the surface tension of the polymer solvent at the liquid-solid interface so that a film of polymer solution will adhere to the substrate as it is withdrawn. Provided that polymer adsorption has occurred, then the amount of polymer deposited increases with increasing withdrawal speed and polymer concentration (*2*). Thus, the film thickness after solvent evaporation and solidification of the film will also increase.

The extent of molecular oxygen quenching of PBN fluorescence in the presence and absence of polymer films is determined by time-resolved methods and used to ascertain the extent of surface coverage. An important aspect of this study is the characterization of PBN photophysics on quartz slides. For this reason it was necessary to devote a separate section to that topic. Following the discussion of PBN photophysics on quartz slides, the photophysics in the presence of polymer films prepared by dip coating is described.

Materials and Methods

Materials Used. PBN was obtained from Molecular Probes, Inc. (Eugene, OR) and used as received. PS (MW = 280,000) and PMMA (MW = 25,000) were obtained from Aldrich and used as received. The polymer solvent used was LC-grade methylene chloride (CH_2Cl_2) containing small amounts of cyclohexene as a stabilizer. Pyrene (Kodak) was purified by liquid-column chromatography using silica gel and cyclohexane.

Equipment Used. Emission spectra were obtained using an SLM Instruments SPF-500C spectrofluorimeter. Absorption spectra were obtained using a Varian Cary-3 UV-Visible spectrophotometer. Emission decays were obtained by exciting at 337.1 nm using a PRA LN-100 N_2 laser (fwhm = 150 ps). Sample emission was focused onto a Bausch and Lomb grating monochromater (1200 grooves/mm) using a series of lenses and detected using a Hammamatsu R1664U multichannel plate PMT. Amplification and digitization of the signal was achieved using a Tektronix 7A29 amplifier and 7912AD digitizer, respectively.

Molecular oxygen was delivered in controlled amounts to the evacuated samples from an O_2 reservoir equipped with a regulator. Oxygen pressures were measured using a Hastings model 760 vacuum gauge and a Hastings DV-760 vacuum gauge tube.

Preparation and Derivatization of Quartz Surfaces. Surface contaminants were removed by exposure of the slides to concentrated H_2SO_4 followed by a thorough rinse in distilled and deionized H_2O. The slides were judged to be free of contaminants if a continuous film of H_2O remained intact for a least one minute on a vertically-oriented slide (*3*). Prior to use, all slides were calcined for 2 hours in air at 600°C.

The PBN cations were adsorbed onto the quartz surfaces by exposure of a slide for a few minutes to a 15mL aliquot of an $11\mu M$ aqueous PBN solution adjusted to pH 8.0 with NH_4OH. Such samples are abbreviated 8-NA (NA meaning no adjustment of the surface concentration after initial adsorption at pH 8.0). When necessary, the surface concentration was adjusted (or in this case lowered) by exposure of an 8-NA slide for a few minutes to 15 mL of H_2O adjusted to pH 8.0; such samples are abbreviated 8-8. A lower surface concentration than sample 8-8 was achieved by exposure of an 8-NA slide to 15 mL of H_2O adjusted to pH 7.0 followed by a second exposure to H_2O adjusted to pH 8.0. Such samples are abbreviated 8-7-8. Prior to use all samples were dried by exposure to air at 150°C for one-half hour followed by room temperature evacuation at mTorr pressures for 1 hour.

Preparation of Polymer Films on Quartz Slides. The polymer films were prepared by dip coating in dilute polymer solutions using a slight modification of a method used to prepare plastic replicas for use in electron microscopy (*4*). This method involves the drainage of a polymer solution in a glass funnel (3 cm diameter, 10 cm length) through a capillary tube (0.20 cm bore, 12 cm length). Slides were submerged in the solution and fixed in position via a clamp attached to a two-holed rubber stopper acting as a lid. After two minutes exposure, the stopcock was opened and the solution was withdrawn by gravity at a rate of 0.2 in/s. A low pressure and non-agitating stream of dry N_2 was used to purge the funnel and to expedite solvent evaporation from the film after withdrawal.

Sample films were evacuated at mTorr pressures for 1 hour immediately prior to analysis in order to remove any excess solvent and atmospheric gases. When necessary, samples were stored in the dark in a sealed, evacuated chamber containing activated $CaSO_4$.

Sample Measurement. During the collection of emission decays, slides were positioned tightly along the diagonal of the cuvette and the laser pulse positioned at 45° to the surface normal. In order to reduce the amount of scattered excitation light on the PMT, an excitation cutoff filter was placed at the monochromater entrance and the cuvette positioned so that specular reflection was away from the

PMT. A 1mm slit was placed 15 mm from the sample at the excitation port of the cuvette holder in order to limit excitation to the middle portion of the slide and to avoid excitation of and/or scatter from the cut edges of the slide. Annihilating pulses were eliminated by placing glass slides between the sample and laser, which reduces the pulse energy via reflective losses.

Acquisition of steady-state fluorescence emission spectra utilizing a 150 Watt Xe arc lamp was more difficult due to much higher levels of scattered excitation light. A similar sample orientation and setup to the time resolved analysis was ineffective in reducing scattered light intensities. Rotational adjustments of cuvettes that may reduce scattered light were also ineffective. In order to obtain fluorescence spectra of PBN on quartz surfaces, a special dimension and orientation of the slide was necessary. For the time-resolved analysis, a slide dimension of 0.45 x 1.5 inches was used; such slides fit tightly along the diaganol of the cuvette. For the steady-state analysis, a dimension of 0.39 x 1.5 inches was used. Such slides could be rotated and positioned inside the cuvette so that scattered excitation was reduced to small levels. However, the use of such slides involved excitation of corners and cut edges of slides. From a time-resolved analysis of oxygen quenching of the PBN fluorescence we were able to show that the coverage of polymer films on the roughened, cut edges was less than on the smooth planar surface. Therefore, under evacuated and controlled oxygen conditions, the steady-state fluorescence spectral analysis was limited to the analysis of PBN adsorbed on quartz slides in the absence of films.

Results And Discussion

Photophysics of PBN on Quartz Slides. Figure 1a illustrates the concentration dependence of the fluorescence of adsorbed PBN. Each spectrum was collected at a different voltage applied to the PMT. The monomer fluorescence is represented by the structured region from 370 to 400 nm, which is characteristic of the vibronic transitions of the pyrenyl group. The sharp, intense peak at 376.5 nm represents the forbidden 0-0 transition. The intensity of forbidden transitions in pyrene is strongly enhanced by covalently-attached alkyl groups at the 1-position, which reduce the electronic symmetry via electron withdrawal. Such derivatives (e.g., 1-methylpyrene) exhibit a weak dependence of forbidden transitions on solvent polarity and thus, unlike pyrene, are not very useful as probes of microenvironment polarity. This is the case for PBN which possesses a covalently attached butyltrimethylammonium ion.

The curvature in Figure 1a from 350 to 370 nm is from scattered excitation light; an excitation cutoff filter was not used. The broad, structureless band centered at 475 nm represents the PBN excimer. Due to ionic bonding with the surface silanols, the PBN cations are immobile on the quartz surface. Thus, excimer formation is possible only because the butyl group of PBN allows the motion necessary for closely-spaced pyrenyl groups to interact (5).

An important feature in Figure 1a is the decrease in the monomer to excimer ratio (M/E), or the ratio of the fluorescence intensity at 385 nm to the maximum intensity of the excimer fluorescence, with increasing concentration of

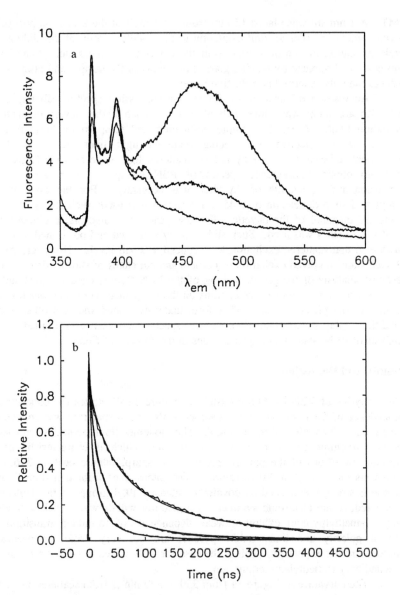

Figure 1. The dependence of the PBN fluorescence spectrum at room temperature (a), and the normalized PBN monomer fluorescence decay profile (b), on the concentration of PBN adsorbed on freshly-prepared fused-quartz slide surfaces. a) at 475 nm from top to bottom and at 376 nm from bottom to top: 8-NA, 8-8, and 8-7-8 (λ_{ex}=337nm, emission and exciation slit width = 1 and 10 nm, respectively). b) at 100 ns from top to bottom: 8-7-8; 8-8; 8-NA.

adsorbed PBN. The surface concentration corresponding to sample 8-NA results in significant excimer formation and a relatively low M/E value. A lower surface concentration can be achieved by exposure of the sample 8-NA to H_2O, when desorption of PBN cations occurs until equilibrium is reached. The equilibrium constant depends on pH since the number of adsorption sites, or ionized surface silanols, increases with pH. For example, sample 8-8 exhibits less excimer and a much greater M/E value than sample 8-NA, while sample 8-7-8 exhibits a higher M/E value than sample 8-8. The increase in M/E with decreasing concentration is readily explained by an increase in spacing between adsorbed PBN molecules, which inhibits excimer formation.

The normalized intrinsic fluorescence decay profiles of the monomeric species in Figure 1a are illustrated in Figure 1b. The heterogeneous decay profiles were modeled using the Gaussian approach (7,8), which yields approximate values of the average decay rate (k_{avg}, s^{-1}). Fits to the decay profiles using the Gaussian model are reasonable. The decay rate is concentration-dependent; higher concentrations exhibit greater decay rates. Although a growing-in of the excimer fluorescence was observed for samples 8-NA and 8-8 (data not shown), the timescale of the growing-in was rapid, such that it could not be entirely responsible for the increase in the monomer fluorescence decay rate. Previous studies of PBN adsorbed on clay surfaces indicated that the ammonium group of adsorbed PBN cations was responsible for a decrease in fluorescence quantum yield (5). Such quenching reactions may be intra- or intermolecular. It is possible that crowding among molecules could enhance such reactions. An important point is that the decay rate of the most highly dispersed PBN distribution, corresponding to sample 8-7-8, is not significantly lowered by further reductions in surface concentration. This implies that for sample 8-7-8, PBN molecules are dispersed on the surface such that crowding is no longer responsible for any fluorescence quenching that may still be present.

Figure 2a illustrates the dependence of the PBN singlet decay profiles on added O_2, at room temperature. Figure 2b illustrates the affect of the O_2 pressure on I_o of the profiles in Figure 2a, expressed as the ratio of the initial intensity at a given pressure, $I_{o,p}$, to the intensity under evacuated conditions, I_o. The data in Figure 2b indicate that there is significant prompt quenching at high O_2 levels. This prompt quenching can be described as a type of static quenching, that results from rapid decay inside the timescale of the laser pulse width ($\tau_{pulse} \approx 1ns$). A similar result was found for the other two decay profiles in Figure 1b (i.e., the level of prompt quenching at 760 Torr is ca. 10%). The plots of k_{avg}, versus O_2 pressure are given in Figure 3a for the decay profiles in Figure 2a, and also at higher PBN concentrations. The prompt quenching limits the decay analysis to low O_2 pressures. In this region, the response is linear and conforms to the Stern-Volmer relation for a bimolecular quenching reaction: $k_{obs} = k_o + k_q[Q]$, where k_o is the intrinsic decay rate and k_{obs} is the observed decay rate at some concentration [Q] of quencher species. The slope of each plot in Figure 3 yields the quenching rate constant (or in this case the average quenching rate constant), k_q ($s^{-1}Torr^{-1}$), which at the three concentrations is not significantly different.

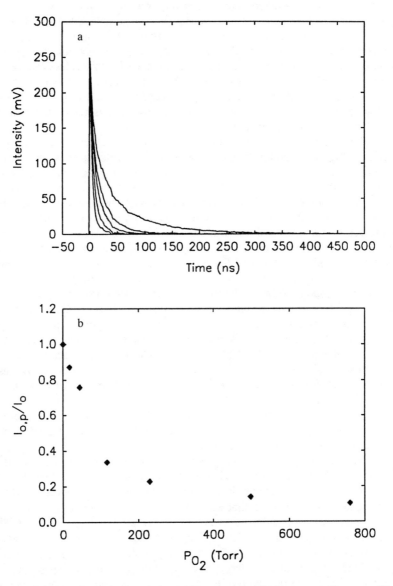

Figure 2. The dependence of the PBN monomer fluorescence decay profile (a), and values of $I_{o,p}/I_o$ (b), on O_2 pressure at room temperature. The PBN concentration corresponds to sample 8-8. a) at 50 ns from top to bottom: 0, 2.3, 4.6, 9.1, and 17 Torr O_2.

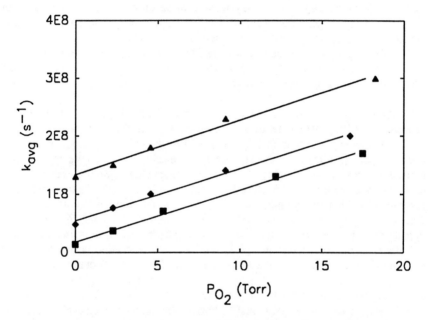

Figure 3. The dependence of k_{avg} of the PBN monomer fluorescence decay on O_2 pressure at room temperature. Filled triangles: 8-NA; filled diamonds: 8-8; filled squares: 8-7-8.

Previous studies of fluorescence quenching by gas-phase O_2 of pyrene adsorbed on silica gel (8) and alumina (9) at the gas-solid interface indicate that quenching proceeds by a dynamic, surface-mediated O_2 bombardment mechanism. It was also shown that shielding of adsorbates from O_2 bombardment by either the adsorbant itself or coadsorbates (such as hexadecanol) substantially decreases the quenching rate. The average value of k_q from Figure 4a is $1.7 \cdot 10^{11}$ $s^{-1} M^{-1}$, which is 3 and 1.7 times greater than values reported in references 8 and 9, respectively. The difference in k_q may indicate a more open surface on the planar substrate relative to the porous substrate, but it may also reflect differences in oxygen adsorption, surface preparation, and the different probe molecule used in each study. Additional experiments may resolve these differences, but they are not necessary at this stage. The details of the PBN photophysics on SiO_2 surfaces is the subject of a separate investigation underway in our laboratory.

The following section describes O_2 quenching of PBN monomer fluorescence in the presence of PMMA and PS films, where a PBN concentration corresponding to sample 8-8 was used.

PBN Photophysics in the Presence of Polymer Films. Figure 4a illustrates the dependence of the PBN singlet decay rate on the O_2 pressure in the presence of PS films. At coverages corresponding to 0.10M PS, the difference in k_{avg} between 0 and 760 Torr (1 atm) O_2 indicates that the quenching rate relative to the open surface is reduced by roughly 3 orders of magnitude. Similar results were observed by using 0.010M PS. The large reduction in k_q implies that quenching is controlled by the mobility and/or solubility of O_2 in the film. At lower coverages, corresponding to 0.001M and 0.0025M PS regions, k_q is much higher and is indicated by the steep rise in the region 0 to 100 Torr. At O_2 pressures exceeding 100 Torr, a gradual decline in k_{avg} is observed. The decline is due to prompt quenching which emphasizes the longer decaying (or shielded) components in the tail region.

Figure 4b illustrates the dependence of $I_{o,p}/I_o$ on the O_2 pressure for the data in Figure 4a. At 0.001M and 0.0025M PS, there is a gradual increase in prompt quenching with increasing O_2; prompt quenching is not significant at higher PS concentrations. The prompt quenching is less in the region 500-800 Torr for 0.0025M PS relative to 0.001M, indicating a higher degree of coverage with increasing PS concentration. Due to a greater proportion of shielded components, k_{avg} in the region 0-100 Torr rises to lower values for 0.0025M PS relative to 0.001M PS.

The observation of prompt quenching for 0.001M and 0.0025M PS indicates open regions where the quenching rate is similar to that on the open surface. At low O_2 pressures, where prompt quenching is not significant, the rate of quenching can be determined. Plots of k_{avg} versus O_2 pressure in the region 0-20 Torr are illustrated in Figure 5. Values of k_q for 0.001M and 0.0025M PS are 75% and 55% of k_q on the open surface. This indicates a greater proportion of shielded PBN molecules at higher PS concentrations.

Figure 6a illustrates the dependence of k_{avg} on the O_2 pressure in the presence of PMMA films. At coverages of 0.10M and 0.010M PMMA, the

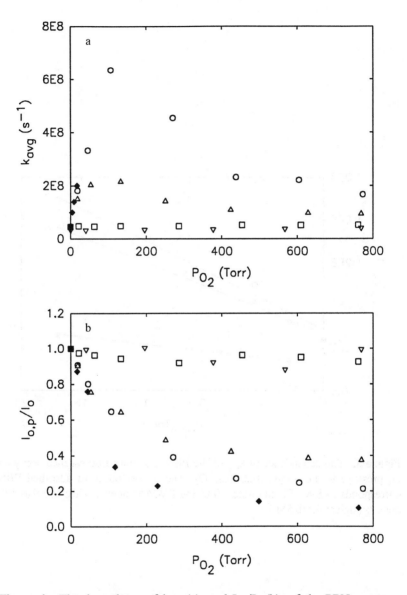

Figure 4. The dependence of k_{avg} (a), and $I_{o,p}/I_o$ (b), of the PBN monomer fluorescence decay on O_2 pressure in the presence of PS films. The concentration of adsorbed PBN corresponds to 8-8. Open circles: 0.001M PS; open triangles: 0.0025M PS; open squares: 0.010M PS; open inverse triangles: 0.10 M PS; filled diamonds: open surface.

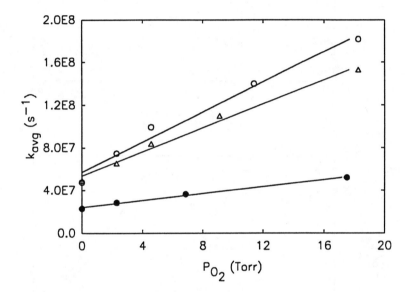

Figure 5. The dependence of k_{avg} of the PBN monomer fluorescence decay on O_2 pressure in the region 0-20 Torr O_2. The concentration of adsorbed PBN corresponds to 8-8. Filled circles: 0.001M PMMA; open circles: 0.001M PS; open triangles: 0.0025M PS.

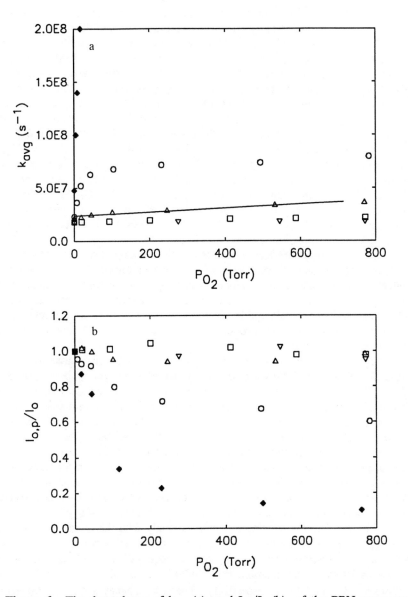

Figure 6. The dependence of k_{avg} (a), and $I_{o,p}/I_o$ (b), of the PBN monomer fluorescence decay on O_2 pressure in the presence of PMMA films. The concentration of adsorbed PBN corresponds to 8-8. Open circles: 0.001M PMMA; open triangles: 0.0025M PMMA; open squares: 0.010M PMMA; open inverse triangles: 0.10M PMMA; filled diamonds: open surface.

difference in k_{avg} between 0 and 1 atm O_2 is small and prompt quenching is not significant. At 0.0025M PMMA prompt quenching is not significant, as observed at 0.010M PMMA, but unlike 0.010M PMMA, there is measureable quenching. For this reason, a regression analysis was performed (k_q of 0.0025M PMMA is 0.2% of k_q on the open surface). The increase in the quenching relative to 0.010M PMMA indicates the existence of partially exposed regions, or regions where the polymer density in the vicinity of the PBN molecules is less relative to higher polymer coverages. At 0.001M PMMA, an initial fast rise in k_{avg} is observed, but k_{avg} rises to a much lower value than in the presence of either 0.001M or 0.0025M PS (note the difference in scale of the y axis in Figures 4a and 6a). A plot of k_{avg} versus the O_2 pressure in the region 0-20 Torr (Figure 5) indicates a value of k_q which is 18% of k_q on the open surface. The k_q value in the low pressure region and the level of prompt quenching in the region 500-800 Torr O_2 (Figure 6b) are much smaller than either 0.001M or 0.0025M PS, which indicates a higher proportion of shielded components. For 0.001M PMMA some degree of open surface quenching is probably present.

The above results reflect differences in the coverage between PMMA and PS when using polymer concentrations of 0.0025M or lower. The differences at identical polymer concentrations can be explained by differences in the amount of polymer deposited on the surface. The presence of oxygenated functionalities in PMMA may increase its adsorption potential on polar quartz surfaces relative to the more hydrophobic PS. This could affect both the equilibrium constant as well as the kinetics of adsorption.

Another contribution to the difference in coverage may arise from the relative affinities of each polymer for the surface after solvent evaporation. If a uniform film of polymer solution coats the surface prior to withdrawal, then the removal of solvent from the film by evaporation constitutes a severe change in the solvation parameters. The resulting morphology adopted by the polymer may depend on the degree of interaction between it and the surface. PS possesses no hydrophilic functionality and in the absence of solvent should be repelled by the polar surface. This could result in polymer aggregation and open surface regions between polymer aggregates. On the other hand, the oxygenated functionalities in PMMA can hydrogen bond with the quartz surface hydroxyls. This may lead to less shrinkage on the surface and a more continuous film morphology.

The idea that PMMA interacts more strongly with the quartz surface than PS is supported by measurements of k_{avg} before and after coverage. At PBN concentrations corresponding to 8-8, the average decay time, τ_{avg}, prior to coverage is 18 ns. After exposure to 0.10M PS and 0.10M PMMA the lifetime increases to 33 ns and 60 ns, respectively. The larger change in τ_{avg} for PMMA is due to a greater degree of interaction with the quartz surface, which inhibits intra- or intermolecular quenching reactions of PBN presumably by the interaction of PMMA segments with quartz surface regions between the PBN molecules.

Measurements of PBN Spectra in the Presence of Films. Of further interest is the excimer level in the presence of films, and the thickness of continuous films. As stated in the Materials and Methods, the steady-state fluorescence analysis under evacuated conditions was restricted to the open surface. However, as indicated in Figures 4a and 6a, oxygen quenching is significantly reduced at coverages of 0.01M or greater. Therefore, at higher coverages, it is not essential to evacuate samples in order to obtain a representative spectrum. The fluorescence spectrum can be collected in air outside the cuvette, which makes it possible to position the sample such that the influence of scattered light is negligible when the excitation beam is limited to the center portion of the slide, as necessary.

In the presence of either 0.010M (or higher) PMMA or PS films, the excimer of sample 8-8 is eliminated; the spectra (in terms of the excimer level) closely resemble that of sample 8-7-8 in Figure 1a. Shifts in the emission and the excitation bands are also observed (data not shown). The ^1L_a absorption band, inferred from excitation scans, shifts from 342.5 nm on the open surface, to 344.0 nm and 345.5 nm in the presence of PMMA and PS films, respectively. The 0-0 emission band shifts from 376.0 nm on the open surface to 377.0 nm and 378.0 nm in the presence of PMMA and PS films, respectively. The observed trend in emission and excitation frequency (quartz > PMMA > PS) is consistent with the trend in dielectric constant and refractive index of each system (*10*).

Measurements of Film Thickness of Continuous Films. Sample thickness of 0.010M and 0.10M films was estimated using absorption spectroscopy of pyrene at a concentration of 50 mM (mol pyrene/L polymer) in the film. The peak absorbance of the ^1L_a band of pyrene was corrected by subtraction of the absorbance due to the slide and polymer by use of a blank. Films prepared from 0.10M solutions of either PS or PMMA produced corrected absorbance readings of ca. 0.025. Using the Beer-Lambert Law and the peak molar extinction coefficient of the pyrene ^1L_a band ($\varepsilon_{max} \approx 50,000$ M^{-1}cm^{-1}), a film thickness of 500Å is calculated by considering that both slide surfaces are coated. Films prepared by dip coating in 0.01M solutions produced absorbance readings of ca. 0.003, or a film thickness of 50Å. Solutions containing 100mM pyrene exhibited higher absorbance readings but the same calculated thickness.

Conclusions

Measurements of O_2 quenching of the fluorescence of PBN adsorbed on planar, fused-quartz substrates in the presence of ultrathin polymer films are useful to assess the extent of surface coverage by PS and PMMA. Our results indicate that withdrawal from 0.010M or 0.10M PS solutions yields continuous films; withdrawal from 0.001M and 0.0025M PS solutions results in patchy films. These results are similar to those determined by STM and AFM studies of PS spin-coated on silicon wafers, i.e., continuous films at 0.009M PS; patchy films below 0.004M PS (*1*). These similarities are surprising since the molecular weights and the method of film preparation are different.

Our results also indicate that withdrawal from 0.0025M PMMA results in a continuous film with regions of lowered polymer density relative to higher coverages (i.e., 0.010M or higher). At a given concentration of polymer at or below 0.0025M, the extent of surface coverage by PMMA is greater than PS.

Future experiments will evaluate the effects of molecular weight, surface derivatization, surface geometry and heat treatments on polymer coverage and morphology.

Acknowledgments

The authors wish to thank The National Science Foundation for support of this work.

Literature Cited

(1) Stange, T. G.; Mathew, R.; Evans, D. F.; Hendrickson, W. A. *Langmuir* **1992**, *8*, 920.
(2) Yang, C.; Josefowicz, J. Y.; Alexandru, L. *Thin Solid Films* **1980**, *74*, 117.
(3) Licari, James J.; Hughes, Laura, A. *Handbook of Polymer Coatings for Electronics: Chemistry, Technology and Applications*; Noyes Publications: Park Ridge, NJ, 1990; p 161.
(4) Bradley, D. E. In *Techniques for Electron Microscopy*; Desmond Kay, Ed., Blackwell Publishing: Oxford, 1965; p 58.
(5) Viaene, K.; Schoonheydt, R. A.; Crutzen, M.; Kunyima, B.; De Schryver, F. C. *Langmuir*, **1988**, *4*, 749.
(6) Albery, J. W.; Bartlett, P. N.; Wilde, C. P.; Darwent, J. R. *J. Phys. Chem.* **1985**, *106*, 1854.
(7) Thomas, J. K. *Chem. Rev.* **1993**, *93*, 301.
(8) Krasnansky, R.; Koike, K.; Thomas, J. K. *J. Chem. Phys.* **1990**, *94*, 4521.
(9) Pankasem, S.; Thomas, J. K. *J. Phys. Chem.* **1991**, *95*, 7385.
(10) Kavanaugh, R. J.; Thomas, J. K. *Langmuir,* **1992**, *8*, 3008.

RECEIVED February 2, 1995

Chapter 25

Photophysical Approaches to Characterization of Guest Sites and Measurement of Diffusion Rates to and from Them in Unstretched and Stretched Low-Density Polyethylene Films

Jawad Naciri[1], Zhiqiang Hé[2], Roseann M. Costantino[3], Liangde Lu,
George S. Hammond, and Richard G. Weiss[4]

Department of Chemistry, Georgetown University,
Washington, DC 20057-2222

Various luminescence and photochemical techniques have been developed to probe the natures and accessibilities of the sites available to guest molecules in low-density polyethylene films. The methods, which include covalent modification of interior sites with fluorophores, are described. The results indicate that a distribution of site types, each with its characteristic shape, free-volume, and accessibility, is present. Furthermore, this distribution is changed drastically when a film is stretched. Activation energies for diffusion of a series of N,N-dialkylanilines in unstretched and stretched, modified and "native" films are reported, compared, and analyzed.

Polyethylene is the name given to literally thousands of polymer formulations, derived in some cases from mixtures of monomers which include 1-alkenes in addition to ethylene. Molecular weight average and distribution, degree of chain branching, mode of processing, degree of crystallinity, types of amorphous regions, density etc. differentiate the various polyethylenes(1). A combination of density, melting point, and degree of crystallinity are used to classify and distinguish low density polyethylene (LDPE) samples(1).

Two general site types in LDPE --those near points of chain-branching and those along the interfaces between amorphous and crystalline regions -- have been suggested for guest (dopant) molecules(2). Recent ^{129}Xe NMR studies with

[1]Current address: Naval Research Laboratory, 4555 Overlook Avenue, Code 6090, Washington, DC 20375
[2]Current address: King Industries, Science Road, Norwalk, CT 06852
[3]Current address: Office of Premarket Approval, Chemistry Review Branch, Food and Drug Administration, 200 C Street, SW, HFS-247, Washington, DC 20204
[4]Corresponding author

0097-6156/95/0598-0425$12.25/0

polyethylene support this hypothesis(3). Guest molecules are unable to enter the crystalline portions of <u>LDPE</u> below the melting transition(2b).

To characterize further the microscopic environments provided by <u>LDPE</u> to guest molecules, we have probed the shapes and free volumes of the sites and the dynamics of diffusion to and from them in several different ways(4). The results obtained thus far indicate that there is a distribution of site types which can be changed drastically by macroscopic stretching of polymer sheets. A review is presented of our efforts to develop new methods for the measurement of diffusional rates to and from guest sites in <u>LDPE</u> and to gain insights into the natures of the sites(5) at which guest molecules reside.

Materials and Procedures

Films of Sclairfilm (76 mμ thick, 0.92 g/cm^3, M$_w$ 112600, from DuPont of Canada) have been employed throughout our investigations. This allows results from different experiments to be compared directly. The films were soaked in chloroform before being used in order to remove plasticizers, antioxidants, and other additives. From the heat of the melting transition (measured by differential scanning calorimetry), Sclairfilm is ca. 50% crystalline(6). Reagents, solvents, and methods are as described in our cited publications.
Cold-stretching to 4-6X the original film length was accomplished with a device designed for this purpose or by hand; experiments employing the two methods led to indistinguishable results.

Preparation of Modified LDPE Films

There are many recipes for preparation of polymeric films containing low concentrations of lumophoric or other "reporter" groups. The vast majority of these involve either copolymerization of two (or more) different monomers, one of which is initially in large excess with respect to the other(s), or the physical mixing of tagged and untagged chains. Such films have two important problems if they are to be used to probe the physical or microscopic properties of their unmodified analogues: (1) the reporter groups are located at surfaces as well as at interior positions of the films; (2) the natures of the sites at which the reporters reside is determined in large part by them rather than by the intrinsic nature of an unmodified film. The modified films whose preparations are described below diminish greatly the seriousness of both problems: (1) the reporter groups can be excluded almost completely from surface-accessible sites; (2) it is the reporter group which must adapt to the demands of the sites offered by the native film.

Py-LDPE (7). Films strips were soaked in chloroform containing 0.24 $\underline{M}$ pyrene until an appropriate concentration of the lumophore (ca. 10^{-2} $\underline{M}$ usually, as measured by UV/vis absorption spectroscopy) had been embibed. After standing in air for 15-30 min (to allow chloroform to evaporate) and being washed with methanol (a non-swelling solvent which removes surface pyrene), doped films were irradiated with the pyrex-filtered output of a 450 W medium pressure Hg lamp (Hanovia) for various periods (usually 1 h) which were adjusted empirically to provide the desired concentrations of 1-pyrenyl groups covalently attached to the polymethylene chains. The amount actually attached was determined by washing the irradiated films exhaustively in baths of chloroform and them measuring the UV/vis absorption spectra. We have sought to produce films with ca. 10^{-3}-10^{-4} M pyrenyl groups (designated Py-LDPE). The emission spectra of a Py-LDPE film and an LDPE film doped with pyrene are presented in Figure 1. The absorption and emission spectra of Py-LDPE are typical of 1-pyrenyl groups(8).

Evidence that the pyrenyl groups of Py-LDPE are isolated from film surfaces was provided by the inability of 0.95 M 2-(dimethylamino)ethanol in methanol to quench more than 6% of the fluorescence intensity from even stretched films. The same quencher decreased the fluorescence intensity of pyrene in methanol by >90%(7).

An-LDPE (9). After doping a film by immersing it in an ether solution of 9-anthryldiazomethane(10), the solvent was removed by evaporation in air, the film surfaces were washed with methanol, and the dopant concentration (commonly $(1-5) \times 10^{-3}$ M) was determined as described above. Each film was sealed in a glass tube and immersed in boiling methanol for 1 h. The unreacted 9-anthryldiazomethane was removed from the films by exhaustive washing in ether baths. A spectroscopic history of the preparation of an anthryl-modified film, An-LDPE, is shown is Figure 2. Typical concentrations of covalently attached anthryl groups were 10^{-3}-10^{-4} M.

Spectroscopic Investigations

Michl and Thulstrup(11) and Yogev et al.(4d), especially, have shown that doped, stretched films of LDPE can be used to align guest molecules according to a laboratory frame of axes. Linear dichroic spectra of the dopant molecules provide information concerning the directions of transition dipoles of the electronic transitions.

Fluorescence spectra of both Py-LDPE and An-LDPE films gave no evidence for excimer emission(7,9). Thus, we assume that virtually all of the occupied sites have no more than one lumophoric group. Much higher concentrations of non-covalently linked pyrene in the films did provide emission spectra with detectable excimer emission (Figure 1).

Since it has been conjectured that dopant molecules tend to translocate from one site type to another when LDPE films are stretched(2), we sought to devise an experiment in which the spectroscopic properties of dopant molecules which are forced to remain at their original locations, regardless of the applied stress on the material, can be compared with the properties of similar species which are free to migrate. To accomplish this goal, an An-LDPE film was stretched. For comparison purposes, another piece of film was doped with an equal concentration of 9-methylanthracene (MA, a non-covalently attached guest) and stretched by the same amount.

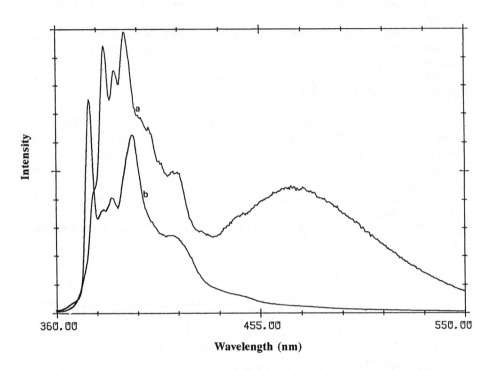

Figure 1. Room temperature emission spectra in air (a) from an LDPE film after immersion in a 0.24 M pyrene in chloroform solution for ca. 12 h (Py/LDPE) and (b) from a film, doped as above, after being irradiated for 1 h and exhaustively extracted with chloroform (Py-LDPE). The spectra are not normalized; λ_{ex} 343 nm(7a).

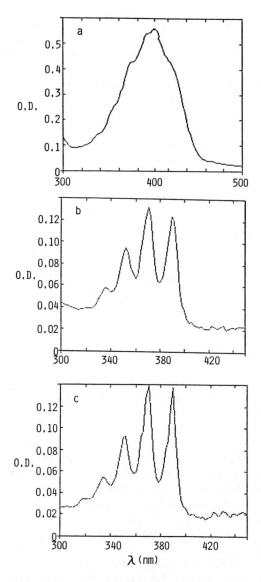

Figure 2. Representative UV/vis absorption spectra: (a) an unstretched LDPE film doped with 9-anthryldiazomethane before heating; (b) film in (a) after heating and exhaustive extraction with ether and chloroform (An-LDPE); (c) an unstretched LDPE film doped with 9-methylanthracene. All spectra were recorded with an undoped film as reference.

The emission anisotropy of an unstretched An-LDPE film is ca. 0.15-0.18(*9a*), indicating that the anthryl groups are not able to move rapidly within a film. When swelled by ca. 17% by cyclohexane, the films become more tolerant of anthryl motions and the emission anisotropy decreased to ca. one-third of its original value(*9a*).

Unstretched An-LDPE films and MA/LDPE films showed no linear dichroism (d): d =$OD_\parallel/OD_\perp$ was unity at 260 and 390 nm, wavelengths at which anthryl transition dipoles are polarized almost exclusively along the long (Z) and short (Y) in-plane axes. However, the same films, after being stretched to 5X their original lengths, displayed non-unity dichroic ratios from which orientation factors (O_f =$d_f/[d_f$ +2] where f =X,Y,Z are the principal axes of anthracene) can be calculated.

As shown in Table I, the values calculated for MA differ somewhat from those reported by Michl and Thulstrup(*12*). The disparities can be attributed to different degrees of stretching, different LDPE sources, and different concentrations of MA in the two experiments. More importantly, an internal comparison between our results from An-LDPE and MA/LDPE, with about equal chromophore concentrations, reveals that the non-covalently and covalently attached anthryl groups reside in nonequivalent site types after film stretching. Although these results do not confirm the hypothesis that guest molecules translocate when films are stretched(*2*), they are at least consistent with it.

Distributions of Guest Site Sizes

Information presented thus far indicates that different site types may exist in LDPE, but does not address quantitatively their free volumes. Since pyrene excimer emission can be detected from LDPE films which were swollen by chloroform at the time of doping(*7*), some of the sites must have a minimum of ~645 $Å^3$ (i.e., the sum of the van der Waals volumes of two pyrene molecules(*13*)) of available free volume. However, this volume may not be representative of sites in LDPE which has not been swelled during doping.

In other experiments, Py-LDPE (and An-LDPE) films were placed in methanolic solutions of N,N-dialkylanilines (DAA) until the pyrenyl (or anthryl) fluorescence intensity no longer changed. Since methanol is a non-swelling solvent for LDPE, DAA molecules had to enter unswelled films and occupy sites more like those of the native polymer. The concentrations of DAA in the films were ascertained from the film volumes and the quantity of DAA which could be extracted from them. Some of the pertinent data for Py-LDPE are presented in Table II(*14*).

F, the fraction of film fluorescence quenched by a DAA homologue, is expressed in equation 1(*7*) where I_o and I_{eq} are the fluorescence intensities observed in the absence of DAA and after equilibration with it. Since DAA are known to be very efficient (diffusion controlled) quenchers of pyrenyl and

Table I. Dichroic ratios and orientation factors from anthryl groups in stretched <u>LDPE</u> films(*9a*)

dopant	d_X	d_Z	O_X[b]	O_Y	O_Z
<u>MA</u>	0.77	2.8	0.14	0.28	0.58
<u>MA</u>[a]			0.20	0.29	0.51
An-<u>LDPE</u>	0.89	1.94	0.21	0.30	0.49

a) Data from Michl and Thulstrup(*12*).   b) Calculated assuming
   $O_X = 1 - (O_Y + O_Z)$(*11*).

Table II. Data related to fluorescence quenching in <u>Py</u>-<u>LDPE</u> films by <u>DAA</u> homologues at 25 °C(*14*)

<u>DAA</u>	van der Waals volume, $Å^3$	unstretched film [DAA]$_{film}$ **M**	**F**	F_M M^{-1}	stretched film [DAA]$_{film}$ **M**	**F**	F_M M^{-1}
<u>DMA</u>	128.7	0.19	0.35	1.6	0.14	0.15	0.9
<u>DEA</u>	162.8	0.12	0.25	1.7	0.09	0.10	0.9
<u>DPA</u>	196.9	0.09	0.19	1.6	0.08	0.04	0.3
<u>DBA</u>	231.0	0.07	0.16	1.6	0.09	<0.02	<0.1

anthryl excited singlet states(15), the values of F represent the fraction of pyrenyl or anthryl occupied sites which also contain at least one DAA molecule at the moment of fluorophore excitation; quenching is static in these experiments. From the volumes of pyrenyl, anthryl, and the DAA, the distribution of the minimum free volumes at the sites containing fluorophores can be calculated.

$$F = (I_o - I_{eq})/ I_o \qquad (1)$$

The values of F for one DAA homologue are larger in unstretched than in stretched films. However, the dependence of F on the molecular volume of DAA is greater for the stretched films. Since the total concentration of DAA in a film, $[DAA]_{film}$, depends upon its initial concentration in the methanolic doping solution and its partition coefficient, the values of F were normalized with respect to DAA (F_M in equation 2). The F_M values are more indicative of the ability of a DAA to enter a site which is already occupied by a covalently attached fluorophore and of the distribution of free-volumes at such sites.

$$F_M = F/[DAA]_{film} \qquad (2)$$

From the absence of change in F_M using unstretched films, we conclude that the total free volume of pyrenyl-occupied sites which are quenchable by any DAA must be at least ~ 555 Å^3, the sum of the van der Waals volumes of a pyrenyl group and one DBA molecule(17) (the largest DAA employed). Furthermore, the access route to those sites must not have a constriction point whose diameter is smaller than that of a DAA molecule. These data, presented graphically in Figure 3, demonstrate the striking difference between unstretched and stretched films. In the latter, the accessible sites can be bracketed to have average free volumes between ~ 485 and 519 Å^3. As judged from the large decreases in F_M between DEA and DPA or between DPA and DBA ($\Delta V \simeq 34$ Å^3)(17), the distribution of volumes within these sites is rather narrow. Comparatively, DMA is less than twice as efficient a quencher in unstretched as in stretched Py-LDPE films, but DBA is more than 10 times more efficient in the unstretched film(14). In essence, a much smaller *fraction* of DBA than DMA molecules enter sites already occupied by a fluorophore.

Since the information on F_M that this method provides is limited only to sites with covalently bound fluorophores, we also examined the influence of their concentration in An-LDPE films(9). Thus, using DMA as the quencher, the F_M values were found to remain constant over a five-fold change of concentration of anthryl groups in unstretched films (Table III). However, the corresponding stretched films exhibited F_M values which increase as the anthryl concentration decreases. These results are totally consistent with those in Table II. They provide additional insights, suggesting that DMA is distributed rather evenly in sites with at least minimally adequate free volumes in the amorphous regions of the polymer.

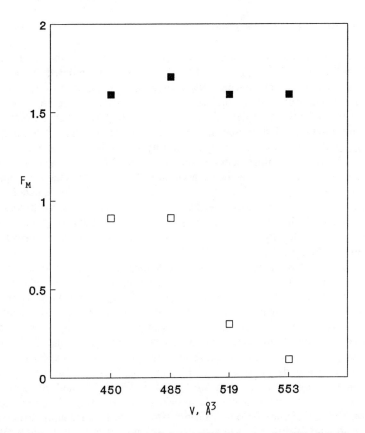

Figure 3. F_M versus the sum of the calculated van der Waals volumes, V, of one pyrenyl group and one <u>DAA</u> molecule; from out-diffusion data using unstretched (■) and stretched (□) <u>Py</u>-<u>LDPE</u> films(*14*).

Table III. Data related to fluorescence quenching of <u>An</u>-<u>LDPE</u> films by <u>DMA</u> at 25 °C(*9*)

[An],<u>M</u>	unstretched film[a]		stretched film[b]	
	F	F_M,<u>M</u>$^{-1}$	F	F_M,<u>M</u>$^{-1}$
7.8×10^{-4}	0.34	1.8	0.12	0.9
6.5×10^{-4}	0.34	1.8	0.15	1.1
1.6×10^{-4}	0.35	1.8	0.21	1.5

a) [<u>DMA</u>]$_{film}$ =0.19 <u>M</u>.   b) [<u>DMA</u>]$_{film}$ =0.14 <u>M</u>.

Rates of Diffusion in LDPE Films

As alluded to in the discussion above, an impediment to the quenching of covalently attached fluorophores in LDPE films may be imposed by restriction of the motions of polymethylene chains which are necessary to allow diffusion of the DAA molecules along the pathway from a film surface to an occupied site. If a Py-LDPE or An-LDPE film is placed in a DAA/methanol solution, the decreasing intensity of fluorescence from the film (by exciting at a wavelength where DAA does not absorb and by monitoring the emission at one wavelength) can be followed with time until a minimum (equilibrium) value is reached. When the DAA-doped film is then placed in methanol (or another non-swelling liquid in which DAA is soluble), the increasing intensity of fluorescence can be monitored with time. Provided the methanol is in extremely large excess of the film volume, virtually all of the DAA will diffuse eventually from the film and the fluorescence intensity will attain a maximum value(7). Typical intensity plots are shown in Figures 4 and 5.

Analysis of these data as being controlled by competing first-order rate processes(7) or by Fickian diffusion (early-time integrated form of the second law(17)) allows characteristic values for rates of DAA migration from the films to be calculated. The two expressions, for the rate constant (k) and the diffusion coefficient (D), are presented in their integrated forms in equations 3 and 4, respectively. Plots of the raw data are shown in Figures 4 and 5; the inserts show the same data plotted in the form indicated by equations 3a and 3b. A typical fluorescence decay (in-diffusion) curve using DMA and an An-LDPE film is presented in Figure 6a; the data are treated in Figures 6b and 6c according to equations 3a and 4a, respectively. In principle (and practice), k (or D) from in-diffusion (initial doping) or out-diffusion (from film to methanol) experiments may lead to slightly different values because the liquids in contact with a film are not exactly the same in the two cases(7). For the purposes of the present discussions, this factor will be ignored. In eqs. 3a and 4a, I_∞, I_t, and I_o are the fluorescence intensities from the film after complete in-diffusion, after a time = t, and before entry by DAA into a film; in equations 3b and 4b, I_∞ is the fluorescence intensity after complete loss of DAA from the film and I_o is the intensity from the film in its maximally-doped state; 2ℓ is the film thickness.

$$\ln[(I_o-I_t)/(I_o-I_\infty)] = -kt \tag{3a}$$

$$\ln[(I_\infty-I_t)/(I_\infty-I_o)] = -kt \tag{3b}$$

$$(I_o-I_t)/(I_o-I_\infty) = (4/\ell)(D/\pi)^{1/2}t^{1/2} \tag{4a}$$

$$(I_\infty-I_t)/(I_\infty-I_o) = (4/\ell)(D/\pi)^{1/2}t^{1/2} \tag{4b}$$

By performing experiments over a temperature range in which no phase change of LDPE occurs, it is possible to calculate the activation energies, E_a and E_D, associated with k and D (equations 5 and 6(18)). Thus, the values of E_a or

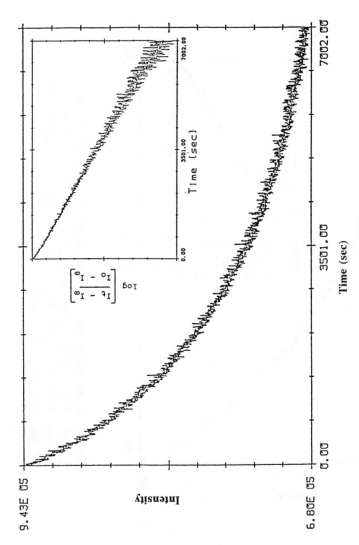

Figure 4. Fluorescence decay (in-diffusion) curve from an unstretched Py-LDPE film in contact with 0.5 M DMA and 0.1 M 2-(dimethylamino)ethanol (a quencher of fluorescence from surface pyrenyl groups) in methanol at 7 °C. λ_{ex} 343 nm; λ_{em} 395 nm. The inset is a plot of the data according to equation 3a(7).

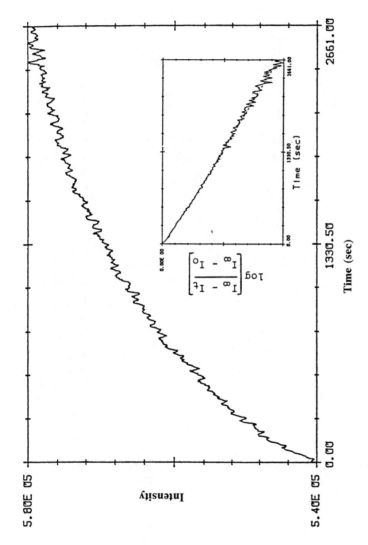

Figure 5. Fluorescence growth (out-diffusion) curve from a stretched Py-LDPE film, previously doped with DMA and in contact with methanol, at 22 °C. λ_{ex} 343 nm; λ_{em} 395 nm. The inset is a plot of the data according to equation 3b(7).

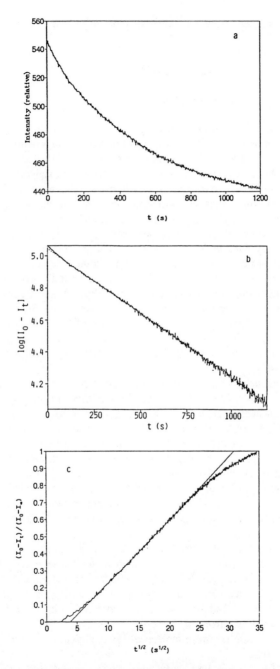

Figure 6. (a) Fluorescence decay curve from an unstretched An-LDPE film in contact with 0.95 M DMA in methanol at 22 °C. λ_{ex} 369 nm; λ_{em} 415 nm. (b) Data plotted according to equation 3a. (c) Data plotted according to equation 4a with a best linear fit to points between 0.2 and 0.7 along the ordinate(*9*).

E_D can be determined from the slopes of plots of ln k or ln D versus the inverse of temperature.

$$k = k_o exp(-E_a/RT) \qquad (5)$$

$$D = D_o exp(-E_D/RT) \qquad (6)$$

E_a values from fluorescence growth curves (i.e., DAA out-diffusion experiments) using one set of Py-LDPE films and a temperature interval which is far from the glass and melting transitions(19) are collected in Table IV(14). They lead to the contra-intuitive conclusion that the resistance to DAA motion *increases* as molecular size *decreases*! Furthermore, it is more difficult to remove DAA molecules from sites in a stretched film than in an unstretched one. It should be remembered that the physical process being observed in these experiments is removal of the last (and probably only) DAA molecules from sites occupied by pyrenyl groups. Only if that process is fast with respect to motion of DAA molecules within the bulk should the activation energies in Table IV be related to diffusion. That they are is indicated by the similarity between the E_a from fluorescence decay (in-diffusion) and fluorescence growth (out-diffusion) experiments: for DMA as quencher, the E_a from in-diffusion experiments for the films of Table IV are 13.5 ± 0.5 (unstretched) and 17.4 ± 0.8 kcal/mol (stretched)(14); from fluorescence decay curves in An-LDPE and using DMA as quencher, we have found that E_D is also ca. 4 kcal/mol higher in the stretched films(9).

Any conclusion that it is easier for larger DAA molecules than for smaller ones to move within LDPE films must be qualified further by the previously mentioned data which shows that the fraction of pyrenyl-occupied sites which participate in the DAA-induced quenching decreases as the volume of a DAA molecule increases. Thus, it is more reasonable to ascribe the bizarre changes in activation energies to the hypothesis that since the *number* of sites accessed becomes smaller as the DAA becomes larger; the relatively higher activation energies for removal of the smaller quencher molecules reflects the fact that some come from sites for which transport is more difficult. Again, we conclude that there appears to be a distribution of sites, each with its characteristic free volume and accessibility.

In order to monitor the diffusional characteristics represented by *all* guest-occupied site types, we have developed methods which allow k and D values for diffusion in native LDPE films to be measured(20). They are complementary to methods used by Wang *et al.*(21), but differ in a very important respect: instead of monitoring fluorescence intensity from molecules (which have diffused from a film) in a receiving liquid, the intensity of fluorescence is monitored from molecules remaining in a film but in contact with the receiving liquid.

In the first step, a fluorescent molecule is doped into LDPE. It is then allowed to diffuse from LDPE and into a non-swelling liquid which contains an efficient quencher which does not enter LDPE. Thus, out-diffusion in these experiments leads to decay of total fluorescence intensity (Figure 7). Provided the quencher and liquid do not absorb at wavelengths where the dopant molecule is excited, treatment of the decrease in fluorescence intensity versus time according to equations 3a and 4a (where I_o and I_∞ now represent the maximum and minimum intensities, respectively) allows k and D to be calculated. In Table V, some characteristic values for DAA diffusion from LDPE into 2 N HCl(22) are compared to data obtained using covalently modified films.

The activation energies, E_a and E_D, derived from rate data using native unstretched LDPE according to equations 3 and 4, are collected in Table VI(22). They lead to a very different conclusion from that made on the basis of data in Table IV. The activation energies are higher and nearly constant throughout the series. The small decrease between DEA and DBA values can be rationalized by assuming that the lower overall concentration of the larger molecule in the film is due to its smaller partition coefficient between the film and liquid(14); if all doping procedures employ the same reservoir concentration of DAA, the film doped with DBA will have the lowest guest concentration at time =0 in out-diffusion experiments. As a result, it will also have the smallest fraction of sites occupied. If these are the more accessible sites, it is reasonable that escape from them should be easier, also.

The decrease in the *differences* among the activation energies reported in Table VI (as compared to Table IV) can be explained, therefore, on the basis of site selection. The Py-LDPE and An-LDPE films allow DAA molecules to enter only those sites at which a relatively voluminous lumophore is covalently attached to the polymer and which also retains sufficient additional free volume to accommodate quencher molecules. In the native films, all doped sites (i.e., potentially those with free volume equal to or greater than the van der Waals volume of one DAA molecule) contribute to the kinetics of diffusion. In fact, at early out-diffusion times, we have observed systematic deviations from single exponential decays of fluorescence intensity(7,9,14,22) which may be due to loss of DAA molecules from very large sites with relatively unencumbered access routes to the film-liquid interface, to a lack of complete temperature equilibration, to film inhomogeneities, or to a combination of factors. Alternatively, it may be a consequence of the expected multiexponential character of Fickian diffusion(17b).

To explore the latter possibility, we have attempted to fit decay curves from DAA-doped LDPE films using a more precise "all-time" integrated form of Fick's second law (equation 7)(23). Theoretical decay curves using the first 8 terms of the infinite series of equation 7 and different values of D are shown in Figure 8 along with a representative data set for DMA diffusion at 25.5 °C(22). Although the experimental data do not coincide with any of the theoretical curves, they can be made to do so quite well by correcting empirically for the fact that our experiments do not measure the first ca. 30 sec of real diffusion(7,22). Although the validity of this manipulation may be questioned,

Table IV. Activation energies from fluorescence growth curves for <u>DAA</u> out-diffusion from <u>Py</u>-LDPE films into methanol between 15 and 35 °C(*14*)

<u>DAA</u>	E_a(kcal/mol)	
	unstretched	stretched
<u>DMA</u>	12.7±0.5	16.9±0.7
<u>DEA</u>	9.2±0.4	11.7±0.5
<u>DPA</u>	8.0±0.5	9.1±0.6
<u>DBA</u>	6.0±0.8	a

a) No fluorescence intensity changes detected.

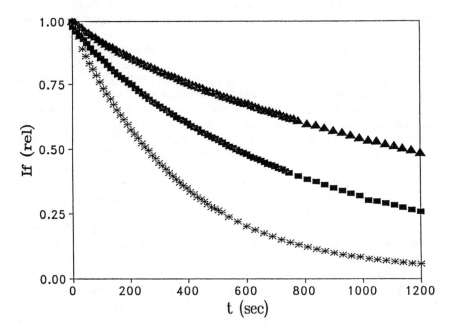

Figure 7. Fluorescence decay (out-diffusion) spectra from a native unstretched <u>LDPE</u> film, doped with <u>DMA</u>, in contact with 2 <u>N</u> HCl at 15.0 (▲), 25.5 (■), and 34.0 °C (*)(*22*).

Table V. Representative values of k and D for out-diffusion of DAA from native(22) and modified(7,9,14) unstretched LDPE films

DAA	T(°C)	native LDPE[a] $10^4 k$ s^{-1}	$10^8 D$ cm^2/s	Py-LDPE[b] $10^4 k$ s^{-1}	$10^8 D$ cm^2/s	An-LDPE[b] $10^4 k$ s^{-1}	$10^8 D$ cm^2/s
DMA	25	12	0.90		0.8	9.6	1.8
	34	28	2.0				
	35			12	1.8	18.0	3.7
DEA	24	3.8	0.22				
	34	8.8	0.53				
	35			7.1			
DPA	35			5.9			
DBA	24.5	2.3	0.14				
	33	3.5	0.23				
	35			4.7			

a) Receiving liquid is 2 N HCl. b) Receiving liquid is methanol.

Table VI. Activation energies for DAA out-diffusion from native unstretched LDPE films into 2 N HCl(22)

DAA	E_D, kcal/mol	E_a, kcal/mol
DMA	15.1 ± 0.3	14.5 ± 0.1
DEA	15.2 ± 0.4	14.7 ± 0.1
DBA	13.4 ± 0.1	13.7 ± 0.1

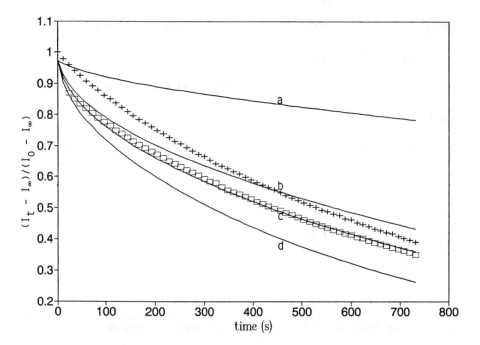

Figure 8. Theoretical fluorescence intensity decay curves from the first 8 terms of the infinite series in equation 7 and setting $D = 10^{-9}$ (a), 5×10^{-9} (b), 6.5×10^{-9} (c), and 9×10^{-9} cm^2/s (d). Also included are a data set from diffusion of <u>DMA</u> in native, unstretched <u>LDPE</u> at 25.5 °C (+) and the same data set with each point divided by a constant (1.15) and offset by 24 s (□)(22).

the best fit (D =6.5x10^{-9} cm^2/s) is extremely close to the value calculated using the simpler "early time" equation 4. Consequently, most analyses were performed using the latter.

$$\frac{I_t - I_\infty}{I_0 - I_\infty} = 1 - \sum_{n=0}^{\infty} \frac{8}{(2n+1)^2\pi^2} e^{-[D(2n+1)^2\pi^2t/\ell^2]} \tag{7}$$

The magnitudes of the E_D for the DAA in native LDPE are surprisingly large. The flow activation energy for an LDPE with M_w similar to that of Sclairfilm is only about one-half of $E_D(24)$. Flow activation is a measure of the energy necessary to produce reptant motions of intertwined polymethylene chains(25). That the E_D values are significantly larger than flow activation energy indicates that migration of DAA molecules within a film requires extraordinary chain motions(26) by the local host. Even the highest energy (α) relaxation processes in LDPE are no more than 10 kcal/mol(27). Thus, movement of the DAA within LDPE involves processes not common to the native (undoped) film. Since the width of an aromatic ring is nearly twice that of a methylene chain, it is reasonable to assume that DAA motion requires the synchronous relaxation of at least *two* vicinal polymethylene chain segments. Alternatively, but less likely, the chain motions associated with bulk relaxation modes do not emanate from the regions where the guest molecules reside (N.B., the amorphous regions and the interfaces between crystalline lamellae and amorphous domains(2,12,28)).

Regardless, it is clear that activation energies for diffusion should not be viewed as the result of motions across single, well-defined barriers. DAA molecules are confronted with a distribution of sites which differ in their abilities to accept guest molecules of fixed volume and shape and in the ease with which those guests can migrate to and from them. Within the two major families of site types, there must be subgroups which are more of less accessible than others. In essence, the diffusional rate parameters and associated activation energies being measured are the ensemble averages of many stochastic events. They are changed when the distribution of events being followed is altered by the population of the diffusing species. Also, there should be a limit beyond which activation energies for diffusion in LDPE will be dependent upon the molecular volumes of the guests.

Summary

Some experimental approaches to the characterization and, in some cases, exploitation of the dopant sites available in low density polythylene have been described. Although a full understanding of the distributions of shapes and free volumes of sites in LDPE films and the dynamics associated with them cannot be claimed yet, several conclusions can be made: (1) the sites provided to dopant molecules are reduced in volume and the guests may change their location during film stretching; (2) the distribution of free volumes for sites in stretched films is much narrower than in unstretched ones; (3) the observed activation energies for diffusion of guest molecules is a convolution of their distribution in the film and the fraction of the total number of sites which are occupied. Obviously, the techniques described here are applicable to the investigation of many types of polymers.

Acknowledgments. We thank the U.S. National Science Foundation (Grant No. CHE-9213622) for its support of this research and Dr. Changxing Cui for technical assistance.

Literature cited

1. (a) Axelson, D.E.; Levy, G.C.; Mandelkern, L. *Macromolecules* **1979**, *12*, 41. (b) Glenz, W.; Peterlin, A. *Macromol. Sci. Phys.* **1970**, *B4*, 473. (c) Hadley, D.W. In *Structures and Properties of Oriented Polymers*; Ward, I.M., Ed.; Wiley: London, 1975, Chapter 9. (d) Nordmeier, E.; Lanver, U.; Lechner, M.D. *Macromolecules* **1990**, *23*, 1072, 1077. (e) Renfrew, A.; Morgan, P. *Polyethylene: The Technology and Uses of Polyethylene Polymers*; Interscience: New York, 1957.

2. (a) Phillips, P.J. *Chem. Rev.* **1990**, *90*, 425. (b) Jang, Y.T.; Philips, P.J.; Thulstrup, E.W. *Chem. Phys. Lett.* **1982**, *93*, 66.

3. Kentgens, A.P.M.; van Boxtel, H.A.; Verveel, R.-J.; Veeman, W.S. *Macromolecules* **1991**, *24*, 3712.

4. Others, using different approaches, have made important contributions to our understanding of the micromorphology of LDPE. See for example: (a) Aggarwal, S.L.; Tilley, G.P.; Sweeting, O.J. *J. Polym. Sci.* **1961**, *51*, 551. (b) Meirovitch, E. *J. Phys. Chem.* **1984**, *88*, 2629. (c) Hentschel, D.; Sillescu, H.; Spiess, H.W. *Macromolecules* **1981**, *14*, 1605. (d) Yogev, A.; Riboid, J.; Marero, J.; Mazur, Y. *J. Am. Chem. Soc.* **1969**, *91*, 4559. (e) Read, B.E. In *Structure and Properties of Oriented Polymers*; Ward, I.M., Ed.; Wiley: London, 1975, Chapter 4. (f) Gottlieb, H.E.; Luz, Z. *Macromolecules* **1984**, *17*, 1959. (g) Yogev, A.; Margulies, L.; Amar, D.; Mazur, Y. *J. Am. Chem. Soc.* **1969**, *91*, 4558. (h) Margulies, L.; Yogev, A. *Chem. Phys.* **1978**, *27*, 89.

5. (a) Weiss, R.G.; Ramamurthy, V.; Hammond, G.S. *Acc. Chem. Res.* **1993**, *26*, 530. (b) Rammamurthy, V.; Weiss, R.G.; Hammond, G.S. In *Advances in Photochemistry*; Volman, D.H., Neckers, D., Hammond, G.S., Eds.; Vol. 18; Wiley-Interscience: New York, 1993; pp 67-234.

6. (a) Brennan, W.P. *Thermal Analysis Application Study 24*; Perkin-Elmer Instrument Division: Norwalk, CT; 1978. (b) Gray, A.P. *Thermochimica Acta* **1970**, *1*, 563.

7. (a) Naciri, J.; Weiss, R.G. *Macromolecules* **1989**, *22*, 3928. (b) Naciri, J. Ph.D. Thesis, Georgetown University, Washington, DC, 1989.

8. (a) Lamotte, M.; Pereyre, J.; Joussot-Dubien, J.; Lapouyade, R. *J. Photochem.* **1987**, *38*, 177. (b) Lamotte, M.; Joussot-Dubien, J.; Lapouyade, R.; Pereyre, J. In Photophysics and Photochemistry above 6 eV; Lahmani, F., Ed; Elsevier: Amsterdam, 1985; p 577.

9. (a) He, Z.; Hammond, G.S.; Weiss, R.G. *Macromolecules* **1992**, *25*, 1568. (b) He, Z. Ph.D. Thesis, Georgetown University, Washington, DC, 1991.

10. (a) Nakaya, T.; Tomomoto, T.; Imoto, M. *Bull. Chem. Soc. Jpn.* **1967**, *40*, 691. (b) Barker, S.A.; Monti, J. A.; Cristian, S.T.; Benington, F.; Morin, R.D. *Anal. Biochem.* **1980**, *107*,116.

11. Michl, J.; Thulstrup, E.W. *Spectroscopy with Polarized Light*; VCH: Deerfield Beach, FL, 1986.

12. Thulstrup, E.W.; Michl, J. *J. Am. Chem. Soc.* **1982**, *104*, 5594.

13. Camerman, A.; Trotter, J. *Acta Crystalogr.* **1965**, *18*, 636.

14. Jenkins, R.M.; Hammond, G.S.; Weiss, R.G. *J. Phys. Chem.* **1992**, *96*, 496.

15. Birks, J.B. *Photophysics of Aromatic Molecules*; Wiley-Interscience: London, 1970, Chapter 9.

16. Calculated from: Bondi, A. *J. Phys. Chem.* **1964**, *68*, 441.

17. (a) Comyn, J. In *Polymer Permeability*; Comyn, J., Ed.; Elsevier: Barking, Essex, 1985; Chapter 1. (b) Crank, J. *The Mathematics of Diffusion*; Oxford University Press: Oxford, 1956.

18. (a) Moisan, J.Y. In *Polymer Permeability*; Comyn, J., Ed.; Elsevier: Barking, Essex, 1985, Chapter 4. (b) Moisan, J.Y. *Eur. Polym. J.* **1980**, *16*, 979.

19. Aggarwal, S.L. In *Polymer Handbook*; 2nd ed; Brandup, J., Immergut, E.H., Eds.; Interscience: New York, 1975, p V-13.

20. He, Z.; Hammond, G.S.; Weiss, R.G. *Macromolecules* **1992**, *25*, 501.

21. (a) Howell, B.F.; McCrackin, F.L.; Wang, F.W. *Polymer* **1985**, *26*, 433. (b) Wang, F.W.; Howell, B.F.; *Polymer* **1984**, *25*, 1626. (c) Wang, F.W.; Howell, B.F. *Org. Coat. Appl. Polym. Sci. Proc.* **1982**, *47*, 41.

22. Lu, L.; Weiss, R.G. *Macromolecules* **1994**, *27*, 219.

23. Ref 17b, p 45.

24. (a) Mendelson, R.A.; Bowles, W.A.; Finer, F.L. *J. Polym. Sci., Polym. Phys. Ed.* **1970**, *8*, 105. (b) Raju, V.R.; Smith, G.G.; Marin, G.; Knox, J.R.; Graessley, W.W. *J. Polym. Sci., Polym. Phys. Ed.* **1979**, *17*, 1183. (c) Pearson, D.S.; Ver Strate, G.; von Meerwall, E.; Schilling F.C. *Macromolecules* **1987**, *20*, 1133.

25. Klein, J. *Nature* **1978**, *271*, 143.

26. Atvars, T.D.Z.; Sabadini, E.; Martins-Franchetti, S.M. *Eur. Polym. J.* **1993**, *29*, 1259.

27. (a) Boyer, R.F. *J. Polym. Sci.* **1966**, *C14*, 3. (b) Matsuoka, S. *Relaxation Phenomena in Polymers*; Hanser: Munich, 1992.

28. Radziszewski, J.G.; Michl, J. *J. Phys. Chem.* **1981**, *85*, 2934.

RECEIVED February 2, 1995

Chapter 26

Photophysical Behavior of Phenylene—Vinylene Polymers as Studied by Polarized Fluorescence Spectroscopy

M. Hennecke and T. Damerau

Federal Institute for Materials Research and Testing, Unter den Eichen 87, Berlin 12205, Germany

Polarized fluorescence spectroscopy is used to investigate the photophysical behavior of phenylene-vinylene polymers in THF solution and in polystyrene matrix in comparison with oligomeric model compounds. To explain various fluorescence properties, poly(p-phenylphenylenevinylene) can be modelled by a chain consisting of a distribution of independent oligomeric segments. Poly(1,4-phenylene-1-phenylvinylene) and poly(1,4-phenylene-1,2-diphenylvinylene) show a different fluorescence behavior.

Poly(phenylenevinylene) (PPV) and its derivatives are promising materials for future applications, e.g. electro-luminescent diodes (*1*), non-linear optical devices (*2*), photoconductive electrophotographic recording materials or electroactive battery electrodes (*3*). Therefore, the photophysical behavior of these polymers is of great fundamental interest and has been extensively studied (*4-10*). As PPV is insoluble in common solvents, various derivatives have been synthesized to improve its solubility. This is achieved by substituents along the phenylene-vinylene chain that decrease intermolecular interactions.

Optical Properties of Conjugated Polymers

Band Model and Exciton Model. It is a controversial subject whether the optical properties of conjugated polymers are more adequately described by the semiconductor band model or the exciton model (*4, 5, 7, 9, 11*). Within the band model, the polymer has a quasi one-dimensional band structure along the chain and a photoexcitation across the band gap causes the creation of polarons, bipolarons, exciton polarons and/or polaron pairs (*4, 9, 10*). Luminescence is attributed to the recombination of exciton polarons. Within the exciton model (*5, 11*), the backbone consists of conjugated segments separated from each other due to conformational distortions. Both, photoexcitation and emission, are attributed to a localized molecular transition of a single segment. The excited state, called exciton, has a high mobility and can be transfered to other segments very efficiently, preferentially to those of lower potential energy (7). Usually, the exciton mobility is known as excitation energy transfer (EET). Bässler et al. found strong evidence for the exciton model (5, 11) but Heeger et al. argue against this interpretation and confirm the validity of the band model (4).

Fluorescence Spectroscopy.   Fluorescence spectroscopy is one method to investigate optical properties. By tuning excitation (λ_{ex}) and emission (λ_{em}) wavelengths, properties of species with different transition energies can be observed selectively. Experiments with polarized light can give additional information. The stationary degree of polarization r depends on the fundamental molecular anisotropy of the chromophore r_0 (-0.2 < r_0 < 0.4), the rotational mobility within fluorescence lifetime and the orientational decorrelation by EET among the chromophores. Therefore, r is sensitive to intermolecular interaction with increasing chromophore concentration because the orientation distribution function (odf) of the transition moments for the excited chromophores is broadened by EET which causes a decrease of r.

In this contribution, an investigation on the fluorescence behavior of phenylene-vinylene polymers and related oligomers is presented. The polymers differ from each other with respect to the main-chain substituents. The influence of the substituents is discussed. Special attention is drawn to the point whether the properties can be described within the exciton model and in comparison with oligomeric model compounds. For this purpose, the degree of polarization is analyzed quantitatively as a function of chromophore concentration and of excitation wavelength. Not only the degree of polarization but also fluorescence spectra of some of the compounds under investigation are sensitive to the microscopic environment. This is verified by using solvents of different viscosities. Polarized fluorescence experiments with chromophores incorporated in stretched polyethylene films provide information on the molecular orientation coefficient and on conformational effects.

In the first part of this paper attention is focussed on a detailed description of poly(1,4-phenylphenylenevinylene) in comparison with oligomeric model compounds, in the second part, this compound is discussed in comparison with other phenylene-vinylene polymers.

Experimental

Polymers under investigation are poly(1,4-phenylphenylenevinylene) (PPPV, **1**) (*12*) (M_W = 7100 g/mol), poly(1,4-phenylene-1-phenylvinylene) (MPPPV, **2**) (M_W = 43000 g/mol) and poly(1,4-phenylene-1,2-diphenylvinylene) (DPPPV, **3**) (M_W = 60000 g/mol) (*3*). A series of oligomers consists of compounds with n repetitive styrene-units plus one terminal phenyl ring (n = 1 - 6) and with tertiary butyl-groups to improve solubility (*13*). 1,4-Distyrylbenzene (DSB) (*14*) is similar to the element n = 2 of this series.

Fluorescence experiments are carried out in a commercial spectrometer (Spex-Fluorolog) with an angle of 90° between the excitation and the fluorescence beam and polarization equipment. In isotropic specimens, polarized excitation and emission spectra consist of spectral recording of the normalized total intensity $I = I_{VV} + 2 \cdot I_{VH}$ and of the degree of polarization $r = (I_{VV} - I_{VH})/(I_{VV} + 2 \cdot I_{VH})$. The indices V and H denote the vertical and horizontal position of the polarizer in the excitation and the fluorescence beam, respectively. The intensities have to be corrected for instrumental polarization sensitivity.

     To avoid rotational mobility within fluorescence lifetime, the compounds are incorporated in films of polystyrene. Films of different weight ratios chromophore to polystyrene are prepared, either by casting or by spin-coating on glass plates. The oligomers with n = 2, 3, and 4 are incorporated in LD-polyethylene films (15) which are subsequently stretched to 300% of their initial length. Within the presumption that stretched polyethylene films show uniaxial symmetry around the stretching direction, the orientation coefficients $<P_2>$ and $<P_4>$ can be determined from the polarized fluorescence intensities according to (16). The quantum yields Φ^* are relative to DSB and are determined from comparison of the total (integral) fluorescence intensity of the specimen under investigation in tetrahydrofuran (THF) or in polystyrene relative to DSB at $\lambda_{ex} = 366$ nm. In THF, an absolute quantum yield of 0.86 has been determined for DSB. All solutions are saturated with oxygen.

Results and Discussion

PPPV Modelled by a Distribution of Segments. Figure 1 displays the long-wavelength part of the UV/Vis-absorption spectrum of PPPV, 1, dissolved in tetrahydrofuran at room temperature. If a distribution of segments of various conjugation length is responsible for absorption of PPPV, the absorption spectrum of the polymer should be a linear combination of spectra of the segments. Using the absorption spectra of oligomers (n = 1 to 6) for such a linear combination, a least-square fit is calculated and displayed as broken line. The relative contribution x_n of normalized oligomer spectra to the normalized PPPV spectrum is shown and is as follows: $x_1 = 0.19$, $x_2 = 0.41$, $x_3 = 0.01$, $x_4 = 0.31$, $x_5 = 0$, $x_6 = 0,74$. Obviously, the linear combination does not fit the polymer spectrum very well, probably because phenyl substituents in PPPV influence spectra of the segments compared to those of the oligomers. Nevertheless, we determine the contribution of the oligomers in order to describe the spectral course of the degree of polarization (see below).

The idea that PPPV consists of a distribution of segments is supported by a significant dependence of the excitation spectrum in dilute solutions on the wavelength of emission (especially for $\lambda_{em} < 480$ nm which is the peak wavelength of the emission spectrum). The peak of the excitation spectrum shifts from 330 nm to 405 nm, when λ_{em} is chosen between 390 and 480 nm. These results can easily be explained by rapid and efficient EET from the absorbing segment (the length depending upon λ_{ex}) to the longest and therefore energetically lowest segment on each chain prior to emission. Regarding the whole sample, the emitting segments from each chain form a narrow distribution of conjugation lengths, leading to the narrow emission spectrum peaking at 480 nm. Upon selecting short wavelengths for the emission (e.g. $\lambda_{em} = 450$ nm), one monitors preferentially a subset of chains with shorter length of emitting segments, leading to the observed shift in the excitation spectrum. The effect vanishes when interchain EET becomes significant, e.g. in concentrated solutions.

Intramolecular EET in PPPV. For a conjugated polymer, EET along an isolated chain should be dispersive because of the energetic distribution of segments. Furthermore, it has to be considered that the mutual orientation between segments along a rigid polymer chain can be correlated. For a semi-quantitative discussion of EET in this system, we refer to the partial alignment of segments by a rough approximation:

$$r / r_0 = k_D \cdot \tilde{g}^S + (1 - \tilde{g}^S) \cdot < {}^2P_2 > . \tag{1}$$

$\tilde{g}^S$ represents the probability that the excitation resides at the site where it has been formed (*17*). $<{}^2P_2>$ is a second order pair orientational order coefficient of a pair-type distribution function, taken in the spatial range of EET ($0 < <{}^2P_2> < 1$). In perfectly isotropic solids, where even neighbored sites are uncorrelated, $<{}^2P_2> = 0$. r_0 is the fundamental molecular anisotropy of the segments (oligomers) which is constant for all oligomers in the S_1-S_0-absorption band ($r_0 = 0.35$).

For PPPV, r_0 increases linearly with increasing λ_{ex} (Figure 2, solid line). By using equation 1 the magnitude of $<{}^2P_2>$ can be estimated. The wavelength dependence of $\tilde{g}^S$ is derived from a simple model based on the relative contribution (i.e. absorption probability) of the oligomers n = 1 - 6 to the UV/Vis-spectrum of PPPV, taken from Figure 1. If any segment with $1 \leq n \leq 5$ absorbs the exciting light, efficient EET should take place, finally to a segment with n = 6 which is energetically favored. From segments with n = 6, EET is not taken into account, neither between themselves nor to shorter segments. As a consequence, only those segments with n = 6 that are excited directly by the exciting light, contribute to the probability that the excitation energy resides at the site where it has been formed.

The dotted line in Figure 2 is calculated for $<{}^2P_2> = 0$. From comparison with the curve for $<{}^2P_2> = 1$ (dash-dotted line), it is concluded that only small values of

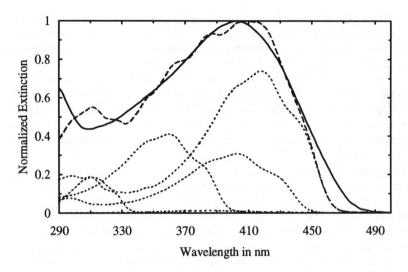

Figure 1.   Normalized absorption spectrum of PPPV in toluene at a concentration of $2.5 \cdot 10^{-3}$ g/l (———). The spectrum is modelled by a linear combination (— — —) of spectra of the oligomers (n = 1 to n = 6). The individual oligomer spectra (· · · · ·) are shown with their contribution to the linear combination.

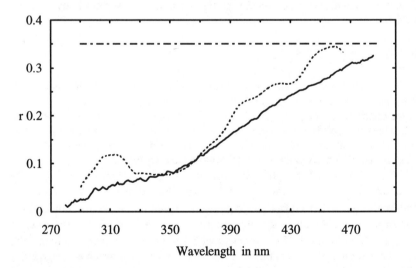

Figure 2.   Degree of polarization r versus wavelength of excitation for PPPV incorporated in a cast film of polystyrene at a weight ratio of 10^{-5} (———). The course of r is modelled by equation 1 for the extreme values $\langle {}^{2}P_{2} \rangle = 0$ (· · · · ·) und $\langle {}^{2}P_{2} \rangle = 1$ (—·—·—·—).

$<^2P_2>$ are allowed in PPPV. The measured values of r are smaller throughout the major part of the spectrum; this systematic difference is explained by effects that have not been taken into account in the model calculation.

Intermolecular EET in PPPV. To discuss the properties of intermolecular EET, fluorescence depolarization as a function of chromophor concentration is analyzed by GAF theory (*18, 19*) for PPPV and the oligomer with n = 4, although presumptions for GAF theory are not strictly fulfilled. When EET is described by Förster mechanism, GAF theory only contains one parameter, the critical concentration c_{OD} which determines the course of r/r_0 versus chromophore concentration. As shown in Figure 3, the critical concentration c_{OD} increases significantly for excitation at long wavelength (known as 'Weber's red edge effect'), and this is more pronounced for PPPV than for the oligomer. This behavior has to be explained by an energetic distribution of the chromophores caused by the glassy polystyrene matrix. Additionally, for PPPV a distribution of segment lengths has to be taken into account.

PPPV in Stretched PE-Films. In polyethylene-films stretched to 300% of their initial length, the orientation parameter $<P_2>$ for the incorporated oligomers with n = 2, 3, 4 and for PPPV is determined to be 0.19, 0.38, 0.59 and 0.35, respectively. It is worth questioning whether PPPV, having a $<P_2>$-parameter in the range of the oligomer with n = 3, maintains its conformation during stretching (as expected for the oligomers) or whether it undergoes conformational changes like a flexible polymer. In the first case, emission spectra should be independent of the stretching ratio. Possible conformational changes are rotations around C-C single bonds of the vinylene units or around single bonds at chemical defects. The first type would influence the effective conjugation length, the second the pathway of EET between the segments. Changes in the emission spectra of stretched films of PPV (Bässler, (*5*)) and in the absorption spectra of a derivate of PPV blended with polyethylene (Heeger, (*4*)) are interpreted as an extension of the effective conjugation length of the polymer during stretching. PPPV in a PE film, as used in this work, does not show significant differences in the unpolarized emission spectra before and after stretching. Using polarized light, the emission spectrum of the component I_{VV} for a film with stretching direction vertical shows less intensity at wavelengths above 500 nm than the same component for horizontal stretching direction (Figure 4), i.e. the energetic distribution of the fluorescent segments is shifted to higher energies. It is assumed that all molecules that orientate parallel to the stretching direction undergo more conformational changes than the less oriented fraction which is preferentially monitored by I_{VV} at stretching direction horizontal. If stretching increases the effective conjugtion length, I_{VV} at vertical stretching direction should be red shifted to I_{VV} at horizontal stretching direction. As this is not the case and in contrast to (*4, 5*), another effect has to be more significant under experimental conditions used here. In a more elongated overall conformation of the macromolecule, the preferred pathway of EET is along the chain. With a random sequence of conjugated segments of various length this implies that the excitation can be immobilized at energetically higher segments from which EET is improbable, i.e. the energetic distribution of the emission spectrum is shifted to higher energies.

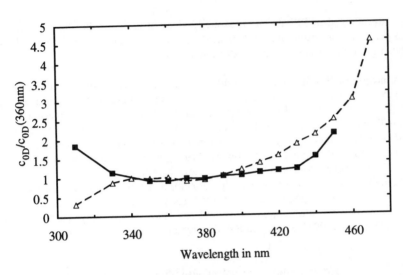

Figure 3. c_{OD} normalized to c_{OD} at $\lambda_{ex} = 360$ nm versus wavelength of excitation for oligomer with n = 4 (■) and for PPPV (Δ).

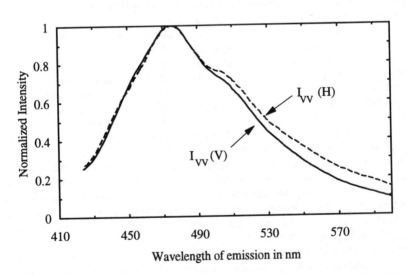

Figure 4. Normalized emission spectra of the component I_{VV} of a polyethylene film doped with PPPV, stretched to 300% of the initial length for vertical (V) and horizontal (H) position of the stretching direction ($\lambda_{ex} = 400$ nm, slit width: 5.4 nm/5.4 nm). The ratio of the absolute intensities at $\lambda_{em} = 475$ nm is about $I_{VV}(V)/I_{VV}(H) = 3$.

Comparison of Different Phenylene-Vinylene-Polymers. Compared to PPPV which has been discussed so far, a pure film of PPV has a red shifted emission spectrum (cf. Table I). This indicates a more extended conjugation length and efficient EET in PPV. In PPPV the phenyl-substituents impose steric hindrance. It is worth noting the high degree of polarization in the PPV-film (*11*). This has to be explained by a high value of $<^2P_2>$ for the segments involved in EET and is contrary to the situation in PPPV.

   In comparison with PPPV, MPPPV and DPPPV show significant differences in polarized fluoresescence spectra (Figure 5 and 6, Table I). For MPPPV and DPPPV the S_1-S_0 band of the excitation spectrum is blue shifted. Again, this is typical for chromophores with a less extended or less planarized π-electronic system. The emission spectra of MPPPV and DPPPV are broad and structureless bands. The Stokes shift, which is the difference between the long wavelength absorption maximum and the short wavelength emission maximum, is significantly higher than for PPPV and the overlap integral J (*20*) is rather small. Both, excitation and emission spectra of MPPPV and DPPPV only depend slightly on λ_{em} and λ_{ex}, respectively. This indicates that chromophores contributing to emission have a very narrow distribution of transition energies. This is in contrast to PPPV where a strong dependence of the excitation spectra on λ_{em} was found.

Behavior in Solvents of Different Viscosity. Comparing the behavior in solvents of different viscosity, further aspects are revealed. Usually, spectra should not change and the degree of polarization is lowered in the solvent with low viscosity due to rotational mobility. This is true for PPPV but not for DPPPV (comparing Figure 6 and 7). In PS, the emission spectrum of DPPPV is blue shifted by 16 nm relative to its spectral position in THF. As we exclude solvatochromic

Table I. Fluorescence properties of phenylene-vinylene-polymers. Maximum of the long wavelength absorption band (λ_{max}(abs)), maximum of the emission spectrum λ_{max}(em), degree of polarization r (λ_{em} is chosen as λ_{max}(em)) and quantum yield Φ^* relative to the oligomer DSB

Specimen	solvent, concentration	λ_{max} (abs) [nm]	λ_{max} (em) [nm]	r(λ_{ex}[nm]) absorption maximum	r(λ_{ex}[nm]) absorption edge	Φ^*
PPV cf. (*11*)	pure film	400	554	0.17 (400)	0.27 (515)	-
PPPV (1)	pure film	400	500	0.03 (400)	0.17 (490)	-
PPPV (1)	THF 2.5 mg/l	400	481	0.16 (400)	0.27 (470)	0.97
PPPV (1)	PS 0.01 mg/g	-	477	0.18 (400)	0.33 (470)	0.4
MPPPV (2)	THF 2 mg/l	371	474	0.11 (370)	0.27 (460)	0.09
MPPPV (2)	PS 0.1 mg/g	-	490	0.05 (370)	0.22 (460)	0.3
DPPPV (3)	THF 3 mg/l	356	528	0.11 (370)	0.26 (450)	0.01
DPPPV (3)	PS 0.1 mg/g	-	512	0.02 (360)	0.20 (450)	0.7
DSB	PS 0.1 mg/g	360	417	0.35(360)	0.35 (400)	1

effects, we favor an interpretation given by Castel et al. (*21*) for a similar phenomenon observed with 1,2-diarylethylenes that are distorted around the C-C-bond in the ground state due to steric reasons: The S_0 and S_1 states of these compounds have their minima of potential energy at different combinations of twist-angles around C-C- and C=C-bonds. In low viscous solvents, a conformative relaxation of the excited state takes place and leads to a high Stokes shift. In viscous media the relaxation is hindered within fluorescence lifetime and emission occurs significantly from those S_1 states which are unrelaxed with respect to these angles, i.e. the emission spectrum changes and the Stokes shift is reduced. This interpretation is supported by ^{13}C-NMR-experiments for DPPPV (*3*) which prove that a steric interaction between the phenyl substituents and the phenyl rings of the main chain leads to a distortion in the ground state.

Surprisingly, the degree of polarization of DPPPVin rigid PS matrix is lower than in fluid THF (Figure 6 and 7). Additionally, r changes its course as a function of λ_{ex} from convex in PS to concave in THF. We exclude the formation of aggregates in the films. Therefore, the only explanation for the unusual result r(3, THF) > r(3, PS) is that r_0(3, THF) has to be greater than r_0(3, PS) in the whole range of λ_{ex}. As the components of both, the absorption and the emission tensor, contribute to r_0, the emission tensor of the rotationally relaxed form may have other components, in such a way that r_0 may be increased compared to the unrelaxed form. This increase may not be compensated by rotational depolarization in fluid media.

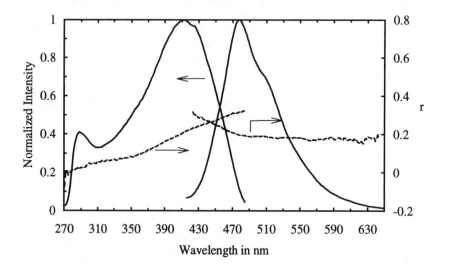

Figure 5. Polarized excitation (λ_{em} = 510 nm) and emission spectrum (λ_{ex} = 400 nm) of PPPV in solution of polystyrene (weight ratio 10^{-5}): total intensity (——, left hand scale) und degree of polarization (– – –, right hand scale); slit widths: 2.7 nm/3.6 nm.

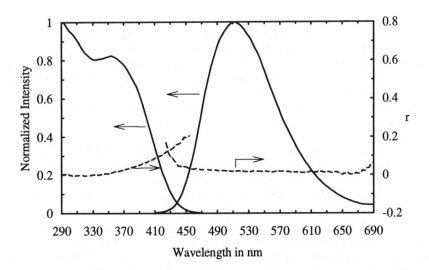

Figure 6. Polarized excitation (λ_{em} = 510 nm) and emission spectrum (λ_{ex} = 360 nm) of DPPPV in solution of polystyrene (weight ratio 10^-4): total intensity (———, left hand scale) and degree of polarization (— — —, right hand scale); slit widths: 2.7 nm/2.7 nm.

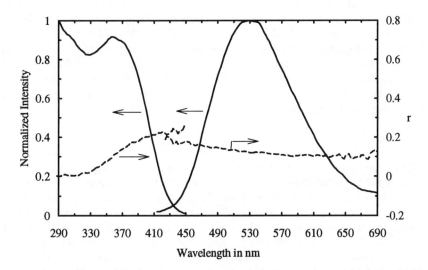

Figure 7. Polarized excitation (λ_{em} = 526 nm) and emission spectrum (λ_{ex} = 360 nm) of DPPPV in solution of THF (concentration 3·10^-3 g/l): total intensity (———, left hand scale) and degree of polarization (— — —, right hand scale); slit widths: 3.6 nm/6.3 nm.

Quantum Yield. The significance of rotational relaxation is confirmed by the quantum yields relative to DSB determined in PS and THF. DSB has an absolute quantum yield of 0.86 in THF. It is obvious from Table I that relative quantum yields of DPPPV and MPPPV are increased by one order of magnitude in polystyrene films. Within the model of rotational relaxation, this can be explained by the relationship between the internal conversion and the flexibility of a molecular skeleton (22). With increasing viscosity non-fluorescent desactivation processes caused by changes in the molecular geometry are reduced, i.e. the quantum yield is increased.

Photophysically Effective Unit in DPPPV and MPPPV. It has to be checked whether the model used for PPPV (effective transfer of excitation energy between segments of different length) is also applicable to DPPPV or MPPPV. Contrary to the behavior of PPPV, the excitation spectrum of DPPPV is independent of λ_{em}. To explain this within the model, one has to assume that the polymer chain of DPPPV predominantly contains short or distorted segments which are responsible for the blue shifted absorption. On the other hand, each chain has to contain a few very long segments that are exclusively responsible for the long wavelength emission. However, the increase of r as a function of λ_{ex} for $\lambda_{ex} > 330$ nm (Figure 3) indicates, that this assumption is not fulfilled: within the model the increase of r has been attributed to an increasing probability of direct excitation of long segments. But for DPPPV, segments with a fluorescence peaking at 530 nm should not contribute preferentially to absorption at 330 nm. Therefore, the model of segments and EET along the polymer chain should not be used for DPPPV.

Most probably, the photophysically effective unit consits only of a small number of monomers but the high degree of polarization in THF solution indicates that there is no significant EET between these nearly isoenergetic chromophores. As the product of quantum yield Φ and the overlap integral J is proportional to the sixth power of the Förster distance R_0 (representing the efficiency of EET), one can conclude that R_0 for PPPV is larger by a factor 5500 than for DPPPV. The increase of r as a function λ_{ex} appears to be a molecular property of the chromophore units built by the conformative distortions.

Conclusion

Soluble derivatives of PPV discussed here show a strong dependence of their behavior on the substituent. In comparison with a PPV-film, emission is blue shifted for all derivates, since the substituents impose steric distortions and reduce the effective conjugation length. The chain of PPPV can be modelled by oligomeric segments which form the photophysically effective units. Intermolecular EET is dispersive, first because of the glassy polystyrene matrix and second because of the energetic distribution of the oligomeric segments. The results for MPPPV and DPPPV could not be described within this model; a geometric distortion and a small overlap integral are made responsible.

Acknowledgments. Financial support by the Fonds der Chemischen Industrie is gratefully acknowledged. The authors thank K. Müllen, A. Greiner, H.H. Hörhold, and H. Meier who supplied the compounds. Measurements were performed at the Institut für Physikalische Chemie, TU Clausthal, courtesy of J. Fuhrmann.

References

1. Bradley, D.D.C. *Synth. Met.* **1993**, *54*, 401.
2. Bradley, D.D.C. *Makromol. Chem. Macromol. Symp.* **1990**, *37*, 247.
3. Hörhold, H.H.; Helbig, M. *Makromol. Chem., Macromol. Symp.* **1987**, *12*, 229.
4. Hagler, T.W.; Pakbaz, K.; Voss, K.F.; Heeger, A.J. *Phys. Rev. B.* **1991**, *44*, 8652.
5. Heun S.; Mahrt, R.F.; Greiner, A.; Lemmer, U.; Bässler, H.; Halliday D.A.; Bradley, D.D.C.; Burn, P.L.; Holmes, A.B. *J. Phys. Cond. Matter* **1993**, *5*, 247.
6. Vestweber, H.; Greiner, A.; Lemmer, U.; Mahrt, R.F.; Richert, R.; Heitz, W.; Bässler, H. *Adv. Mater.*, **1992**, *4*, 661.
7. Woo, H.S.; Graham, S.C.; Halliday, D.A.; Bradley, D.D.C.; Friend, R.H.; Burn, P.L.; Holmes, A.B. *Phys. Rev. B* **1992**, *46*, 7379.
8. Stubb, H.; Punka, E.; Paloheimo, J. *Mat. Sci. Engineering* **1993**, *10*, 85.
9. Heeger, A.J.; Kivelson, S.; Schrieffer, J.R.; Su, W.P. *Rev. Mod. Phys.* **1988**, *60*, 781.
10. Hsu, J.W.P.; Yan, M.; Jedju, T.M.; Rothberg, L.J. Hseih, B.R. *Phys. Rev. B* **1994**, *49*, 712.
11. Rauscher, U.; Bässler, H.; Bradley, D.D.C.; Hennecke, M. *Phys. Rev. B* **1990**, *42*, 9830.
12. Martelock, H.; Greiner, A.; Heitz, W. *Makromol. Chem.* **1991**, *192*, 967.
13. Tian, B.; Zerbi, G.; Müllen K. *J. Chem. Phys.* **1991**, *95*, 3198.
14. Oelkrug, D.; Rempfer, K.; Prass, E.; Meier, H. *Z. Naturforsch.* **1988**, *43a*, 583.
15. Hennecke, M.; Damerau, T.; Müllen, K. *Macromolecules* **1993**, *26*, 3411.
16. Nobbs, J.H.; Ward, I.M. In *Polymer Photophysics*; Phillips, D., Ed.; Chapman and Hall, London 1985; pp 159 - 220.
17. Fredrickson, G.H.; Frank, C.W. *Macromolecules* **1983**, *16*, 1198.
18. Gochanour, C.R.; Andersen, H.C.; Fayer, M.D. *J. Chem. Phys.* **1979**, *70*, 4254.
19. Gochanour, C.R.; Fayer, M.D. *J. Phys. Chem.* **1981**, *85*, 1989.
20. Berlman, I.B. *Energy Transfer Parameters of Aromatic Compounds;* Academic Press, New York 1973.
21. Castel, N.; Fisher, E.; Bartocci, G.; Masetti, F.; Mazzucato, U. *J. Chem. Soc. Perkin Trans. II* **1985**, 1969.
22. Klessinger, M.; Michl, J. *Lichtabsorption und Photochemie organischer Moleküle*, VCH Publishers, Weinheim 1989.

RECEIVED February 2, 1995

Chapter 27

Free-Volume Hole Distribution of Polymers Probed by Positron Annihilation Spectroscopy

J. Liu, Q. Deng, H. Shi, and Y. C. Jean[1]

**Department of Chemistry, University of Missouri,
Kansas City, MO 64110**

Positron annihilation spectroscopy (PAS) has been developed to characterize the free-volume properties of polymers and polymer blends. Positron annihilation lifetime (PAL) measurements give direct information about the dimension, content, and hole-size distributions of free volume in amorphous polymeric materials. The free-volume hole distribution in epoxy, polypropylene, polycarbonates, and polystyrene is presented as a function of pressure, molecular relaxation, and miscibility. The unique capability of PAS to probe free-volume properties derives from the fact that the positronium atom (Ps) is preferentially trapped in atomic-scale holes ranging in size from 1 to 10 Å.

The concept of free volume in a liquid, proposed four decades ago [1], has been applied to explain the free-volume hole properties of polymers in recent years [2]. In 1959, Cohen and Turnbull [3] used free-volume hole size to explain diffusion in liquids as a function of temperature and pressure and defined free volume to be the volume difference between the space of the molecular cage made by the surrounding molecules and the van der Waals volume of the molecule in the cage.

Free-volume parameters of polymeric materials have recently been calculated using theories such as molecular dynamics and kinetics [4, 5]. Despite the existence of various free-volume theories, only limited experimental data about free volumes in polymers have been reported due to the intrinsic difficulties of probing free volume, which has a size of a few angstroms and a duration as brief as a few 10^{-10} s. However, a great deal of effort has been put into measuring free-volume properties in polymeric materials. Small-angle X-ray and neutron diffractions have been used to determine density fluctuations and then to deduce free-volume size distributions [6-8]. A photochromic labeling technique by site-

[1]Corresponding author

0097–6156/95/0598–0458$12.00/0

specific probe has also been developed [9-12].   In recent years, a new microanalytical probe, Positron Annihilation Spectroscopy (PAS), has been developed to probe the free-volume properties in polymeric materials. In 1986, we reported [13] the mean free-volume size in an epoxy by measuring its positron annihilation lifetime (PAL) based on a spherical model developed for the free-volume bubbles in liquids [14]. In recent experiments [15-17], using the PAL method we have determined the free-volume hole size in polymers as a function of pressure, temperature, and physical aging. We have also extended our study of free-volume properties to polymer blends [18].

The unique sensitivity of PAS in probing free-volume properties is due to the fact that positronium (a bound atom consisting of an electron and a positron) is found to be preferentially localized in the free-volume region of polymeric materials. Evidence of Ps-localization in free volumes has been found from temperature-, pressure-, and crystallinity-dependent experiments [13, 15, 16]: (1) o-Ps (triplet Ps) lifetime undergoes a dramatic change as $T > T_g$ (glass transition temperature) and $T < T_g$; (2) the lifetime temperature coefficient ($\approx 10^{-3}$ K^{-1}) is one order of magnitude larger than the volume expansion coefficients ($\approx 10^{-4}$ K^{-1}); (3) a large variation of positron lifetime has been observed when a polymer is under a static pressure; and (4) o-Ps formation is found only in amorphous regions where free volume exists.   In contrast to other techniques, PAS probes the free-volume properties directly without significant interference from bulk properties.

Mean Free-Volume Hole Sizes

In PAL measurements, the experimental measured positron annihilation rate λ is defined by the integration of the overlap between the positron density $\rho_+(r)$ and the electron density $\rho_-(\mathbf{r})$:

$$\lambda = constant \times \int \rho_-(r)\, \rho_+(r)\, dr \qquad (1)$$

where the constant is a normalization constant related to the number of electrons involved in the annihilation process. The annihilation lifetime τ is the reciprocal of the annihilation rate λ. In polymeric materials, the positron lifetime spectrum reveals a function containing a multiexponential:

$$N(t) = \sum_{i=1,n} I_i e^{-\lambda_i t} \qquad (2)$$

where n is the number of exponential terms, and I_i and λ_i represent the number of positrons (intensity) and the positron annihilation rate, respectively, for the annihilation from the ith state. It is customary to fit PAL spectra into three or four lifetime components by using a computer program (PATFIT) [19]. The longest lifetime component is contributed from o-Ps (ortho-positronium) annihilation. Ps is considered to be confined in spaces between molecules.

In order to determine the size of free-volume holes from the o-Ps lifetime, generally we assume the free-volume hole is in a spherical shape under isotropic conditions.   By developing equation (1) based on a simple particle-in-spherical-box quantum mechanical model [20, 21], a correlation between the o-Ps annihilation lifetime (τ) and hole radius (R) has been obtained:

$$\tau = \frac{1}{2} \left[1 - \frac{R}{R_o} + \frac{1}{2\pi} \sin \left(\frac{2\pi R}{R_o} \right) \right]^{-1} \qquad (3)$$

where $R = R_o - \Delta R$ is the radius of the free-volume hole and $\Delta R = 1.656$ Å is an empirical parameter which was obtained by fitting the measured annihilation lifetimes of cavities with known sizes [22]. The universal correlation between o-Ps and free-volume is shown in Fig. 1.

The mean free-volume hole size of polymers can be easily determined by measuring the o-Ps lifetime and converting the mean lifetime of o-Ps obtained by conventional methods of data analysis into hole size according to the semi-empirical equation (3). In the temperature study on a series of epoxy polymers DGEBA/DAB/DDH in different chemical compositions, we found that the o-Ps lifetime increases dramatically at T_g [13]. One of these lifetime/temperature variations is shown in Fig. 2.

It is seen in Fig. 2 that the largest change of o-Ps lifetime coincides with T_g and the temperature coefficient is on the order of 10^{-3} K^{-1}, which is about one order of magnitude larger than the volume expansion coefficient. Furthermore, the temperature dependence also occurs at $T < T_g$. This effect has been used to probe the sub-T_g annealing and the physical aging effects on the polymeric structures [23, 24].

However, it is known that the free-volume hole is not exactly spherical, especially under anisotropic conditions. Therefore, we need to consider the shape factor. Under anisotropic external forces, such as anisotropic pressure, stretching, etc., the free-volume hole shape is more likely to be ellipsoidal. We have recently developed a new relationship between the o-Ps lifetime and the anisotropic dimension of an ellipsoidal free-volume hole using an ellipsoidal hole model [25, 26].

In an ellipsoidal hole model, an ellipsoid is defined by the dimensions a and b with respect to the semi-major and the semi-minor elliptic axes and by the eccentricity, $\varepsilon = (a^2-b^2)^{1/2}/a$. Solving the Ps wave function, Φ_{Ps}, inside the ellipsoid ($\Phi_{Ps} = 0$ outside the ellipsoid) from the Schrödinger equation by a numerical method, we obtain the o-Ps lifetime in a manner similar to that used for the spherical model. In practice, we fit the numerical results into a variety of mathematical forms, including polynomials and exponentials. The following equation is obtained for the ratio of o-Ps lifetimes as a function of the eccentricity (ε):

$$\frac{\tau_{ell}}{\tau_{sph}} = 1 + 0.400\varepsilon - 4.16\varepsilon^2 + 2.76\varepsilon^3 \qquad (4)$$

The above equation shows that the o-Ps lifetime in an ellipsoid is always shorter than in a sphere with the same hole volume. An ellipsoid with a larger

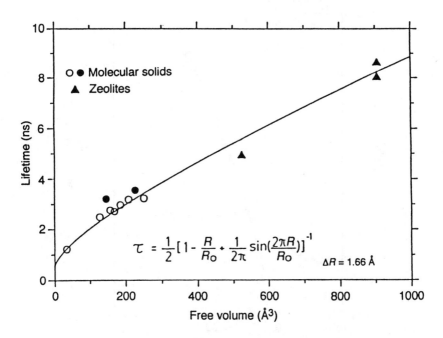

Fig. 1. A universal correlation between o-Ps lifetime and free volumes in polymers. The solid line is the best fitted equation (3) to known cavity volumes [22]. Reprinted by permission of Ref. 22 (© 1988 World Scientific).

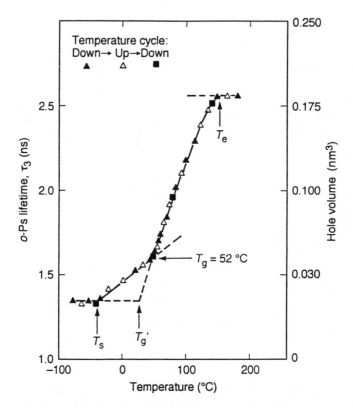

Fig. 2. Temperature dependance of o-Ps lifetime and free volume in an amine-cure epoxy polymer ($T_g = 52$ °C) [13]. Four onset temperatures were observed: T_s (solidified), T_g' (real glass transition), T_g (apparent glass transition), T_e (Ps-bubble formation). Reprinted by permission of Ref. 13 (© 1986 *J. Polym. Sci. B*).

eccentricity gives a larger reduction of o-Ps lifetime. Eq. (4) can be used to correct the shape factor from PAL results.

Free-Volume Hole Distributions

In reality there is a range of free-volume hole sizes in polymeric materials. Therefore, it is more adequate to express o-Ps lifetime in a distribution form rather than as a discrete value, τ_3. Hence, we employ the expression of positron lifetime spectra in the form originally suggested by Tao and later by Schrader [27] as:

$$N(t) = \int_0^\infty \lambda \alpha(\lambda) e^{-\lambda t} d\lambda + B \qquad (5)$$

where $\lambda\alpha(\lambda)$ is an annihilation probability density function and B is the background of the spectrum. The exact solution of $\lambda\alpha(\lambda)$ is a very difficult problem because the actual experimental spectra are convoluted with a resolution function which cannot be exactly measured. However, if one measures a reference spectrum $Y_r(t)$, which has a known single positron decay rate (λ_r), the solution of $\lambda\alpha(\lambda)$ can be obtained by using $Y_r(t)$ to deconvolute sample spectra. Here, $Y_r(t)$ can be expressed as:

$$Y_r(t) = R(t) * N_r \lambda_r e^{-\lambda_r t} \qquad (6)$$

where N_r is the normalized total count for a reference spectrum and R(t) is the resolution function. Lifetime spectra of reference samples were obtained from the following materials: a well-annealed single Al (τ = 162 ps, 98%) and a [207]Bi radioisotope which emits two γ-rays at τ = 183 ps with energies very close to [22]Na. The computer program CONTIN [28, 29] was used to perform the deconvolution procedure through a Laplace inversion technique. In this continuous data analysis method, a result of $\lambda\alpha(\lambda)$ vs λ is obtained from spectra. All reference spectra give consistent $\lambda\alpha(\lambda)$ vs λ results.

Following the correlation Eq. (3) between τ_3 and hole radius R, and considering the difference of o-Ps capture probability in different hole size with a linear correction K (R) (=1+8R) and with a spherical approximation of free-volume holes, the free-volume hole volume probability density function, $V_f$pdf is expressed as [30]:

$$V_f pdf = -3.32 \frac{[\cos(\frac{2\pi R}{R+1.66}) - 1]\alpha(\lambda)}{(R+1.66)^2 K(R) 4\pi R^2} \qquad (7)$$

The free-volume distributions in polymers obtained by PAL measurements using continuous data analysis are compared quantitatively with the Simha-Somcynsky theory [31]. A good agreement between our experimental results and predictions of the theory is a further indication that PAS is a reliable and sensitive probe to determine the free-volume distributions in polymeric materials.

Pressure Dependence of Free-Volume Hole Distributions

The investigation of pressure/volume behavior of polymers is essential to the fundamental understanding of material properties. In the microscopic point of view, the changes in macro-physical properties due to applied pressure are related to the changes in free-volume properties. In the study of pressure dependence of free-volume properties in polymers, we have used a thermosetting polymer, epoxy, and a thermoplastic polymer, polypropylene [30, 32]. The positron lifetime distributions of epoxy and polypropylene are plotted in Fig. 3 under different pressures. Each spectrum contains a total statistics of 40 x 10^6 counts. The good agreements between the mean lifetime results from a least-squares fit method (PATFIT) and the peak lifetimes from a Laplace inverting method (CONTIN) as shown in Fig. 3 confirm the accuracy of our new method of data analysis. The new information here is a continuous lifetime distribution instead of discrete lifetimes.

The free-volume hole radius distributions are then obtained by converting the o-Ps lifetimes (right peaks) into R according to the τ-R correlation of Eq. (3), then converting to volume, V, according to Eq. (7). The results of hole volume distributions are shown in Fig. 4.

As expected, an increase of pressure results in collapse and compression of the free-volume holes. In epoxy the hole distributions shift from maxima 50 Å^3 to 25 Å^3, 8 Å^3, and 1 Å^3 for 1.8, 4.9, and 14.0 kbar respectively, while in polypropylene the maxima shift from 100 Å^3 to 25 Å^3 and 7 Å^3 for 4.2 and 14.7 kbar. The width of the distribution is also correspondingly narrowed as P increases. We found that the experimental hole distributions can be fitted into gaussian function or with a sum of gaussians. The compressibility of free volume, β_f, calculated from the peak values of free-volume distributions under different pressures, is in the range of 10^{-5} to 10^{-6} at low pressures. It is found to be a few times larger than the results obtained from a volumetric method [2]. We believe that our β_f value is more accurate than those obtained by the conventional methods because PAS is probing the free volume directly. It is worthwhile to mention that a distorted distribution of very small holes, clearly seen in Fig. 3 for P of ca. 14 kbar, is a result of using a crude theoretical quantum mechanical model for Ps-localization in the free-volume holes. It also shows a limiting radius of holes, ~ 0.5 Å, that Ps can probe properly.

Free-Volume Hole Distributions and Molecular Relaxation

A non-equilibrium state often exists in glassy polymers resulting from the thermal or mechanical treatment history. Glassy polymers exhibit a tendency toward equilibrium state in order to minimize the overall energy through molecular relaxation of polymer chains over a period of time. The chain mobility of macromolecules in a closely packed system is primarily determined by and inversely proportional to the degree of packing of the system. At the molecular level, the chain mobility is restricted by the free-volume hole size and hole size distribution. On the other hand, the free-volume properties are changed through molecular relaxation. Hence, a very fruitful approach to understanding the

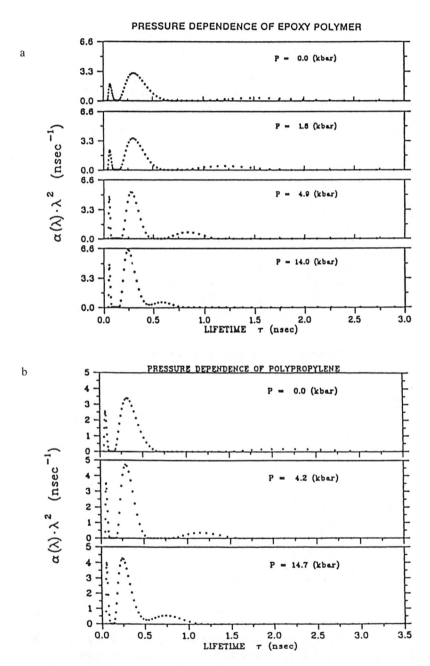

Fig. 3.   Positron lifetime distribution functions of   (a) epoxy,   (b) polypropylene.   The applied pressure is quasi-isotropic [30, 32].   The distributions of positron lifetime were obtained by using the Laplace inverting program CONTIN.   Reprinted by permission of Refs. 30 (© 1992 *J. Polym Sci. B)* and 32 (© 1992 ACS).

a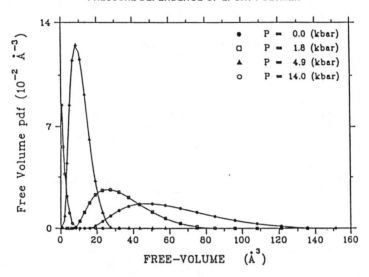

b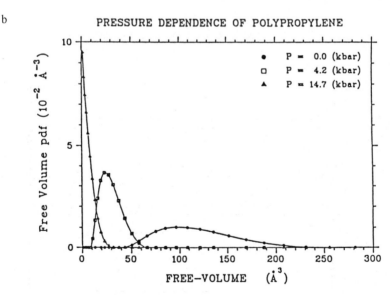

Fig. 4. Free-volume hole volume distribution functions of (a) epoxy, (b) polypropylene, under different pressures. Smooth lines were drawn through data for eye-guide purposes only [30, 32]. Reprinted by permission of Refs. 30 (© 1992 *J. Polym Sci. B)* and 32 (© 1992 ACS).

molecular relaxation mechanism is to investigate the microscopic free-volume properties as a function of time after polymeric materials are thermally or mechanically treated [1].

Recently, we have studied the epoxy and polypropylene polymers for free-volume distribution as a function of time after polymers are released from a static pressure [33]. The free-volume hole distributions of epoxy and polypropylene are shown in Fig. 5 before and after press.

As shown in Fig. 5, the free-volume hole volume distributions are found to be observably different for polymers before and after press. The distributions become narrower after applying 14.7 kbar of pressure for one week. The distribution in epoxy can be approximately expressed by gaussian type functions with FWHM (Full Width at Half Maximum) of 60 $Å^3$ and 50 $Å^3$ before and after press respectively, *i.e.* 17 % narrower after press. In polypropylene, FWHM are found to be 105 $Å^3$ and 65 $Å^3$ before and after press respectively, *i.e.* 38 % narrower after press.

These results show that applying an external pressure has significantly changed the microstructure of polymers. Increasing pressure causes a compression or collapse of free-volume holes. Upon the release of pressure, a major fraction of the molecular chains, where the free volume is at a maximum distribution (*i.e.* the peak of the distribution), relaxes spontaneously back to the original size. Part of the molecular chain relaxes at a slower rate and is detectable from the current PAS method. As shown in Fig. 5, in epoxy, after 75 days the free-volume distribution recovers gradually back to the original distribution. In polypropylene, we found the difference of hole distribution before and after press is much greater than that in epoxy. On the other hand, we found the result of the distribution is not changed as a function of time up to 75 days of aging, *i.e.* the distributions are all the same as that at 10 days, as shown in Fig. 5. This indicates that a fraction of molecular chains in polypropylene relaxes at a rate much slower than in epoxy. This slower molecular relaxation in polypropylene than in epoxy after press may be due to the difference in the polymer structures: polypropylene is a semi-crystalline sample and epoxy is 100% amorphous; polypropylene is a thermoplastic polymer and epoxy is a thermosetting polymer. These differences in thermal and mechanical properties between these polymers result in a different response in the molecular relaxation after the release of pressure. It appears that deformation of free-volume holes can be easily recovered in thermosetting polymers but not in thermoplastic materials.

Free-Volume Hole Distributions and Miscibility

In comparison with their neat polymer components, the morphology of polymer blends is more complicated due to the process of mixing. A variety of interesting new features in polymer blends can be generated according to the equilibrium thermodynamics and kinetics of mixing. Based on thermodynamics, a negative enthalpy of mixing, ΔH_{mix}, is often required for polymer blends to contribute to lower the free energy of mixing, ΔG_{mix}, due to the small and negligible entropy of mixing, ΔS_{mix} [34]. As a result, the majority of polymer blends are immiscible. The phase separation of immiscible polymer blends creates different domains as well as interfacial regions. The heterogeneity of the morphology resulting from the phase separation has a significant effect on physical properties of blends. In our

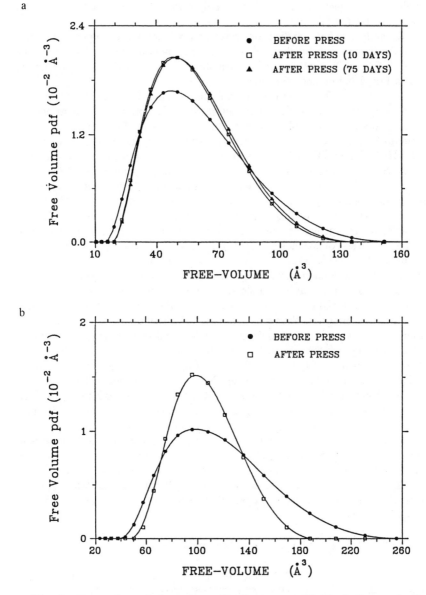

Fig. 5. Free-volume hole volume distributions at 25 °C before and after press (10 and 75 days after releasing pressure) (a) epoxy, (b) polypropylene [33]. Reprinted by permission of Ref. 33 (© 1992 ACS).

study of the change of free-volume distributions corresponding to the miscibility of polymer blends, we investigated two different types of polymer blends: a miscible blend of bisphenol-A polycarbonate (PC) and tetramethyl bisphenol-A polycarbonate (TMPC), and an immiscible blend of PC and polystyrene (PS) [18].

As shown in Fig. 6, the differences in free-volume distributions can be seen for the two types of blends. A larger difference between the blends and the pure polymers in the distribution of free-volume holes is observed in the immiscible blend (PS/PC) than in the miscible blend (TMPC/PC). In the immiscible blend, the distribution is broader than in the pure polymers due to the free volumes formed in the interfacial regions. Further systematic investigation on the free-volume effect using the PAL method on blends is in progress at our laboratory.

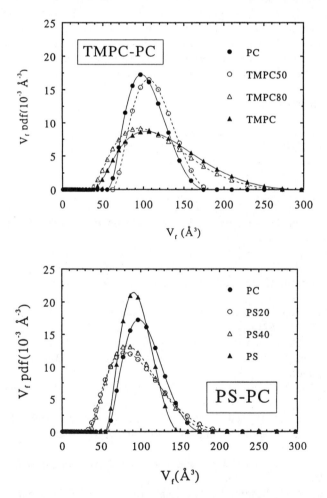

Fig. 6.   Free-volume hole volume distributions in the TMPC/PC (miscible) and PS/PC (immiscible) polymer blends [18]. Reprinted by permission of Ref. 18 (© 1994 Materials Research Society).

Conclusion

We have presented the PAS technique as a new microanalytical probe for the characterization of free-volume properties in polymers and polymer blends. The unique behavior of Ps localization in free-volume holes of polymers enables us to use PAS to determine the hole size, hole size distribution function, and concentrations of free volumes at atomic scales in polymeric materials. PAS is a novel probe with applications not only to polymers but also to other technologically important materials, such as pores in catalytic materials and surface states of solids. Four major developments in PAS will be very beneficial in the future: (1) improvement of PAL data analysis into a lifetime vs. amplitude spectrum as in conventional spectroscopy, so that a hole size distribution function can be obtained more accurately; (2) development of two-dimensional ACAR spectroscopy for polymeric applications, so that a detailed three-dimensional hole structure may be mapped out; (3) development of a monoenergetic slow positron beam, so that a direct application to thin film polymers can be made; and (4) development of a universal correlation equation including chemical quenchings.

Acknowledgement

This research was supported by a grant from the National Science Foundation (DMR-90040803).

Literature Cited

1.   Doolittle, A. K. *J. Appl. Phys.*, **1951**, *22*, 1471.
2.   For example, see Ferry, J. D. **Viscoelastic Properties of Polymers**, 3rd. ed. Wiley, N.Y. 1980.
3.   Cohen, M. H.; and Trunbull, D. *J. Chem. Phys.* **1959**, *31*, 1164.
4.   Takenchi, H.; and Roe, R. J. *J. Chem. Phys.* **1991**, *94*, 7446.
5.   Robertson, R. E.; Simha, R.; and Curro, J. G. *Macromolecules* **1988**, *18*, 2239.
6.   Flouda, G.; Pakula, T.; Stamm, M.; and Fisher, E. W. *Macromolecules* **1993**, *26*, 1671.
7.   Song, H. H.; and Roe, R. J. *Macromolecules* **1987**, *20*, 2723.
8.   Nojima, S.; Roe, R. J.; Rigby, D.; and Han, C. C. *Macromolecues* **1990**, *23*, 4305.
9.   Hooker, J. C.; Royal, J. S.; and Torkelson, J. M. *Polym. Prepr.* **1993**, *34*, 498.
10.   Winudel, M. B.; and Torkelson, J. M. *Polym. Prepr.* **1993**, *34*, 500.
11.   Royal, J. S.; Victor, J. G.; and Torkelson, J. M. *Macromolelues* **1992**, *25*, 729.
12.   Royal, J. S.; Victor, J. G.; and Torkelson, J. M. *Macromolelues* **1992**, *25*, 4792.
13.   Jean, Y. C.; Sandreczki, T. C.; and Ames, D. P. *J. Poly. Sci. B.* **1986**, *24*, 1247.
14.   Ferrell, R. A. *Phys. Rev.* **1957**, *108*, 167.

15. Nakanishi, H.; Jean, Y. C.; Smith, E. G.; and Sandreczki, T. C. *J. Poly. Sci. B.* **1989**, *27*, 1419.

16. Wang, Y. Y.; Nakanishi, H.; Jean, Y. C.; and Sandreczki, T. C. *J. Poly. Sci. B.* **1990**, *28*, 1431.

17. Jean, Y. C.; Zandiehnadem, F.; and Deng, Q. in **Proc. of MRS Symp. Structure, Relaxation, and Physical Aging of Glassy Polymers**, vol. 215, pp. 163-174 (R.J. Roe and J.M. O'Reilly, Eds.) MRS Pub., Pittsburgh, PA (1991).

18. Liu, J.; Jean, Y. C.; and Yang, H. in **Proc. of MRS Symp. Crystallization and Related Phenomena in Amorphous Materials**, vol. 321 (Libera, M.; Cebe, P.; Dickinson Jr., J. E.; Eds.) MRS Pub., Pittsburgh, PA (1994), p. 47.

19. A PATFIT package was purchased from Risø National laboratory, Roskilde, Denmark.

20. Tao, S. J. *J. Chem Phys.* **1972**, *56*, 5499.

21. Eldrup, M.; Lightbody, D.; and Sherwood, J. N. *Chem Phys.* **1981**, *63*, 51.

22. Nakanishi, H.; and Jean, Y. C. in **Positron and Positronium Chemistry**, (D.M. Schrader and Y. C. Jean, Eds.) Elsevier Pub. Amsterdam (1988), Chapter 5.

23. Sandreczki, T. C.; Nakanishi, H.; and Jean, Y. C. in **Proc. of Int. Symp. in Positron Annihilation Studies of Fluids**, (S. C. Sharma, Ed.) World Sci. Pub., Singapore (1988) p. 200.

24. Kobayashi, Y.; Zeng, W.; Meyer, E. F.; McGervey, J. D.; Jamieson, A. M.; and Simha, R. *Macromolecules*, **1989**, *22*, 2302.

25. Jean, Y. C.; Shi, H.; Dai, G. H.; Huang, C. M.; and Liu, J. in **Proc. of 10th Conf. on Positron Annihilation**, (He, Y. J.; Cao, B. S.; Jean, Y. C. Eds) Trans Tech pub., Beijing, China (1994) p. 691.

26. Jean, Y. C.; and Shi, H. *J. Non-Crys. Solid.*, in press (1994).

27. Tao, S. J. *IEEE Trans. Nucl. Sci.*, **1968**, 15, 175; Schrader, D. M. in **Positron Annihilation**, pp. 912-914 (P. G. Coleman, S. C. Sharma and L. M. Diana, Eds.) North-Holland, Amsterdam (1982).

28. Provencher, S. W. CONTIN Users Manual, EMBL Technical Report DA05, European Molecular Biology Laboratory, Heidelberg, Germany

29. Gregory, R.B.; and Zhu, Y. *Nucl. Instr. Meth. Phys.*, **1990**, *A290*, 172.

30. Jean, Y. C.; and Deng, Q. *J. Polym. Sci. B, Polym. Phys.* **1992**, *29*, 1359.

31. Liu, J.; Deng, Q.; and Jean, Y. C. *Macromolecules* **1993**, *26*, 7149.

32. Deng, Q.; and Jean, Y. C. *Macromolecules* **1993**, *26*, 30.

33. For example, see Deng, Q., in **Ph.D. Dissertation: Characterization of Free-Volume Properties in Polymers by Positron Annihilation Spectroscopy**, University of Missouri-Kansas City, Kansas City, Missouri, 1993.

34. Paul, D. R.; and Sperling, L. H.; Eds. **Multicomponent Polymeric Materials**; American Chemical Society: Washington, D. C. 1986.

RECEIVED February 2, 1995

Chapter 28

Monitoring Degree of Cure and Coating Thickness of Photocurable Resins Using Fluorescence Probe Techniques

J. C. Song and D. C. Neckers[1]

Center for Photochemical Sciences, Bowling Green State University, Bowling Green, OH 43403

In this paper, we report recent progress in developing fluorescence probe technology for photocurable coatings. Intramolecular charge transfer (ICT) fluorescence probes based on derivatives of dansyl amide were found to be useful for monitoring cure of various photocurable resins the polymerization of which was initiated with UV or visible initiators. Fluorescence emission spectra of chosen ICT probes exhibited spectral blue shifts due to the increases in microviscosity surrounding the probe molecules as the photocuring process proceeded. A fluorescence intensity ratio method was used to follow the probe spectral changes as a function of the degree of polymerization. Linear correlation plots between the intensity ratios as measured from the probe fluorescences and the degree of cure were obtained for varieties of photocurable resins with UV or visible initiators. In addition, a fluorescence probe technique for measuring coating thickness based on the nascent fluorescence from coating substrate is also described.

Photopolymerization, a process of converting liquid monomers or oligomers into solid polymers using UV or Visible light, has found increasing applications especially in the coating industries because of its speed, efficiency and solvent-free nature. The ultimate chemical and physical properties of the photopolymer coatings formed by photopolymerization processes depend on various factors, such as monomer structure, forms of initiation, reaction conditions and the rate and degree of polymerization. However, the industrial processing cycles of these coating resins have generally been developed empirically, by "trial and error", because of the lack of proper techniques to monitor the changes in chemical and physical properties as they occur during the processing. Resins with significant "batch to batch" variations in composition are processed using the same general curing cycle, resulting in products with significantly differing properties. Thus off-line as well as on-line cure monitoring techniques are essential for property control and processing optimization of the final products produced by photopolymerization. Several methods have been previously developed to study the kinetics of photopolymerization. Among these are differential scanning calorimetry (DSC) and photo-differential scanning calorimetry (PDSC) (*1-4*), laser interferometry, (*5*)

---

[1]Corresponding author

photoacoustic spectroscopy, (*6*) and UV and FTIR spectroscopy. (*7-9*) DSC and PDSC are useful for studies of the kinetics of photopolymerization. The degree of cure of the resin systems can be determined from the residual heat of reaction of the unreacted functionalities. However, both DSC and PDSC are destructive techniques and therefore are not suitable for on-line cure monitoring applications. FTIR is the most utilized spectroscopic method for degree of cure measurement by monitoring changes of the characteristic monomer or polymer absorption bands. Recently, a real time infra-red (RTIR) technique was developed to monitor the polymerization process continuously and rapidly in real time. (*7,8,10*) Both the rate and degree of polymerization can be measured. However, IR techniques can only be used to analyze thin films (less than 20 mm thick). They are not useful for monitoring polymerization of coatings on opaque substrates.

Fluorescence spectroscopy has been recognized as a powerful analytical technique for polymer analysis because of its sensitivity, selectivity and non-destructive characteristics. Remote and on-line sensing can be readily achieved via a fiber-optic fluorimeter with today's fiber optic technology. The performance of a fluorescence technique depends to a large extent on the properties of the fluorescence probes employed. Generally, there are two types of fluorescence probes according to their reactivities: reactive probes and non-reactive probes. Reactive fluorescence probes can be aromatic monomers or curing agents that can take part in a polymerization process. The fluorescence characteristics of a reactive fluorescence probe changes as it reacts with other components in the polymerizing system. For instance, the fluorescent excitation spectra of aromatic diamine curing agents such as 4,4'-diaminodiphenyl sulfone (DDS) and 4,4'-diaminodiphenylmethane (DDM) were found to exhibit significant spectral red shifts during their curing reactions with epoxides, as a consequence of the conversion of the primary amines to the secondary and tertiary amines. (*11*) The fluorescence excitation spectral peak positions of these aromatic diamine curing agents were found to correlate well with the degree of cure. Reactive probes can also be extrinsic probes added to the polymerization systems. (*12*) Non-reactive fluorescence probes are useful for polymer cure characterization because of their viscosity-dependent characteristics which arises from the dependence of the internal rotations of the probe molecules to the local viscosity or microviscosity. (*13*) The photophysical properties of the probes can be correlated to the degree of polymerization which affects the microviscosity surrounding the probe molecules. Several cure monitoring methods have been developed including excimer fluorescence, (*14*) fluorescence polarization, (*14,15*) fluorescence quenching, (*16*) time-resolved fluorescence (*17*) and fluorescence recovery after photobleaching. (*18*)

Fluorescence Probes for Photopolymerization Studies

During the last few years, our laboratories has devoted considerable efforts to develop fluorescence probe techniques for monitoring photoinitiated polymerization. Our approach has been to use viscosity-sensitive probes such as excimer forming molecules and intramolecular charge transfer complexes in which the fluorescence quantum yield and/or the fluorescence emission peak maximum are related to an intramolecular rotation-dependent relaxation process.

Excimer forming molecules such as pyrene and its derivatives exhibit dual fluorescence emissions in which the shorter wavelength band corresponds to the monomeric emission and the longer wavelength band to that of the excimer. The fluorescence intensity ratio (I_m/I_e) of the monomer emission and the excimer emission is sensitive to the changes in the microviscosity surrounding the probe and thus can be correlated with the degree of polymerization. However, a major disadvantage in using excimer probes is that large concentrations of the probes are required to form excimers, and this can limit the possible depth of observation and interfere with the photoinitiating process. (*19*)

Certain aromatic compounds with donor-acceptor structures such as 4-(N,N-dimethylamino)benzonitrile (DMABN) and 4-(N,N-dimethylamino) benzoate exhibit dual fluorescence emission bands corresponding to two different excited state conformations due to internal twisting. (*20*) The short wavelength band is due to a coplanar (parallel) excited state conformation and the long wavelength band originates from a perpendicular conformation which is the so-called twisted intramolecular charge transfer (TICT) state. The perpendicular conformation usually exhibits an energy minimum at the excited singlet state due to a larger degree of charge separation. In contrast, the ground state exhibits an energy maximum at the perpendicular conformation. Thus, upon excitation, a molecule with a coplanar ground-state conformation will spontaneously twist toward the more stable perpendicular excited-state conformation. (*21*) The rate of twisting and the population of the TICT state is expected to be strongly dependent on the microenvironment of the probe molecule. Our interests were in using TICT molecules as fluorescence probes for photopolymerization studies. A number of intramolecular charge transfer probes have been evaluated for probing the degree of polymerization of polyolacrylates photopolymerized using visible initiators. (*21-25*) Among them are 5-dimethylamino-1-naphthalene sulfonamide (dansylamide, DA), 2-dimethylamino-5-naphthalene-n-butyl-sulfonamide (2,5-DASB), N-(4'-cyano-phenyl)-carbazole (CBC), N-(4'-butyl benzoate)-carbozole (BBC), N-(1'-naphthyl)-carbazole (NNC) and 9,9'-dianthryl (see Chart 1). Dansyl amide and 2,5-DASB were successfully employed as fluorescence probes for visible light initiated polymerization. Among other probes, N-(1'-naphthyl)-carbazole was found to be the most sensitive while 9,9'-dianthryl was the least sensitive probe. (*25*)

To search for intramolecular charge transfer probes with large Stokes shifts and high fluorescence quantum yields, we investigated several probe compounds which were 1,5-derivatives of dansyl amide. The 1,5-isomers of dansyl amide were found to exhibit larger Stokes shift than the 2,5-isomers. For example, the Stokes shift value of 5-dimethylaminonaphthalene-1-sulphonyl-n-butylamide (152 nm) is twice of that of 5-dimethylaminonaphthalene-2-sulphonyl-n-butylamide (76 nm) when measured in THF at room temperature. (*26*) Thus we focused on the 1,5-isomers of dansyl amide derivatives. Among them are 5-dimethylaminonaphthalene-1-sulphonyl-n-butylamide (DASB), 5-dimethylaminonaphthalene-1-sulphonyl-di-n-butylamide(DASD),5-dimethylaminonaphtha-lene -1-sulphonyl-aziridine (DASA) and 5-dimethylaminonaphthalene-1-sulphonyl-pyrolidone (DASP) (see Chart 2). All four of the probes exhibited large values of Stokes shift of about 150-160 nm (in THF solution), as shown in Figure 1. Intramolecular charge transfer fluorescence probes with large Stokes shift values are advantageous to use for cure monitoring applications since they can provide wider spectrum windows in which to operate as well as higher sensitivities. DASB and DASD were found to be more fluorescent that DASA and DASP. (*27*) In this paper, we will focus on studies of the applicability of DASB probe for cure monitoring applications. Evaluation of other probes is now in progress and will be reported soon.

Cure Monitoring Device

Since it usually takes time to acquire a fluorescence spectrum using a scanning emission monochromator and PMT detecting system, a conventional fluorimeter is not suitable for monitoring spectral changes of ICT fluorescence probes during a rapid photopolymerization process. Conventional fluorimeters are comprised of, mainly, a light source, an excitation monochromator, an emission monochromator,

**Chart 1. Fluorescence Probes Used Previously For Photopolymerization
Studies**

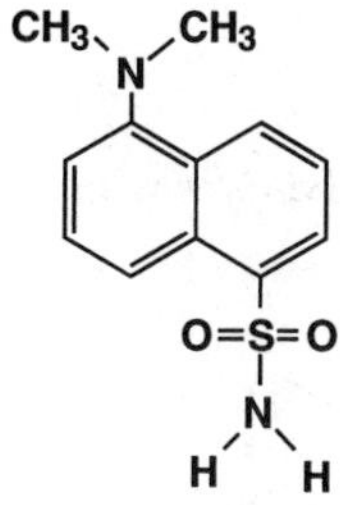

(Dansyl amide)

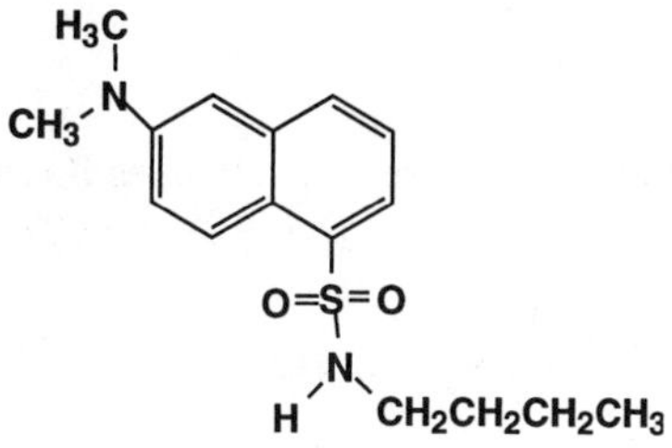

(2,5-DASB)

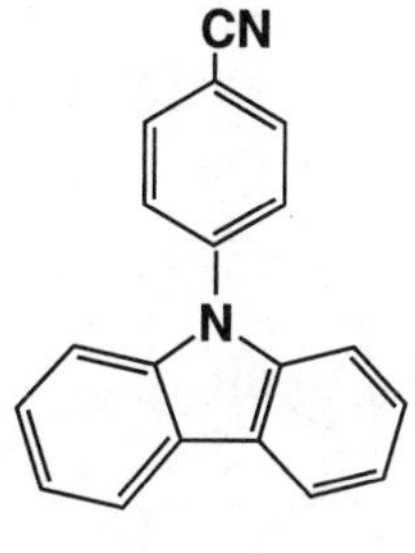

(CBC)

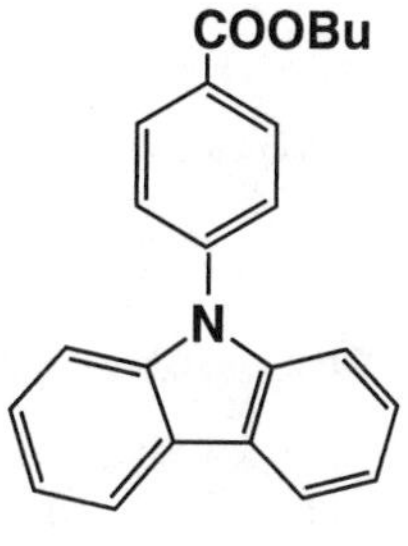

(BBC)

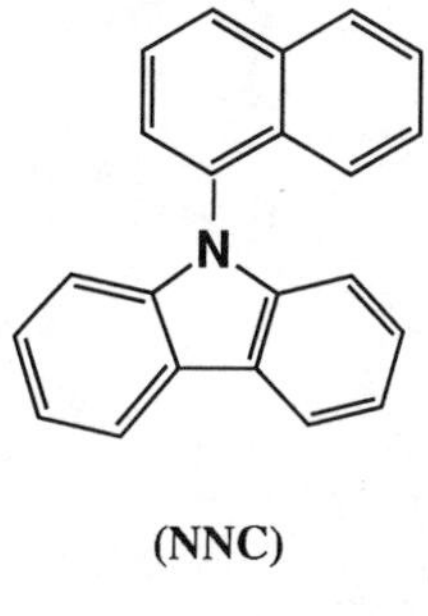

(NNC)

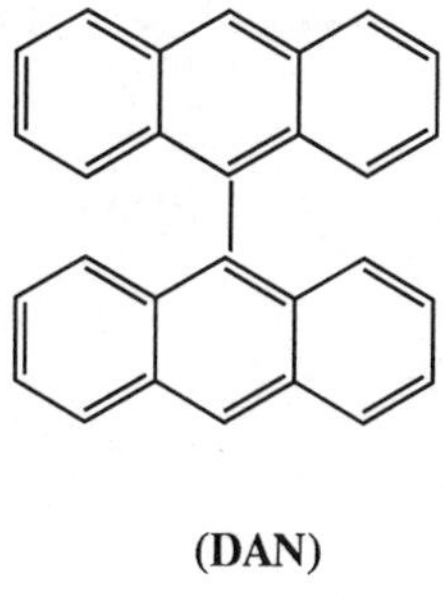

(DAN)

Chart 2. Fluorescence Probes Based on Dansyl Amide Derivatives with Large Values of Stokes Shifts

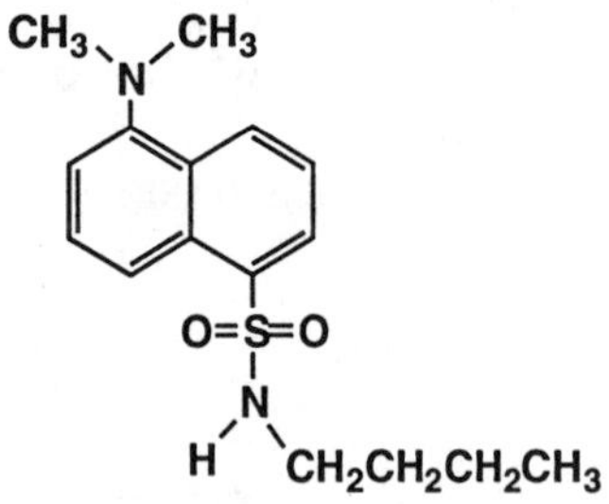

(DASB)

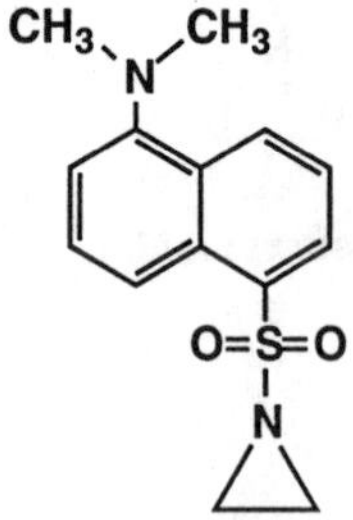

(DASA)

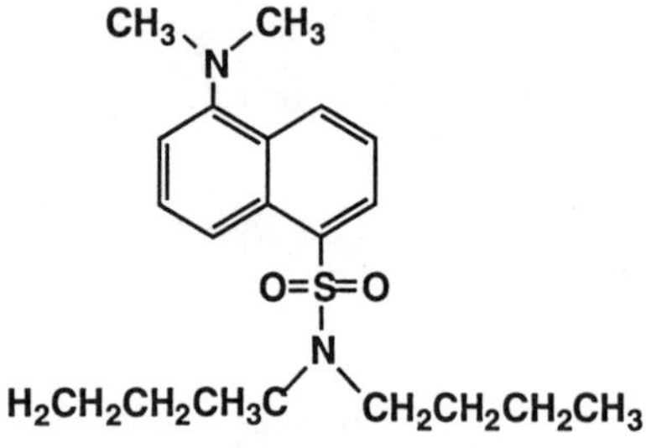

(DASD)

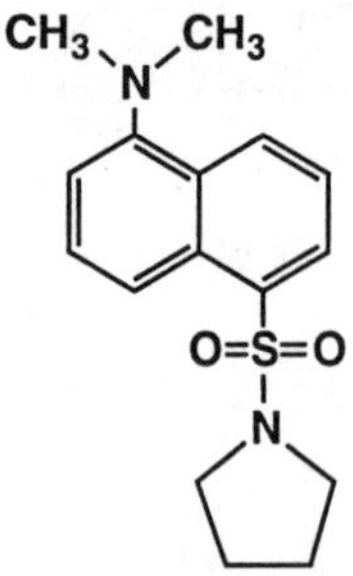

(DASP)

a photomultiplier tube (PMT), a data station and a recorder. Remote sensing can be achieved by using a bifurcated fiber-optic cable. One way to shorten the spectral acquisition time is to use multiple detection heads at various emission wavelengths. For example, by using a pair of photomultiplier tubes with narrow-band interference filters and trifurcated fiber optic bundles, Neckers et al. developed a real time cure monitoring device for laser-induced photopolymerization. (*22*) The fluorescence intensity ratio of two selected emission wavelengths can be monitored continuously, in real time, as the polymerization occurs. However, this system is not capable of recording a full spectrum rapidly. To overcome this difficulty, a fast detection system featuring a multichannel analyzer and a CCD detector was developed by SGL/Oriel Instruments for radiation cure monitoring applications. With this system, the spectral acquisition time can be as short as 25 ms, so a full emission spectrum can be obtained instantaneously. Using a bifurcated optic-fiber sampling technique, this device can be readily used for both on-line and off-line cure monitoring needs.

Fluorescence Intensity Ratio Method for Photopolymerization Studies

When using intramolecular charge transfer fluorescence probes to monitor photopolymerization, the probe molecules are doped into the uncured resin systems at small concentrations (0.01-0.02 wt%). As the curing reaction proceeds, the probe fluorescence emission spectral peak position generally exhibits spectral blue shifts due to the increasing microviscosity surrounding the probe. For example, a total spectral shift of about 26 nm was observed from 5-dimethylamino-1-naphthalene-sulphonyl-n-butylamide (DASB) in a acrylic coating formulation when the degree of cure increases from 0% to 87%. (Figure 2). To quantitatively monitor the spectral changes as a function of the degree of cure, we used a fluorescence intensity ratio method in which fluorescence intensities at two different emission wavelengths were ratioed. Thus, any variations in probe concentration, sample thickness and lamp intensity are internally calibrated. We found that the intensity ratio method was more sensitive and accurate than monitoring the emission peak position since fluorescence emission spectra are generally broad and structureless.

To utilize the fluorescence probe information acquired, one needs to establish a correlation between the fluorescence response and polymer properties. One example, the measure of the degree of cure which has been determined independently using FTIR or DSC, has been particularly investigated. (*28*) For acrylic resins, we used FTIR to monitor the rate of disappearance of the acrylic C=C functionality. The degree of cure (a) was calculated using the following equation.

$$\alpha = 1 - \frac{A_{acry}(t) * A_{ref}(0)}{A_{acry}(0) * A_{ref}(t)} \tag{1}$$

where $A_{acry}(t)$ and $A_{acry}(0)$ are the absorbances at 810 cm^{-1} due to the acrylate double bond, after curing times t and zero, respectively. $A_{ref}(t)$ and $A_{ref}(0)$ are the reference peak absorbances at 2945 cm^{-1} due to the CH groups, after curing times t and zero, respectively. This reference peak was used as an internal standard to calibrate any thickness fluctuation during the curing process.

For unsaturated polyester resins, we found it was difficult to use FTIR to monitor the curing processes due to peak overlapping in the infrared. Thus we used differential scanning calorimetry (DSC) instead. The degree of cure (a) was calculated using the following equation.

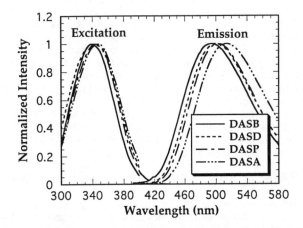

Figure 1. Fluorescence excitation and emission spectra of several dansyl amide derivatives, 5-dimethylaminonaphthalene-1-sulphonyl-n-butylamide (DASB), 5-dimethylaminonaphthalene-1-sulphonyl-di-n-butylamide (DASD),   5-dimethylaminonaphthalene-1-sulphonyl-aziridine (DASA) and 5-dimethylaminonaphthalene-1-sulphonyl-pyrolidone (DASP). (Solvent: THF).

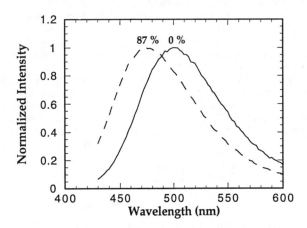

Figure 2. Comparison of the fluorescence emission spectra of the DASB probe in an UV curable acrylic coating formulation (Durethane low gloss coating,from the PPG Industries) as a function of the degree of cure. (Excitation at 380 nm).

$$a = 1 - DH_{residue}/DH_{total} \qquad (2)$$

where $DH_{residue}$ is the residual heat of reaction measured for the sample after curing for a period of time. DH_{total} is the total heat of reaction obtained when benzoyl peroxide (1.0 wt%) was used as the thermal initiator.

Figure 3 compares the fluorescence and FTIR cure monitoring profiles of an acrylic coating formulation (Durethane low gloss coating from the PPG Industries) as a function of irradiation time using a medium pressure mercury lamp. It can be seen the two cure profiles are about the same and a linear correlation curve between the fluorescence intensity ratios and the degree of cure was obtained as shown in Figure 4.

We have examined the applicability of DASB probe to a number of commercial UV coating formulations including urethane acrylates, epoxy acrylates, silicone acrylates and polyester acrylates. (26-28) In all cases, linear correlation curves between the fluorescence intensity ratios and the degree of cure were obtained. The slopes of the correlation curves were found to vary from 2.3 to 4.8, while the intercepts ranged from 1.2 to 1.6 (see Table I). These differences probably reflected the differences in network structures of these resins.

DASB has also been successfully used to monitor photopolymerization processes initialized by visible light initiators such as fluorone dye/amine or fluorone dye/onium salt/amine systems developed in our laboratories. (29) A photocurable multifunctional acrylate resin system made of trimethylopropane triacrylate (40 wt%), dipentaerythrytol monohydroxyl pentaacrylate (40 wt%) and polyethylene glycol diacrylate (20 wt%) was employed as the standard resin. The visible photoinitiator was 5,7-diiodo-3-butoxy-6-fluorone (5X10^{-4} M) with N-phenyl glycine (5X10^{-2} M) as the coinitiator. The photoinitiating mechanism (Scheme 1) is believed to be involving an electron transfer process in which the amine coinitiator donates an electron to the triplet excited state of the dye molecule. A subsequent proton transfer process will then generate an amine radical which initializes radical polymerization and a dye radical which is further reduced by hydrogen abstraction or termination. (30) So the dye is bleached to leuco form. Before the bleaching process was completed, the fluorone dye, DIBF, was found to exhibit weak fluorescence emissions which could interfere with the measurement of the probe fluorescence signals. However, because of the probe's high fluorescence quantum yield, no significant interference was noticed. In fact, the cure profiles observed by fluorescence probe and FTIR techniques were very similar as shown in Figure 5. The probe fluorescence intensity ratio (I470/I560) of DASB was correlated linearly with the extent of C=C conversion determined by FTIR as shown in Figure 6.

DASB was also used to monitor visible light curing of commercial resins such as polyester acrylate and unsaturated polyesters. For photoinitiation, a "hybrid" three-component initiating system consisting of a fluorone dye initiator, 5,7-diiodo-3-butoxy-6-fluorone (DIBF), an onium salt, 4-octyloxyphenyliodonium hexafluoro antimonate (OPPI) and N,N'-dimethylamino-2,6-diisopropylaniline (DIDMA) was used with a molar ratio of 1/2/3. Even though the presence of onium salt was found to have some quenching effect on the fluorescence emission intensity of the DASB probe, linear correlation between the fluorescence intensity ratio and the degree of cure was obtained. (28) This showed one of the advantages of using the fluorescence intensity ratio method which is independence of the probe concentration.

Figure 7 shows an example of linear correlation curve for one of the unsaturated polyester resins (from Owens-Corning Fiberglass) cured with the three-component "hybrid" visible initiator (DIBF/OPPI/DIDMA). The degree of cure in this case

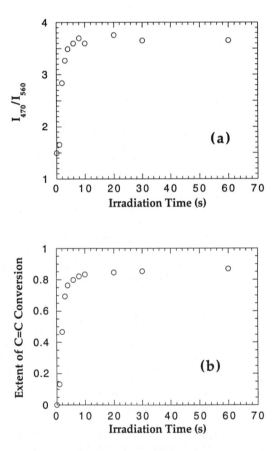

Figure 3. Comparisons of fluorescence (a) and FTIR (b) cure monitoring profiles as a function of irradiation time using a medium pressure mercury lamp for an UV curable acrylic coating formulation (Durethane low gloss coating, from the PPG Industries).

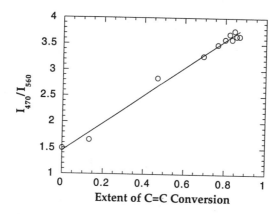

Figure 4. Linear correlation curve between the fluorescence intensity ratio (I_{470}/I_{560}) of the DASB probe and the extent of C=C conversion for an UV curable acrylic coating formulation (Durethane low gloss coating, from the PPG Industries).

Table I. Slopes and intercepts of the correlation curves between the probe fluorescence intensity ratio and the degree of cure for different UV coatings using DASB as the fluorescence probe

Resin System	Intercept	Slope
Acrylic Coating for Fibers (Primary)[a]	1.380	2.631
Acrylic Coating for Fibers (Secondary)[b]	1.628	2.320
Acrylic Coating for Vinyls (Alpha Gloss)[c]	1.447	2.654
Acrylic Coating for Vinyls (Low Gloss)[d]	1.241	2.714
Silicone Acrylate[e]	1.338	4.817

Notes: (a). An UV curable acrylic formulation (Desotech 950-076) from DSM Desotech Inc., Elgin, Illinois.
(b). An UV curable acrylic formulation (Desolite 950-044) from DSM Desotech Inc., Elgin, Illinois.
(c). An UV curable acrylic formulation (Durethane 602Z70) from PPG Industries, Inc., Allison Park, PA.
(d). An UV curable acrylic formulation (Durethane 509Z70) from PPG Industries, Inc., Allison Park, PA.
(e). An UV curable silicone acrylic formulation (Trade Code: UV 8550-D1) from GE Silicones, Waterford, NY.

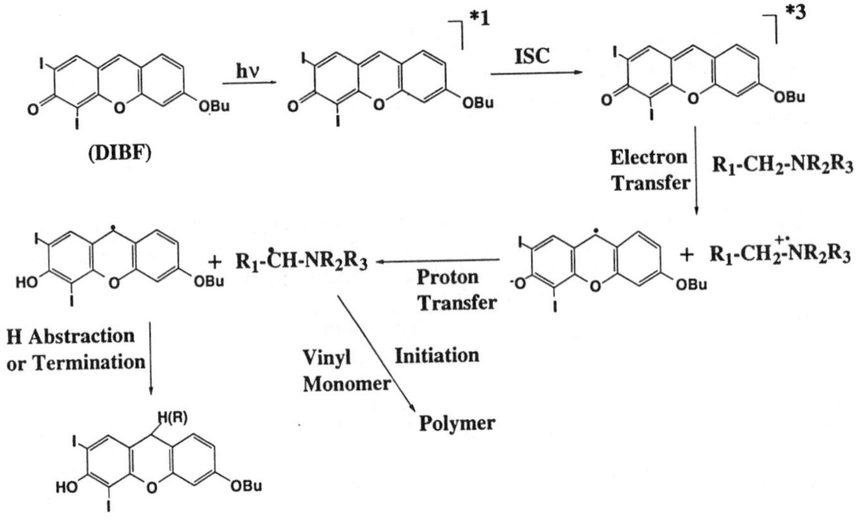

Scheme 1.   Initiating Mechanism of a Fluorone Dye (DIBF)/Amine Radical
Initiating System (from Ref. 32)

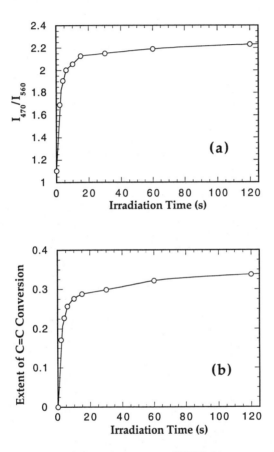

Figure 5. Comparisons of fluorescence (a) and FTIR (b) cure monitoring profiles as a function of irradiation time for a multifunctional acrylate formulation containing TMPTA (40 wt%), DPHPA (40 wt%) and PEGA (20 wt %) with a visible initiating system. ([DIBF] = 5×10^{-4} M; [NPG]= 5×10^{-2} M; [DASB] = 0.015 wt %; Excitation at 380 nm).

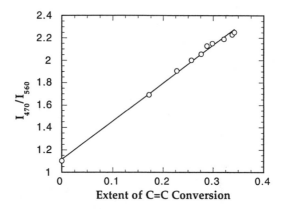

Figure 6. Linear correlation plot between the fluorescence intensity ratio
(I_{470}/I_{560}) of the DASB probe and the extent of C=C conversion for a
multifunctional acrylate formulation containing TMPTA (40 wt%), DPHPA (40
wt%) and PEGA (20 wt %) with a visible initiating system. ([DIBF] = $5X10^{-4}$
M; [NPG]= $5X10^{-2}$ M; [DASB] = 0.015 wt %; Excitation at 380 nm).

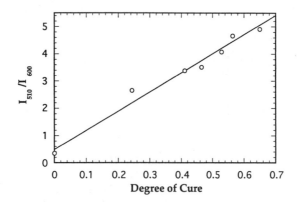

Figure 7. Linear correlation plot between the fluorescence intensity ratio
(I_{510}/I_{600}) of the DASB probe and the extent of C=C conversion for an
unsaturated polyester formulation (E711 resin, from Owens Corning
Fiberglass) with a visible initiator system consisting of 5,7-diiodo-3-6-fluorone
(DIBF), 4-octyloxyphenyliodonium hexafluoroantimonate (OPPI) and N,N'-
dimethylamino-2,6-diisopropylaniline (DIDMA) with a molar ratio of
DIBF/OPPI/DIDMA = 1/2/3. The concentration of DIBF is 0.01 wt%.

[10] Kutal, C.; Grutsch, P.A.; Yang, D. B. *Macromolecules*, **1991**, *24*, 6872.

[11] Song, J.C.; Sung, C.S.P. *Macromolecules,* **1993**, *26*, 4818.

[12] (a) Sung, C.S.P.; Pyun, E.; Sun, H.L. *Macromolecules* **1986**, *19*, 2922. (b) Sung, C.S.P. in *Photophysics of Polymers, ACS Symposium Series 358* (Hoyle, C. E.; Torkelson, J. M. eds.) Am. Chem. Soc., Washington, D.C., 1987, P404. (c) Sung, C.S.P. ; Mathisen, R. *Polymer* **1987**, *28*, 941. (d) Yu, W.C.; Sung, C.S.P. *Macromolecules* **1990**, *23*, 386. (e) Pyun, E.; Sung, C.S.P. *Macromolecules* **1991**, *24*, 855.

[13] (a) Loufty, R.O.; Arnold, B.A. *J. Phys. Chem.* **1982**, *86*, 4205. (b) Loufty, R.O. *J. Polym. Sci., Polym. Phys.* **1982**, *20*, 825.

[14] Wang, F.W.; Lowry, R.E.; Fanconi, B.M. *Polymer* **1986**, *27*, 1529. (b) Scalata, S.F. Ors, J. A. *Polymer Communications* **1986**, *27*, 41.

[15] Noel, C.; Laupetre, F.; Fredrich Leonard, C.; Halary, J.L.; Monnerie, L. *Macromolecules* **1986**, *19*, 202.

[16] Dousa, P.; Konak, C.; Fidler, V.; Dusek, D. *Polymer Bulletin* **1989**, *22*, 585.

[17] Strehmel, B.; Stremel V.; Timpe, H. J.; Urban, K. *Eur. Polym. J.* **1992**, *28*, 525.

[18] Wang, F. W.; Wu, E.-S. *Polymer Communications* **1987**, *28*, 73.

[19] Paczkowski, J.; Neckers, D.C.*Chemtracts-Macromolecular Chem .* **1992**, *3*, 75.

[20] (a) Lipinski, J.; Chojnacki, H.; Rotkiewicz, K.; Grabowski, Z.R. *Chem. Phys. Lett.* **1980**, *70*, 449.

[21] Lippert, E.; Rettig, W.; Bonacic-Koutecky, V.; Heisel, F.; Miehe, J.A. *Advances in Chemical Physics*, John Wiley: New York, **1987**.

[22] Paczkowski, J.; Neckers, D. C. *Macromolecules* **1992**, *25*, 548.

[23] Zhang, X.; Kotchetov, I, N.; Paczkowski, J.; Neckers, D. C. *J. Imaging Sci. Tech.* **1992**, *36*, 322.

[24] Kotchetov, I. N.; Neckers, D. C *J. Imaging Sci. Tech.* **1993**, *37*, 156.

[25] Paczkowski, J.; Neckers, D. C. *J. Polym. Sci., Polym. Chem. Ed.* **1993**, *31*, 841.

[26] (a) Song, J.C.; Neckers, D.C. *Radtech Conference Proceedings*, **1994**, *1*, 338. (b) Song, J.C.; Neckers, D.C. *Polym. Mater. Sci. Eng.* **1994**, *2*. 71.

[27] Song, J.C.; Wang, Z.J.; Bao, R.; Neckers, D.C. To be published.

[28] Song, J.C.; Neckers, D.C. *Macromolecules* **1994**, (submitted).

[29] (a) Valdes-Aguilera, O.; Pathak, C.P.; Shi, J.; Watson, D.; Neckers, D.C. *Macromolecules* **1992**, *25*, 541. (b) Torres-Filho, A.; Neckers, D.C. *J. Appl. Polym. Sci.* **1994**, *51*, 931. (c) Torres-Filho, A.; Neckers, D.C. *RadTech Conference Proceedings*, **1994**, *1*, 259. (d) Bi, Y.; Neckers, D.C. *Macromolecules*, **1994**, *27*, 3683.
(e) Marino, T, Martin, D. and Neckers, D.C. *RadTech Conference Proceedings*, **1994**, *1*, 169.

[30] Bi, Y. *Ph.D. Thesis*, Bowling Green State University, **1993**.

RECEIVED February 2, 1995

MULTIDISCIPLINARY APPROACHES

Chapter 29

Spectroscopic Characterization of Unimer Micelles of Hydrophobically Modified Polysulfonates

Yotaro Morishima

Department of Macromolecular Science, Faculty of Science, Osaka University, Toyonaka, Osaka 560, Japan

Polysulfonates modified with bulky hydrophobic groups of cyclic structure form unimolecular micelles (unimers) in aqueous solution owing to intramolecular self-organization. This chapter deals with the characterization of the unimers of random copolymers of sodium 2-(acrylamido)-2-methylpropanesulfonate and hydrophobic methacrylamides carrying cyclododecyl, 1-adamantyl, or 1-naphthylmethyl substituent group, by light scattering, small-angle X-ray scattering, fluorescence, NMR, and IR techniques. The absence of intermolecular association is indicated by the absence of non-radiative energy transfer from naphthalene to pyrene which are labeled respectively on the individual polymer molecules. Light scattering studies show that the unimers have extremely large ratios of mass to dimension indicating an extremely compact conformation. Local motions of the hydrophobes in the unimers are highly restricted as indicated by ^{1}H-NMR relaxation times. FTIR indicates that hydrogen bonding between the spacer amide bonds contributes to the stabilization of the compact unimer structure.

Amphiphilic polymers, a class of water-soluble polymers consisting of both hydrophobic and hydrophilic segments in a polymer chain, have a tendency to self-organize in aqueous solution. Water molecules surrounding the hydrophobes in the amphiphilic polymers in aqueous solution form an ice-like structure (1) because the water molecules tend to form the largest possible number of hydrogen bonds on the surface of the hydrophobes. The hydrophobes would associate such that they have the smallest possible area of contact with the aqueous phase by releasing the structured water molecules into free water. The principal driving force for this hydrophobic association is a large increase in entropy which is sufficient to overcome a positive enthalpy change due to the breakup of the hydrogen bonds (2,3). The hydrophobic association is the main cause for the self-organization of amphiphilic polymers in aqueous solution.

In amphiphilic polyelectrolytes, the hydrophobic interaction competes with electrostatic repulsion. Therefore, the balance of the contents of the hydrophobes and charged segments in the polymers are a critical factor to determine whether or not the self-organization occurs. For example, amphiphilic polycarboxylic acids, which adopt an extended chain conformation at high pH, would collapse to form compact conformation upon decreasing pH (4-9). This transition from an extended to a compact

0097–6156/95/0598–0490$13.75/0

structure, a typical of cooperative processes that can be viewed as a two-state transition, takes place sharply within a narrow range of pH. On the other hand, amphiphilic polysulfonic acids, strong polyelectrolytes, show no such pH dependence of their conformation because they are fully ionized even at very low pH (*10,11*).

The compact conformation of amphiphilic polyelectrolytes shows an analogy to surfactant micelles. In 1951, Strauss and Jackson (*12*) first synthesized a "polysoap" by quaternization of poly(2-vinylpyridine) with *n*-dodecyl bromide. They demonstrated that the polymer adopted a highly compact conformation which had an ability to solubilize hydrophobic small molecules in dilute aqueous solution (*12*). Later, Strauss *et al.* (*13,14*) synthesized a series of alternating copolymers of maleic acid and alkyl vinyl ethers, and showed that the copolymer with *n*-octyl vinyl ether adopted a hypercoiled compact structure even at high degrees of ionization, while the copolymers with *n*-butyl and *n*-hexyl vinyl ethers existed in compact conformations only at low degrees of ionization.

In spite of the pioneering studies of Strauss and Jackson (*12*) over forty years ago, relatively little work on amphiphilic polyelectrolytes seems to have followed. In fact, it was not many years ago when the volume of published work on this subject rapidly increased. Recent interest in the amphiphilic polymers stems not only from their biological relevance but also from the potential for utilizing them in applications such as coatings, paints, drugs, and personal care goods. The class of amphiphilic polymers is broad and has a variety of interesting features worthy of study from a scientific as well as a technological point of view. Moreover, the functionalization of amphiphilic polymers has also been the recent subject of intensive investigation.

Over the decade, a variety of functionalized amphiphilic polyelectrolytes have been synthesized (*15-21*). For example, if a small amount of hydrophobic chromophore dyes are covalently incorporated into polysulfonates modified with bulky hydrophobes, the dyes can be "compartmentalized" in the clusters of the hydrophobes in aqueous solution (*22-24*). The compartmentalized dyes are completely isolated from one another and "protected" from the aqueous phase, leading to a large modification of their photophysical and photochemical behavior (*22-24*).

An advantage of the functionalization of amphiphilic polyelectrolytes with fluorescent dyes is that fluorescence from the incorporated dyes provides a useful tool to investigate the self-organization of the polymers. To this end, polycyclic aromatic chromophores, such as naphthalene and pyrene can be successfully employed as fluorescence labels to be tagged on the amphiphilic polyelectrolytes (*15,17*).

As illustrated schematically in Figure 1, amphiphilic polyanions consisting of bulky hydrophobes, for example, may first form hydrophobic clusters (micelle units) within a polymer chain due to intramolecular associations of the hydrophobes in dilute aqueous solution. However, this conformation, which may be viewed as a "second-order structure", may not be stable conformation because a significant portion of the surface of the hydrophobic clusters should inevitably be exposed to water. Consequently, the micelle units may conglomerate to form a higher-order conformational structure which may be referred to as a unimolecular micelle (unimer). If, however, the associative interactions between the hydrophobes and/or between the micelle units are an intermolecular process, then multimolecular aggregates would be formed instead of the unimers. It depends primarily on the chemical structure, the first-order structure, of the amphiphilic polyions whether the intramolecular association predominates over the intermolecular association. Thus, it is important to establish structural requirements for polymers that would form unimers as compared to polymers that would form multimolecular aggregates.

Various types of amphiphilic polyelectrolytes have been synthesized so far, including random copolymers (*19,20,22-26*), block copolymers (*27-30*), and alternating copolymers (*31-37*). The sequence distribution of charged and hydrophobic units along the polymer chain is an important structural factor to determine whether the hydrophobic

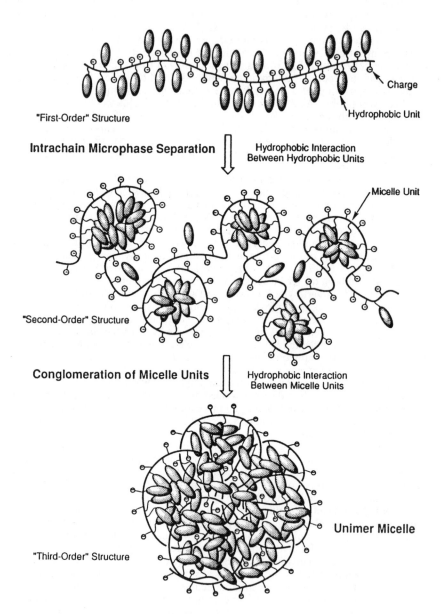

Figure 1. Schematic illustration of the self-organization of a hydrophobically modified polyanion in aqueous solution.

self-association is an intra- or intermolecular event. Block sequences have a tendency for intermolecular association, whereas random and alternating sequences tend to associate intramolecularly (*21,28,30,32,38,39*).

Random copolymers of sodium 2-(acrylamido)-2-methylpropanesulfonate (AMPS) and methacrylamides carrying bulky hydrophobic groups with cyclic structures, such as cyclododecyl (Cd), 1-adamantyl (Ad), and 1-naphthyl (1-Np) groups, form unimers in aqueous solutions of a wide range of concentrations if the contents of the hydrophobes are higher than a certain critical value (*40*). This phenomenon is due to predominant intramolecular self-association of the hydrophobes with cyclic structure (*23,24,41-43*). These unimers are very different from the classical surfactant micelles in that (a) all charged and hydrophobic groups are covalently linked to the polymer backbone, (b) the unimers are "static" in nature as oppose to the "dynamic" nature of the surfactant micelles which exist in equilibrium between association and dissociation, (c) the micellar structure is retained even at very low concentrations, and (d) the unimers remain as such even at very high concentrations.

This chapter describes the self-organization of these random copolymers in aqueous solution by fluorescence, ^{1}H-NMR, FTIR, static light scattering (SLS), dynamic light scattering (DLS), and small-angle X-ray scattering (SAXS) techniques. For the fluorescence studies, the polymers were labeled with a small amount of 1-pyrenyl (Py) or 2-naphthyl (2-Np) groups by terpolymerization techniques with use of the respective methacrylamide monomers (Schemes 1 and 2). In the text, the terpolymers are denoted as poly(A/R/P), where A, R, and P represent the AMPS, hydrophobe, and label units, respectively.

Synthesis of Hydrophobically Modified Polysulfonates

Amphiphilic polysulfonates with various bulky hydrophobes were prepared by free-radical copolymerization of AMPS and hydrophobic methacrylamides, such as *N*-Laurylmethacrylamide (LaMAm) (*26*), *N*-cyclododecylmethacrylamide (CdMAm), *N*-(1-adamantyl)methacrylamide (AdMAm), and *N*-(1-naphthylmethyl)methacrylamide (1NpMAm) (*23*). To incorporate a fluorescence label, a small mole fraction of *N*-(1-pyrenylmethyl)methacrylamide (1PyMAm) (*23*) or *N*-(2-naphthylmethyl)methacrylamide (2NpMAm) (*44*) was terpolymerized with the above monomers. These co- and terpolymerizations were carried out in vacuum-sealed glass ampules containing respective monomers and 2,2'-azobis(isobutyronitrile) (AIBN) (0.5 mol% based on the total monomers) in *N,N*-dimethylformamide (DMF) solution at 60 °C for 12 h (*23*). The polymers were purified by reprecipitation from methanol into ether followed by dialysis against pure water for a week. The compositions of the co- and terpolymers were determined by N/C and S/C ratios and also by absorption spectroscopy for the fluorescence-labeled polymers.

These co- and terpolymers are characterized as "ideal copolymers"; the monomer reactivity ratios are practically unity, giving copolymer compositions equal to monomer feed compositions (*23,44*). Therefore, one can determine the terpolymer compositions directly by the molar ratios of the monomers in feed, which permits one to prepare terpolymers with well defined compositions and completely random distributions of the monomeric units along the polymer chain.

The ability of the charged segments of AMPS to solubilize the sequences of the hydrophobic monomer units into water is very high. The terpolymers with up to ca. 60 mol% hydrophobic monomer units are completely soluble in water. The contents of the hydrophobes in the Py-labeled polymers, poly(A/La/Py), poly(A/Cd/Py), poly(A/Ad/Py), and poly(A/1-Np/Py) (Scheme 1), are sufficiently high for the self-association of the hydrophobes to occur in aqueous solution. In these terpolymers, the Py labels are expected to be incorporated into the hydrophobic clusters of the respective hydrophobes.

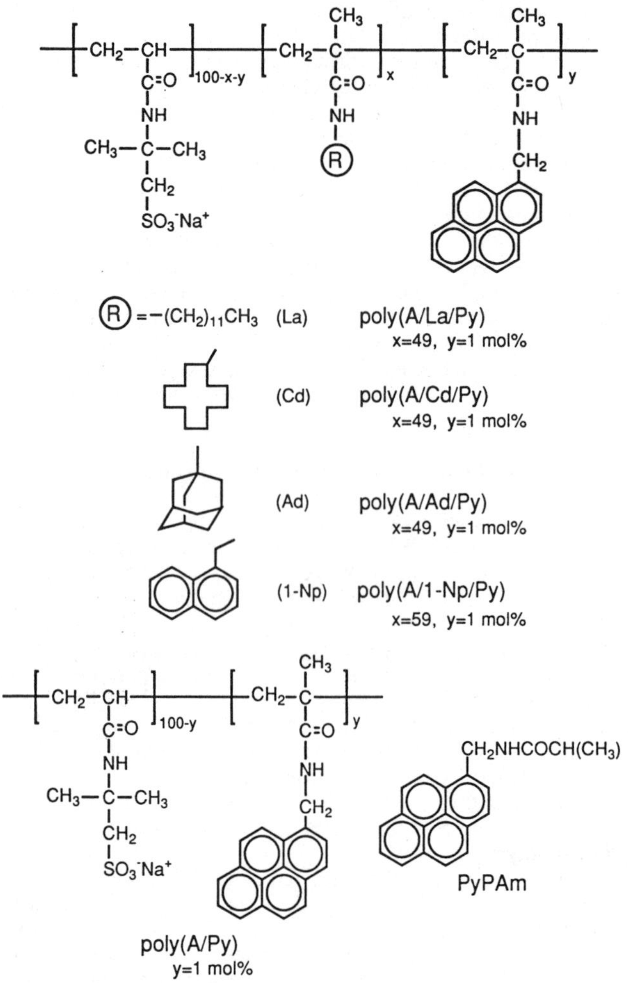
(La)
poly(A/La/Py)
x=49, y=1 mol%
(Cd)
poly(A/Cd/Py)
x=49, y=1 mol%
(Ad)
poly(A/Ad/Py)
x=49, y=1 mol%
(1-Np)
poly(A/1-Np/Py)
x=59, y=1 mol%
PyPAm
poly(A/Py)
y=1 mol%

Scheme 1

Light and X-Ray Scattering Studies

A characteristic feature of the unimers of the amphiphilic polysulfonates in aqueous solution is that the hydrodynamic volumes are extremely small for their high molecular weights. In Table I are listed static and dynamic light scattering data for poly(A/La/Py), poly(A/Cd/Py), poly(A/Ad/Py), and poly(A/1-Np/Py) in aqueous solution containing 0.1 M NaCl. Weight average molecular weights (M_w), determined by SLS, of these polymers are reasonably high; the M_w values of poly(A/La/Py), poly(A/Cd/Py), and poly(A/1-Np/Py) are on the order of 10^5, and that of poly(A/Ad/Py) is on the order of 10^4. In contrast, the Stokes radii, determined by DLS, are considerably small for their molecular weights. This is an experimental manifestation of the compact conformation, which, in turn, indicates that the terpolymers adopt the unimers in aqueous solution. The unimer of poly(A/Cd/Py) is particularly compact, as indicated by the largest ratio of mass to dimension ($M_w=5.1\times10^5$ against $R_s=5.5$ nm).

Table I. Light scattering and SAXS data for the La-, Cd-, and Ad-containing terpolymers in aqueous solution

polymer	$M_w{}^a$	$R_s{}^b$ (nm)	d^c (nm)
poly(A/La/Py)	1.2×10^5	7.0	...
poly(A/Cd/Py)	5.1×10^5	5.5	11
poly(A/Ad/Py)	3.5×10^4	6.2	7.3
poly(A/1-Np/Py)	1.3×10^5	...	...

a. Weight average molecular weight determined by SLS.
b. Average Stokes radius determined by DLS.
c. Spacing calculated from the scattering angle in SAXS.

Figure 2a shows SAXS data for poly(A/Cd/Py) and poly(A/Ad/Py) in concentrated (ca. 10 wt %) aqueous solutions. Scattering peaks were observed at $2\theta=0.8°$ and $1.2°$ for the Cd- and Ad-containing terpolymers, respectively, which correspond to the spacings of 11 and 7.3 nm, respectively. A spacing of 11 nm for the Cd-containing terpolymer agrees with the Stokes diameter determined by DLS (Table I). These scattering peaks can be interpreted to be due to closely packed unimer particles in the concentrated aqueous solutions. The Cd- and Ad-containing terpolymers remain as unimers with no interpenetration even at the high concentration (ca. 10 wt%) employed for the SAXS measurements. As is evident in Figure 2b, no scattering peaks were observed in ca. 10 wt% methanol solutions. This is because these terpolymers adopt random coils in methanol, and the random coils interpenetrate one another at such a high concentration.

In the case of aqueous solutions of poly(A/La/Py), however, SAXS shows no scattering peaks because the La-containing terpolymer can only exist as unimer at very low concentrations and the intermolecular association is predominant at higher concentrations, as will be described in the following section.

Fluorescence Studies

Figure 3 compares fluorescence spectra of the Py labels in the hydrophobic terpolymers and the reference copolymer without hydrophobes, poly(A/Py) (Scheme 1), in aqueous solution (*40*). All the polymers show only monomeric Py fluorescence because the content of the Py labels in the polymers is as low as 1 mol%. The ratio of the third to the first vibronic bands (I_3/I_1) in fluorescence spectra of pyrene is known to depend on the polarity in media where pyrene exists (*45*), I_3/I_1 ratio being larger in less polar

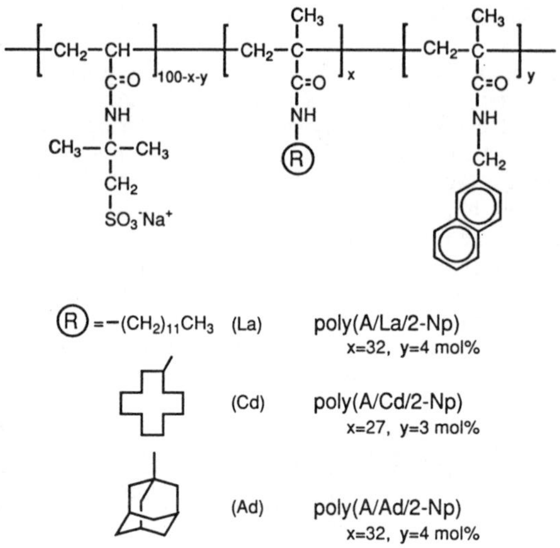

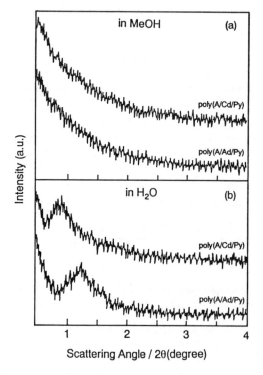

Scheme 2

Figure 2. SAXS patterns for poly(A/Cd/Py) and poly(A/La/Py) in methanol (a) and in aqueous solution (b); concentration of the polymers, ca. 10 wt%.

media. In a parallel experiment, this empirical rule was confirmed to apply to the model compound, PyPAm (Scheme 1), which has a substituent group similar to the Py label in the polymers (46). The I_3/I_1 ratio for poly(A/Py) is 0.59, reflecting the polarity of water (PyPAm shows the same value in water). This is an expected observation because the reference copolymer assumes an open-chain conformation in water, and the Py labels are exposed to the aqueous phase. In contrast, the I_3/I_1 ratios of the terpolymers are all larger than that of the reference copolymer as compared in Table II. This is an indication that the Py labels are experiencing lower polarities in the terpolymers because the Py labels are confined within the clusters of the hydrophobes in the unimers.

Table II. Steady-state fluorescence and decay parameters for the fluorescence of the Py-labeled co- and terpolymers in aqueous solution

polymer	$I_3/I_1{}^a$	fitting parameters $\tau_i(ns)/\alpha_i{}^b$		$<\tau>^c$ (ns)
poly(A/Py)	0.59	29/0.08	148/0.916	136
poly(A/La/Py)	0.80	55/0.258	207/0.742	168
poly(A/Cd/Py)	0.83	57/0.284	238/0.716	187
poly(A/Ad/Py)	0.76	41/0.336	219/0.664	159

a. Intensity ratio of peak 3 to peak 1 in the fluorescence spectra.
b. Fitting function; $I(t)=\Sigma\alpha_i\exp(-t/\tau_i)$.
c. Average lifetime defined by $<\tau>=\Sigma\alpha_i\tau_i$.

The hydrophobic microenvironments about the Py labels are also reflected in fluorescence decay profile; a fluorescence decay curve for poly(A/Cd/Py) in aqueous solution is compared with that for poly(A/Py) in Figure 4 (23). The decay for poly(A/Cd/Py) is significantly slower than that for the reference copolymer. Similarly slower decays were also observed for the La- and Ad-containing terpolymers.

The decay curves for all these polymers are best-fitted with double-exponential functions: the values of the relative weight of the pre-exponential factors α_i and the lifetimes τ_i are listed in Table II. The presence of the shorter life component suggests that the microenvironments about the Py labels are not uniform. The lifetimes for the longer life component, which is the main component, indicate that the Py labels are encapsulated in the hydrophobic clusters in the unimers.

Among the hydrophobic groups in the terpolymers, there are significant differences in the local polarity that the Py labels experience in the clusters of the hydrophobes. The steady-state fluorescence and fluorescence decay data consistently indicate that the microenvironmental polarity decreases in the order of Ad > La > Cd groups. A reason for the lowest micropolarity of the Cd cluster may be that the Cd groups can intimately be packed together with the Py labels in the hydrophobic cluster, leading to an effective protection of the Py moieties from the aqueous phase.

As discussed above, self-organization of an amphiphilic polyelectrolyte can occur, in general, via intramolecular or intermolecular association of hydrophobes. Non-radiative energy transfer between polymers tagged with an energy donor and an energy acceptor on separate polymer molecules is a useful tool to look into intermolecular associations of polymers (47-49). Naphthalene and pyrene labels can be used as a singlet-excited energy donor and acceptor, respectively, because they have a large spectral overlap and naphthalene can be selectively excited at a wavelength near 290 nm.

This technique was applied to the La-, Cd-, and Ad-containing polymers to examine whether the hydrophobic association is an intra- or intermolecular event. For this particular experiment the La-, Cd-, and Ad-containing polymers labeled with the 2-Np groups (Scheme 2) were employed together with the corresponding Py-labeled

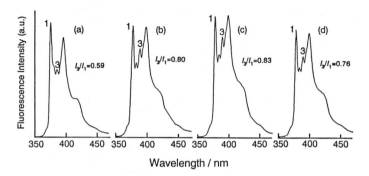

Figure 3.   Fluorescence spectra of poly(A/Py) (a), poly(A/La/Py) (b), poly(A/Cd/Py) (c), and poly(A/Ad/Py) (d) in aqueous solution; excitation wavelength, 347 nm. The I_3/I_1 ratios are indicated in the figure.

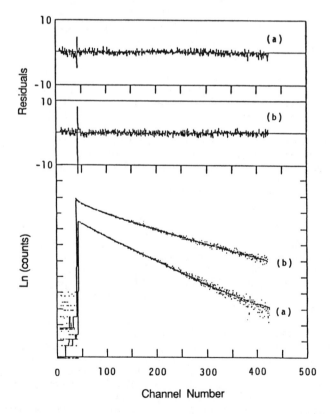

Figure 4. Fluorescence decay profiles monitored at 400 nm for poly(A/Py) (a) and poly(A/Cd/Py) (b) in aqueous solution; excitation at 355 nm; 1 channel=1.15 ns.

polymers. If the self-association of the hydrophobes is an intramolecular event, the terpolymers form unimers, and the fluorescence labels are confined within the hydrophobic clusters. Consequently, in an aqueous solution of a mixture of the 2-Np-labeled and Py-labeled polymers having the same aliphatic hydrophobes, the 2-Np and Py labels should be isolated from each other in separate unimers. Furthermore, each unimer should exist in separation because of electrostatic repulsion between the unimers. In these circumstances, the possibility of the energy transfer from the singlet-excited 2-Np label in one unimer to the Py label in another unimer is precluded. On the other hand, if the hydrophobic association is an intermolecular event, it should be possible for the 2-Np and Py labels to come close to each other within the Forster radius (R_0=2.86 nm for transfer from 2-methylnaphthalene to pyrene(*50*)), which permits the energy transfer to occur. If the intermolecular energy transfer takes places, then fluorescence from the Py label is observed when the 2-Np label is selectively excited.

Aqueous solutions of the mixtures of poly(A/La/2-Np) plus poly(A/La/Py), poly(A/Cd/2-Np) plus poly(A/Cd/Py), and poly(A/Ad/2-Np) plus poly(A/Ad/Py) were excited at 290 nm, and emissions from the 2-Np and Py labels were monitored at 340 and 395 nm, respectively. Figure 5 shows the ratios of the intensities of the fluorescence emitted by the Py and 2-Np labels plotted as a function of the total polymer concentration for the La-, Cd-, and Ad-containing terpolymer systems. In these systems, the 2-Np labels can be predominantly excited at 290 nm, but a slight contribution of direct excitation of the Py labels cannot completely be eliminated. The polymer concentrations were varied by adding the non-labeled copolymers carrying the same hydrophobes (Scheme 3), keeping the concentration of the labeled polymers constant. In the case of the La-containing polymer system, a much larger increase in the Py fluorescence intensity was observed with increasing polymer concentration than in the Cd- and Ad-containing polymer systems. An increase in the Py fluorescence for the La-containing polymer was recognized even at a concentration of ca. 0.2 wt%, and above this concentration, significant intermolecular interactions are indicated by an increase in Py fluorescence. In the Cd- and Ad-containing polymer systems, by contrast, there was no increase in Py fluorescence until the concentration was increased to ca. 4 and 7 wt% , respectively. These observations indicate that the Cd and Ad groups have much stronger tendency for intermolecular association than does the La group. In the La-containing terpolymer, intramolecular hydrophobic association occurs at concentrations below 0.2 wt%, and the polymers can exist as unimers only at these low concentrations. By contrast, in the Cd- and Ad-containing terpolymers, intramolecular association is predominant at concentrations below 4 and 7 wt%, respectively. These polymers can exist as unimers at much higher concentrations than the La-containing polymer. An important conclusion from this experiment is that the hydrophobes with cyclic structure (Cd and Ad) show much stronger tendency for the intramolecular association than that with linear structure (La), although the numbers of the carbon atom are approximately the same (C_{12} for La and Cd, and C_{11} for Ad).

If polycyclic aromatic groups are employed as hydrophobes in the amphiphilic polysulfonates in place of the bulky aliphatic hydrocarbons, the clusters of the aromatic chromophores in the unimers show characteristic photophysical behavior. Figure 6a shows fluorescence spectra of the copolymers of AMPS and 1NpMAm, poly(A/1-Np(x)) (Scheme 3), with varying 1-Np contents in aqueous solution at room temperature (*44*). The fluorescence spectra depend markedly on the copolymer composition. As the 1-Np content increases from 13 to 28 mol%, fluorescence due to the monomeric 1-Np chromophore peaking at 337 nm decreases, and excimer fluorescence peaking at 390 nm increases systematically. The copolymers with >41 mol% 1-Np content emit strong excimer fluorescence which dominates over the monomer fluorescence. Figure 7 shows a plot of the ratio of the excimer to monomer fluorescence intensities (I_E/I_M) as a function of the 1-Np content in the copolymer. The I_E/I_M ratio sharply increases at a 1-Np content between 28 and 41 mol%, indicating that

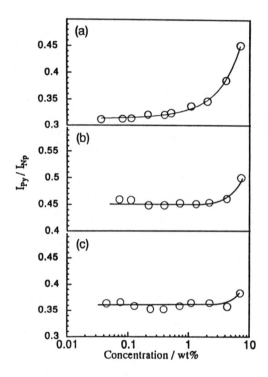

Figure 5. Ratio of the fluorescence intensities for the Py and 2-Np labels (I_{Py}/I_{Np}) as a function of the total concentration of the polymers in aqueous solution. The concentrations were varied by adding the non-labeled polymers to the aqueous solutions containing 0.01 wt% of the respective 2-Np-labeled and Py-labeled polymers: (a), poly(A/La/Py)+poly(A/La/2-Np)+poly(A/La); (b), poly(A/Cd/Py)+poly(A/Cd/2-Np)+poly(A/Cd); (c), poly(A/Ad/Py)+poly(A/Ad/2-Np)+poly(A/Ad).

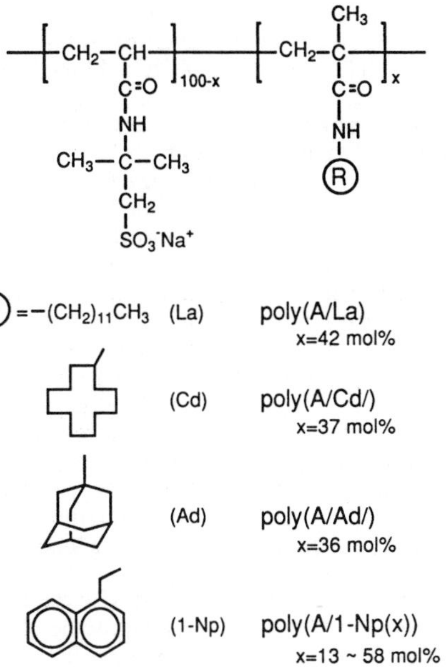

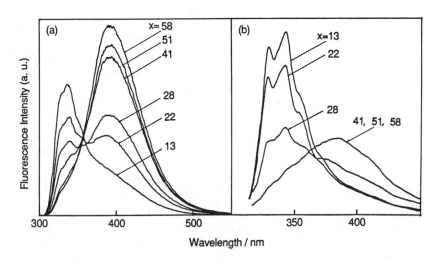

Scheme 3

Figure 6. Fluorescence spectra of poly(A/1-Np(x)) with varying mol% content (x) of the 1-Np unit in aqueous solution at room temperature (a) and at 77 K (b); excitation wavelength, 290 nm.

the critical content of the 1-Np group for the unimer formation lies between 28 and 41 mol%. The copolymers with >41 mol% 1-Np content show practically only excimer emission even in rigid solution at 77 K (Figure 6b), although the intensity is lower than that at room temperature. The excimer at 77 K is blue-shifted by ca. 10 nm from the excimer peak at room temperature. This blue-shifted excimer emission is attributable to a preformed excimer with configurational constraints leading to a deviation from the full overlap between the two naphthalene rings (44).

Since the fluorescence spectra shown in Figure 6a were measured at the same molar concentration of the 1-Np residue (20 μM), the relative fluorescence intensity can approximately be related to the relative fluorescence quantum yield. The total fluorescence quantum yield increases as the excimer component increases with increasing 1-Np content at room temperature; e.g., the copolymers with 13 and 58 mol% 1-Np units showing quantum yields of 0.09 and 0.16 at room temperature, respectively. This is rather unusual as compared with the tendency shown by polymer-bound aromatic chromophores studied so far (15,18). Unlike conventional random copolymers with comparable 1-Np contents, there exists little or no self-quenching site in poly(A/1-Np(x)), despite the fact that the 1-Np residues are tightly packed in the cluster in aqueous solution. From studies of fluorescence decay, it has become evident that extremely rapid migration of singlet-excited energy occurs throughout the cluster of the 1-Np chromophores in the unimer, and that the migrating singlet energy is thoroughly trapped by the preformed excimer sites which exist in a very small amount within the cluster.

In the 1-Np-containing polysulfonates labeled with the Py chromophores, poly(A/1-Np/Py) (Scheme 1), the Py labels are confined within the clusters of the 1-Np groups in the unimer in aqueous solution. Figure 8 compares fluorescence spectra of poly(A/1-Np/Py) with excitation of the 1-Np chromophore in methanol and in water (23). In methanol, in which poly(A/1-Np/Py) assumes a random coil conformation, the 1-Np containing polymer exhibits both 1-Np fluorescence and Py fluorescence, the former being predominant over the latter. The random coil allows energy transfer from the singlet-excited 1-Np chromophore to the Py label to some extent, as can be seen from the significant Py fluorescence (Figure 8a). In contrast, poly(A/1-Np/Py) emits entirely Py fluorescence in aqueous solution when the 1-Np chromophore is selectively excited at 290 nm (Figure 8b). This is a clear indication that the singlet-excited energy migrating over the 1-Np chromophores in the unimer is thoroughly trapped by the Py labels. This observation suggests that, even if there are a number of cluster units in a unimer (cf. Figure 1), all the cluster units are in contact with one another such that the 1-Np singlet-excited energy can migrate from one cluster unit to another.

A comparison of the fluorescence decay of the 1-Np chromophore with that of the Py trap provides insight into the rate of energy transfer. Figure 9 compares fluorescence rise and decay curves for poly(A/1-Np/Py) in aqueous solution monitored at 330 nm and at 400 nm, respectively, on a picosecond time scale. The data were satisfactorily fitted with a three-exponential function (Figure 9). The lifetimes τ_i and the relative weight of the preexponential factors α_i for the three-exponential function are listed in Table III. As can be seen in Figure 9a, the 1-Np fluorescence at 330 nm decays extremely rapidly; a large portion of the fluorescence decays with a lifetime of 19 ps. Importantly, a rapid rise time of 20 ps was observed in fluorescence monitored at 400 nm, and no other slower rise times than 298 ps were found. This clearly indicates that extremely rapid energy transfer occurs from the singlet-excited 1-Np residues to the Py traps in the unimers of poly(A/Np/Py).

Figures 10a and 10b show decays of 400-nm fluorescence on a nanosecond time scale for poly(A/1-Np/Py) on excitation of the 1-Np chromophore at 290 nm and on direct excitation of the Py chromophore at 355 nm, respectively. Both the decay data are fitted with a double-exponential function and the decay times are almost identical for both the decay curves as listed in Table IV. These facts indicate that the fluorescence

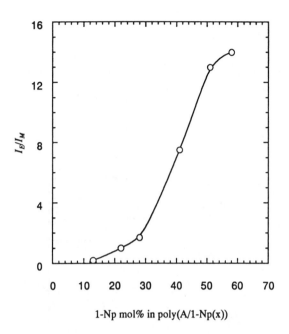

1-Np mol% in poly(A/1-Np(x))

Figure 7. Ratio of the excimer to monomer fluorescence intensities as a function of the 1-Np content in poly(A/1-Np(x)) in aqueous solution at room temperature.

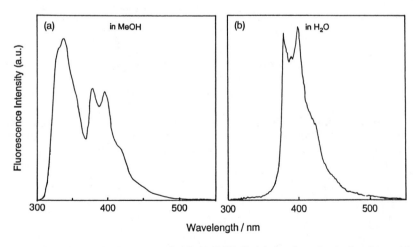

Figure 8. Fluorescence spectra of poly(A/1-Np/Py) in methanol solution (a) and in aqueous solution (b) at room temperature; excitation wavelength, 290 nm.

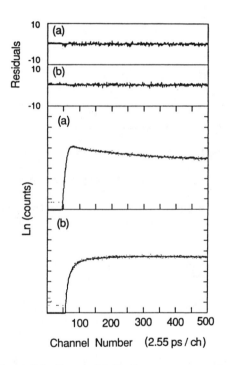

Figure 9. Comparison of the decay of 1-Np fluorescence monitored at 330 nm (a) and the rise of Py fluorescence monitored at 400 nm (b) for poly(A/1-Np/Py) in aqueous solution at room temperature; excitation wavelength, 290 nm.

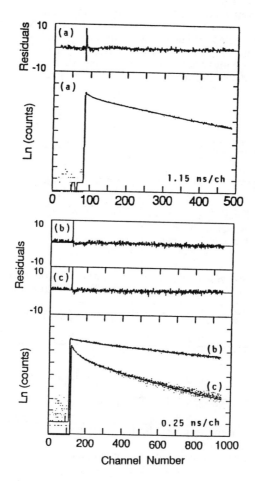

Figure 10. Comparison of the fluorescence decay profiles monitored at 400 nm on the nanosecond time scale: (a), poly(A/1-Np/Py) excited at 290 nm; (b), poly(A/1-Np/Py) excited at 355 nm; (c), poly(A/Np(58)) excited at 290 nm.

observed at 400 nm for poly(A/1-Np/Py) is emitted entirely from the Py traps regardless of whether they are directly excited at 355 nm or indirectly excited at 290 nm. On the other hand, the 1-Np excimer emission of the copolymer poly(A/1-Np(58)) monitored at 400 nm contains faster decay times as shown in Figure 10c and Table IV; a 49-% portion of the excimer decay has a lifetime of 4.5 ns. Since the 400-nm fluorescence of poly(A/1-Np/Py) excited at 290 nm contains no such fast decay component, the possibility of the contribution from the 1-Np excimer emission can be ruled out. Therefore, the steady-state fluorescence spectrum for poly(A/1-Np/Py) in aqueous solution shown in Figure 8 is essentially due to the singlet-excited Py moieties and no 1-Np excimer emission band lies beneath the Py fluorescence. This is supported by the observation that the steady-state fluorescence spectra of poly(A/1-Np/Py) in aqueous solution excited at 290 and 345 nm were essentially the same in their emission maxima and spectral profiles.

Table III. Fluorescence decay and rise times on the picosecond time scale for poly(A/1-Np/Py) in aqueous solution at room temperature[a]

λ_{ex}[b] (nm)	λ_{em}[c] (nm)	fitting parameters τ_i(ps)/α_i	χ^2
290	330	19/0.771, 167/0.175, 2550/0.054	1.03
290	400	20[d]/-0.300, 298[d]/-0.160, 25500/1.43	1.04

a. Fitting function is the same as indicated in Table II.
b. Excitation wavelength.
c. Wavelength at which fluorescence was monitored.
d. Rise time.

Table IV. Fluorescence decay times on the nanosecond time scale for poly(A/1-Np(58)) and poly(A/1-Np/Py) in aqueous solution at room temperature[a]

polymer	λ_{ex}[b] (nm)	λ_{em}[c] (nm)	fitting parameters τ_i(ps)/α_i	χ^2
poly(A/ 1-Np(58))	290	400	4.5/0.490, 19.8/0.256, 69.3/0.254	1.64
poly(A/ 1-Np/Py)	290	400	24.2/0.323, 187/0.677	1.66
poly(A/ 1-Np/Py)	355	400	24.0/0.253, 188/0.747	1.81

a. Fitting function is the same as indicated in Table II.
b. Excitation wavelength.
c. Wavelength at which fluorescence was monitored.

The encapsulation of the Py labels in the hydrophobic clusters in the unimer and their protection from the aqueous phase are clearly demonstrated by a sharp suppression of fluorescence quenching by Tl+ ions. Fluorescence of pyrene is known to be quenched by Tl+ ions only through short range interaction due to a heavy atom effect (51). In the reference copolymer, poly(A/Py), the Py labels can come into contact with Tl+ ions because the reference copolymer assumes an open chain conformation in aqueous solution, and the Tl+ ions are electrostatically concentrated in the vicinity of the anionic polymer chain. Thus, the Py fluorescence in the reference copolymer is efficiently quenched by Tl+ (Figure 11). In sharp contrast, as the Py labels in the terpolymers are confined within the hydrophobic clusters in the unimers, Tl+ ions are prohibited from

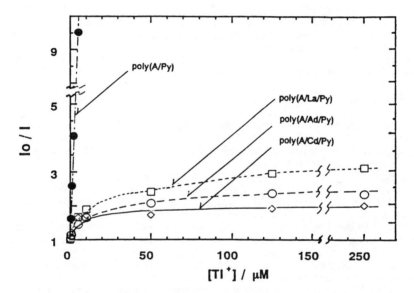

Figure 11. Stern-Volmer plots for fluorescence quenching by Tl+ for the Py labels in poly(A/Py), poly(A/La/Py), poly(A/Cd/Py), and poly(A/Ad/Py) in aqueous solution.

coming into contact with the Py labels, leading to a sharp suppression of the fluorescence quenching (Figure 11).

In the case where there are two fluorophore sites, one is accessible to quenchers and the other is not, the Stern-Volmer equation can be modified as

$$(I_0/(I_0-I))=(1/fK[Tl^+])+(1/f) \qquad (1)$$

where I and I_0 are the fluorescence intensity in the presence and absence of Tl^+, respectively, K is the quenching constant, and f is the fraction of the Py labels accessible to Tl^+.

Figure 12 shows the plots of the quenching data according to eq 1 for the La-, Cd-, and Ad-containing terpolymers. Apparently, the data for all the terpolymers follow eq 1. From the intercepts, f values were estimated to be 0.76, 0.49, and 0.55 for poly(A/La/Py), poly(A/Cd/Py), and poly(A/Ad/Py), respectively. These values imply that 51 and 45 % of the Py labels are completely protected from the access of Tl^+ for the Cd- and Ad-containing terpolymers, respectively, while only 24 % of the Py labels are protected in the La-containing terpolymer.

A plausible model for the unimer with the Py label confined within the hydrophobic clusters is illustrated in Figure 13.

NMR Studies

^{1}H-NMR spectroscopy provides useful information on the self-organization of amphiphilic polyelectrolytes. In D_2O, resonance bands for the protons on the hydrophobes become extremely broad when the unimers are formed because motions of the hydrophobes are highly restricted.

In Figure 14 are compared NMR spectra for the 1-Np protons in poly(A/1-Np(x)) with varying 1-Np contents (44). These spectra were measured in DMSO-d_6 and in D_2O at a constant polymer concentration (7 wt%) under identical instrumental conditions. In DMSO-d_6, the peak intensity increases systematically with increasing 1-Np content in the copolymer. In D_2O, however, the peak intensity sharply decreases with increasing 1-Np content at room temperature when the 1-Np content exceeds 28 mol%. This is due to the motional line broadening of the 1-Np proton resonance, indicating that the motion of the naphthalene ring becomes highly restricted when the copolymer forms unimer in aqueous solution.

A marked line broadening occurs at a 1-Np content between 28 and 41 mol% in the copolymer, indicating that the critical content of the 1-Np group for the unimer formation lies between 28 and 41 mol%. This is consistent with the tendency observed in the intensity ratio of the excimer to monomer fluorescence described above. When the temperature was raised to 85 °C, the NMR peak in D_2O became apparent as shown in Figure 14. At this temperature the structured water diminishes, and thereby hydrophobic interaction is destabilized. Thus, the packing of the 1-Np groups in the cluster is loosened, and the motional restriction of the hydrophobes is alleviated.

The NMR relaxation times give more quantitative information about the self-organization of these polymers. Figure 15 shows ^{1}H-NMR spectra measured with the CPMG pulse sequence (52) for the aliphatic protons in poly(A/La/Py) and poly(A/Cd/Py) in D_2O at 25 °C, from which spin-spin relaxation times (T_2) were estimated. The peaks at 0.9 and 1.4 ppm for the La-containing terpolymer are attributable to the methyl and methylene protons in the La group, respectively. In the Cd-containing terpolymer, on the other hand, the methylene protons in the Cd group give no peaks but a broad shoulder at ca. 1.2 ppm, despite the fact that the number of the Cd methylene protons is larger than the total of the AMPS methyl and main chain methylene protons, all giving a broad peak at ca. 1.5 ppm. This is because the resonance line of the Cd methylene in the terpolymer is extremely broad because of

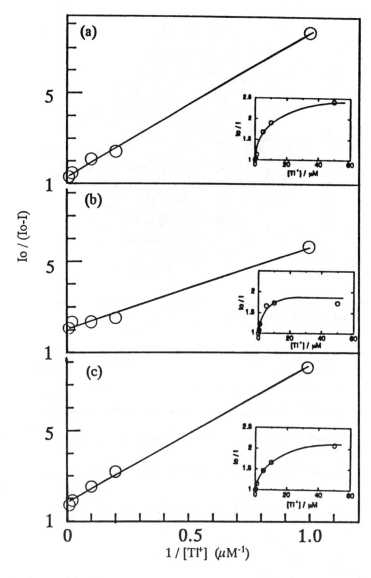

Figure 12. Modified Stern-Volmer plots according to eq 1 for the fluorescence quenching by Tl+ for the Py labels in poly(A/La/Py) (a), poly(A/Cd/Py) (b), and poly(A/Ad/Py) (c) in aqueous solution.   Inserts: Stern-Volmer plots for the corresponding data.

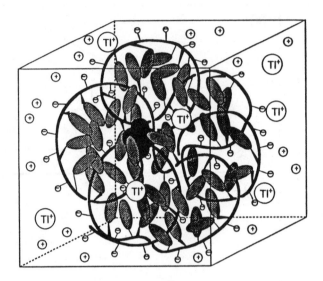

Figure 13. Conceptual illustration of a model for the unimer in which the Py label is confined and protected from Tl$^+$ ions.

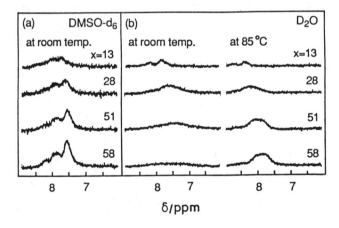

Figure 14. ^{1}H-NMR spectra of poly(A/1-Np(x)) with varying contents of the 1-Np unit measured in DMSO-d_6 (a) at room temperature and in D$_2$O (b) at room temperature and at 85 °C.

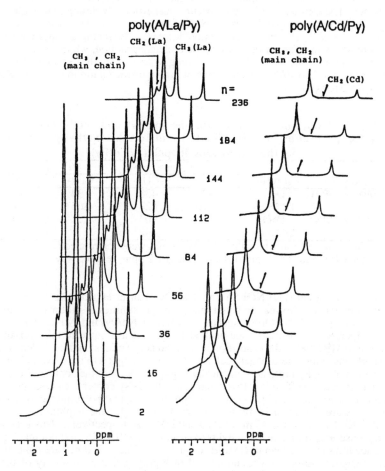

Figure 15. ¹H-NMR stack plots for poly(A/La/Py) and poly(A/Cd/Py) in D₂O at 25 °C; *n* represents the repeat number of the 180° pulse.

restricted motions. The values of T_2 are listed in Table V together with the values of spin-lattice relaxation times (T_1). The T_2 value for the Cd methylene protons in the Cd-containing terpolymer estimated at 1.2 ppm (shoulder) is significantly smaller than those for the La methyl and La methylene protons in the La-containing terpolymer. This indicates that the motion of the Cd group in the hydrophobic cluster is more restricted than that of the La group. The T_1 value for the Cd methylene in the Cd-containing terpolymer is much larger than that for the La methylene in the La-containing terpolymer. The spin-lattice relaxation occurs most efficiently through molecular motion whose frequency is comparable to the NMR frequency (53). Therefore, T_1 decreases concurrently with T_2 as a molecular motion decreases. Reaching a minimum value, T_1 then increases with a further decrease in the molecular motion, while T_2 remains as the minimum value. The large T_1 value, along with the small T_2 value, for the Cd methylene in poly(A/Cd/Py) are indicative of highly restricted motions of the Cd groups in the hydrophobic cluster.

Table V. Values of T_1 and T_2 for the methyl and methylene protons of the La groups in poly(A/La/Py) and for the methylene protons in the Cd groups in poly(A/Cd/Py)[a]

terpolymer	solvent	T_1 (ms)		T_2 (ms)	
		CH$_3$-	-CH$_2$-	CH$_3$-	-CH$_2$-
poly(A/La/Py)	D$_2$O	472	438	34	15
	DMF-d_7	2916	627	805	211
poly(A/Cd/Py)	D$_2$O	...	904	...	4
	DMF-d_7	...	442	...	13

a. Chemical shifts at which the relaxation times were determined for the CH$_3$- and -CH$_2$- protons are 0.9 and 1.2 ppm, respectively.

Two-dimensional (2D) NOESY (nuclear Overhauser enhancement spectroscopy) provides qualitative information about the spatial proximities of protons in a molecule. Figure 16 compares 2D-NOESY spectra of poly(A/1-Np/Py) in DMF-d_7 and in D$_2$O at room temperature. The resonance peak at 1.7 ppm is attributed to the overlap of the methyl (in AMPS and at the α-position in the methacrylamides) and methylene (in main chain) protons, while the broader peak at 7.3 ppm is assigned to the 1-Np protons. The contribution of the Py protons is practically negligible because the content of the Py unit in the terpolymer is sufficiently low compared to the 1-Np content. In D$_2$O, cross peaks between the aromatic and aliphatic protons appear owing to dipolar interactions, whereas cross peaks are absent in DMF-d_7. The values of T_2 for the naphthyl resonance in poly(A/1-Np/Py) are 4 and 45 ms in D$_2$O and DMF-d_7, respectively. This is consistent with the observation that the naphthyl protons in poly(A/1-Np/Py) exhibit much broader resonance lines in D$_2$O than in DMF-d_7 (Figure 16). These findings are indicative of a highly compact unimer conformation and of highly restricted mobility of the 1-Np residue owing to the cluster formation in the unimer in aqueous solution.

FTIR Studies

In the co- and terpolymers, the hydrophobes are connected to the main chain via amide spacer bonds. Figure 17 compares FTIR spectra of the La-, Cd-, and Ad-containing terpolymers measured as KBr pellets. The KBr pellets were prepared with these polymers recovered from their aqueous solutions by freeze-drying. The self-organized structures of the terpolymers in aqueous solution should be retained in the KBr pellet. An IR spectrum for the reference copolymer, poly(A/Py), is also presented in Figure 17 for comparison. All the polymers show two characteristic IR absorption bands in the

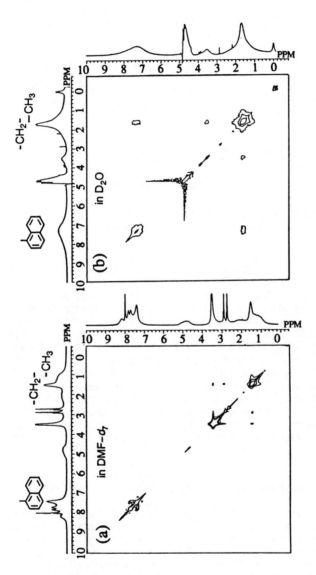

Figure 16. 2D-NOESY contour plots for poly(A/1-Np/Py) in DMF-*d*₇ (a) and in D₂O (b) at room temperature.

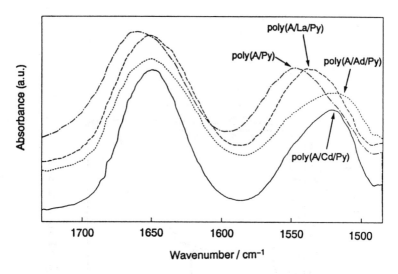

Figure 17.   Expanded-scale FTIR spectra for poly(A/La/Py), poly(A/Cd/Py), poly(A/Ad/Py), and poly(A/Py) measured as KBr pellets.

region of 1500-1700 cm^{-1} due to the amide bonds, the peaks at higher and lower wavenumbers attributable to ν(C=O) and δ(N-H), respectively. These bands for the terpolymers are shifted toward lower wavenumber as compared with those for the reference copolymer. More significant shifts were observed in the δ(N-H) band than in the ν(C=O) band. These lower-wavenumber shifts are attributable to the hydrogen bonding of the amide spacer bonds. An important point is that there are differences in the extent of the lower-wavenumber shift in the δ(N-H) band among the terpolymers. The Cd- and Ad-containing terpolymers show a lower-wavenumber shift of ca. 30 cm^{-1} in the δ(N-H) band as compared with that of the reference copolymer, while the La-containing terpolymer shows a much smaller shift of 8 cm^{-1}. These observations indicate that hydrogen bonds are formed between the spacer amide bonds much more extensively in the Cd- and Ad-containing terpolymers than are in the La-containing terpolymer. This is an indication that hydrogen bonding between the spacer amide bonds contributes to the reinforcement of the unimer structure.

Conclusions

Random copolymers of AMPS and a methacrylamide N-substituted with La, Cd, Ad, or 1-Np groups undergo self-organization in aqueous solution. The absence of the intermolecular singlet-energy transfer between Py-labeled and 2-Np-labeled polymer molecules in aqueous solutions of a wide range of concentrations revealed that the hydrophobic self-association was primarily an intramolecular event leading to the formation of unimers. The unimer formation was also evidenced by rapid migration of singlet-excited energy among the 1-Np groups and extremely rapid energy transfer of the migrating 1-Np singlets to the Py traps in the 1-Np containing terpolymer labeled with a small amount of the Py groups. Among the bulky aliphatic hydrophobes, the Cd and Ad groups show a much stronger tendency for the intramolecular association than does the La group. The Cd- and Ad-containing terpolymers exist as unimers even at high polymer concentrations up to ca. 4 and 7 wt%, respectively. In contrast, the La-containing terpolymers can exist as unimers only at concentrations <0.2 wt%, the unimers reorganizing into multimolecular aggregates at higher concentrations. The unimers are extremely compact; e.g., the Cd-containing terpolymer with a weight average molecular weight of 5.1×10^5 forms unimers with a mean hydrodynamic radius of 5.5 nm. ^{1}H-NMR relaxation times indicated that the local motions of the Cd groups in the unimers were much more restricted than those of the La groups. FTIR showed hydrogen bonding between the spacer amide bonds contributing to the stabilization of such compact unimer structures.

Literature Cited

(1) Nemethy, G.; Scheraga, H. A. *J. Chem. Phys.* **1962**, *36*, 3382.
(2) Nemethy, G.; Scheraga, H. A. *J. Phys. Chem.* **1962**, *66*, 1773.
(3) Jencks, W. P. *Catalysis in Chemistry and Enzymology*; McGraw-Hill: New York, 1969; pp 393.
(4) Kotin, L.; Nagasawa, M. *J. Chem. Phys.* **1962**, *36*, 873.
(5) Nagasawa, M.; Murase, T.; Kondo, K. *J. Phys. Chem.* **1965**, *69*, 4005.
(6) Joyce, D. E.; Kurucse, T. *Polymer* **1981**, *22*, 415.
(7) Morcellet-Sauvage, J.; Morcellet, M.; Loucheux, C. *Makromol. Chem.* **1981**, *182*, 949.
(8) Chu, D.; Thomas, J. K. *J. Am. Chem. Soc.* **1986**, *108*, 6270.
(9) Arora, K.; Turro, N. J. *J. Polym. Sci., Polym. Phys. Ed.* **1987**, *25*, 243.
(10) Morishima, Y.; Itoh, Y.; Nozakura, S. *Makromol. Chem.* **1981**, *182*, 3135.
(11) Guillet, J. E.; Wang, J.; Gu, L. *Macromolecules* **1986**, *19*, 2793.
(12) Strauss, U. P.; Jackson, E. G. *J. Polym. Sci.* **1951**, *6*, 649.
(13) Dubin, P.; Strauss, U. P. *J. Phys. Chem.* **1967**, *71*, 2757.
(14) Dubin, P.; Strauss, U. P. *J. Phys. Chem.* **1970**, *71*, 2842.
(15) Morishima, Y. *Prog. Polym. Sci.* **1990**, *15*, 949.
(16) Morishima, Y. *Adv. Polym. Sci.* **1992**, *104*, 51.
(17) Morishima, Y. *Trends Polym. Sci.* **1994**, *2*, 31.
(18) Webber, S. E. *Chem. Rev.* **1990**, *90*, 1469.
(19) Chatterjee, P. K.; Kamioka, K.; Batteas, J. D.; Webber, S. E. *J. Phys. Chem.* **1991**, *95*, 960.
(20) Nowakowska, M.; Guillet, J. E. *Macromolecules* **1991**, *24*, 474.
(21) McCormick, C. L.; Chang, Y. *Macromolecules* **1994**, *27*, 2151.
(22) Morishima, Y.; Furui, T.; Nozakura, S.; Okada, T.; Mataga, N. *J. Phys. Chem.* **1989**, *93*, 1643.
(23) Morishima, Y.; Tominaga, Y.; Kamachi, M.; Okada, T.; Hirata, Y.; Mataga, N. *J. Phys. Chem.* **1991**, *95*, 6027.
(24) Morishima, Y.; Tsuji, M.; Kamachi, M.; Hatada, K. *Macromolecules* **1992**, *25*, 4406.
(25) Morishima, Y.; Kobayashi, T.; Nozakura, S. *J. Phys. Chem.* **1985**, *89*, 4081.
(26) Morishima, Y.; Kobayashi, T.; Nozakura, S. *Polym. J.*, **1989**, *21*, 267.
(27) Morishima, Y.; Itoh, Y.; Hashimoto, T.; Nozakura, S. *J. Polym. Sci., Polym. Chem. Ed.* **1982**, *20*, 2007.
(28) Kamioka, K.; Webber, S. E.; Morishima, Y. *Macromolecules* **1988**, *21*, 972.
(29) Ramireddy, C.; Tuzar, Z.; Prochazka, K.; Webber, S. E.; Munk, P. *Macromolecules* **1992**, *25*, 2541.
(30) Prochazka, K.; Kiserow, D.; Ramireddy, C.; Tuzar, Z.; Munk, P.; Webber, S. E. *Macromolecules* **1992**, *25*, 454.
(31) Morishima, Y.; Kobayashi, T.; Nozakura, S.; Webber, S. E. *Macromolecules* **1987**, *20*, 807.
(32) Morishima, Y.; Lim, H. S.; Nozakura, S.; Strutevant, J. L. *Macromolecules* **1989**, *22*, 1148.
(33) Bai, F.; Chang, C. H.; Webber, S. E. *Macromolecules* **1986** *19*, 588.

(34) Bai, F.; Chang, C.-H.; Webber, S. E. *Macromolecules* **1986**, *19*, 2484.
(35) Burkhart, R. D.; Haggquist, G. W.; Webber, S. E. *Macromolecules* **1987**, *20*, 3012.
(36) Bai, F.; Webber, S. E. *Macromolecules* **1988**, *21*, 628.
(37) Itoh, Y.; Webber, S. E.; Rodgers, M. A. J. *Macromolecules* **1989**, *22*, 2766.
(38) Chang, Y.; McCormick, C. L. *Macromolecules* **1993**, *26*, 6121.
(39) McCormick, C. L.; Salazar, L. C. *Polymer*, **1992**, *33*, 4617.
(40) Morishima, Y.; Nomura, S.; Ikeda, T.; Seki, M.; Kamachi, M. *Macromolecules*, to be published.
(41) Morishima, Y.; Tominaga, Y.; Nomura, S.; Kamachi, M.; Okada, T. *J. Phys. Chem.* **1992**, *96*, 1990.
(42) Morishima, Y.; Tsuji, M.; Seki, M.; Kamachi, M. *Macromolecules* **1993**, *26*, 3299.
(43) Aota, H.; Morishima, Y.; Kamachi, M. *Photochem. Photobiol.* **1993**, *57*, 989.
(44) Morishima, Y.; Tominaga, Y.; Nomura, S.; Kamachi, M. *Macromolecules* **1992**, *25*, 861.
(45) Kalyanasundaram, K.; Thomas, J. K. *J. Am. Chem. Soc.* **1977**, *99*, 2039.
(46) Morishima, Y.; Seki, M.; Tominaga, Y.; Kamachi, M. *J. Polym. Sci., Polym. Chem. Ed.* **1992**, *30*, 2099.
(47) Ringsdorf, H.; Simon, J.; Winnik, F. M. *Macromolecules* **1992**, *25*, 7306.
(48) Ringsdorf, H.; Simon, J.; Winnik, F. M. *Macromolecules* **1992**, *25*, 5353.
(49) Winnik, F. M. *Polymer* **1990**, *31*, 2125.
(50) Berlman, I. B. *Energy Transfer Parameters of Aromatic Compounds*; Academic Press: New York, 1973.
(51) Hashimoto, S.; Thomas, J. K. *J. Am. Chem. Soc.* **1985**, *107*, 4655.
(52) Meiboom, S.; Gill, D. *Rev. Sci. Instrum.* **1958**, *29*, 688.
(53) Pake, G. E. In *Solid State Physics*; Seitz, F., Turnbull, D., Eds,; Academic Press: New York, 1965; *Vol. 2*, pp. 1-92.

RECEIVED February 2, 1995

Chapter 30

Fourier Transform IR and NMR Observations of Crystalline Polymer Inclusion Compounds

N. Vasanthan[1], I. D. Shin[2], and A. E. Tonelli[1]

[1]Fiber and Polymer Science Program, College of Textiles, and
[2]Department of Chemistry, North Carolina State University, P. O. Box 8301, Raleigh, NC 27695–8301

Several small-molecule hosts form clathrates or inclusion-compounds (ICs) with polymers. In these polymer-ICs the guest polymer chains are confined to occupy narrow channels in the crystalline matrix formed by the host. The walls of the IC channels are formed entirely from the molecules of the host, and they serve to create a unique solid state environment for the included polymer chains. Each polymer chain included in the narrow, cylindrical IC channels(ca. 5.5Å in diameter) is highly extended and also separated by the host matrix channel walls from neighboring polymer chains. The net result is a solid state environment where extended, stretched (as a consequence of being squeezed) polymer chains reside in isolation from their neighbors inside the narrow channels of the crystalline matrix provided by the small-molecule, host. Comparison of the behavior of isolated, stretched polymer chains in their crystalline ICs with observations made on ordered, bulk samples of the same polymer are beginning to provide some measure of the contributions made by the intrinsic nature of a confined polymer chain and the pervasive, cooperative, interchain interactions which can complicate the behavior of bulk polymer samples. Just as dilute polymer solutions at the Θ-temperature have been effectively used to model disordered, bulk polymer phases (both glasses and melts), polymer-IC's may be utilized to increase our understanding of the behavior of polymer chains in their ordered, bulk phases as found in crystalline and liquid-crystalline samples. Solid state FT-IR and NMR observations of polymer-IC's can provide detailed conformational and motional information concerning the included, stretched, and isolated polymer chains.

The long-chain nature of polymers, coupled with their ability to adopt an almost inexhaustible variety of different conformations permits them to adjust their sizes and shapes in response to both external stresses and internal (intrachain)

0097–6156/95/0598–0517$12.00/0

interactions. It is this additional internal degree of freedom that confers upon polymers many of their unique physical properties. In disordered bulk polymers (melts and glasses) each polymer chain is in contact with or influences many (ca. 100) neighboring polymer chains, solely because of their large sizes and pervasive, randomly-coiling shapes. By comparison, the number of neighboring polymer chains influenced by any given polymer chain in bulk ordered samples (semi-and liquid-crystalline) are much reduced as a consequence of the highly extended polymer chain conformations adopted there. Though fewer in number, the interactions between polymer chains in bulk, ordered samples are stronger, more intimate, and extend over longer portions of their chain contours. To understand the physical properties of all bulk polymer systems requires two distinct types of information: first, the inherent or intrinsic characteristics (conformations and mobilities) of individual polymer chains, which depend solely upon intramolecular or intrachain interactions, and secondly, the effects of cooperative interchain interactions between polymer chains. Observation of isolated polymer chains in dilute solutions at the θ-temperature (1), where the chains are free from both excluded-volume self-intersections and cooperative interchain interactions, provides the means to secure a useful description (2) of the inherent intramolecular behavior of disordered polymer chains. Comparison with the behavior of molten polymers, for example, can provide some measure of the nature of cooperative, interchain interactions occurring in disordered, bulk polymers.

We have no dilute solution analogue for the conformationally, orientationally, and/or positionally ordered polymer chains found in liquid-crystalline and crystalline polymer samples. How then might we attempt to understand the properties of ordered, bulk polymers? We believe that certain inclusion compounds (ICs) formed between small-molecule hosts and guest polymers can serve as solid-state analogs/model systems/touchstones useful for separating the inherent, single-chain behavior from the cooperative interchain interactions occurring in ordered, bulk polymer phases.

Certain small molecules, such as urea (U) and perhydrotriphenylene (PHTP), are able to form crystalline ICs with polymers (3-7) via a cocrystallization process. The guest polymer chains in these ICs are included in and confined to occupy narrow channels provided by the crystalline matrix formed by the small-molecule host. Two such polymer-ICs are illustrated in Figure 1 and are based on X-ray diffraction analysis (8,9) performed on single crystals of polyethylene (PE)-U-(10,11) and trans-1,4-polybutadiene (TPBD)-PHTP-ICs (9). The channels occupied by the included polymer chains are nearly cylindrical and quite narrow (D = channel diameter = ca. 5.5Å), and each included polymer chain is separated, and therefore decoupled, from all neighboring chains by the walls of the IC channels, which are constructed exclusively of the host matrix molecules. Polymer chains included in the narrow channels of their ICs with U and PHTP clearly are both highly extended (12) and isolated from neighboring polymer chains. It is precisely these two features of polymer-ICs which serve to

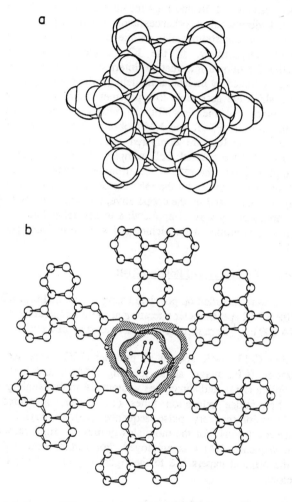

Figure 1 - (a) Space-filling drawing of a channel in the urea-n-hexadecane clathrate (10,11) and (b) schematic drawing of a trans-1,4 polybutadiene chain in the channel of its inclusion compound with PHTP (9).

recommend them as model systems whose study can lead to an assessment of the inherent, single-chain behavior of ordered, bulk polymer solids.

Principal among the experimental probes used to study polymer-ICs are FT-IR and NMR spectroscopy. FT-IR and high resolution NMR [C-13 (CPMAS/DD)] report individual vibrational absorbances and nuclear magnetic resonances for each structurally and environmentally unique constituent group of atoms and carbon nucleus, respectively, in the IC sample and can provide both conformational and motional information. Interpretation of the FT-IR and NMR observations is facilitated by molecular modeling, which attempts to define those conformations which are accessible to a polymer chain confined to occupy its narrow IC channel and the feasibility of interconverting between channel conformers. In addition, FT-IR and NMR observations and modeling analysis performed on polymer-ICs are compared with identical FT-IR and NMR observations recorded for ordered, bulk samples of the same polymers. In this way we attempt to gauge the relative contributions made to the properties of ordered, bulk polymer samples by the inherent behavior of single, isolated, and stretched polymer chains and by the cooperative, interchain interactions. We will illustrate this approach by way of application to several polymer-U-ICs, where the conformations, mobilities, and stoichiometries of the included polymer chains are estimated.

FT-IR and NMR Observations of Polymer-U-ICs

Infrared spectra were recorded on powdered samples pressed into KBr pellets on a Nicolet 510P FT-IR spectrometer operating at room temperature. Spectra in the region 400 - 4,000 cm^{-1} were recorded with a resolution of 2 cm^{-1}.

High resolution C-13 NMR spectra of polymer-U-ICs were recorded under cross-polarization (CP), magic angle spinning (MAS), and high power H-1 dipolar decoupling (DD), and sometimes without CP in the single-pulse mode at 50.3 MHz. Spin-lattice relaxation times (T_1) were measured using the 180°-τ-90° inversion-recovery pulse sequence (13) employing a 5T delay between pulse repetitions. For the more slowly moving, rigid carbon nuclei the CP-T_1 pulse sequence (14) was employed. For further details the reader is referred to the original papers and to the figure captions of the NMR spectra presented below.

Modeling Polymer Chains in IC Channels

Two approaches have been taken when modeling polymers confined to the channels of their ICs with U and PHTP. Mattice and coworkers (15-19) have employed molecular dynamics simulation techniques. They place a polymer chain fragment inside a single IC channel formed by the host clathrate crystalline lattice, whose molecules are positioned according to their X-ray determined (9) structures. For example (15), a 10-repeat unit fragment of TPBD was placed in the clathrate channel which is formed by 6 stacks containing 15 PHTP molecules each [see Figure 1 (b)], which results in a channel that is 60% longer than the fully extended TPBD fragment. Each of the 4,422 atoms of the TPBD fragment and PHTP lattice channel was explicitly considered, and the CHARMm potential was employed to calculate the energy of the system. Their dynamic simulation was carried out for a 70 ps trajectory using the Verlet algorithm with 0.5 fs integration steps.

We have adopted another approach to model (12, 20-26) the behavior of polymers in their ICs, which utilizes a complete search of rotational isomeric states (RIS) (2) conformations of polymer chain fragments confined to occupy a structureless cylinder whose D = 5.5Å mimics the narrow U- and PHTP-IC channels. Once a population of channel conformers has been established, a test is performed to determine if interconversion between channel conformers is possible within the confines of the channel cylinder.

Both modeling methods have particular advantages and shortcomings. The complete RIS conformational search generates all possible channel conformers, but, aside from indicating the potential for interconversion of channel conformers, does not provide detailed information concerning the dynamics of the included polymer motions. The molecular dynamics simulations, on the other hand, must assume a starting channel conformation to begin each trajectory. In a sense, one must know ahead of time the channel conformers. Even if one "guesses" a potential channel conformer, the molecular dynamics approach will never reveal alternative channel conformers unless they are readily obtained by interconversion of the starting conformer during the course of the dynamics trajectory. A practical way out of this dilemma would seem to be afforded by beginning a molecular dynamics trajectory from each of the channel conformers found in a preliminary, yet complete, RIS conformational search performed on the included polymer fragment as outlined below.

Consider the PE (21) chain fragment drawn in Figure 2. The statistical weight matrix U given below the PE fragment embodies the RIS conformational description of the PE chain (27), and its elements are appropriate to room temperature. We adopt C-C and C-H bond lengths of 1.53 and 1.10Å, respectively, along with C-C-C and H-C-H valence angles of 112 and 110°. A Cartesian coordinate system is affixed to the middle of the first C-C bond at o and is used to express the x,y,z coordinates of each C and H atom in the PE chain fragment. These atomic coordinates depend on the set of 7 ϕ-rotation angles, resulting in 3^7 (2,187) total PE fragment conformers when each backbone bond is restricted to adopt only the staggered trans (t) ($\phi = 0°$) and gauche (g±) ($\phi = \pm 120°$) rotational states. For each of these conformations the x, y, z coordinates of all the atoms are calculated and then transformed to the Cartesian coordinate system x',y',z' whose z'-axis connects the mid-points of the terminal bonds in the PE fragment (o and • in Figure 2). The radius of the corresponding cylindrical coordinate system with a coincident z'-axis is $r = (x'^2 + y'^2)^{1/2}$.

In our selection of channel conformers, we simply determine if $r = (x'^2 + y'^2)^{1/2} < r_c$, where r_c is the radius of the cylindrical channel. If each atom in the PE fragment passes this test, then the conformation is considered to be a channel conformer.

Matrix multiplication techniques (2) are used to calculate various properties of the PE chain fragment, averaged over all conformations and also averaged over just the set of channel conformers found. This is made possible by the RIS conformational model developed for PE (27). Average probabilities, or populations, of channel conformers and fragment bond conformations averaged over all 2,187 conformations, and for just the set of channel conformers, are obtained in this manner for channels (cylinders) of various radii.

Summation of the elements in the matrix product $(U)^7$ yields the configurational partition function $Z(PE)$ of the PE fragment in Figure 2. The probability that the PE fragment adopts the ttg+tg-tt conformation, for example, is obtained from $[U(1)U(1,1)U(1,2)U(2,1)U(1,3)U(3,1)U(1,1)]/Z(PE)$, where $U(1)$ is the statistical weight matrix U of Figure 2 with all elements in columns 2 and 3 replaced by zeros. To determine the probability of finding the fourth C-C bond of the PE fragment in the $\phi = g-$ rotational state (or conformation), we simply divide the matrix product $(U)^3 U(g-)(U)^3$ by $Z(PE)$, where $U(g-)$ is the statistical weight matrix U with first and second column elements, which correspond to the t and g+ conformations, replaced by zeros.

As a measure of the mobility of PE chains confined to occupy the channels of its ICs with urea and PHTP, we have attempted to determine the feasibility of interconverting between the channel conformers without any part (atom) of the chain fragment leaving the channel during any step in the interconversion process. One of the channel conformers is selected as the starting conformation. As each rotation angle is incremented ($\Delta\phi| = 20°$), the x' , y' , z' coordinates of each atom in the PE fragment are calculated and checked to see that all atoms remain inside the cylinder of the starting channel conformer. This procedure is repeated until one or more atoms passes through the cylinder wall, or another channel conformer is reached. If the former occurs, then another channel conformer is selected as the starting conformer, and the interconversion procedure is repeated. After reaching another channel conformer, all rotation angles are reinitialized to the values of the new starting channel conformer, and the interconversion process is restarted. The test for interconversion between channel conformers is complete after each channel conformer has been used as the starting fragment conformation.

The channel conformers found (21) for PE are partially characterized in Table 1, where their numbers, probabilities, and bond rotation state probabilities are presented for cylinders of various diameters (D). Because in the search for channel conformers each atom is considered to be a volumeless point, a channel conformer found to fit into a cylinder of diameter $D' = 2r_c$, would actually fill a cylinder with a diameter D = D' + 1Å, if van der Waals spheres of 0.5Å radius are assigned to each hydrogen atom. The steric requirements of 0.5Å van der Waals spheres placed on each hydrogen atom are reflected in the cylinder diameters given in Table 1.

It should be noted that only ca. 1% of the 2,187 total PE fragment conformers are slim enough to fit into those cylinders (D=5.5Å) corresponding to the channels in the polymer-ICs with urea and PHTP. Among these channel conformers is the lowest energy, all-trans, planar zigzag conformation which is found in bulk PE crystals. Furthermore, it is not possible to interconvert between these channel conformers until the cylinder diameter is increased to D > 6.5Å. The inability of PE channel conformers to interconvert parallels the behavior observed when modeling poly(ethylene oxide) (21), aliphatic polyesters and polyamides (22), polypropylenes (23), and trans-1,4-poly(isoprene) (TPIP) (20), when they are also confined to cylinders with D=5.5Å. On the other hand, two other trans-1,4-polydienes, namely,trans-1,4-poly(butadiene) (TPBD) and isotactic trans-1,4-poly(penta-1,3-diene) (TPPD), do exhibit (20) the ability to interconvert between channel conformers when confined to occupy cylinders with D=5.5Å (see below).

Results and Discussion

To illustrate the utility of FT-IR observations of polymer-U-ICs we will focus on the U-IC formed with poly(ε - caprolactone) (PEC) (28). The FT-IR spectra obtained for urea, bulk, semicrystalline PEC, and PEC-U-IC are presented in Figure 3a, b, and c, respectively. The assignments for uncomplexed, tetragonal urea are well established (29). Tetragonal urea has strong absorption bands at 1682 cm^{-1} due to the C=O stretching vibration, at 1628 and 1599 cm^{-1} due to N-H bending vibrations, and at 1467 cm^{-1} due to the N-C-N stretching vibration. In the spectrum of PEC, we can assume contributions from both crystalline and amorphous regions. It has been demonstrated previously that the crystalline vibrations can be isolated by digital subtraction of amorphous contributions from semicrystalline samples (30). Though normal coordinate calculations have not as yet been performed on PEC, Coleman et al. (31) have assigned some bands using a group frequency approach. The bands at 1724 and 1737 cm^{-1} were assigned to C=O stretching in the crystalline and amorphous phases, respectively. The FT-IR spectrum of bulk PEC in Figure 3b shows the band at 1724 cm^{-1} with a shoulder at 1738 cm^{-1}, indicating contributions from both crystalline and amorphous regions. Three bands at 1192, 1178, and 1161 cm^{-1} are observed and were previously assigned (31) to coupled modes variously associated with C-C-H and O-C-H bending, and C-C and C-O stretching vibrations. The band at 1161 cm^{-1} was identified as an amorphous band and the other two bands derive from the crystalline conformation. In order to confirm these band assignments made for amorphous and crystalline PEC, FT-IR spectra were recorded at a temperature well above the melting temperature (59° C) of PEC.

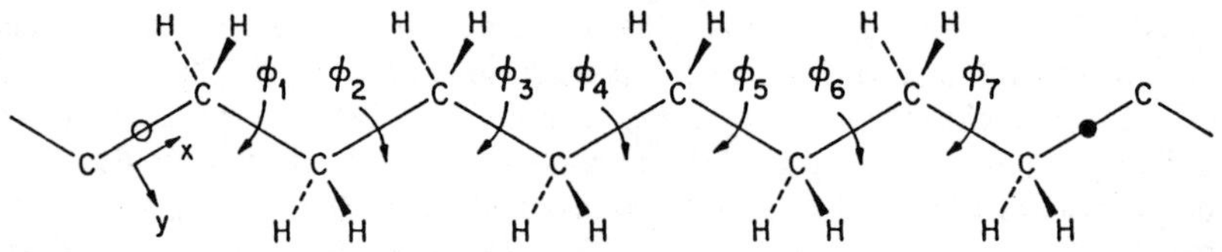

$$U = \begin{array}{c} \\ t \\ g^+ \\ g^- \end{array} \begin{array}{ccc} t & g^+ & g^- \\ \left[\begin{array}{ccc} 1 & .54 & .54 \\ 1 & .54 & .034 \\ 1 & .034 & .54 \end{array}\right] \end{array}$$

Figure 2 - A four repeat unit fragment of PE used to derive (21) the conformations and motions of its channel-bound chains.

Table 1. Polyethylene Channel Conformers (21)

D,Å	channel conformers	probability of channel conformers	$P(\phi_n=t)$ for n=			
			1,7	2,6	3,5	4
4.0	1	0.012	1.0	1.0	1.0	1.0
4.5	3	0.014	0.855	1.0	0.855	1.0
5.0	5	0.016	0.746	1.0	0.746	1.0
5.2	11	0.032	0.591	1.0	0.591	1.0
5.5	25	0.058	0.575	0.750	0.640	0.870
6.0	101	0.164	0.481	0.756	0.648	0.751
7.0	365	0.349	0.519	0.618	0.695	0.650
8.0	805	0.659	0.522	0.605	0.663	0.674
9.0	1597	0.865	0.525	0.608	0.617	0.641
11.0	2165	0.998	0.539	0.605	0.596	0.599
free chain	2187	1.0	0.540	0.605	0.596	0.598

The PEC-U-IC spectrum shows bands characteristic of uncomplexed urea, PEC included in the IC channel, and channel-forming, hexagonal urea in the IC. New bands that are not observed in the spectra of pure urea and bulk PEC are observed at 1658 and 1491 cm^{-1} in the PEC-U-IC spectrum (Figure 3c) and are assigned to channel-forming, hexagonal urea. The bands originating from channel-included PEC chains do not show appreciable differences from those of bulk crystalline PEC, except for the following: the C=O stretching band at 1724 cm^{-1} shifts to 1738 cm^{-1} and a single band is observed for the C-O-C stretching vibration at 1178 cm. It is known from X-ray diffraction (32) that crystalline PEC adopts a nearly all trans conformation, and similarities in both spectra (b and c) indicate that PEC chains in the channels of PEC-U-IC adopt a conformation similar to bulk crystalline PEC. The PEC carbonyl group in PEC-U-IC vibrates 14 wave numbers higher than in bulk, crystalline PEC. This is opposite to the behavior expected if the PEC carbonyl group were hydrogen-bonded to the amide hydrogen of urea in the inclusion compound, and so we suggest that the host (U)-guest(PEC) interactions in PEC-U-IC are principally of the vander Waals type.

The inclusion compounds of n-hexadecane (H-U-IC) (33) and poly(L-lactic acid) (PLLA-U-IC) (34) with urea were formed and characterized using X-ray diffraction, DSC, FT-IR, and solid state C-13 NMR. The FT-IR spectra of H-U-IC and PLLA-U-IC are compared in Figure 4 with that of PEC-U-IC. Hexagonal urea lattice bands at 1658 and 1491 cm^{-1} are observed for all three U-ICs indicating similar overall crystal structures. The crystal structures of H-U-IC (8) and PLLA-U-IC (34) were studied using x-ray diffraction and showed that urea forms a hexagonal lattice in the presence of H and PLLA in the IC channel. Also a very similar X-ray diffraction pattern was observed recently (35) for PEC-U-IC. Consequently it seems reasonable to assign the 1658 and 1491 cm^{-1} bands to C=O and N-C-N stretching vibrations of urea in its hexagonal IC lattice structure.

Analysis of the FT-IR spectrum of a polymer-U-IC may provide information concerning the types and strengths of interactions between the guest polymer and host urea matrix. To carry out such an analysis a clean polymer-U-IC sample, free from uncomplexed urea and/or polymer, is required. Though not shown here, the DSC thermogram of our PEC-U-IC sample indicated very little uncomplexed urea and PEC. The FT-IR spectrum for PEC-U-IC (see Figure 4c) shows small bands at 1682 and 1467 cm^{-1} consistent with the presence of a small amount (ca. 5%) of free, uncomplexed, tetragonal urea. By subtraction of the spectrum observed for tetragonal urea (Figure 3a) from our PEC-U-IC spectrum, which is achieved by monitoring the disappearance of the 1682 and 1467 cm^{-1} bands due to tetragonal urea, the FT-IR spectrum of 100% crystalline PEC-U-IC can be obtained, as shown in Figure 5. Here only the vibrational bands characteristic of hexagonally arranged host urea and channel-included PEC chains are observed.

The stoichiometry of the PEC-U-IC (moles of urea/moles of PEC repeat units)

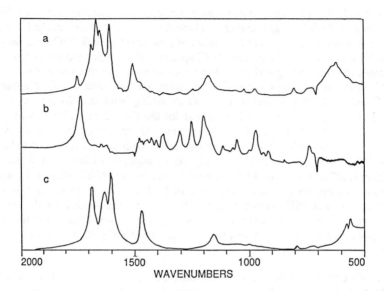

Figure 3 - FT-IR spectra in the region between 500 and 2000 cm^{-1} for (a) urea, (b) PEC, and (c) PEC-U-IC.

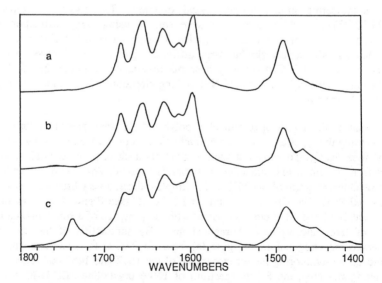

Figure 4 - A comparison of FT-IR spectra in the region between 1400 and 1800 cm^{-1} for the urea based inclusion compounds (a) H-U-IC, (b) PLLA-U-IC, and (c) PEC-U-IC.

can be determined by taking the ratio of urea carbonyl (1658 cm^{-1}) to PEC carbonyl (1738 cm^{-1}) intensities if the relative absorption coefficients are known for these two vibrations. Though absorption coefficients are generally not easily obtained, in this case when the integrated carbonyl absorbance is normalized with respect to a PEC-U-IC internal standard band (1491 cm^{-1}) a ratio of one is obtained. Thus it appears that the C=O stretching vibrations of both carbonyl groups have approximately equal absorption coefficients. On this basis, a stoichiometry of 4 is obtained through comparison of the areas of both carbonyl stretching bands. This FT-IR estimate of PEC- U-IC stoichiometry agrees closely with that obtained previously (35) from DSC measurements.

We close our discussion of the FT-IR observation of polymer-U-ICs with a comparison of the FT-IR spectra observed (36) for polytetrahydrofuran (PTHF) and PTHF-U-IC as presented in Figure 6. Focusing on the 745 cm^{-1} CH$_2$ rocking band observed in the bulk polymer, but absent in the IC, we can draw a conclusion concerning the conformation of PTHF chains in the PTHF-U-IC. Imada et al. (37) have demonstrated that the 745 cm^{-1} CH$_2$ rocking band only appears in those poly(alkane oxides) like PTHF that assume an all trans, planar zig zag conformation in their crystals (38). Those poly(alkane oxides) adopting helical crystalline conformations do not evidence this CH$_2$ rocking band, because the CH$_2$ rocking mode couples with other skeletal stretching and bending modes when the chain conformation is distorted from the planar symmetry of the all trans conformation. The absence of the 745 cm^{-1} CH$_2$ rocking band in PTHF-U-IC points to a nonplanar, helical conformation for the included PTHF chains. An X-ray diffraction analysis of PTHF-U-IC single crystals (39) reached a similar conformational conclusion based on the reduced fiber repeat distance observed for PTHF chains in the U-IC compared with the fiber repeat obtained (38) from bulk PTHF fibers.

The CPMAS/DD C-13 NMR spectrum (28) of PEC crystallized from solution is presented in Figure 7b. Five resonances at 172.9, 64.8, 32.6, 28.5, and 25.1 ppm are evident. A single resonance is obtained for each carbon though two methylene carbons in the PEC repeat unit resonate at identical frequencies. Since the spectrum was recorded with cross polarization (CP), the rigid crystalline carbons, which cross polarize efficiently, are enhanced relative to the mobile carbons in the amorphous phase, which do not cross polarize efficiently at temperatures well above the glass transition temperature. Therefore these resonances are assigned to crystalline carbons. The crystal structure of PEC was studied by Chatani et al. (40) who proposed a nearly all trans, planar zig zag conformation for the crystalline PEC chains. This is consistent with the observations of single resonances for the methylene carbons in the CPMAS/DD C-13 NMR spectrum (Figure 7b). Recording the C-13 NMR spectrum without cross polarization (not shown) also yielded five resonances at 172.0, 63.7, 33.3, 28.2, and 25.0 ppm corresponding to the amorphous carbons in our sample of PEC. The chemical shifts observed for carbons in both the crystalline and amorphous phases are listed in Table 2.

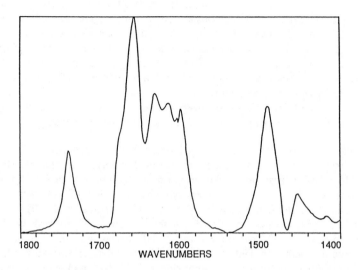

Figure 5 - FT-IR spectrum of 100% PEC-U-IC obtained by subtraction of the FT-IR spectrum of free urea [Figure 3(a)] from that of our PEC-U-IC sample [Figure 3(c)].

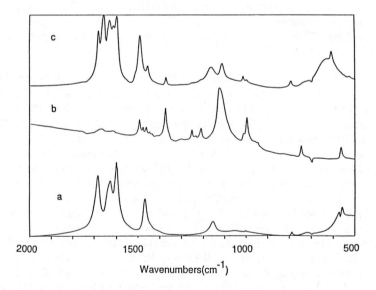

Figure 6 -   FT-IR spectra of (a) bulk PTHF and (b) PTHF-U-IC.

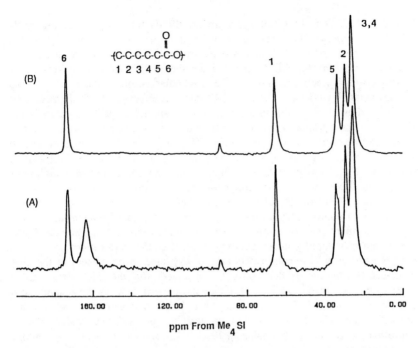

Figure 7 - The CPMAS/DD C-13 NMR spectra at 50.3 MHz measured (28) for (a) PEC-U-IC and (b) bulk PEC.

Table 2. Chemical Shift Values for Poly (ϵ-Caprolactone) in Bulk and in the IC Channel

Carbon	Crystalline chemical shift(ppm)	Amorphous chemical shift(ppm)	PEC-U-IC
1	64.8	63.3	64.7
2	28.5	28.0	28.8
3	25.1	25.0	24.9 26.2
4	25.1	25.0	24.9 26.2
5	32.6	33.3	32.9 34.0
6	172.9	172.0	172.6

The CPMAS/DD C-13 NMR spectrum of PEC-U-IC is shown in Figure 7a, where it is observed that three of its methylene carbons (carbons 3, 4, and 5) show multiple resonances. A similar spectrum is observed when recorded without cross polarization (MAS/DD) and reflects the motion of PEC chains in the narrow U-IC channels which render the cross polarization of magnetization from the abundant H-1 nuclei to the rare C-13 nuclei ineffectual. Multiple resonances can be attributed to more than a single conformational or packing environment. Our molecular modeling (22) of aliphatic polyesters suggested that the all trans bond conformation accounts for at least 40% of all channel conformers found for PEC. The remaining channel conformers found were of the g±tg+ kink variety. The bulk PEC chains crystallize in a nearly all trans conformation, which is the most stable for isolated chains, though the conformations about the CH_2 -O bonds deviate somewhat from 0 . The modeling of PEC in the narrow IC channels also indicated that conformational interconversions among and between kink and all trans conformers was not possible. Because the increased entropy attendant upon the presence of rapidly interconverting channel conformers cannot be realized, the higher energy kink conformers are not expected to be present in the U-IC channels. Furthermore, the splittings observed for the methylene carbon resonances are not consistent with either the pattern or with the magnitude of splitting expected (41) for a population of rigid all trans and kink conformers in the U-IC channels. These observations lead us to suggest a nearly all trans conformation for the PEC chains in PEC-U-IC, and we attribute the multiple resonances observed for methylene carbons 3, 4, and 5 to distinct packing environments in the IC channels.

Spin-lattice relaxation times (T_1) were determined for bulk PEC and PEC-U-IC via the CP-T_1 method (14) and the results for each of the carbon nuclei are presented in Table 3 along with the results for several other bulk and U-IC polymers. T_1's observed for the carbons of U-IC included PEC chains are reduced by factors of 2 to 5 compared with the bulk polymer and provide some measure of the motional constraints produced by the cooperative, interchain interactions occurring in the ordered bulk polymer. The T_1 's of PEC in the urea channels can be compared with the T_1's of n-hexadecane (H) in H-U-IC and PE in its PHTP-IC (42). The T_1's of CH_2 carbons in PEC-U-IC are 4-10 times longer than the T_1 observed for PE CH_2 carbons in PE-PHTP-IC, while the central carbons of H in H-U-IC have T_1 = 15s, intermediate between 6s for PE in PE-PHTP-IC and 20-50s for PEC in PEC-U-IC, though all three types of included chains adopt the nearly all trans conformation. Variable temperature, spin-lattice relaxation measurements (42) demonstrated that PE in its PHTP-IC is on the fast motion side of its T_1 minimum. If we assume that n-hexadecane is also on the fast motion side of its T_1 minimum, then the 16 carbon n-alkane in H-U-IC is moving faster than PE in its PHTP-IC. On the other hand, it seems most likely that PEC in its U-IC is on the slow motion side of its T_1 minimum, and is therefore moving more slowly than either n-hexadecane or PE in their ICs. T_1's two to three times longer are observed for PEC compared with n-hexadecane in the urea channels, reflecting the differences in the interactions between guest

Table 3. C-13 Spin Lattice Relaxation Times (T_1)Observed for
Polymers in Their Ordered IC and Bulk Crystalline Phases

T_1, s

Polymer	IC-Host	IC	Bulk
PEC	U	16-51 (CH_2), 45 (c=o), 16 (U)	93-111 (CH_2), 146 (c=o)
n-hexa-decane	U	15, 16 (U)	
PE	PHTP	6	320
TPBD	PHTP	8 (CH_2), 10 (CH=)	10.5 (CH_2), 12.2 (CH=), Form II crystals (42)
TPIP[a]	PHTP	80 (1), 70 (2), 50 (3), 60, 10 (2')	80 (1), 50 (2), 50 (3), 90 (4), 40, 10 (2') β - form crystals (44)
PTHF	U	1.5, 10.4 (U)	75
PLLA	U		17-26 (CH), 32-36 (C=O), 1-1.5 (CH_3)
PEO[b]	U	4.2, 35 (U)	8.2

a- $-C^1$-$C^2 = C^3$-C^4-
$\quad\quad |$
$\quad\quad C^{2'}$

b - PEO-U-IC adopts a trigonal crystal structure (39) and not the more prevelent
hexagonal structure (8).

polymer and urea host. PEC contains the bulky, rigid, and polar carbonyl group which may interact with urea amide groups forming the channel, and this may be the reason for the longer T_1's observed for PEC compared with n-hexadecane, which cannot interact specifically with urea in the channel. T_1 measurements of the urea C=Os were also carried out for H-U-IC and PEC-U-IC. Approximately the same T_1 (16s) is observed for both ICs. Differences in the mobilities of the included n-hexadecane and PEC chains apparently do not effect the mobilities of urea molecules forming the IC channels.

Aside from the two diene polymers [trans-1,4-polybutadiene (TPBD) and isoprene (TPIP)] (43,44), the spin-lattice relaxation times observed for included polymers are generally substantially shortened compared with the T_1's for their bulk crystalline samples. This serves to emphasize the important contribution made by cooperative, interchain interactions to the mobilities of ordered bulk polymers. The geometrical constraints imposed on included polymer chains by the narrow IC channels (D~5.5Å) are very comparable to those imposed by nearest neighbor chains in bulk crystalline polymer samples. [See for example the interchain distances observed in the crystal structures of polymers (38,45).] The substantial shortening of spin-lattice relaxation times observed for isolated polymer chains residing in their IC channels must be a consequence of the removal of coordinated polymer-polymer motions which impede the mobilities of bulk crystalline polymer chains.

In order to determine the stoichiometry of PEC-U-IC by C-13 NMR, we recorded the MAS/DD spectrum without cross polarization, but with a sufficiently long delay time between signal accumulations to obtain quantitative spectra (13). This estimate of PEC-U-IC stoichiometry was obtained by measuring the intensities of the C=O resonances for PEC and urea. The C=O resonance of PEC had the longer T_1 of 45s, so the delay time of 200s (5 x the longest T_1) was employed. The areas of both C=O resonances were then measured and resulted in a 4:1 mole ratio (U:PEC) estimate of PEC-U-IC stoichiometry. This is in close agreement with the stoichiometry expected for PEC-U-IC based on the all trans conformation for included PEC chains and 1.83Å channel length per urea molecule observed (8) for H-U-IC by single crystal X-ray diffraction and the stoichiometry determined previously by DSC (35) and by FT-IR spectroscopy (28).

Acknowledgment

We are grateful to the National Science Foundation (DMR-9201-094), North Carolina State University,and the College of Textiles for supporting this work.

References

1. Flory, P. J. "Principles of Polymer Chemistry", Cornell University Press, Ithaca, NY, 1953.

2. Flory, P. J. "Statistical Mechanics of Chain Molecules", Wiley-Interscience, New York, 1969.

3. Fetterly, L. C. in "Non-Stoichiometric Compounds", L. Mandelcorn, Ed., Academic Press, New York, 1965, Ch. 8.

4. Farina, M. in "Proceedings of the International Symposium on Macromolecules, Rio de Janeiro, 1974", E. B. Mano, Ed., Elsevier, New York, 1975, p. 21.

5. Farina, M. in "Inclusion Compounds", J. L. Atwood et al., Eds., Academic Press, New York, 1984, Vol. 2, p. 69.

6. Farina, M. in "Inclusion Compounds", J. L. Atwood et al., Eds., Academic Press, New York, 1984, Vol. 2, p. 297.

7. Di Silvestro, G and Sozzani, P. in "Comprehensive Polymer Science", G. C. Eastman et al., Eds., Pergamon, London, 1988, Vol. 4, Ch. 18, p. 303.

8. Smith, A. E. Acta Crystallogr. 1952, 5, 224.

9. Colombo, A. and Allegra, G. Macromolecules 1971, 4, 579.

10. Harris, K. D. M. Ph. D. Thesis, Cambridge University, 1988.

11. Harris, K. D. M. and Joneson, P. Chem. Phys. Lett. 1989, 154, 593.

12. Tonelli, A. E. Comp. Polym. Sci. 1991, 1, 22.

13. Farrar, T. C. and Becker, E. D. "Pulse and Fourier Transform NMR", Academic Press, New York, 1971.

14. Torchia, D. A. J. Magn. Reson. 1978, 30, 613.

15. Dodge, R. and Mattice, W. L. Macromolecules 1991, 24, 2709.

16. Zhan, Y. and Mattice, W. L. Macromolecules 1992, 25, 1554.

17. Zhan, Y. and Mattice, W. L. Macromolecules 1992, 25, 3439.

18. Zhan, Y. and Mattice, W. L. Macromolecules 1992, 25, 4078.

19. Zhan, Y. and Mattice, W. L. Trends in Polym Sci. 1993, 1(11), 343.

20. Tonelli, A. E. Macromolecules 1990, 23, 3129.

21. Tonelli, A. E. Macromolecules 1990, 23, 3134.

22. Tonelli, A. E. Macromolecules 1991, 24, 1275.

23. Tonelli, A. E. Macromolecules 1991,24, 3069.

24. Tonelli, A. E. Macromolecules 1992, 25, 3581.

25. Tonelli, A. E. Makromol. Chem. Symp. Ser. 1993, 65, 133.

26. Tonelli, A. E. Polymer(British) 1994, 35, 573.

27. Abe, A.; Jernigan, R. L. and Flory, P. J. J. Am. Chem. Soc. 1966, 88, 631.

28. Vasanthan, N.; Shin, I. D. and Tonelli, A. E. Macromolecules 1994, 27, 6515.

29. Bhoopathy,T. J.; Baskaran, M. and Mohan, S. Ind. J. Phys. 1988, 62B(1), 47.

30. Vasanthan, N.; Corrigan, J. P. and Woodward, A. E. Polymer 1993, 34, 2270.

31. Coleman, M. M. and Zarian, J. J. Polym. Sci. 1979, 17, 837.

32. Hu, H. and Dorset, D. L. Macromolecules 1990, 23, 4604.

33. Vasanthan, N.; Nojima, S.; and Tonelli, A. E. Macromolecules 1994, 27, 7220.

34. Howe, C.; Vasanthan, N.; Sankar, S.; Shin, I. D.; Simonsen, I. and Tonelli, A. E. Macromolecules 1994, 27, 7433.

35. Choi, C. Davis, D. D. and Tonelli, A. E. Macromolecules 1993, 26, 1468.

36. Vasanthan, N; Shin, I. D. and Tonelli, A. E. J. Polym. Sci., Polym. Phys. Ed., in press.

37. Imada, K.; Miyakawa, T.; Chatani, Y. and Tadokoro, H. Makromol. Chem. 1965, 83, 113.

38. Tadokoro, H. "Structure of Crystalline Polymers," John Wiley, New York, 1979.

39. Chenite, A. and Brisse, F. Macromolecules 1992, 25, 783.

40. Chatani, Y.; Okita, Y.; Tadokoro, H. and Yamashita, Y. Polym. J. 1970, 1, 555.

41. Tonelli, A. E. "NMR Spectroscopy and Polymer Microstrusture: The Conformational Connection", VCH, New York, 1989.

42. Sozzani, P.; Schilling, F. C. and Bovey, F. A. Macromolecules 1991, 24, 6764.

43. Sozzani, P.; Bovey, F. A. and Schilling, F. C. Macromolecules 1989, 22, 4225.

44. Schilling, F. C. ; Sozzani, and Bovey, F. A. Macromolecules 1991,24, 4369.

45. Wunderlich, B. "Macromolecular Physics, " Academic Press, New York, 1973, Vol. 1.

RECEIVED March 7, 1995

Chapter 31

Solvent-Induced Changes in the Glass Transition Temperature of Ethylene–Vinyl Alcohol Copolymer Studied Using Fourier Transform IR and Dynamic Mechanical Spectroscopy

Marsha A. Samus and Giuseppe Rossi

Ford Research Laboratory, Ford Motor Company, P.O. Box 2053, Mail Drop 3198, Dearborn, MI 48121–2053

Results of dynamic mechanical spectroscopy and Fourier-transform infrared spectroscopy are presented for a glassy polymer (ethylene vinyl alcohol copolymer, EVOH) exposed to a plasticizing solvent (methanol), a non-plasticizing fluid (toluene), and mixtures of the two. Quantitative FTIR measurements on thin films of EVOH are used to analyze the sorption and diffusion characteristics of the systems. The presence of even minute quantities of methanol in toluene allows toluene to penetrate EVOH to a significant extent. We interpret the observed large increase in toluene uptake in the presence of small amounts of methanol as due to the fact that the methanol present in the solvent mixture is preferentially absorbed by the EVOH and plasticizes it, thereby lowering its glass transition temperature and eliminating the kinetic constraints preventing penetration by toluene.

Polymer films are often used as barrier layers to prevent or reduce the transport of gases or liquids from one system to another (1). For example, poly(vinyl chloride) and polyethylene are used as barriers to water migration or evaporation, while nylon is an effective hydrocarbon barrier (2). Barrier properties are often due to the glassy nature of the polymer material. In these instances they may be compromised if the fluid being confined is either contaminated or intentionally modified to include an ingredient which plasticizes the barrier.

A glassy polymer in contact with a low molecular weight fluid can be plasticized by the fluid if the molecular interactions are sufficiently favorable (3). In the other

0097–6156/95/0598–0535$12.00/0
© 1995 American Chemical Society

extreme, the interactions between the polymer and the fluid
may be so unfavorable that liquid molecules will enter the
polymer only in very minute amounts not usually detectable
by weight gain measurements.   In these instances, the
polymer will remain glassy (4-9). In the present work we
consider an ethylene-vinyl alcohol copolymer (EVOH) (which
is glassy at room temperature) in contact with two liquids,
methanol and toluene, which are examples, respectively, of
the first and second type of behavior.

Recently, we used dynamic mechanical measurements and
standard sorption and desorption experiments to study the
diffusion behavior in the methanol/EVOH binary system at
temperatures from about 10 to 50 °C below the glass transition temperature (T_g) of the EVOH. (Samus, M. A. and Rossi,
G., submitted to *Polymer*, 1994.) We found that, although
sorption curves taken in this temperature range exhibit
sigmoidal behavior (5), the main features of the sorption
and desorption process can be accounted for by Fickian
diffusion with a diffusion coefficient whose dependence on
concentration exhibits a drastic increase at the methanol
concentration at which the polymer is plasticized.

In the present work we first briefly summarize the
results of our dynamic mechanical analysis on the effect of
solvation on T_g for the EVOH/methanol system: these data
are central to our interpretation of the behavior of
toluene-methanol mixtures in contact with EVOH.   We then
use Fourier-transform infrared (FTIR) spectroscopy to show
that the sorption (and diffusion) behavior of the non-
plasticizing liquid (toluene) is itself critically affected
by the polymer glass transition temperature.   Finally, we
examine in some detail the toluene-methanol-EVOH ternary
system.   We exploit the fact that quantitative FTIR
spectroscopy on thin EVOH films allows detection of small
amounts of penetrant fluid and makes it possible to
differentiate between dissimilar fluid species.   In this
way we are able to obtain separate sorption curves for
methanol and toluene. We find that raising the temperature
above T_g and solvating the polymer with the plasticizing
liquid (methanol) have similar effects on the solubility
(and diffusivity) of the non-plasticizing liquid.

Experimental

The depression in the T_g with increasing methanol concentration is measured by dynamic mechanical spectroscopy.
The dynamic shear storage modulus (G') and loss tangent
(tan δ) were obtained for "neat" and solvent-soaked specimens on a Rheometrics RMS 800 mechanical spectrometer in
torsion rectangular geometry.   Isochronal spectra were
obtained from -70 or -50 °C to +100 °C for determination of
T_g.   The spectra were obtained at .2% strain (verified to
be within the linear viscoelastic region), 240 grams
tension and 1 Hz frequency.

Thin (.7 mm) molded slabs of EVAL F101 EVOH (a random copolymer of ethylene and vinyl alcohol containing 32% ethylene, T_g = 68 °C), G110 EVOH (44% ethylene, T_g = 58 °C) and L101 EVOH (27% ethylene, T_g = 73 °C) were equilibrated in sealed vials for a time sufficiently long to assure that all concentration gradients had disappeared. The equilibration time was chosen on the basis of a diffusion coefficient determined from standard sorption measurements. However, the dynamic mechanical measurements themselves provide a check for whether equilibrium has been reached. Indeed, in samples where concentration gradients (solvated shell, unsolvated core) are present, the observed dynamic mechanical spectra exhibit two peaks in the loss tangent measurement (G''/G'), as shown in Figure 1. The presence of two well separated peaks in non-equilibrated specimens indicates a sharp diffusion front through the EVOH: a uniform change in solvent concentration as a function of penetration distance would result in a very broad, flattened tan δ.

Infrared spectra of .012 mm F101 extruded polymer films held between two KBr plates were recorded with a nitrogen purged Mattson Galaxy Series 5000 FTIR spectrometer with a room temperature DTGS detector. Absorbance spectra were produced by ratioing the single beam spectra of the samples to the single beam spectra of the background. Each ratioed absorbance spectrum consisted of 16 background and 16 sample scans recorded using triangular apodization with approximately 1 wavenumber resolution. Baseline corrections were made as necessary. The peak absorbance of the internal non-reacting (secondary) C-OH stretching mode at 1091 cm^{-1} associated with the EVOH film was used to correct for potential changes in sample thickness due to swelling.

Results and Discussion

Examples of our dynamic mechanical results are shown in Figure 2, which gives storage moduli and loss tangents (tan δ) as functions of temperature for a series of EVOH (F101) specimens equilibrated at different levels of methanol uptake. The transitions of the amorphous phase from the glassy low temperature regime to the rubbery high temperature behavior are smooth and relatively narrow (*10*). As the amount of methanol within the sample increases, the transition region (e.g., T_g) is shifted towards lower and lower temperatures. Figure 3 summarizes the relation between methanol concentration within the polymer and change in T_g for the three copolymers that we examined. The values of T_g reported in the Figure correspond to the temperature at which the maximum in the loss tangent occurs. Although T_g is different for different (ethylene/vinyl alcohol) ratios, the shifts in T_g fall approximately on the same curve. A relatively small amount of methanol is sufficient to drastically lower T_g: for example a 3% uptake induces a drop in T_g of about 40 °C.

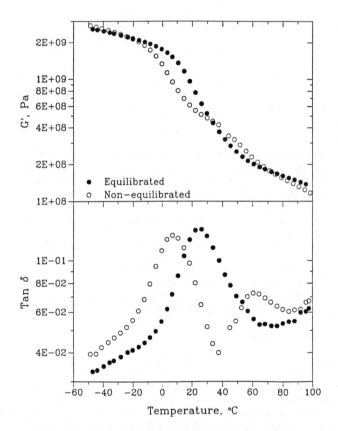

*Figure 1. Equilibrated and non-equilibrated dynam-
ic-mechanical specimen response. The storage
modulus (G') and the loss tangent (tan δ) are given
as a function of temperature.*

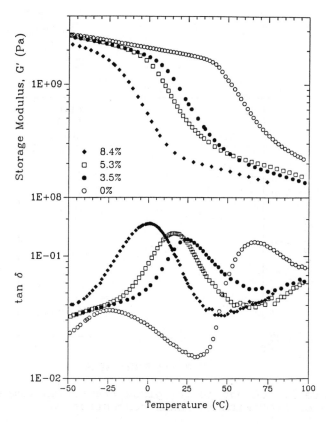

Figure 2. Effect of methanol concentration on the equilibrated dynamic-mechanical specimen response of EVOH F101 polymer. The storage modulus (G') and the loss tangent (tan δ) are given as a function of temperature.

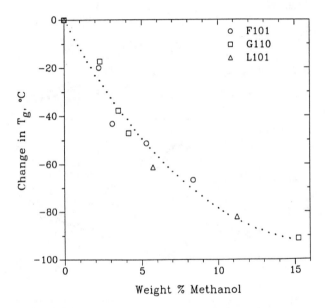

Figure 3.   Change in the glass transition tempera-
ture of EVOH with methanol concentration.

These results show that methanol behaves as a plasticizer for EVOH: they are similar to results reported in the literature for polyvinyl alcohol in the presence of water or glycerol (a standard plasticizer for PVA) (*11*).

Figure 4 shows FTIR spectra obtained from a pure EVOH film (curve a), from methanol (curve b) and from toluene (curve c) in the region of 1200 to 675 cm^{-1}. In order to determine the amount of methanol present in soaked EVOH samples we chose the peak at 1031 cm^{-1} in the methanol spectrum as the characteristic (signature) absorbance: this peak is attributed to the C-OH stretch of the primary alcohol. To identify toluene, the absorbance at 696 cm^{-1}, which is characteristic of the C-H out-of-plane deformation of a monosubstituted benzene, was selected. This choice allows us to detect toluene in EVOH film by FTIR measurements to concentrations as low as .3% by weight.

Using a combination of FTIR spectroscopy and long term weight gain experiments we tried to ascertain the dependence of the solubility and diffusion coefficient on the state of aggregation (glassy or rubbery) of the polymer in the toluene-EVOH binary system. We found that, unlike methanol, toluene does not solvate EVOH under ambient conditions to any detectable extent. Evidence supporting this result is shown in curve (a) of Figure 5: it gives the spectrum of a 12 μm thick EVOH (F101) film soaked in toluene at room temperature for 24 hours and does not exhibit any sign of the signature toluene absorbance. Indeed a series of long term room temperature sorption experiments on thicker films produced no measurable weight gain. On the other hand, the signature toluene absorbance is clearly evident in the FTIR spectrum of a film held at 98°C in the presence of toluene (shown in curve (b) of Figure 5). From weight uptake measurements we determined that the corrected intensity of the toluene absorbance in this instance corresponds to 0.8 wt% toluene in the EVOH film. In other words increasing the temperature from 23 to 98 °C increases the equilibrium volume fraction of toluene in EVOH from near zero to about 1%. In order to correlate precisely this increase in solubility with the effects of the glass transition region, we obtained FTIR spectra from a series of EVOH films held in toluene at different temperatures for several hours. This allowed us to narrow the temperature window where the increase in solubility takes place: there is no evidence of toluene after a 22 hour soak at 50 °C (about 18 °C below T_g), while the toluene signature is clearly present at 78 °C (only 10 °C above T_g).

We also found that for EVOH films which were first exposed to toluene at temperatures well above T_g and then quenched to room temperature, the absorbed toluene remains "trapped" in the film. Specifically, the characteristic toluene absorbance is essentially the same in spectra taken immediately after quenching and several days later. This suggests that, even for these very thin films, little

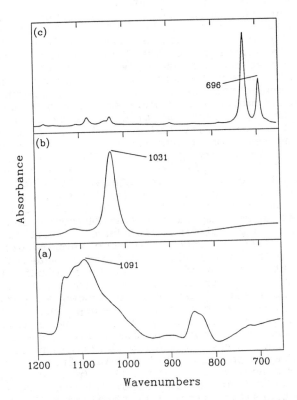

Figure 4.   Fourier-transform infrared spectra showing characteristic absorbances used for identification of components: (a) pure EVOH F101 film, (b) methanol, and (c) toluene.

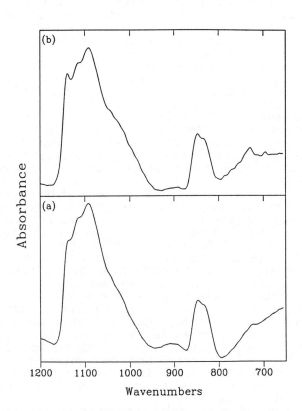

Figure 5.   FTIR spectra of (a) EVOH F101 film exposed to toluene at room temperature for 24 hours, and (b) EVOH F101 film exposed to toluene at 98 °C for 23 hours.

toluene loss occurs during this period. This behavior can be rationalized assuming that the diffusion coefficient of toluene in EVOH changes drastically at the glass transition temperature of the soaked polymer. According to this assumption the diffusion coefficient is relatively large above T_g but extremely small below T_g, so that when the temperature of the soaked film is lower than T_g, toluene diffusion out of the film is severely inhibited. Similar assumptions on the diffusion coefficient of methanol in EVOH can account for many of the observed features of sorption and desorption in the methanol-EVOH system. (Samus, M. A., and Rossi, G., loc. cit.)

The results presented above suggest that the glass transition plays a central role in determining both the equilibrium (solubility) and kinetic (diffusion coefficient) properties of the toluene-EVOH system. In particular, penetration of toluene into EVOH appears to be inhibited as long as EVOH is in its glassy state. These results also indicate that the "apparent" solubility (determined from the observed long-tern uptake) of toluene in EVOH at a given temperature may well depend on the temperature history to which the sample is subjected.

On account of these ideas, it should be expected that if EVOH is plasticized (for example, by exposing it to a solvent such as methanol) and then exposed to toluene, toluene will be able to penetrate the plasticized polymer just as it can diffuse into pure EVOH when the temperature is raised above T_g. The crucial feature in both cases is that the polymer is in its rubbery state. Consequently, if EVOH is exposed to a mixed toluene-methanol solution, both species should be able to penetrate EVOH even at temperatures well below the T_g of the pure polymer. Evidence for this behavior is shown in Figure 6: here spectrum (a) is the same as that reported in Figure 5(b) (e.g., it refers to a sample soaked in toluene at 98 °C for 23 hours); spectrum (b), on the other hand, was obtained from a film soaked at room temperature in a 97% toluene/3% methanol solution by volume. Both spectra exhibit the toluene signature at 696 cm^{-1}.

This kind of qualitative observation on the toluene-methanol-EVOH ternary system can be made quantitative if the characteristic methanol and toluene absorbances are correlated to a given weight uptake of methanol or toluene in the film. This is done through the study of the methanol-EVOH and toluene-EVOH binary systems. The results of such a calibration procedure are reported in Figure 7.

We use the methanol calibration data of Figure 7 to correlate the observed intensity of the methanol absorbance at 1031 cm^{-1} to the amount of methanol present in the EVOH film for the methanol-toluene-EVOH ternary system. Referring again to Figure 6(b), a concentration of 3% methanol in the methanol-toluene fluid mixture in contact with the EVOH film results in an equilibrium concentration within the EVOH film of .11 grams methanol/gram polymer. Accord-

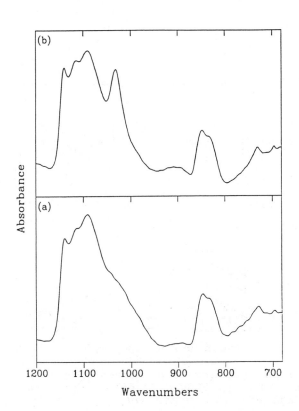

Figure 6. FTIR spectra of (a) EVOH F101 film exposed to toluene at 98 °C for 23 hours, and (b) EVOH F101 film exposed to a 97% toluene/3% methanol solution at 23 °C for 22 hours.

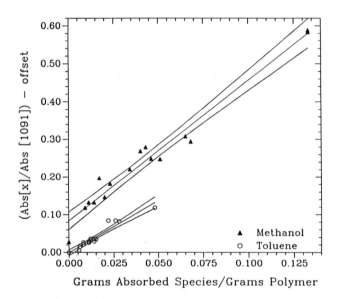

*Figure 7.   Calibration curves for methanol and
toluene in 12 μm EVOH F101 film. Error bars represent the 95% confidence interval for the regression.*

ing to the data reported in Figure 3, this level of methanol uptake depresses T_g by approximately 80 °C, reducing it to -12 °C. Therefore, a room temperature soak in the mixed solvent actually corresponds to immersion in toluene at about 35 °C above the "new" T_g. Using the toluene calibration data of Figure 7, the toluene absorbance in Figure 6(b) corresponds to .011 grams toluene/gram polymer. This is an amount comparable to the .008 grams toluene/gram polymer of curve 6(a), which was obtained by raising the temperature of the binary toluene/EVOH system 30 °C above the T_g of the pure polymer.

The FTIR analysis discussed above provides a useful tool to study the equilibrium concentrations in the ternary system as a function of the composition of the mixed two component liquid to which the polymer is exposed. Furthermore, the same kind of FTIR analysis can be used to obtain information on the kinetic properties (fluid transport) in the ternary toluene-methanol-EVOH system. Preliminary examples of this kind of study are shown in Figures 8(a-c) which give the amounts of methanol and toluene (determined from the FTIR absorbances) as functions of $\sqrt{t}/l$, where t is the time in seconds and l the film thickness in mm, in polymer films immersed at room temperature in three different toluene-methanol mixed solvent compositions: 10% methanol (by volume) in (a), 3% methanol in (b), and .5% methanol in (c). The data of Figure 8(a-c) are in effect the ternary system analog of the usual sorption curves obtained by weight uptake measurements in binary systems. These data show that both the final equilibrium concentration of methanol and toluene within the polymer and the time required to reach equilibrium depend strongly on the composition of the mixed fluid.

The data for the final equilibrium concentrations of methanol and toluene are summarized in Figure 9, which shows that they depend in a very non-linear way on the amount of methanol present in the liquid. For example, the equilibrium uptake of methanol in the polymer grows from approximately 3% to over 12% as the volume fraction of methanol in the liquid goes from .5% to 3%. However, a further increase of the amount of methanol in solution has little effect on the equilibrium amount of methanol taken up by the polymer. At 10% methanol in solution, the uptake nears 15%, a value comparable to the amount taken up by polymer immersed in pure methanol (Samus, M. A. and Rossi, G., *loc. cit.*). For a given solvent-glassy polymer pair, there should be a threshold concentration of solvent below which the solvent is unable to plasticize the polymer (8,9). The data of Figure 9 show that such a threshold for the amount of methanol present in solution falls below the lowest methanol concentration (.5%) that we studied.

The shape of the sorption curves for methanol in Figures 8(a,b) is similar to that of the sorption curves obtained for EVOH in pure methanol (Samus, M. A. and Rossi, G., *loc. cit.*). However, the time required for methanol

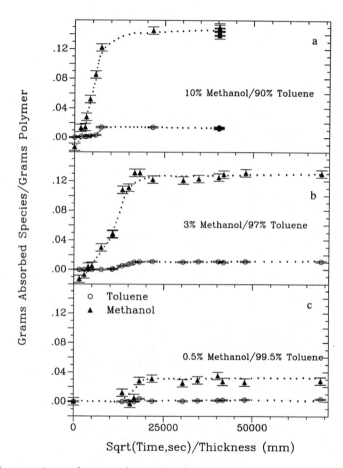

Figure 8. Absorption of methanol and toluene into EVOH F101 films from mixed-fluid systems of three different compositions: (a) 90% toluene/10% methanol; (b) 97% toluene/3% methanol; (c) 99.5% toluene/0.5% methanol.

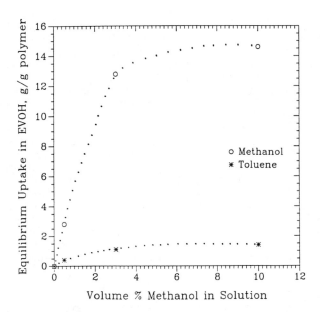

Figure 9. Effect of methanol concentration on the equilibrium uptake of methanol and toluene by EVOH.

absorption to reach equilibrium (e.g., to be within about 20% of the final equilibrium value) increases substantially as the amount of methanol in the mixed liquid decreases. For 10% methanol in the mixed solvent, this time is about four times larger than the corresponding time in pure methanol and it increases by another factor of four as the concentration of methanol in solution falls to .5%. It appears reasonable to rationalize this kind of behavior as due to the dependence of the methanol diffusion coefficient on concentration. The data of Figure 8 also indicate that toluene uptake lags somewhat behind methanol. For example, in 8(b) the toluene uptake becomes sizeable only when about half of the final amount of methanol has already been taken up. Presumably this result reflects the fact that diffusion of toluene and methanol in the ternary system are controlled by different transport parameters.

Conclusions

In summary, we have studied the relation between the polymeric glass transition and the equilibrium behavior and transport properties in the toluene-methanol-EVOH ternary system when EVOH is exposed to a methanol-toluene blend. We have also reported data on the toluene-EVOH and methanol-EVOH binary systems which clarify the behavior of the three component system. The main result of this study is that taking into proper consideration the effect of the glass transition is central to understanding and predicting

550 MULTIDIMENSIONAL SPECTROSCOPY OF POLYMERS

the behavior of the mixed system. In particular, the presence of a plasticizing fluid (methanol) changes the state of the polymer to rubbery, and allows non-solvents such as toluene to penetrate the polymer at temperatures where ordinarily there would be no uptake.

Our study of the ternary system is based on the results of a detailed FTIR analysis. This method makes it possible to obtain separate sorption curves for each low molecular component in the mixed system, so that information on both the equilibrium properties and the transport behavior can be obtained. This method is relatively easy to use and flexible enough to be applicable to most situations where a polymer is put in contact with two or more fluid components. It has the potential of providing reliable information in the area of diffusion of mixed solvent into polymers. Although this is a field of considerable technological importance, it appears to have received little attention in the past precisely due to the difficulty in obtaining reliable experimental information on these systems.

Literature Cited

1. *Barrier Polymers and Structures*; Koros, W. J., Ed.; ACS Symposium Series; ACS: Washington, D. C., 1990.
2. Pauly, S. In *Polymer Handbook*; Brandrup, J.; Immergut, E. H.; Eds.; Wiley: New York, NY, 1989; pp VI/435.
3. Sears, J. K.; Touchette, N. W. In *Encyclopedia of Polymer Science and Technology*; Wiley: New York, NY, 1988, Supplement Volume; pp 335.
4. Flory, P. J.; *Principles of Polymer Chemistry*; Cornell University Press: Ithaca, NY, 1953.
5. Kokes, R. J.; Long, F. A.; Hoard, J. L. *J. Chem. Phys.* **1952**, *20*, pp 1711.
6. Alfrey, T.; Gurnee, E. F.; Lloyd, W. G. *J. Polym. Sci., Part C* **1966**, *12*, pp 249.
7. Thomas, N. L.; Windle, A. H. *Polymer* **1978**, *19*, pp 255.
8. Hui, C. Y.; Wu, K. C.; Lasky, R. C.; Kramer, E. J. *J. Appl. Phys.* **1987**, *61*, pp 5129.
9. Lasky, R. C.; Kramer, E. J.; Hui, C. Y. *Polymer* **1988**, *29*, pp 673.
10. Ferry, J. D. *Viscoelastic Properties of Polymers*; Wiley: New York, NY, 1980.
11. Toyoshima, K. In *Polyvinyl Alcohol*; Finch, C. A.; Ed.; Wiley: New York, NY, 1980; pp 339.

RECEIVED February 2, 1995

Chapter 32

Photophysical and Rheological Studies of Amphiphilic Polyelectrolytes

Correlation of Polymer Microstructure with Associative-Thickening Behavior

Kelly D. Branham and Charles L. McCormick

Department of Polymer Science, University of Southern Mississippi,
Hattiesburg, MS 39406–0076

The synthesis and aqueous solution properties of two hydrophobically modified polyelectrolyte systems are examined. Micellar polymerization was used provide terpolymers of similar bulk composition but with varied hydrophobic microstructures via alteration of the surfactant to hydrophobic comonomer molar ratio (SMR). This synthetic parameter dictates the initial number of hydrophobic monomers per micelle, which ultimately controls the co- or terpolymer microstructure and resultant solution properties. The first system utilizes the fluorescent N-[(1-pyrenyl-sulfonamido)ethyl]acrylamide (APS) monomer as the hydrophobic component to provide direct evidence of microstructural placement. The second system utilizing N-(4-decyl)phenyacrylamide (DPAM) exhibits associative thickening behavior which is dependent on the SMR used in polymerization. These terpolymer systems indicate how variation of the [SDS] to [hydrophobic comonomer], even over a narrow range, can provide polymer systems with significantly different microstructural characteristics and associative properties.

Controlling co- and terpolymer microstructure is crucial to establishing structure-property relationships and design strategies for water soluble polymers for application as sequestration agents for organic contaminants, stabilizers for suspensions and emulsions, and associative thickeners for rheology modification. Aqueous micellar polymerization, which utilizes an external surfactant to solubilize hydrophobic comonomers, has proven to be a successful technique for the synthesis of amphiphilic polymers(1-5) and polyelectrolytes(6-8) which surpass the properties of polymeric viscosifiers synthesized by homogeneous or solution polymerization. The nature of this microheterogeneous polymerization system and the microstructural placement of the hydrophobic groups is of considerable importance

0097–6156/95/0598–0551$12.00/0
© 1995 American Chemical Society

given the enhanced properties of these co- and terpolymers. Initial copolymerization studies by Valint and coworkers(5) with acrylamide and n-arylacrylamides in sodium dodecyl sulfate (SDS) solution indicated that hydrophobe content varied significantly as a function of conversion with the copolymers being initially rich in hydrophobic comonomer. At higher conversions the arylacrylamide content decreased to that expected from the feed ratio. This decrease in the hydrophobe content with conversion implied that the copolymers were compositionally heterogeneous; polymer generated at higher conversions likely contained little or no hydrophobe. Furthermore, this heterogeneity increased at lower surfactant concentrations. These results were later confirmed by Candau and coworkers(9), who also studied kinetic and mechanistic aspects of micellar polymerization.

Although these studies provided significant insight into the heterogeneous nature of the micellar polymerization process, no evidence of microstructural placement was provided. Peer(10) was the first to suggest a blocky microstructure, but direct evidence of such placement was not obtained until fluorescent hydrophobic comonomers were utilized(11). In the latter studies, copolymers of acrylamide and a fluorescent comonomer synthesized by micellar polymerization exhibited higher local chromophore concentrations than those synthesized by solution polymerization. Excimer to monomer intensity ratios (I_E/I_M) in dilute solution indicated the placement of the labels is in a micro-blocky fashion (i.e. short hydrophobic runs separated by long runs of hydrophilic mers) for the polymer synthesized by micellar polymerization; copolymers synthesized by solution polymerization exhibited a random label distribution.

Recent progress(8,12,13) in our laboratories has led to more fundamental understanding of the micellar polymerization process concerning both how the "micro-blocky" placement of hydrophobic comonomers systems may be controlled and how such placement may affect viscosity enhancement in aqueous systems via interpolymer hydrophobic associations. Terpolymers consisting of approximately 60 mole% acrylamide (AM), 40 mole% acrylic acid and one of two hydrophobic comonomers ($\leq$ 0.5 mole%) were synthesized varying the surfactant (SDS) to hydrophobic comonomer ratio to yield terpolymers of identical bulk composition but with varied "micro-blocky" compositions. The first system utilizes the fluorescent N-[(1-pyrenyl-sulfonamido)ethyl]acrylamide (APS) monomer as the sole hydrophobic component to provide evidence of microstructural placement from steady-state fluorescence emission studies(12,13). In the second system, the non-fluorescent N-(4-decyl)phenyacrylamide (DPAM) monomer is used to impart hydrophobic character and associative properties to the hydrophilic AM/AA backbone(8). The use of the APS and DPAM comonomers in these systems is ideal for purposes of comparison since each has approximately the same solubility in SDS micelles(8,12,13). In conjunction, these studies indicate how variation of [SDS] to [hydrophobic comonomer], even over a narrow range, can provide polymer systems with significantly different microstructural characteristics and associative properties.

Experimental

Terpolymer Synthesis. The synthesis of APS(5) and DPAM(8) have been reported previously. All other materials were purchased from Aldrich. Acrylamide (99+%), AM, and sodium dodecylsulfate (99%), SDS, were used as received. Acrylic acid, AA, was distilled before use and potassium persulfate, $K_2S_2O_8$, was recrystallized from water. The detailed procedure for terpolymer synthesis by micellar polymerization is described elsewhere(8). The SDS to hydrophobic comonomer molar ratio, SMR, was varied in successive polymerizations from 40 to 60, 80 and 100 in the respective systems. A previously published method utilizing elemental analysis and UV spectroscopy was employed to determine terpolymer compositions[7].

Solution Preparation. Stock solutions of the copolymer and terpolymers were prepared in deionized water. After dissolution, the pH values of the polymer solutions were adjusted to 7.1- 7.5 using μL amounts of concentrated HCl or NaOH solutions. For 0.5M NaCl solutions, dry salt was added after the pH was adjusted.

Viscometry. Viscosity measurements were conducted on a Contraves LS-30 low shear rheometer at 25 °C and a shear rate of $6s^{-1}$. An upper limit of 250 centipoise may be obtained on the Contraves LS-30 at this shear rate. This value was arbitrarily assigned to samples which exceeded this upper limit for means of comparison.

Light Scattering. Light scattering studies were performed in 0.5M NaCl (APS terpolymers) or 1.0% SDS solution (DPAM terpolymers) at 25°C. Details of these studies are described elsewhere(8,12).

Steady-State Fluorescence Spectroscopy. Steady-state fluorescence spectra were obtained with a Spex Fluorolog 2 Fluorescence Spectrophotometer equipped with a DM3000F data system. Slit widths were maintained at 1-2 mm. Emission spectra were obtained by excitation at 340 nm while monitoring the emission from 350 to 600 nm. Monomer intensities were recorded at 400 nm and excimer intensities were recorded at 519 nm. Spectra were normalized at 400 nm.

Results and Discussion

Terpolymer Synthesis. The hydrophobically-modified polyelectrolytes of this study were prepared from acrylamide (AM), acrylic acid (AA), and either N-[(1-pyrenylsulfonamido)ethyl]acrylamide (APS) (**P2-P5**) or N-(4-decyl)phenyacrylamide (DPAM) (**P7-P10**). The structures of the resulting terpolymers are shown in Figure 1. The synthesis of APS(5)and DPAM(8) have been reported in earlier studies. The polymerization procedure has been detailed elsewhere(8). AM is utilized as the major hydrophilic component since it is readily polymerized to high molecular

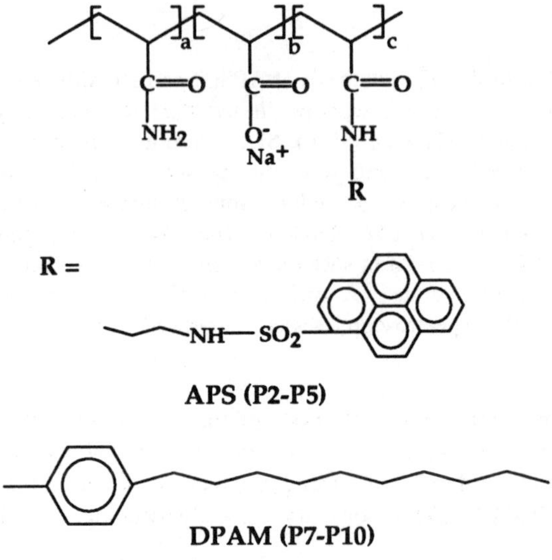

R =

APS (P2-P5)

DPAM (P7-P10)

Figure 1. Terpolymer and Hydrophobe Structures

weights in aqueous media.   AA copolymerizes well with AM and incorporates
ionizable groups along the polymer backbone. The ratio of AM:AA in the feed was
69.5:30.0. A feed content of 0.5 mole% of the hydrophobic comonomer is used to
incorporate covalently attached hydrophobic groups to the hydrophilic polymer
backbone. Potassium persulfate, $K_2S_2O_8$, a water soluble initiator, is used in a ratio
of 3000:1, [total monomer]:[initiator].   These feed ratios provide a series of
moderate charge-density polyelectrolytes which contain very small numbers of
hydrophobic units. Terpolymerizations were carried out in deionized water at 50 °C
under micellar reaction conditions utilizing SDS in the specified ratios to solubilize
the hydrophobic comonomers. The SMR or surfactant to monomer ratio is defined
by Equation 1 below:

$$SMR = \frac{[SDS]}{[H]} \tag{1}$$

where [SDS] is the molar concentration of surfactant and [H] is the hydrophobic
comonomer molar concentration. The SMR may be varied to control the average
number of hydrophobic monomers per micelle, n, as predicted from the Poisson
distribution by:

$$n = \frac{N\,[H]}{[SDS] - CMC}$$

(2)

where CMC is the critical micelle concentration of SDS in this system and N is the aggregation number of SDS(*14,15*).   At the polymerization temperature and monomer concentrations in this study, a CMC value of 6.5×10^{-3} mole/L was obtained for SDS(*8*).   An aggregation number of approximately 60 is generally accepted for SDS.   Equation 2 is appropriate if the aggregation number of the micelle is not significantly altered by the presence of the hydrophobic monomers. This assumption should be quite valid at low values of n.

By setting the SMR values in successive polymerizations (terpolymers **P2-P5** and **P7-P10**) at 40, 60, 80 and 100, the initial number of hydrophobic molecules per micelle can be adjusted as indicated in Tables I and II.   Note that at SDS concentrations well above the CMC, Equation 2 may be approximated by $n = N(SMR)^{-1}$.   Decreasing SMR results in a higher number of hydrophobic comonomers, on the average, per micelle.   An SMR of 60 predicts approximately 1 hydrophobe per micelle initially; below 60 there is an average of less the one hydrophobe/micelle.   Persistent turbidity in polymerization feeds with an SMR of slightly below 40 (approximately 2 monomers per micelle) prohibits the use of lower SMR values for both the APS and DPAM systems.   As indicated in Tables I and II, the range of n in this study is limited to just above and below one hydrophobic monomer per micelle (1.6 to 0.6).   Although this range of n is quite narrow, the data presented here will show how even slight variation of the synthetic conditions can effect differences in terpolymer microstructure and solution properties.

Copolymers of AM and AA, **P1** and **P6** have been synthesized in the presence of SDS (0.132 M) as controls.   A feed ratio of 70:30 (AM:AA) was used in these polymerizations and conditions were identical to those described above.

Terpolymer Characterization.   A previously published method(*7,8*) utilizing elemental analysis and UV spectroscopy was employed to determine the compositions of the AM/AA/APS terpolymers, **P2-P5**, and the AM/AA/DPAM terpolymers, **P7-P10**.   Molar absorptivities for the APS and DPAM model compounds in water were approximately 24,000 (352nm)(*5*) and 11,000 (250nm)(*8*) $M^{-1}cm^{-1}$, respectively.   The compositions of the control AM/AA copolymers, **P1** and **P6**, were determined by elemental analysis.   Compositions for **P1-P5** and **P6-P10** appear in Tables I and II, respectively.   Co- and terpolymers compositions in both systems are very similar in AM and AA content, with approximately 60-65% AM and 35-40% AA.   DPAM content is close to that expected from the feed, while APS incorporation is less than half that expected from the feed concentrations. Decreased hydrophobe incorporation in micellar polymerizations with carboxylate monomers has been attributed to charge effects(*7*).   However conducting the polymerization below the pK_a of the ionizable group can alleviate such problems(*8*). Decreased APS incorporation has been noted in nonionic copolymers prepared by micellar polymerization(*5*), and is likely due to inaccessibility of the acrylamido moiety of APS within the SDS micelle.

Conversions were kept low to avoid drift in polymer composition and to limit heterogeneity. The DPAM terpolymers were polymerized to higher conversions. Note that overall conversion as well as the mole% of hydrophobe incorporated within each respective series appear to be unaffected by SMR within experimental error.

Weight-average molecular weights (M_w) for the APS terpolymers **P2-P5** and copolymer **P1** were determined in 0.5M NaCl. M_w values appear in Table I. Note molecular weights are quite similar for the terpolymers (1.2-1.3 x 10^6 g/mole) while that of the copolymer **P1** is slightly higher (1.7 x 10^6 g/mole). More detailed light scattering studies on **P1-P5** are discussed elsewhere[13].

Table I. Data for AM/AA/APS Terpolymers

Sample	SMR	n	Mole% AM	Mole % AA	Mole % APS	%Conv.	M_w x 10^{-6} (g/mole)
P1	-	-	57	43	0.00	36	1.7
P2	40	1.6	60	40	0.20	23	1.3
P3	60	1.0	61	39	0.17	24	1.2
P4	80	0.8	62	38	0.16	27	1.3
P5	100	0.6	61	39	0.16	28	1.2

M_w values were not determined for the DPAM terpolymers due to aggregation in aqueous salt solutions[8]. Instead, hydrodynamic diameters (d_H) were measured in 1.0% SDS solution for **P6-P10**. Data from these studies are reproduced from Reference 8 in Table II. Similar d_H values were obtained for each polymer under these conditions.

Table II. Data for AM/AA/DPAM Terpolymers[a]

Sample	SMR	n	Mole% AM	Mole % AA	Mole % DPAM	%Conv.	d_H (nm)
P6	-	-	60	40	0.0	25	256
P7	40	1.6	63	37	0.4	53	203
P8	60	1.0	64	36	0.5	53	249
P9	80	0.8	65	35	0.4	57	257
P10	100	0.6	65	35	0.5	52	214

[a] Adapted from ref. 8.

Viscometric Studies - APS Terpolymers. Bulk or macroscopic solution properties of the copolymer **P1** and the terpolymers **P2-P5** were examined using viscometry. Apparent viscosities in deionized water (Figure 2) and 0.5M NaCl (Figure 3) were

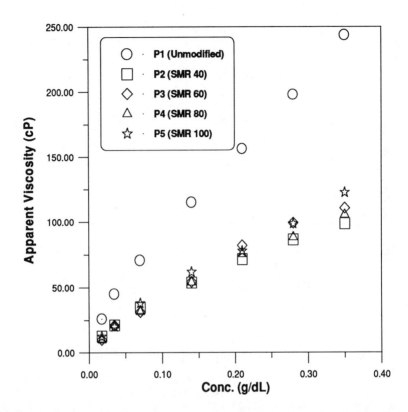

Figure 2: Apparent viscosity as a function of polymer concentration for copolymer **P1** and terpolymers **P2-P5** in water at 25°C and 6s^{-1}.

assessed as a function of polymer concentration. Copolymer **P1** exhibits a linear viscosity increase with concentration in deionized water. Also, the values of the viscosity (in centipoise) are very high, typical of a high molecular weight polyelectrolyte in low ionic strength media. Charge-charge repulsions from adjacent carboxylate groups along the polymer backbone result in an expanded conformation. It can be observed from Figure 2 that the terpolymers **P2-P5** exhibit similar viscosity behavior in deionized water, but differ from **P1** in two ways: the viscosity values are lower over the entire concentration range investigated and the profiles show marked curvature. Lower viscosities for the terpolymers are most likely due to both lower molecular weights and reduced hydrodynamic volume due to intramolecular associations of the APS units. The curvature of the viscosity-concentration profiles for **P2-P5** is likely related to the latter effect as well; the hydrodynamic volumes of the terpolymers decrease at higher polymer concentrations driven by a combination of intramolecular hydrophobic associations and electrostatic screening of atmospherically bound counter-ions. Note that the bulk viscosity continues to increases as the volume fraction of polymer increases.

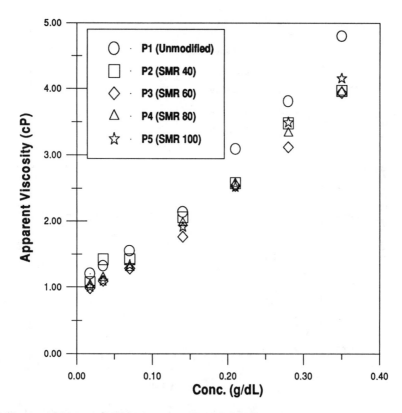

Figure 3: Apparent viscosity as function of polymer concentration for copolymer **P1** and terpolymers **P2-P5** in 0.5M NaCl at 25°C and 6s^{-1}

Copolymer **P1** and terpolymers **P2-P5** exhibit essentially the same viscosity behavior in 0.5M NaCl (Figure 3). Apparent viscosities are an order of magnitude lower than those in deionized water, attributed to collapse of the polymer coil from shielding of the charge-charge repulsions of carboxylate anions along the polymer chain. Viscosities for **P1-P5** essentially lie on the same line and increase in a linear fashion with concentration. No break in the curves or rapid increase in viscosity is evident; therefore, intermolecular associations are not apparent at the macroscopic level.

This behavior differs from earlier work on non-charged AM/APS copolymers of similar label content in which significant intermolecular aggregation was observed (*5,11*). It is believed that the high charge densities ($\approx$ 40 mole% carboxyl groups) of **P2-P5** favor intramolecular associations, especially in deionized water. Studies of hydrophobically-modified polyelectrolytes from our laboratories have correlated decreased associative properties with high polyelectrolyte charge density(*6,7*), particularly at low ionic strength.

Fluorescence Emission Studies - APS Terpolymers. Steady-state fluorescence emission spectra of terpolymers **P2-P5** in water (Figure 4) and in 0.5M NaCl (not shown) for terpolymer concentrations of 0.02g/dL and pH of 7.1-7.5 are qualitatively identical. Each terpolymer exhibits normal or "monomer" fluorescence from approximately 360 to 450 nm as well as excimer fluorescence from 450 to 600 nm (Figure 4, inset).

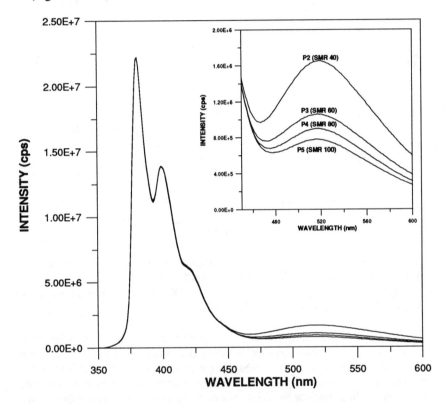

Figure 4. Fluorescence Spectra of the AM/AA/APS terpolymers in water. Inset: Excimer region from 450 to 600 nm for **P2** (SMR 40), **P3** (SMR 60), **P4** (SMR 80), and **P5** (SMR 100).

Emission spectra support the existence of "blocky" microstructures in terpolymers **P2-P5** at ≤ 0.2 mole% of the APS chromophore. The fluorescence emission data are summarized in plots of I_E/I_M in deionized water (O) and 0.5M NaCl ($\bullet$) vs. the SMR used in polymerization (Figure 5). The data in Figure 5 clearly show differences in the amount of excimer formation for each terpolymer. However, before the difference in I_E/I_M may be ascribed to terpolymer microstructure, the contribution of other interactions leading to excimer formation must be considered.

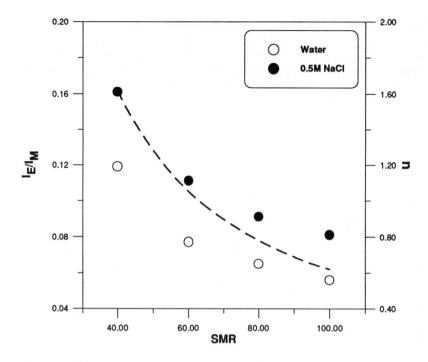

Figure 5. I_E/I_M as a function of the SMR used in polymerzations for the AM/AA/APS terpolymers in water and 0.5M NaCl. The dashed line represents a plot n as a function of SMR calculated from the surfactant and hydrophobe concentrations using Equation 2.

Excimer formation in labeled polymer systems may occur through either intermolecular or intramolecular label interactions. Since interpolymer interactions are not indicated by viscometric measurements in deionized water and 0.5M NaCl, intramolecular interactions appear to dominate. Furthermore, intramolecular label interactions in these systems should be related to polymer conformation (long range label interactions) or neighboring group interactions(16). Differentiation of these two types of interactions has been accomplished by monitoring changes in I_E/I_M in solvents or solutions where the macromolecular dimensions are significantly altered. The dependence of I_E/I_M on thermodynamic quality of the solvent is an indication of long range label interactions(17). In the present case, polyelectrolyte dimensions may be altered by addition of electrolyte or aqueous acid. Comparison of I_E/I_M values for **P2-P5** in deionized water, 0.5M NaCl and at various pH values yields information about the extent of interactions present in aqueous terpolymer solutions.

Fluorescence spectra in deionized water at neutral pH (Figure 4) indicate significant values of I_E/I_M for terpolymers **P2-P5**. At this pH in deionized water , the degree of ionization of the carboxylate groups along the polymer backbone is high and the polymer chains are in an extended conformation. Intramolecular interactions of APS labels or "blocks" of labels from distant parts of the molecule are unfavorable due to the rigidity of the polymer backbone. However addition of electrolytes to the system results in charge shielding allowing a more random coil conformation. Comparison of I_E/I_M for each terpolymer in deionized water and 0.5M NaCl (Figure 5) should reveal additional excimer contribution to the total fluorescence due to the more relaxed polymer conformation in 0.5M NaCl. For terpolymers synthesized at each respective SMR, I_E/I_M is only slightly higher for the terpolymers in 0.5M NaCl, even though the bulk viscosity is an order of magnitude lower than in water because of the decrease in polyelectrolyte hydrodynamic volume in the presence of excess salt. The polymer can adopt more conformations in salt solution and APS interactions within the blocky microstructure which were spatially inaccessible in deionized water may occur in 0.5M NaCl.

The fact that APS chromophore interactions occur on the local level even in 0.5M NaCl is apparent when I_E/I_M for these systems is examined as a function of pH. An example of the pH dependent behavior is shown in Figure 6 for terpolymer **P2** at 0.05g/dL. These data indicate collapse of the polymer coil below pH 6 until phase separation occurs below pH 2.8 (solid symbol). This behavior is the classical behavior for labeled polyacids as a function of pH(*18,19*) and is indicative of long range chromophore interactions in the compact polymer coil at low pH.

Since it is evident that excimer formation at neutral pH in these terpolymer systems arises from a unique microstructure, we may now compare the microstructure and relative label proximity within this series of terpolymers.

Examination of the excimer peaks in the fluorescence spectra of **P2-P5** (inset, Figure 4) reveals that excimer formation scales with the micellar parameters in the polymerization feed. **P2** (SMR 40) has the largest excimer peak followed by **P3** (SMR 60), **P4** (SMR 80) and **P5** (SMR 100), respectively. Also, I_E/I_M vs. SMR for **P2-P5** in deionized water(○) and in 0.5M NaCl (●) (Figure 5) show a striking resemblance to the curve for micellar occupancy number, n, versus SMR (Figure 5, dashed line). Also plots of I_E/I_M as a function n (Figure 7) indicate that label proximity is directly related to the initial number of hydrophobic monomers per micelle. A linear dependence of I_E/I_M with local chromophore composition can be predicted from kinetic and statistical models for random copolymer systems(*17*) and the adherence of the present system to this model will be the subject of further discussion(*13*).

While the APS terpolymers **P2-P5** exhibited no intermolecular aggregation regardless of SMR, the steady-state fluorescence emission data clearly indicate that the initial number of APS monomers per micelle, n, in micellar polymerizations can be used to control microstructure. Studies in the next section with a different hydrophobic group at a higher mole% incorporation address how intermolecular associations may be affected by changes in microstructure.

Viscometric Studies - DPAM Terpolymers. Viscosity studies of **P6-P10** in deionized water and 0.5M NaCl (pH 7.0-7.5) are reproduced from Reference 8. As noted in the preceding sections, the compositions of **P7-P10** are virtually identical and the molecular dimensions in solution are similar. Therefore differences in terpolymer aggregation may be attributed to variation of polymer microstructure based on the SMR during the respective polymerizations.

Viscometric studies were first carried out in deionized water at specified terpolymer concentrations (Figure 8). Plots of apparent viscosity as a function of concentration indicate that the apparent viscosity of the terpolymers and the control copolymer increases in a linear fashion; curvature generally associated with intermolecular hydrophobic associations is not evident in terpolymers **P6-P10**(8). The polymers exhibit high viscosities in deionized water due to the large hydrodynamic volumes typical of polyelectrolytes. **P9** (SMR 80) has the highest hydrodynamic volume in water, followed by **P6** (SMR 40). **P8** (SMR 60) and **P10** (SMR 100) have hydrodynamic volumes approximately equal to the unmodified copolymer **P6**.

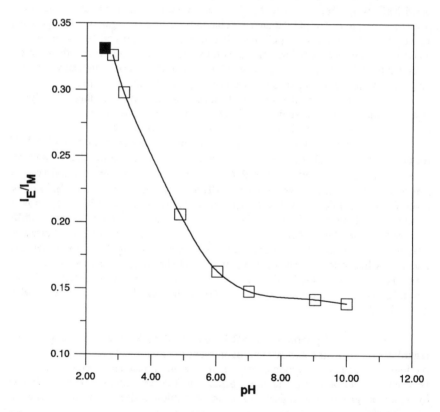

Figure 6. I_E/I_M as a function of pH for **P2** in deionized water. The filled symbol indicates phase separation.

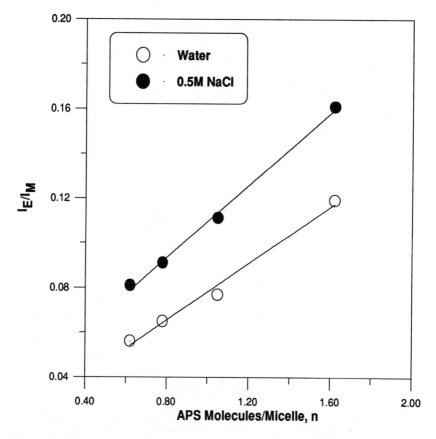

Figure 7. I_E/I_M versus n (Equation 2) for **P2-P5** measured in deionized water and 0.5M NaCl.

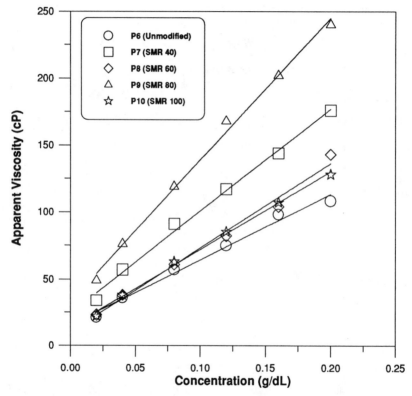

Figure 8. Apparent viscosity as a function of polymer concentration for copolymer **P6** and terpolymers **P7-P10** in water at 25 ° C and 6s[-1]. (Reproduced with permission from reference 8. Copyright 1994 Butterworth-Heinemann.)

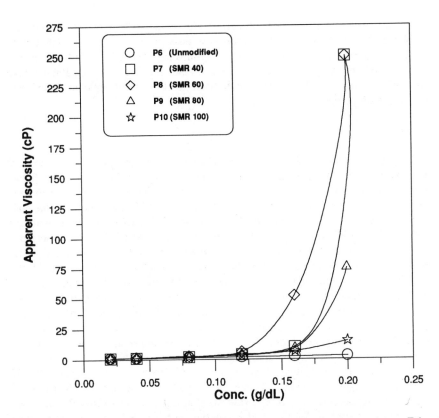

Figure 9. Apparent viscosity as function of polymer concentration forcopolymer **P6** and terpolymers **P7-P10** in 0.5M NaCl at 25 °C and 6s^{-1}. (Reproduced with Permission from reference 8. Copyright 1994 Butterworth-Heinemann.)

Addition of NaCl to hydrophobically-modified polyelectrolytes would be expected to result in: 1) the loss of hydrodynamic volume of the individual polymer coils by shielding of intra-coil ionic repulsions, and 2) the enhancement of hydrophobic associations.   Figure 9 shows plots of apparent viscosity vs. polymer concentration in 0.5 M NaCl(8).   At low polymer concentration, all the polymers exhibit very low viscosities and thus hydrodynamic volumes, indicative of collapse of the individual polymer coils.   Under these conditions, the terpolymers behave identically to the unmodified polymer at low concentration.   At sufficient polymer concentration, three of the four terpolymers exhibit sharp increases in apparent viscosity due to hydrophobic associations of individual polymer coils. At this ionic strength, a correlation between SMR of the polymerization and the associative properties of the resulting terpolymers is noted. Terpolymers **P7** and **P8** with SMR values of 40 and 60 appear to be the most strongly aggregated, with the latter having the lowest critical overlap concentration (C*) (ca. 0.13g/dL). Terpolymers **P7** and **P9** (SMR 80) appear to have approximately the same C* (ca. 0.16g/dL), but **P7** appears to be more strongly aggregated. Terpolymer **P10** (SMR 100) exhibits the lowest viscosity; rheological properties are similar to that of the unmodified copolymer **P6**. Clearly, these salt responsive associations may be correlated with SMR and the initial number of hydrophobic groups per micelle in the polymerization feed. Terpolymers synthesized a low SMR (and thus higher n) exhibit marked associative behavior while ones synthesized at low SMR exhibit diminished associative properties.

Conclusions

Terpolymers of acrylamide, acrylic acid and one of two hydrophobic comonomers were synthesized by micellar polymerization, varying the SDS to hydrophobic comonomer ratio or SMR. In the first system direct examination of microstructural placement was accomplished using the fluorescent APS monomer as the sole hydrophobic constituent. Values of I_E/I_M for the APS terpolymers scale with the initial SDS to APS molar ratio (SMR) and initial number of APS monomers per micelle, n, in the polymerization feed. These studies clearly demonstrate that the initial synthetic conditions control the local chromophore concentration in the resulting terpolymers and that hydrophobic comonomer placement may be controlled by adjusting the SMR. However, no associative-thickening behavior was noted for the APS system due to the low number of hydrophobic groups (≤ 0.2 mole%). Use of a different hydrophobic group in higher mole percentages provided associative-thickening properties. Terpolymers consisting of acrylamide and acrylic acid in the same ratios as the first system, but possessing the DPAM comonomer showed significant associative behavior in 0.5M NaCl which was dependent on the SMR of the polymerization. Terpolymers synthesized at low SMR exhibited extensive associations, while those synthesized at higher SMR exhibited diminished associative properties. Taken together, studies on these two systems demonstrate how "blocky" hydrophobe placement results in enhanced intermolecular hydrophobic associations and how variation of the number of hydrophobic monomers per micelle in micellar polymerization, even over a narrow range, may be used to tailor associative polymer systems.

Acknowledgements. We wish to thank the US Office of Naval Research and the US Department of Energy for financial support of this research.

Literature Cited

1)   Valint, P. L; Bock, J.; Schultz, D. N., *Polym. Mater. Sci. and Eng.,* **1987,** *57,* 482.

2)   Siano, D. B.; Bock, J.; Myer, P.; Valint, P. L., *Polymers in Aqueous Media*, Glass, J. E., Ed., **1989**, Advances in Chemistry Series No. 223; American Chemical Society, Washington D.C, 425

3)   McCormick, C. L.; Nonaka, T.; Johnson, C. B., *Macromolecules*, **1988**, *29,* 731.

4) Valint, P. L.; Bock, J.; Ogletree, J.; Zushuma, S.; Pace, S. J., *Polym. Prepr.*, **1990**, *31*, 67.

5) Ezzell, S. A. and McCormick, C. L., *Macromolecules* **1992**, *25*, 1881.

6) McCormick, C. L.; Middleton, J. C.; Cummins, D. F., *Macromolecules*, **1992**, *25*, 1201.

7) McCormick, C. L.; Middleton, J. C.; Grady, C. E. , *Polymer*, **1992**, *33*, 4184.

8) Branham, K. D.; Davis, D. L.; Middleton, J. C.; McCormick, C. L., *Polymer*, **1994**, *35*, 4429.

9) Biggs, S.; Hill, A.; Selb, J. and Candau, F., *J. Phys. Chem.*, **1992**, *96*, 1505.

10) Peer, W.., *Polymers in Aqueous Media*, Glass, J. E., Ed., Advances in Chemistry Series No. 223; American Chemical Society, Washington D.C,1989, 381.

11) Ezzell, S. A.; Hoyle C. E.; Creed, D.; McCormick, C. L., *Macromolecules* **1992**, *25*, 1887.

12) Branham, K. D.; McCormick, C. L.; Shafer, G. S, *Polym. Mater. Sci. and Eng.*, **1994**, *71*, 423.

13) Branham, K. D.; McCormick, C. L.; Shafer, G. S., *manuscript in preparation.*

14) Thomas, J. K., *The Chemistry of Excitation at Interfaces*, The American Chemical Society, Washington D. C, 1984.

15) Kalyanasundarum, K., *Photochemistry in Microheterogeneous Systems*, **1987**, Academic Press, Orlando, Fla.

16) Soutar, I; Phillips, D., *Photophysical and Photochemical Tools in Polymer Science*, *182*, 97, Winnik, M. A.(Ed.), D. Reidel Publishing Co., 1985.

17) Reid, R. F.; Soutar, I., *J. Polym. Sci., Polym. Physics Ed.*, **1978**, *16*, 231.

18) Turro, N. J.; Arora, K. S., *Polymer*, **1986**, *27*, 783.

19) Arora, K. S;Turro, N. J., *J. Polym Sci., Polym. Chem.*, **1987**, *25*, 259.

RECEIVED February 2, 1995

Chapter 33

Fourier Transform IR and Fluorescence Spectroscopy of Highly Ordered Functional Polymer Langmuir–Blodgett Films

Tokuji Miyashita

Department of Molecular Chemistry and Engineering, Faculty of Engineering, Tohoku University, Aoba Aramaki, Aoba-ku, Sendai 980, Japan

FT-IR spectroscopic measurements (polarized transmission and reflection absorption spectra (RAS) of N-octadecylacrylamide (ODA) LB films indicate that ODA molecules are oriented nearly perpendicular to the surface of the plate, moreover, they are arranged in the plane along the dipping direction (in-plane order). ODA LB films have a biaxial orientation. The emission spectroscopic study for the aromatic chromophore incorporated into the highly ordered N-alkylacrylamide polymer LB films is carried out. The emission spectra for the LB films were measured as a function of the deposited layer and the alkyl chain length of N-alkylacrylamide polymers. The excimer emission intensity increased with increasing number of deposited layers when the effective interlayer energy transfer to the excimer forming site is taking place. The efficiency of vertical energy transfer in the Y-type polymer LB films for naphthalene, pyrene, carbazole and fluorene chromophores was safely judged from the increase in the excimer emission intensity as a function of the deposited layers in the emission spectra. The limiting energy transfer distances for each chromophores are correlated with critical energy transfer distance of Förster type.

Langmuir-Blodgett (LB) method is one of the most useful techniques for fabrication of functional organic ultrathin films. The LB technique makes it possible to prepare thin films with controlled thickness at a molecular size and well-defined molecular orientation (1,2). Because of this superior feature, the LB method has been tried to apply, to polymer materials. We have continued to study the preparation of functional polymer LB films (3). Three approaches, schematically shown in Scheme 1 (A, B, and C) can be considered for preparation of polymer LB films. In the first approach (A), polymerizable amphiphilic monomer is spread onto a water surface and the condensed monomer monolayer is transferred onto a solid support (deposition), and then the polymer LB film is obtained through polymerization by UV light, electron beam, or γ-ray irradiation. In the second approach (B), amphiphilic monomer is polymerized separately by a usual polymerization method and the resulting amphiphilic preformed polymer is spread onto a water, forming the condensed polymer monolayer and then the deposition of the monolayer gives the

0097–6156/95/0598–0568$12.00/0

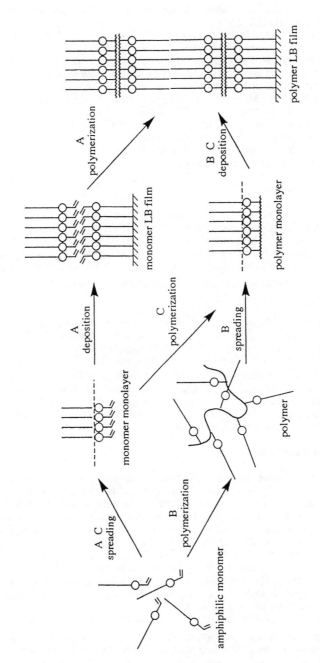

Scheme 1.

desired polymer LB film.   Lastly, the approach (C),   this is just an intermediate of the approach (A) and (B).
We have succeeded in the preparation of fairly uniform polymer LB films using *N*-alkylacrylamide series.   For method A,   *N*-octadecylacrylamide (ODA) monomer gave a stable condensed monolayer on a water surface and a uniform polymer LB film can be obtained by the polymerization of the monomer LB film (*4*).   For method B, preformed poly(*N*-dodecylacrylamide) (PDDA) formed a fairly uniform polymer LB film (*5*).   Moreover, we have proposed a method to incorporate various functional groups as a comonomer of PDDA (*6-8*).   In this paper, we report a study on the molecular orientation of alkyl substituents in ODA and PDDA LB films by FT-IR spectroscopy.   Moreover,   the photophysical behavior of the aromatic chromophores that are incorporated into PDDA LB films by the copolymerization method is discussed from fluorescence spectra.

Experimental Section

Materials. *N*-alkylacrylamide monomers were synthesized by the reactions of the corresponding alkylamines and acryloyl chloride in the presence of trimethylamine at room temperature.   The products were purified by column chromatography and recrystallization.   Poly(*N*-alkylacrylamides) and the copolymers were prepared by usual free radical polymerization.   Chloroform used for spreading monolayers on the water was of spectroscopic grade.   Distilled and deionized water (18 MΩ/cm) by Milli-QII was used for the subphase.

Measurements.   Measurement of surface pressure - area isotherms and the deposition of the monolayer were carried out with automatic working Langmuir trough (Kyowa Kaimen Kagaku HBM-AP using Wilhelmy - type film balance). FT-IR transmission and reflection absorption spectra (RAS) were measured with a JOEL JIR-100 FT-IR spectrophotometer operated at a resolution of 4 cm^{-1}.   CaF$_2$ plate and Au-evaporated slide glass were used for transmission and RAS measurement, respectively.   Fluorescence spectra and UV-visible absorption spectra were measured with a Hitachi 850 spectrofluorophotometer and a Hitachi U-3500 UV-visible spectrophotometer, respectively.

Results and Discussion

Molecular orientation in ODA LB film.   In the previous work, we have showed that ODA and *N*-icosylacrylamide (ICA) form stable condensed monolayers giving LB films and the ODA and ICA LB films are polymerized by UV irradiation, especially ODA LB film can be polymerized completely.   The polymerized ODA LB film was insoluble in usual organic solvents and the possibility of application to a new type of high resolution resist was also indicated (*9*).   An ordered molecular orientation is expected to be formed in ODA LB film.   In this part, a highly ordered structure in ODA (Scheme 2) LB multilayers can be shown by FT-IR measurements (transmission and reflection) absorption spectra (RAS)), especially, the polarized transmission spectra showed a drastic dichroism due to anisotropic orientation of ODA molecule to the dipping direction.
A chloroform solution of ODA was spread onto a water surface.   The surface pressure (π) - surface area (A) isotherms,   which were already studied clearly in the previous paper (*4*),   show the formation of the stable condensed monolayer.   The ODA condensed monolayer was transferred onto solid supports (CaF$_2$ plate for transmission spectra and Au-evaporated glass slide for RAS) at 15 mN/m with a transfer ratio of unity giving Y-type LB multilayers (20 layers).   The transmission

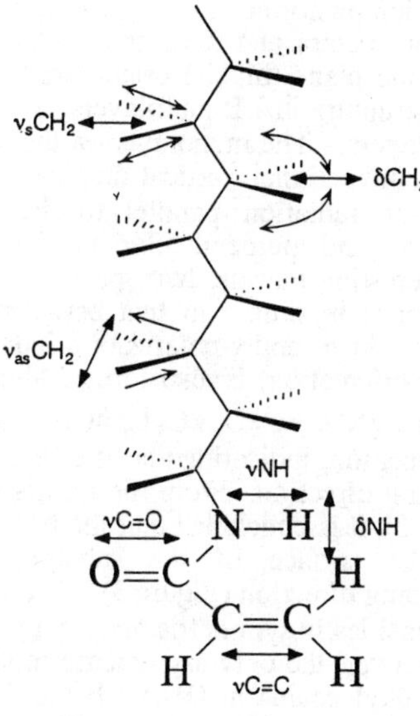

Scheme 2.

spectrum and RAS of the ODA LB multilayers are shown in Figure 1. They are quite different each other. The NH stretching (3300 cm^{-1}), C-H symmetric stretching (2850 cm^{-1}), C=O stretching (1653 cm^{-1}), C=C stretching (1622 cm^{-1}), and CH$_2$ scissoring (1471 cm^{-1}) absorption bands were not observed in RAS in which the electric field vector is perpendicular to the metal surface, whereas those absorption bands appear strongly in a standard transmission spectrum. This indicates that the transition moments of those groups are parallel to the surface of solid supports. The NH, C=O and C=C substituents are highly oriented in parallel and the long alkyl chain is standing up from the surface in the LB multilayers. If the long alkyl chain is completely perpendicular to the surface, the absorption assigned to the C-H asymmetric stretching (2953 cm^{-1}) should not be observed in RAS. However, the ν_{as} CH$_2$ and δNH (1543 cm^{-1}) absorption bands can be observed in RAS, which indicates that the long alkyl chain is tilted regularly.

Although the information on normal plane (uniaxial orientation) can be given by comparison of transmission spectra and RAS, this method is not useful to examine anisotropic orientation in the plane (biaxial orientation). In order to examine the two-dimensional ordered structure in LB multilayers, we measured polarized FT-IR spectra of ODA LB multilayers. The monolayer on the water-surface is transferred onto a CaF$_2$ plate by the conventional vertical dipping method. The transmission spectra are taken with the radiation parallel to the dipping direction of LB multilayers (p-polarization) and perpendicular to the direction (s-polarization) (Figure 2). It is very interesting that the two spectra are drastically different from each other and the difference is similar to that between transmission spectra and RAS. The large dichroism in p- and s-polarization indicates that a highly ordered structure in plane (biaxial orientation) is also formed along the dipping direction in ODA LB multilayers. The νNH, νC=O, νC=C, δCH$_2$ and ν_sCH$_2$ absorption bands appear strongly in s-polarization, indicating that the NH, C=O, and C=C groups are perpendicular to the dipping direction. From the results shown in Figure 1 and the present dichroism (Figure 2), it is concluded that the NH, C=O, and C=C groups are organized parallel to the surface of the substrate and moreover oriented perpendicularly to the dipping direction (Figure 3). On the CH$_2$ stretching vibration bands, the absorption intensities in symmetric and asymmetric modes are reversed in p- and s-polarization, moreover, the only asymmetric mode appears in RAS. These results mean that all the alkyl chains in ODA LB multilayers are packed with the same orientation (tilting with a constant angle). The biaxial molecular orientation shown in Figure 3 is proposed, where the C=C and C=O groups are placed in the same plane and the alkyl chain bends at the bond of CH$_2$ and NH. It can be said that the hydrogen bonding between the amide groups is a driving force for the molecular assembly. This statement is supported by the following results.

Octadecylacrylate (OAc), being the corresponding acrylate derivative to ODA, gives a similar π - A isotherms with ODA monolayer. The condensed OAc monolayer can be transferred onto CaF$_2$ plate with a transfer ratio of unity. The polarized transmission spectra of OAc LB film are shown in Figure 4. The spectra recorded with p- and s-polarized radiation gave nearly the same spectra. The drastic large dichroism shown in Figure 2 for ODA LB film was not observed. Moreover, no dichroism was also observed for the polarized FT-IR spectra of N-octadecylacetamide LB film (Figure 5). The highly ordered structure shown in Figure 3 apparently arises from only the acrylamide structure.

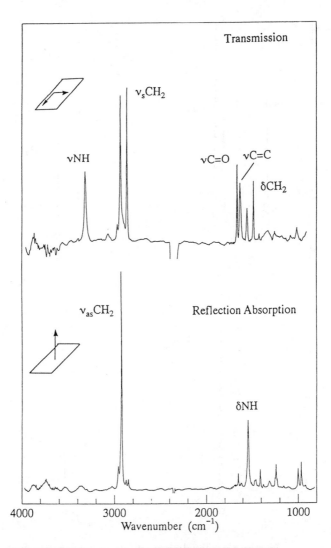

Figure 1. FT-IR spectra of ODA LB film with 20 layers
(upper: transmission, lower:reflection absorption spectra (RAS)).

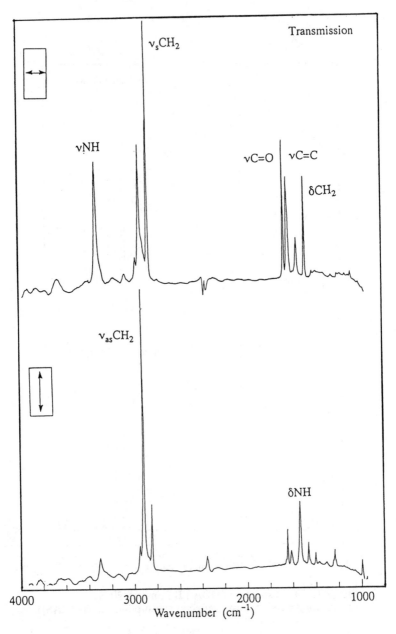

Figure 2. Polarized FT-IR spectra of ODA LB film (upper:s-polarization, lower:p-polarization).

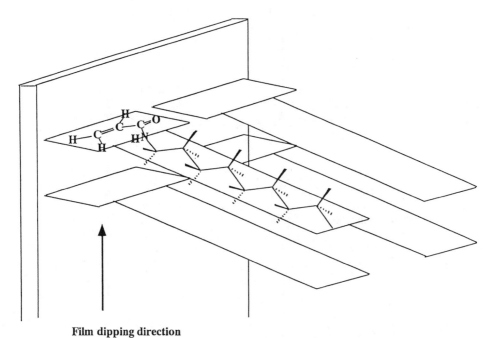

Film dipping direction

Figure 3. A highly two-dimensional ordered structure in ODA LB film

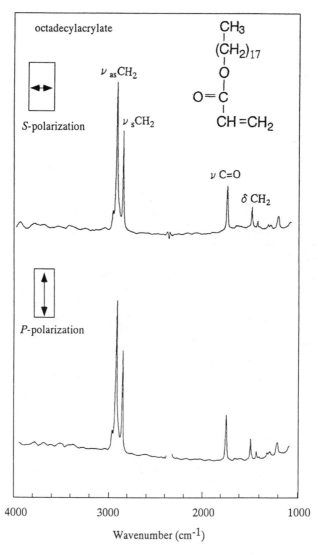

Figure 4. Polarized FT-IR spectra of octadecylacrylate LB film
(upper:s-polarization, lower:p-polarization).

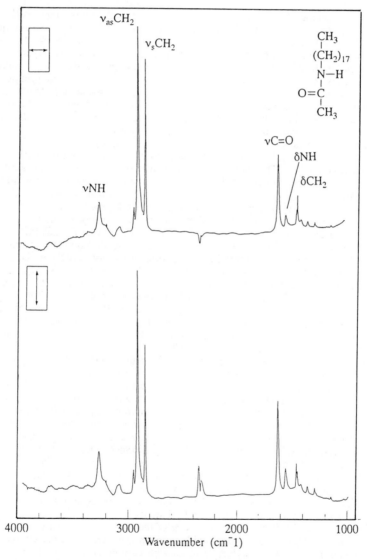

Figure 5. Polarized FT-IR spectra of octadecylacetamide LB film (upper:s-polarization, lower:p-polarization).

Emission spectroscopic study for the chromophores incorporated into polyalkylacrylamide LB films. We have proposed a method to incorporate various functional groups into polymer LB films as a comonomer of N-alkylacrylamides with a help of the excellent ability to form stable LB films (Scheme 3) (*6-8*). In this part, we have prepared LB films containing various aromatic chromophores (Scheme 4) with the purpose of investigating the photochemical behavior of the chromophores placed at a controlled distance in an LB assembly. The emission spectra of the aromatic chromophores were measured as functions of the deposited layers and the alkyl chain length of alkylacrylamides. Effective energy transfer was observed in those LB films owing to a regular molecular orientation of the aromatic groups. The vertical interlayer energy transfer in the LB films, of which efficiency can be estimated from the excimer emission intensity as a function of deposited layers, was controlled by the alkyl chain length of alkylacrylamides (Figure 6). The energy transfer processes are discussed by emission spectroscopy.

Figure 7 shows the emission spectra for the copolymer (DDA/Cz=2/1) of vinylcarbazole (Cz) and N-dodecylacrylamide (DDA) in LB film with various deposited layers (*8*). The excimer emission intensity around 420 nm apparently increases with number of deposited layers. The increment of the intensity, however, is not constant. The excimer emission intensity increases significantly from monolayer to bilayers, whereas the increment from bilayers to three layers is very small. Again the increment from three layers to four layers is larger. The interesting observation is attributable to the Y-type structure of the LB films and the effective interlayer energy transfer to the excimer site through energy migration between Cz chromophores (Figure 8). The interlayer energy transfer in the head-to-head structure occurring for an even number of deposited layers is very effective The fluorescence spectra, however, for the LB multilayers of the VCz copolymer with N-tetradecylacrylamide (TDA), of which alkyl chain is two-carbon longer in length than DDA, showed no change with the number of deposited layers (Figure 9). This result means that vertical energy transfer across the double dodecyl alkyl chains to the excimer forming sites occurs, whereas the transfer across the distance of double tetradecyl alkyl chains is impossible. On the fluorene chromophore, the LB multilayers were fabricated from N-(2-fluorenyl)acrylamide copolymer with TDA or DDA (*10*). The excimer emission intensity even in the TDA copolymer LB multilayers increases with the number of deposited layers (Figure 10), indicating that the vertical energy transfer between the excited fluorene chromophore across the double tetradecyl alkyl chain length occurs. The efficiency of vertical energy transfer can be safely judged from the increase in the excimer emission as a function of deposited layer. The summary on the energy transfer of the chromophores at various alkyl chain lengths is shown in Figure 11. The distance capable of energy transfer for each chromophore is calculated by thickness of the monolayer which is determined by the alkyl chain length and the orientation. The limiting distance for the vertical energy transfer are correlated with the critical distance for Förster type energy transfer. A further study on the detailed structure around the chromophores is necessary to discuss exactly related to the distances.

Conclusively, the fluorescence behavior of the chromophores in LB multilayers described above shows that a fairly stable uniform polymer LB films are fabricated from the copolymers of N-alkylacrylamides with various vinyl aromatic monomers. As a further extension, it is expected that various functional groups can be incorporated into polymer LB films having uniform distribution by this copolymerization method.

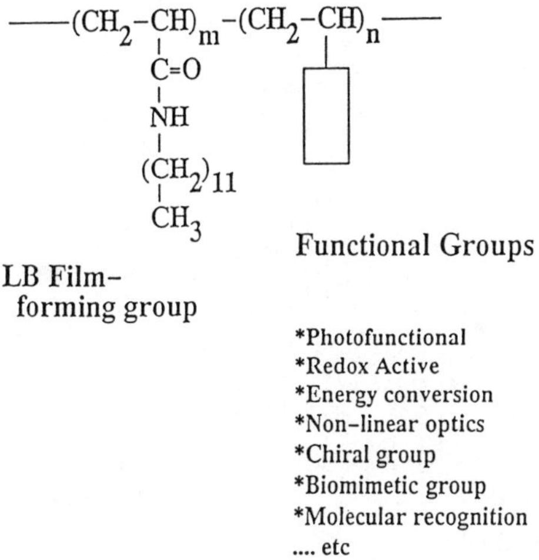

$$—(CH_2\text{-}CH)_m\text{-}(CH_2\text{-}CH)_n—$$
$$\underset{C=O}{|}$$
$$\underset{NH}{|}$$
$$\underset{(CH_2)_{11}}{|}$$
$$\underset{CH_3}{|}$$

LB Film–
forming group

Functional Groups

*Photofunctional
*Redox Active
*Energy conversion
*Non–linear optics
*Chiral group
*Biomimetic group
*Molecular recognition
.... etc

Scheme 3.

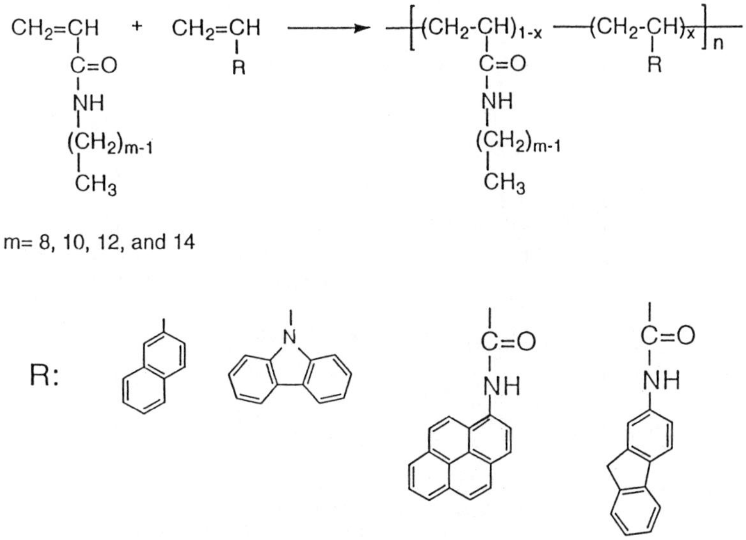

$$CH_2=CH \; + \; CH_2=CH \longrightarrow -\left[(CH_2\text{-}CH)_{1\text{-}x} — (CH_2\text{-}CH)_x\right]_n—$$

with pendant groups:
C=O, NH, $(CH_2)_{m-1}$, CH_3 and R

m= 8, 10, 12, and 14

R:

Scheme 4.

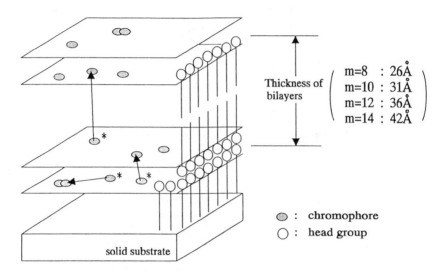

Figure 6. Schematic illustration of energy transfer process in Y-type LB film containing photofunctional groups (Reproduced with permission from ref. 10).

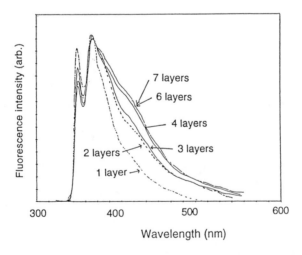

Figure 7. Emission spectra of vinylcarbazole copolymer with *N*-dodecylacrylamide in LB film.

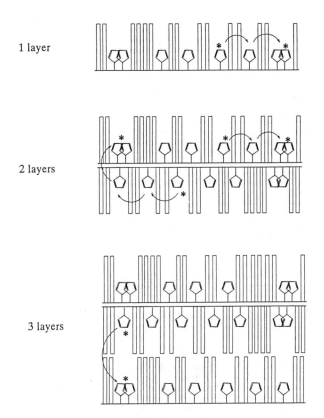

Figure 8. Schematic illustration of energy transfer between Cz chromophores in monolayer, bilayer, and three layers.

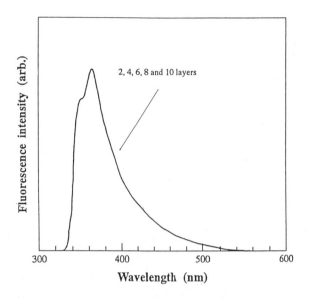

Figure 9. Emission spectra of vinylcarbazole copolymer with N-tetradecylacrylamide in LB film with various deposited layers (Reproduced with permission from ref. 10).

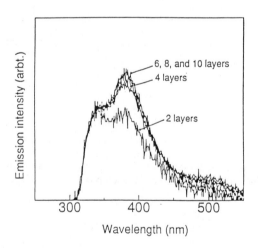

Figure 10. Emission spectra of fluorene copolymer with N-tetradecylacrylamide in LB film with various deposited layers (Reproduced with permission from ref. 10).

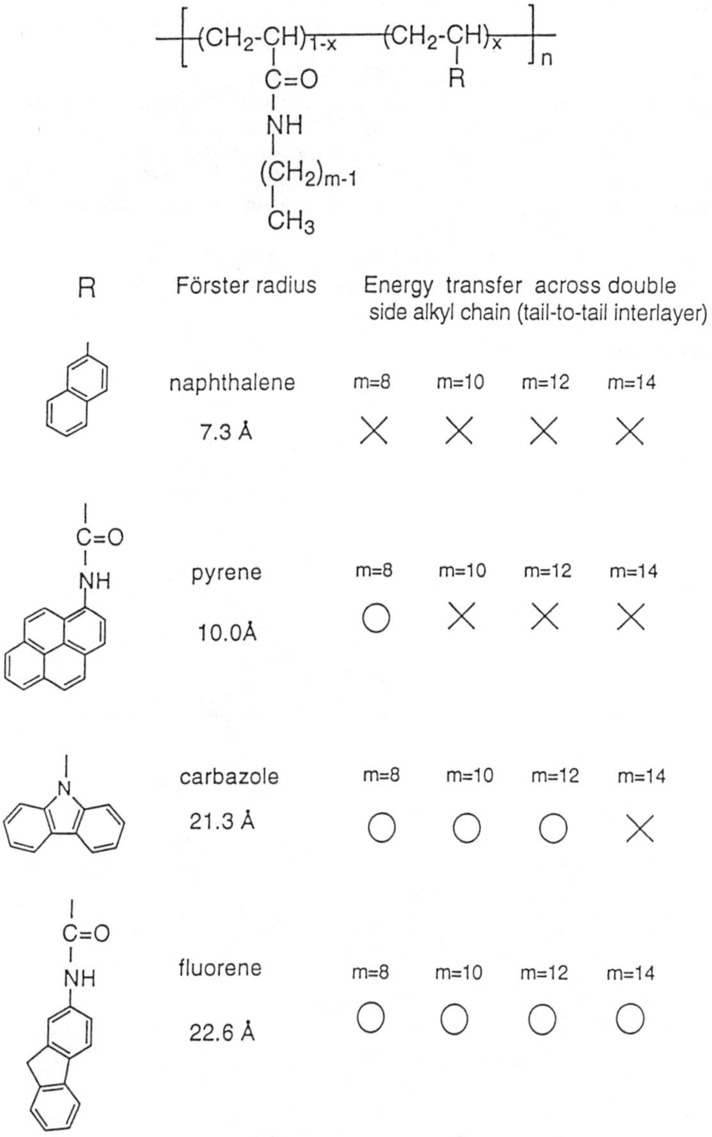

$$-\left[(CH_2\text{-}CH)_{1-x} \quad (CH_2\text{-}CH)_x\right]_n$$

$$C=O \qquad R$$

$$NH$$

$$(CH_2)_{m-1}$$

$$CH_3$$

R		Förster radius	Energy transfer across double side alkyl chain (tail-to-tail interlayer)			
	naphthalene	7.3 Å	m=8	m=10	m=12	m=14
			X	X	X	X
	pyrene	10.0Å	m=8	m=10	m=12	m=14
			O	X	X	X
	carbazole	21.3 Å	m=8	m=10	m=12	m=14
			O	O	O	X
	fluorene	22.6 Å	m=8	m=10	m=12	m=14
			O	O	O	O

Figure 11. Summary of vertical energy transfer across the double alkyl chain length for various aromatic chromophores.

Literature Cited

1. Blodgett, K. B.; Langmuir, I. *Phys. Rev.* **1937**, *51*, 964.
2. Gaines, G. L., Jr. *Insoluble Monolayers at Liquid-Gas Interfaces*; Interscience: New York, **1966**.
3. Miyashita, T. *Prog. Polym. Sci.*, **1993**, *18*, 263.
4. Miyashita, T.; Sakaguchi, K.; Matsuda, M. *Polymer Commun.*, **1990**, *31*, 461.
5. Miyashita, T.; Mizuta, Y.; Matsuda, M. *Br. Polym. J.* **1990**, *22*, 327.
6. Mizuta, Y.; Matsuda, M.; Miyashita, T. *Macromolecules* **1991**, *24*, 5459.
7. Qian, P.; Matsuda, M.; Miyashita, T. *J. Am. Chem. Soc.* **1993**, *115*, 5624.
8. Miyashita, T.; Matsuda, M.; Van der Auweraer, M.; Deschryver, F. C. *Macromolecules*, **1994**, *27*, 513.
9. Miyashita, T.; Matsuda, M. *Thin Solid Films*, **1989**, *168*, L47.
10. Miyashita, T.; Sakai, J.; Mizuta, Y.; Matsuda, M. *Thin Solid Films*, **1994**, *244*, 718.

RECEIVED February 2, 1995

Author Index

Anseth, Kristi S., 116
Attanasio, D., 333
Bandis, Athinodoros, 254
Beckham, Haskell W., 243
Blicharski, J., 290
Boccara, Stephane, 99
Boerio, F. J., 8
Bowman, Christopher N., 116
Branham, Kelly D., 551
Capitani, D., 290,333
Chang, Shih Ying, 147
Chao, James L., 99
Claybourn, M., 41
Cole, Kenneth C., 117
Connors, Laura M., 99
Costantino, Roseann M., 425
Damerau, T., 446
Delgado, Ana H., 117
Deng, Q., 458
Domschke, Angelika, 311
Ellison, E. H., 410
Federici, C., 333
Fina, L. J., 61
Fuji, Akira, 99
Ghiggino, K. P., 363
Gregoriou, Vasilis G., 99
Griffiths, Peter R., 2
Gurau, Mihai, 311
Haines, D. J., 363
Hammond, George S., 425
Harwood, H. James, 215
Hatvany, Gerard S., 215
Hé, Zhiqiang, 425
Hennecke, M., 446
Inglefield, Paul T., 254
Jean, Y. C., 458
Jiang, Eric Y., 99
Jones, Alan A., 254
Kramer, Michael C., 379
Kricheldorf, Hans R., 311

Li, Lan, 215
Liu, J., 458
Lu, Liangde, 425
Ludwig, B. W., 78
Luget, A., 41
Maas, Werner E., 274
McCormick, Charles L., 379,551
Miyashita, Tokuji, 568
Morishima, Yotaro, 490
Naciri, Jawad, 425
Neckers, D. C., 472
Paci, M., 333
Palmer, Richard A., 99
Papini, M., 137
Paroli, Ralph M., 117
Pennington, B. D., 78
Plunkett, Susan E., 99
Probst, Nicolas, 311
Ray, Dale G. III, 215
Rinaldi, Peter L., 215
Rossi, Giuseppe, 535
Samus, Marsha A., 535
Schmidt-Rohr, Klaus, 184,191,243
Segre, A. L., 290,333
Shi, H., 458
Shin, I. D., 517
Smith, T. A., 363
Song, J. C., 472
Soutar, Ian, 356,363,388
Spiess, H. W., 243
Steger, Jamie R., 379
Swanson, Linda, 363,388
Thomas, J. K., 410
Tonelli, A. E., 517
Urban, Marek W., 2,78
Vasanthan, N., 517
Walker, Teri A., 116
Wallace, S. J. L., 363
Wang, Hsin-Ta, 215
Wang, Nam Sun, 147

Weiss, Richard G., 425
Wen, Wen-Yang, 254
Williams, K. P. J., 41

Wutz, Christoph, 311
Young, J. T., 8
Zhao, W. W., 8

Affiliation Index

Bowling Green State University, 472
Bruker Instruments, Inc., 274
Cambridge University, 41
Clark University, 254
Consiglio Nazionale delle Ricerche, 333
Duke University, 99
Federal Institute for Materials Research and Testing—Berlin, 446
Ford Motor Company, 535
Georgetown University, 425
Georgia Institute of Technology, 243
IBM Corporation, 99
ICI Paints, 41
II Università di Roma, 333
Institut für Technische und Makromolekulare Chemie, 311
Istituto Centrale per la Patologia del Libro, 333
Istituto Strutturistica Chimica, 290
Jagellonian University, 290
Lancaster University, 356,363,388

Max-Planck-Institut für Polymerforschung, 191,243
National Research Council of Canada, 117
North Carolina State University, 517
North Dakota State University, 2,78
Osaka University, 490
Renishaw Transducer Systems Ltd., 41
Rutgers University, 61
Tohoku University, 568
Université d'Aix-Marseille I, 137
University of Akron, 215
University of California—Berkeley 191
University of Cincinnati, 8
University of Colorado, 116
University of Idaho, 2
University of Maryland, 147
University of Massachusetts, 184
University of Melbourne, 363
University of Missouri, 458
University of Notre Dame, 410
University of Southern Mississippi, 379,551

Subject Index

A

Absorbance, relationship to concentration, 151
Acrylamide–sodium 11-(acrylamido)undecanoate copolymer, pyrene-labeled, fluorescence studies, 379–386
Adhesion, rheophotoacoustic Fourier transform IR spectroscopy in polymers, 90–97
Alpha processes, merging with beta relaxation in poly(methyl methacrylate), 205–208

Amorphous polymers, dynamics and structure from multidimensional solid-state NMR spectroscopy, 191–213
Amphiphilic polyelectrolytes
advantages of functionalization, 491
examples, 491
hydrophobe content effect on self-organization, 490–491
photophysical and rheological studies, 551–566
previous studies, 491
self-association of hydrophobically modified polyanion, 491,492f
types, 491,493

Amphiphilic polymers
description, 490
self-organization in aqueous
solutions, 490
Ancient paper, characterization, 333–352
Angle-dependent NMR frequencies,
analysis, 192–193,194f
Anionic polymerization of styrene and
isoprene, near-IR spectroscopy, 156
Anisotropy function, definition, 389
Anisotropy measurements, time-resolved,
See Time-resolved anisotropy
measurements
Anthryl-modified low-density polyethylene
films
preparation, 427,429f
See also Photophysical studies,
low-density polyethylene film
characterization
Anti-Stokes scattering, schematic
representation, 3
Applications of luminescence, polymer
science, 356–362
Aqueous micellar polymerization, synthesis
of amphiphilic polymers and
polyelectrolytes, 551
Aromatic polymers, oxygen absorption
using ^{1}H-NMR relaxation, 290–308
Associative-thickening behavior,
correlation with polymer
microstructure, 551–566
Atoms
molecule formation, 2
vibration, 2
Attenuated total reflectance spectroscopy,
description, 4
Azobenzene, photochromic probe studies,
166–180
m-Azotoluene, photochromic probe
studies, 166–180

B

Backbone deuterated polystyrenes,
^{1}H-NMR relaxation of oxygen
absorption, 298–302

Band model, optical properties of
conjugated polymers, 446–447
Beta relaxation in poly(methyl
methacrylate)
coupling of side-group flips to rotations
around local-chain axis, 199–202
function, 196
geometry of large-angle ester group
motions, 196–199
mechanical relaxation relationship,
202,205
merging with alpha processes, 205–208
motional rates, 202,203–204f
Beta sheets of polypeptides and proteins,
example of layer structure, 312
Bulk polymers, chains vs. properties, 518

C

^{13}C cross-polarization–magic-angle
spinning NMR spectroscopy
characterization, 339,343,344f
conformational disorder and dynamics
within crystalline phase of form II
polymorph of isotactic poly(1-butene),
246–248
$\{^{13}C\}^{19}F$ heteronuclear multiple bond
correlation spectroscopy, fluoropolymer
characterization, 228,231–237
$\{^{13}C\}^{19}F$ heteronuclear multiple
quantum coherence spectroscopy,
fluoropolymer characterization,
228,229–230f
$^{13}C/^1H/^{19}F$ triple resonance NMR
spectroscopy for fluoropolymer
characterization, See $^1H/^{19}F/^{13}C$ triple
resonance NMR spectroscopy for
fluoropolymer characterization
^{13}C-NMR spectroscopy, crystalline
polymer inclusion compounds,
520,527,529–533
^{13}C two-dimensional magic-angle
spinning exchange NMR spectroscopy,
conformational disorder and dynamics
within crystalline phase of form II
polymorph of isotactic poly(1-butene),
248,249f

Calcium carbonate loss, weatherability of polyurethane sealants, 126–135

Cellulose, content in paper, 334–352

CH_2 magic-angle spinning lines, structured, conformational information, 208–212

Chain dynamics
smectic poly(ester imides), 325,326f,328–330f
substituted polyaramides, 314,318–323
substituted polyesters, 323,324f,326f

Chain packing, sanidic layer structures, 314,315f,318f

Charge-coupled device detector, description, 43–44

Chemical shift, description, 185

Chemical structure
sanidic layer structures, 314,316–317
smectic poly(ester imides), 323,325,327

Chromophore(s), medium sensitive, study using luminescence spectroscopy, 361

Chromophore incorporation, highly ordered functional polymer Langmuir–Blodgett films, 568–583

Coating thickness, measurement for photocurable resins, 485,486f

Concentration relationship
absorbance, 151
reflectance, 151

Concept, diffuse and specular reflectance measurements of polymeric fibrous materials, 137–138

Confocal measurements, Raman microscopy using charge-coupled device detector, 50–54,55f

Confocal Raman spectroscopy, description, 44

Conformation-dependent NMR frequencies, analysis, 193

Conformational disorder and dynamics
trans-1,4-polybutadiene, 251–252
within crystalline phase of form II polymorph of isotactic poly(1-butene)
^{13}C cross-polarization–magic-angle spinning spectra
chemical shift, 246,248t
various temperatures, 246,247f

Conformational disorder and dynamics—Continued
within crystalline phase of form II polymorph of isotactic poly(1-butene)—Continued
^{13}C two-dimensional magic-angle spinning exchange NMR spectra, 248,249f
conformational energy maps, 251
experimental procedure, 244,246
long-range positional order and dynamic conformational disorder, 248,251
metastable form II to stable form I transformation, 251
origin of conformational exchange, 248,250f

Conformational information from structured CH_2 magic-angle spinning lines
dynamics, 208–212
spectra of aliphatic polymers near glass transition temperature, 208,209f

Conjugated polymers, optical properties, 446–447

Continuous-scan Fourier transform IR spectroscopy, limitations for dynamic spectroscopy, 99–100

Continuous films, film thickness measurements, 423

Copolymer and terpolymer microstructure, importance of control, 551

Correlation time for translational diffusion, definition, 271,272t

Cross-polarization, pulsed NMR spectroscopy, 186

Crystalline polymer inclusion compounds
^{13}C-NMR spectroscopy
procedure, 520
spectrum, 527,529–530
spin–lattice relaxation times, 530–533
stoichiometry, 532
channels occupied by polymer chains, 518–520
formation, 518
Fourier transform IR
chain conformation in bulk polymers vs. inclusion compounds, 527,528f

Crystalline polymer inclusion
compounds—*Continued*
Fourier transform IR—*Continued*
interaction between guest polymer and
host urea matrix, 525,527,528*f*
procedure, 520
spectra, 523,525,526*f*
modeling of polymer chains in inclusion
compound channels, 520–524
study techniques, 520
Cure, measurement for photocurable
resins, 474,477

D

N-(4-Decyl)phenylacrylamide terpolymers,
viscometric studies, 562–566
Degassed polymers, NMR relaxation,
291,292*f*
Degree of cure, measurement for
photocurable resins, 474,477
Desorption diffusion coefficient,
determination, 85–86
Diameter, diffuse and specular reflectance
measurements of polymeric fibrous
materials, 142,145*f*
Diffuse measurements of polymeric
fibrous materials
concept, 137–138
diameter, 142,145*f*
experimental procedure, 138–139
orientation, 142,144–145*f*
surface state effect, 142,143*f*
wavelength effect, 139–141
Diffuse reflectance spectroscopy,
advantages, 6
Diffusion
rheophotoacoustic Fourier transform IR
spectroscopy in polymers, 78–89
study techniques in polyisobutylene, 255
Diffusion rates in low-density
polyethylene films, photophysical
techniques
activation energy calculation,
434,438,440*t*
N,N-dialkylaniline effect on intensity,
434,435–436*f*

Diffusion rates in low-density
polyethylene films, photophysical
techniques—*Continued*
N,N-dialkylaniline migration rate
calculation, 434,437–438
measurement method, 438–443
Dilute solutions, poly(dimethylacrylamide)
behavior studies, fluorescence
techniques, 364–368
4,4'-Diphenylstilbene, photochromic
probe studies, 166–180
Discotic layer structures, schematic
representation, 312,315*f*
Disordered bulk polymers, chains vs.
properties, 518
Dynamic conformational disorder, chain
motions within amorphous polymers
above glass transitions, 243
Dynamic magnitude, determination, 110
Dynamic mechanical spectroscopy
impulse–response mode, 99
modulation–demodulation mode, 99
solvent-induced changes in glass
transition temperature of ethylene
vinyl alcohol copolymer
dynamic mechanical specimen response,
537,538*f*
experimental procedure, 536
glass transition temperature vs.
methanol concentration, 537,540–541
methanol concentration vs. equilibrated
dynamic mechanical specimen
response, 537,539*f*
previous studies, 536
Dynamics, amorphous polymers,
multidimensional solid-state NMR
spectroscopy, 191–213

E

Electric field reorientation, liquid
crystals, 110–115
Electromagnetic enhancement,
definition, 12
Electron paramagnetic resonance
spectroscopy, characterization of
paper, 343,345–352

Electron spin resonance spectroscopy, measurement of local free volume distribution, 168
Emission spectroscopy, description, 4
Energy transfer, See Quenching
Ethylene polymerizations, near-IR spectroscopy, 154,155f
Ethylene vinyl alcohol copolymer, solvent-induced changes in glass transition temperature, 535–550
Excimer formation
process, 405
study using luminescence spectroscopy, 360
water-soluble polymers, 405–407
Excited-state lifetime, determination, 358
Exciton, definition, 447
Exciton model, optical properties of conjugated polymers, 446–447
Extent of isomerization, calculation, 173

F

^{19}F$\}$^{13}C heteronuclear shift correlation spectroscopy, fluoropolymer characterization, 225,227–228
^{19}F$\}$^{1}H insensitive nuclei enhanced by polarization transfer/$\{{}^{13}$C$\}$^{1}H heteronuclear multiple quantum coherence and /$\{{}^{13}$C$\}$^{1}H heteronuclear multiple bond correlation spectroscopy, use for fluoropolymer characterization, 237–241
^{19}F/^{1}H/^{13}C triple resonance NMR spectroscopy for fluoropolymer characterization, See ^{1}H/^{19}F/^{13}C triple resonance NMR spectroscopy for fluoropolymer characterization
Fibers, radiative heat transfer, 137
Fibrous materials, polymeric, diffuse and specular reflectance measurements, 137–145
Film thickness measurements, continuous films, 423
Fluorescence, definition, 358

Fluorescence characterization of unimer micelles of hydrophobically modified polysulfonates
Ad effect, 497,499–501,508–509
Cd effect, 497,499–501,508–509
La effect, 497,499–501,508–509
model, 508,510f
Np effect, 499,501–506
pyrene effect, 495,497,498f
Tl⁺ effect, 506–508
Fluorescence emission quenching
poly(dimethylacrylamide) behavior, 364–365,366f
pyrene-labeled water-soluble polymeric surfactants, 385–386
Fluorescence intensity measurements, poly(dimethylacrylamide) behavior studies, 364,366f,367,369
Fluorescence intensity ratio method for photopolymerization studies
degree of cure calculation, 477,479
description, 477,478f
fluorescence intensity ratio vs. C=C conversion extent, 479,481f,t,484–485
irradiation time effect, 479,480f,483f
photoinitiating mechanism, 479,482
Fluorescence lifetime measurements, poly(dimethylacrylamide) behavior studies, 364,366f
Fluorescence probe techniques
monitoring of photocurable resins
coating thickness measurement, 485,486f
cure monitoring device, 474,477
fluorescence intensity ratio method, 477–485
fluorescence probes, 473–476,478f
photopolymerization studies
examples, 474–476
excitation and emission spectra, 474,476f
types, 473
Fluorescence spectroscopy
advantages for polymer analysis, 473
highly ordered functional polymer
Langmuir–Blodgett films
chromophore incorporation, 578–583
experimental procedure, 570

Fluorescence spectroscopy—*Continued*
  highly ordered functional polymer
    Langmuir–Blodgett films—*Continued*
    molecular orientation in *N*-octadecyl-
      acrylamide film, 570–577
    optical properties of conjugated
      polymers, 447
    poly(dimethylacrylamide) behavior
      complexation with poly(methacrylic
      acid), 375–377
  dilute solutions
    fluorescence intensity and lifetime
      measurements, 364,366*f*
    fluorescence quenching data,
      364–365,366*f*
    time-resolved anisotropy measurements,
      365,367,368*f*
    experimental procedure, 364
    pyrene probe studies, 367
  silica–water interface
    fluorescence intensity measurements,
      367,369
    time-resolved anisotropy measurements,
      369–375
  pyrene-labeled water-soluble polymeric
    surfactants
    experimental procedure, 379–381
    fluorescence emission quenching studies,
      385–386
    instrumentation, 380
    pH effects, 383,385*f*
    polymer syntheses, 381–382
    previous studies, 379
    salt effect, 383,384*f*
    viscosity studies, 382–383,384*f*
Fluorescence techniques, applications, 363
Fluoropolymers, characterization using
  [1]H/^{19}F/^{13}C triple resonance NMR
  spectroscopy, 215–241
Fourier transform IR–attenuated total
  reflectance for surface gradients in
  nylon 11 and nylon 11–poly(vinylidene
  fluoride) bilaminates
  absorption coefficient vs. effective
    depth, 64–66
  experimental coordinate system, 63,65*f*
  experimental procedure, 63–64

Fourier transform IR spectroscopy
  characterization of unimer micelles of
    hydrophobically modified
    polysulfonates, 512,514
  crystalline polymer inclusion compounds,
    520,523,525–528
  diffusion and adhesion in polymers,
    rheophotoacoustic, *See*
    Rheophotoacoustic Fourier transform
    IR spectrocopy of diffusion and
    adhesion in polymers
  highly ordered functional polymer
    Langmuir–Blodgett films
    chromophore incorporation, 578–583
    experimental procedure, 570
    molecular orientation in
      N-octadecylacrylamide
      Langmuir–Blodgett film, 570–577
  polymers, background, 2–7
  solvent-induced changes in glass
    transition temperature of ethylene
    vinyl alcohol copolymer
    calibration of methanol and toluene,
      544,546*f*
    component identification, 541,542*f*
    equilibrium concentration vs.
      two-component liquid composition,
      547,548*f*,549
    experimental procedure, 536–537
    glass transition temperature vs.
      methanol concentration,
      537,540–541
    methanol absorbance intensity vs.
      methanol concentration,
      544,547,549
    previous studies, 536
    solubility and diffusion coefficient on
      state of aggregation, 541,543*f*,544
    toluene absorbance after quenching and
      after time, 541,544,545*f*
Fraction of film fluorescence quenched,
  determination, 430,432–433
Free induction decay, detection, 184–185
Free volume distribution determination
  during multifunctional monomer
  polymerizations, UV–visible
  spectroscopy, 166–180

Free volume hole distributions of polymers, determination using positron annihilation spectroscopy, 458–470

Free volume in liquid, applications, 458

Free volume parameters of polymeric materials
measurement methods, 458–459
theoretical calculations, 458

Fujita theory for diffusion, description, 256

Functional polymer Langmuir–Blodgett films, highly ordered, Fourier transform IR and fluorescence spectroscopy, 568–583

G

Glass transition temperature of ethylene vinyl alcohol copolymer, solvent-induced changes, 535–550

Gold substrates, spin-coated with thin films
(4-mercaptophenyl)phthalimide, 35–38
pyromellitic dianhydride–oxydianiline, 23,25–30

Graphite, example of layer structure, 312

Guest site characterization of low-density polyethylene films, photophysical techniques
determination of fraction of film fluorescence quenched, 430,432–433
procedure, 430

H

^{1}H/^{19}F/^{13}C triple resonance NMR spectroscopy for fluoropolymer characterization
applications, 241
development, 215–216
experimental description, 216
one-dimensional NMR experiments, 216–226
two-dimensional NMR experiments, 225,227–237

^{1}H-NMR relaxation of oxygen absorption on aromatic polymers
best fit of experimental data, 307–308

^{1}H-NMR relaxation of oxygen absorption on aromatic polymers—*Continued*
correlation times, 305–306
experimental description, 291
proton relaxation times vs. temperature, 291,293–295
relaxation in backbone deuterated polystyrenes, 298–302
spin diffusion process, 295–298
spin–lattice relaxation rate equation, 306–307
spin–lattice relaxation theory, 303–305

Helical jump motions
description, 244
occurrence, 244

4,4'-Hexafluoroisopropylidene dianhydride–oxydianiline on silver
reflection–absorption IR spectra, 30,32–35
transmission IR spectra, 30,31f

High-resolution ^{1}H-NMR spectroscopy, characterization of paper, 339,340f,341t,342f

Highly ordered functional polymer Langmuir–Blodgett films, Fourier transform IR and fluorescence spectroscopy, 568–583

Hydrophobic association, cause for self-organization of amphiphilic polymers in aqueous solutions, 490

Hydrophobically modified polysulfonates, spectroscopic characterization of unimer micelles, 490–514

I

Imaging, Raman microscopy using charge-coupled device detector, 54,57–58f

Immiscible polymer–polymer systems, single glass transition temperature for identification, 274

Immobile, definition, 170

Impulse–response mode, dynamic spectroscopy, 99

Inclusion compounds, crystalline polymer, *See* Crystalline polymer inclusion compounds

Interdiffusion, detection using
solid-state NMR spectroscopy of
poly(methyl methacrylate)–
poly(vinylidene fluoride) system,
286–288
Internal conversion, definition, 358
Intramolecular charge transfer
fluorescence probes, monitoring of
degree of cure and coating thickness of
photocurable resins, 472–486
IR absorption scattering
detection method, 3
schematic representation, 3
IR active conditions, vibration, 2
IR spectroscopy
advances, 5
development, 3–5
Isotactic poly(1-butene)
crystalline structure transformation to
more stable structure, 244
wide-angle X-ray diffractograms, 244,245*f*

K

Kubelka–Munk function, definition, 151

L

Langmuir–Blodgett films, highly ordered
functional polymer, Fourier transform
IR and fluorescence spectroscopy,
568–583
Langmuir–Blodgett method, use for
fabrication of functional organic
ultrathin films, 568
Laser-induced sample fluorescence,
problem, 44–45
Layer structures
examples, 312,313*f*,315*f*
formation, 311–312
occurrence, 311
Lecithins, example of layer structure, 312
Light scattering, characterization of
unimer micelles of hydrophobically
modified polysulfonates, 495

Liquid crystal(s)
electric field reorientation, 110–115
step-scan Fourier transform IR
spectroscopy, 99–115
Liquid crystal electroreorientation,
procedure, 102,105,106*f*
Local free volume distribution,
measurement and quantitation
techniques, 168
Low-density polyethylene films
characterization using photophysical
techniques, 425–443
site types, 425
Low-resolution pulsed NMR spectroscopy,
characterization of paper, 335–338
Luminescence
definition, 356
processes, 356–359
Luminescence anisotropy, study using
luminescence spectroscopy, 360
Luminescence quenching, study using
luminescence spectroscopy, 359–360
Luminescence spectroscopy
advantages, 356
applications
excimer formation, 360
luminescence anisotropy, 360
luminescence quenching, 359–360
medium-sensitive chromophores, 361
future, 361–362
processes, 356–359
sensitivity, 356
water-soluble polymers
emission anisotropy measurements,
389–396
energy transfer, 398–405
excimer formation, 405–407
parameters characteristic of intensity,
388–389,390*f*
spectroscopic information, 396–398

M

Macroscopic properties of polymer,
dependence on microstructure of
network, 167

Magic-angle spinning, pulsed NMR spectroscopy, 185

Mean free volume hole sizes, determination, 459–463

Medium-sensitive chromophores, study using luminescence spectroscopy, 361

(4-Mercaptophenyl)phthalimide on gold Raman spectrum, 37,38f
transmission IR spectrum, 35,36f

Methanol, role in glass transition temperature of ethylene vinyl alcohol copolymer, 535–550

Methyl methacrylate polymerization, near-IR spectroscopy, 154–156

Methyl methacrylate–styrene copolymerization, near-IR spectroscopy, 156

Microprobe measurements, Raman microscopy using charge-coupled device detector, 46–50

Microstructure of polymers, correlation with associative-thickening behavior, 551–566

Minimum hole size required for molecule to permit displacement, definition, 270

Miscibility
molecular, detection using solid-state NMR spectroscopy of poly(methyl methacrylate)–poly(vinylidene fluoride) system, 275,277–280
role in positron annihilation spectroscopy of free volume hole distributions of polymers, 467,469

Miscible polymer–polymer systems, single glass transition temperature for identification, 274

Mobile, definition, 170

Modulated mechanical field responses, polymer films, 105,107–110

Modulation–demodulation mode, dynamic spectroscopy, 99

Molecular miscibility, detection using solid-state NMR spectroscopy of poly(methyl methacrylate)–poly(vinylidene fluoride) system, 275,277–280

Molecular orientation, N-octadecylacrylamide Langmuir–Blodgett film, 570–577

Molecular relaxation, role in positron annihilation spectroscopy of free volume hole distributions of polymers, 464,467,468f

Monomer polymerizations, multifunctional, UV–visible spectroscopy for free volume distribution determination, 166–180

Multidimensional NMR spectroscopy
segment identification, 189
use in molecular dynamics studies, 187

Multidimensional solid-state NMR spectroscopy of amorphous polymers
alpha and beta process merging, 205–208
angle-dependent NMR frequencies, 192–193,194f
beta relaxation in poly(methyl methacrylate), 196–205
conformation-dependent NMR frequencies, 193
conformational information from structured CH_2 magic-angle spinning lines, 208–212
experimental procedure, 192
three-dimensional NMR spectroscopy, 195–196
two-dimensional exchange spectroscopy, 193–195

Multifunctional monomer polymerizations, UV–visible spectroscopy for free volume distribution determination, 166–180

Multimode IR analysis of nylon 11
experimental description, 62
Fourier transform IR–attenuated total reflection for surface gradients, 63–66
trichroic IR transmission spectroscopy of thin films, 70,72–76
two-dimensional IR spectroscopy, 66–71

Mutual exclusion rule, description, 2

N

Near-IR spectroscopy
advantages, 5
anionic polymerization of styrene and
isoprene, 156
applications, 148
characteristics, 147
data analysis, 152
ethylene polymerizations, 154,155f
history, 147–148
instrumentation, 148,149t,150f
methyl methacrylate polymerization,
154–156
on-line process analyzer for
polymerization, 156–159,162–163
patent for reaction monitoring
process, 154
polymer characterization
material characterization, 153
polymerization monitoring, 153
spectral deconvolution method,
156,159,160–161f
spectral pretreatment, 148,151
styrene and methyl methacrylate
copolymerization, 156
Negative Poisson's ration phenomenon,
description, 86
NMR characterization of unimer micelles of
hydrophobically modified
polysulfonates
Cd effect, 508,511–512
function, 508
La effect, 508
Np effect, 508,510f,512,513f
NMR spectroscopy
capability to distinguish different
configurations in homopolymers and
different monomer sequences in
copolymers, 290–201
description, 184
fluoropolymer
characterization, See ¹H/¹⁹F/¹³C
triple resonance NMR spectroscopy for
fluoropolymer characterization
oxygen absorption in aromatic polymers,
291,293–308

NMR spectroscopy—*Continued*
polymers
difficulties, 215
multidimensional NMR spectroscopy,
187
outlook, 189
pulsed NMR spectroscopy, 184–186
segment identification, 189
two-dimensional NMR spectroscopy,
186–187,188f
relaxation in degassed polymers,
291,292f
Nuclear Larmor frequency,
determination, 184
Number of hydrophobic monomers per
micelle, calculation, 554–555
Nylon(s), crystallinity vs. properties, 61
Nylon 11
crystal forms, 62
Fourier transform IR–attenuated total
reflectance for surface gradients, 63–66
two-dimensional IR spectroscopy,
66–71
Nylon 11–poly(vinylidene fluoride)
bilaminates, Fourier transform IR–
attenuated total reflectance for
surface gradients, 63–66
Nylon 11 thin films, trichroic IR
transmission spectroscopy, 70,72–76

O

N-Octadecylacrylamide Langmuir–
Blodgett film, molecular
orientation, 570–577
One-dimensional NMR spectroscopy for
fluoropolymer characterization
{¹⁹F}¹³C polarization transfer, 218–223
{¹⁹F}¹H polarization transfer,
222,224–225,226f
spectra, 216–217
Optical properties of conjugated polymers
band model, 446–447
exciton model, 446–447
fluorescence spectroscopy, 447
Ordered bulk polymers, study approach,
518

Orientation, diffuse and specular reflectance measurements of polymeric fibrous materials, 142,144–145f

Oxygen absorption on aromatic polymers, study using ^{1}H-NMR relaxation, 290–308

P

Paper
characterization
^{13}C cross-polarization–magic-angle spinning NMR spectroscopy, 339,343,344f
electron paramagnetic resonance spectroscopy, 343,345–352
high-resolution proton NMR spectroscopy, 339,340f,341t,342f
low-resolution pulsed NMR spectroscopy, 335–338
composition, 334
definition of condition, 334
use as writing material, 333–334

Penetrant diffusion
polyisobutylene, importance, 255
toluene–polyisobutylene solutions, pulsed field gradient NMR spectroscopy, 254–272

Perhydrotriphenylene, formation of crystalline inclusion compounds with polymers, 518

pH, role in fluorescence studies of pyrene-labeled water-soluble polymeric surfactants, 383,385f

Phase separation, detection using solid-state NMR spectroscopy of poly(methyl methacrylate)–poly(vinylidene fluoride) system, 281,283–286

Phenylene–vinylene polymers, photophysical behavior using polarized fluorescence spectroscopy, 446–456

Phosphatides, example of layer structure, 312

Photoacoustic–Fourier transform IR spectroscopy
evaluation of weatherability of polyurethane sealants, 119–135
experimental setup, 78–79

Photoacoustic intensity, determination, 85

Photochromic probe techniques, measurement of local free volume distribution, 168

Photocurable resins, monitoring of degree of cure and coating thickness, fluorescence probe techniques, 472–486

Photophysical behavior of phenylene–vinylene polymers, studies using polarized fluorescence spectroscopy, 446–456

Photophysical studies
low-density polyethylene film characterization
diffusion rates, 434–443
experimental procedure, 426
guest-site size distributions, 430–433
modified film preparation, 426–427,428–429f
spectroscopic investigations, 427,430,431t
microstructure–associative-thickening behavior of amphiphilic polyelectrolytes
N-(4-decyl)phenylacrylamide terpolymers, 562–566
experimental procedure, 553
importance, 551–552
previous studies, 552
N-[(1-pyrenylsulfonamido)ethyl]-acrylamide terpolymers, 556–563
terpolymer synthesis, 553–555,556t
thin films of polystyrene and poly(methyl methacrylate)
experimental materials, 411
instrumentation, 411–412
polymer film preparation on quartz slides, 412
(1-pyrenyl)butyltrimethylammonium bromide
in presence of polymer films, 418,420–422
on quartz slides, 413–419
quartz surface preparation and derivatization, 412
sample measurement procedure, 412–413

Photophysical techniques, advantages, 388
Photopolymer coatings, factors affecting
    properties, 472
Photopolymerization
    applications, 167,472
    description, 472
    methods for kinetic studies, 472–473
Plasticizer loss, weatherability of
    polyurethane sealants, 120–135
Polarized fluorescence spectroscopy of
    photophysical behavior of
    phenylene–vinylene polymers
    behavior in stretched polyethylene
        films, 451–452
    excitation energy transfer, 456
    experimental procedure, 447–448
    fluorescence properties, 453,454–455f
    intermolecular excitation energy
        transfer, 451,452f
    intramolecular excitation energy
        transfer, 449–451
    optical properties of conjugated
        polymers, 446–447
    quantum yield effect, 456
    role of solvents of different viscosity,
        453–455
    segment distribution modeling,
        448–449,450f
Poly(alkylacrylamide) Langmuir–Blodgett
    films, chromophore incorporation,
    578–583
Polyamides
    crystal forms, 62
    number of carbon atoms vs. properties, 61
    prediction of chain arrangement, 61–62
    three-dimensional alignment, 70
Polyaramides, substituted, chain dynamics,
    314,318–323
trans-1,4-Polybutadiene, conformational
    disorder and dynamics, 251–252
Poly(1-chloro-1-fluoroethylene-*co*-
    isobutylene), one-dimensional NMR
    spectroscopy, 216–226
Poly(dimethylacrylamide), behavior studies
    using fluorescence, 363–377
Polyelectrolytes, amphiphilic, photophysical
    and rheological studies, 551–566

Polyesters, substituted, chain dynamics,
    323,324f,326f
Polyethylene, characteristics, 425
Polyimides
    metal substrate–structure relationship
        determination, 12–14
    use as dielectrics in multilevel
        interconnects, 12
Polyisobutylene diffusion, study
    techniques, 255
Polymer(s)
    amorphous, 191–213
    aromatic, oxygen absorption, ^{1}H-NMR
        relaxation, 290–308
    degassed, NMR relaxation, 291,292f
    Fourier transform IR and Raman
        spectroscopy, 2–7
    free volume hole distributions using
        positron annihilation spectroscopy,
        458–470
    NMR spectroscopy, 184–189
    Raman microscopy and imaging, 41–58
    rheophotoacoustic Fourier transform IR
        spectroscopy of diffusion and
        adhesion, 78–97
    size and shape adjustment for external
        stresses and internal interactions,
        517–518
    step-scan Fourier transform IR
        spectroscopy, 99–115
    water soluble, luminescence spectroscopy,
        388–407
    with layer structures, chain
        packing and chain dynamics,
        311–331
Polymer characterization, near-IR
    spectroscopy, 153
Polymer films
    modulated mechanical field responses,
        105,107–110
    photophysics of (1-pyrenyl)butyl-
        trimethylammonium bromide,
        418,420–422
    use as barrier layers, 535–536
Polymer inclusion compounds,
    crystalline, *See* Crystalline polymer
    inclusion compounds

Polymer Langmuir–Blodgett films
highly ordered, functional,
Fourier transform IR and
fluorescence spectroscopy, 568–583
preparation approaches, 568–570
Polymer microstructure, correlation with
associative-thickening behavior,
551–566
Polymer network structure, reaction
conditions vs. final properties, 167
Polymer science, applications of
luminescence spectroscopy, 356–362
Polymeric fibrous materials, diffuse and
specular reflectance measurements,
137–145
Polymeric surfactants, pyrene-labeled,
water-soluble, fluorescence studies,
379–386
Polymerization behavior of multifunctional
monomers, problems with characterizing
network structure, 167
Polymerization reactions, monitoring by
near-IR spectroscopy, 147–163
Poly(methacrylic acid), complexation
with poly(dimethylacrylamide), 375–377
Poly(methyl methacrylate)
beta relaxation, 196–205
luminescence spectroscopy, 388–407
photophysical study, 410–424
Poly(methyl methacrylate)–poly(vinylidene
fluoride) system, miscibility, phase
separation, and interdiffusion detection
using solid-state NMR spectroscopy,
274–288
Poly(1,4-phenylene-1,2-diphenylvinylene),
excitation energy transfer, 456
Poly(1,4-phenylene-1-phenylvinylene),
excitation energy transfer, 456
Poly(phenylenevinylene), applications, 446
Poly(1,4-phenylphenylenevinylene)
excitation energy transfer, 456
photophysical behavior using polarized
fluorescence spectroscopy, 448–452
Polystyrene(s), backbone deuterated,
^{1}H-NMR relaxation of oxygen
absorption, 298–302

Polystyrene thin films, photophysical
study, 410–424
Polysulfonates, hydrophobically modified,
spectroscopic characterization of
unimer micelles, 490–514
Poly(tetrafluoroethylene)
negative Poisson's ration phenomenon, 86
permeability, 86,88–89
Polyurethane sealants
formulation, 117–118
production, 117
weatherability, 117–135
weathering analytical techniques, 118–119
Poly(vinylidene fluoride), polarization
characterization techniques, 63
Poly(vinylidene fluoride) diffusion
diffusion coefficient
vs. diffusion parameter, 86,87f
vs. percent crystallinity, 86,88f
diffusion parameters, 85,87t
elastic deformation effect, 82,83f
plastic deformation effect, 82,84f
procedure, 81–82
quantitative analysis, 82,85
Poly(vinylidene fluoride)–poly(methyl
methacrylate) system, miscibility,
phase separation, and interdiffusion
detection using solid-state NMR
spectroscopy, 274–288
Positron annihilation lifetime
spectroscopy
free volume hole distributions of polymers
determination, 463
future research, 470
mean free volume hole sizes, 459–463
miscibility effect, 467,469
molecular relaxation effect,
464,467,468f
pressure effect, 464,465–466f
previous studies, 459
local free volume distribution
measurement, 168
sensitivity, 459
Pressure, role in positron annihilation
spectroscopy of free volume hole
distributions of polymers, 464,465–466f

Probe molecule
absorbance vs. trans and cis
concentrations, 171
mobile fraction, 171–172
reactions during polymerization, 170–171
Pulsed field gradient NMR spectroscopy
advantages, 255
cross-polarization, 186
description, 255
dipolar decoupling, 185
free induction decay, 184–185
interactions, 185
magic-angle spinning, 185
nuclear Larmor frequency, 184
spin diffusion, 185–186
toluene–polyisobutylene solutions
behavior vs. that of other polymers,
270–271,272*t*
correlation time
for segmental motion, 271
for translational diffusion, 271,272*t*
experimental procedure, 258
fractional free volume
of solution vs. volume fraction of
toluene, 268,269*f*
of toluene vs. entanglement in
polyisobutylene, 268
free volume analysis of viscosity, 268
interpretation, 258–267,269
polymer segmental motion
model, 264
temperature dependence,
264,266–267,269*t*
self-diffusion coefficient
concentration dependence,
258,260–264,265*f*
temperature dependence, 258,259*f*
N-[(1-Pyrenylsulfonamido)ethyl]-
acrylamide terpolymers
fluorescence emission studies, 559–563
viscometric studies, 556–558
Pyrene-labeled acrylamide–sodium
11-(acrylamido)undecanoate copolymer,
fluorescence studies, 379–386
Pyrene-labeled water-soluble polymeric
surfactants, fluorescence studies,
379–386

Pyrene-modified low-density polyethylene
films
preparation, 426–427,428*f*
See also Photophysical studies,
low-density polyethylene film
characterization
Pyrene probes, poly(dimethylacrylamide)
behavior studies, 367
Pyromellitic dianhydride–oxydianiline
on gold, reflection–absorption IR
spectroscopy, 23,25–30
on silver
reflection–absorption IR spectroscopy,
18,19–21*f*
surface-enhanced Raman scattering,
18,22–23,24*f*
transmission IR spectra, 16–18

Q

Quadrupolar coupling, determination, 185
Quantum mechanical selection rules,
description, 2
Quantum yield of fluorescence,
determination, 359
Quartz slides, photophysics of (1-pyrenyl)-
butyltrimethylammonium bromide,
413–419
Quenching
process, 398–399
water-soluble polymers, 399–405

R

Raman active conditions, vibration, 2
Raman mapping, Raman microscopy with
charge-coupled device detector, 54,56*f*
Raman microscopy
advances in instrumentation, 43–44
applications, 43
problem with laser-induced sample
fluorescence, 44–45
system, 43,47*f*
with charge-coupled device detector
confocal measurements, 50–54,55*f*
experimental procedure, 45
imaging, 54,57–58*f*

Raman microscopy—*Continued*
with charge-coupled device detector—
Continued
micropolar measurements, 46–50
Raman mapping, 54,56*f*
Raman scattering, detection method, 3
Raman spectroscopy
advances, 6–7
advantages for analysis and
characterization of polymers, 41
analytical capabilities, 41,42*f*
background, 2–7
development, 3,6
structural ordering effect, 41–43
Raman spectrum, (4-mercaptophenyl)-
phthalimide on gold, 37,38*f*
Ratio of molar volume of solvent jumping
unit to molar volume of polymer
jumping unit, determination, 268,270
Rayleigh, scattering, schematic
representation, 3
Reflectance, relationship to
concentration, 151
Reflection–absorption IR spectroscopy
description, 4
polyimide–metal interface structure
determination
absorbance
vs. angle of incidence and
polarization, 9,10*f*
vs. film thickness, 9
experimental procedure, 14–16
4,4'-hexafluoroisopropylidene
dianhydride–oxydianiline on silver,
30,32–35
(4-mercaptophenyl)phthalimide on
gold, 35–37
perpendicularity of electric field
vector to metal surface, 9
previous studies, 12–13
pyromellitic dianhydride–oxydianiline
on gold, 23,25–30
on silver, 18,19–21*f*
spectral dependence on real and complex
parts of refractive index of
absorbing species, 9,11*f*

Reorientational motion of penetrant/
solvent, determination, 257
Resonance Raman scattering,
occurrence, 6
Rheological studies of microstructure–
associative-thickening behavior of
amphiphilic polyelectrolytes
N-(4-decyl)phenylacrylamide terpolymers,
562–566
experimental procedure, 552–553
importance, 551–552
previous studies, 552
N-[(1-pyrenylsulfonamido)ethyl]-
acrylamide terpolymers, 556–563
terpolymer
characterization, 555–556
synthesis, 553–555,556*t*
Rheooptical studies, procedure, 102
Rheophotoacoustic Fourier transform IR
spectroscopy of diffusion and adhesion
in polymers
adhesion
analytical results, 93–95
bilayer interaction measurements,
95–97
vibrational energy changes, 90,92–93
work of adhesion, 93
cell
description, 78*f*,80
with photoacoustic umbrella, 79*f*,80
diffusion
diffusion coefficient
vs. diffusion parameter, 86,87*f*
vs. percent crystallinity, 86,88*f*
poly(tetrafluoroethylene), 86,88–89
poly(vinylidene fluoride) diffusion,
81–88
quantitative analysis of cross-linking
processes, 89–90,91*f*

S

Salt, role in fluorescence studies of
pyrene-labeled water-soluble polymeric
surfactants, 383,384*f*

Sanidic layer structures
chain dynamics
substituted polyaramides, 314,318–323
substituted polyesters, 323,324*f*,326*f*
chain packing, 314,315*f*,318*f*
chemical structure, 314,316–317
experimental procedure, 325,331
schematic representation, 312,313*f*
Segment(s), identification by NMR
spectroscopy, 189
Segmental motion
determination, 257
toluene–polyisobutylene solutions, pulsed
field gradient NMR spectroscopy,
254–272
Self-diffusion coefficients of penetrant
into polymer matrix, measurement
techniques, 255
Self-diffusion of small molecules into
polymer matrices, concentration
dependence, 255–256
Semicrystalline polymers
helical jump, 244
molecular motions within crystalline
regions, 243–244
Silica–water interface, poly(dimethylacrylamide) behavior studies using
fluorescence techniques, 367,369–375
Silver substrates, spin-coated with thin
films
4,4'-hexafluoroisopropylidene
dianhydride–oxydianiline, 30–35
pyromellitic dianhydride–oxydianiline,
16–24
Single glass transition temperature,
criterion for distinguishing between
miscible and immiscible polymer–
polymer systems, 274
Small-angle X-ray scattering,
characterization of unimer micelles of
hydrophobically modified polysulfonates,
495,496*f*
Small molecule diffusion into polymers,
importance, 255
Smectic layer structures
poly(ester imides), 323,325–331
schematic representation, 312,313*f*

Smectic poly(ester imide)s
chain dynamics, 325,326,328–330
chemical structure, 323,325,327
experimental procedure, 325,331
Sodium 11-(acrylamido)undecanoate–
acrylamide copolymer, pyrene-labeled,
fluorescence studies, 379–386
Solid-state ^{13}C-NMR spectroscopy
angle-dependent NMR frequencies,
192–193,194*f*
conformation-dependent NMR
frequencies, 193
three-dimensional NMR spectroscopy,
195–196
two-dimensional exchange NMR
spectroscopy, 193–195
Solid-state NMR spectroscopy
amorphous polymers, multidimensional,
See Multidimensional solid-state NMR
spectroscopy of amorphous polymers
poly(methyl methacrylate)–
poly(vinylidene fluoride) system
experimental procedure, 274–275,276*f*
interdiffusion, 286–288
molecular miscibility detection,
275,277–280
phase separation, 281,283–286
quantification of degree, 280–281,282*f*
Solvent-induced changes in glass
transition temperature of ethylene vinyl
alcohol copolymer, study using Fourier
transform IR and dynamic mechanical
spectroscopies, 535–550
Solvent loss, weatherability of
polyurethane sealants, 120–135
Spectroscopic characterization of unimer
micelles of hydrophobically modified
polysulfonates
examples, 493
experimental description, 493–494,496
fluorescence studies, 495,497–510
Fourier transform IR studies, 512,514
light scattering studies, 495
NMR studies, 508,510–513
small-angle X-ray scattering studies,
495,496*f*
synthesis, 493

Specular reflectance measurements of polymeric fibrous materials
concept, 137–138
diameter, 142,145*f*
experimental procedure, 138–139
orientation, 142,144–145*f*
surface state effect, 142,143*f*
wavelength effect, 139–141
Spin diffusion
description, 295–298
pulsed NMR spectroscopy, 185–186
Spin–lattice relaxation
rate equation, 306–307
theory, 303–305
Step-scan Fourier transform IR spectroscopy
advantages for dynamic spectroscopy, 100–101
electric field reorientation of liquid crystals, 110–115
instrumentation, 101–102,103–104*f*
liquid-crystal electroreorientation procedure, 102,105,106*f*
microrheometer, 102,104*f*
modulated mechanical field responses of polymer films, 105,107–110
rheooptical study procedure, 102
two-dimensional analysis, 101
variable-temperature liquid crystal IR cell, 105,106*f*
Stilbene, photochromic probe studies, 166–180
Stokes scattering, schematic representation, 3
Stretched low-density polyethylene film, characterization using photophysical techniques, 425–443
Structure
amorphous polymers using multi-dimensional solid-state NMR spectroscopy, 191–213
polyimide–metal interfaces, determination using reflection–absorption IR spectroscopy and surface-enhanced Raman scattering, 8–39

Structured CH_2 magic-angle spinning lines, conformational information, 208–212
Styrene and methyl methacrylate copolymerization, near-IR spectroscopy, 156
Substituted polyaramides, chain dynamics, 314,318–323
Substituted polyesters, chain dynamics, 323,324*f*,326*f*
Surface electromagnetic wave spectroscopy, description, 4
Surface-enhanced Raman scattering, occurrence, 6
Surface-enhanced Raman spectroscopy for polyimide–metal interface structure determination
experimental procedure, 15–16
mechanisms of enhancement, 12
(4-mercaptophenyl)phthalimide on gold, 37–39
pyromellitic dianhydride–oxydianiline on silver, 18,22–23,24*f*
surface selectivity, 12
Surface gradients in nylon 11 and nylon 11–poly(vinylidene fluoride) bilaminates, Fourier transform IR–attenuated total reflectance spectroscopy, 63–66
Surface-sensitive IR techniques, examples, 4
Surface state effect, diffuse and specular reflectance measurements of polymeric fibrous materials, 142,143*f*
Surfactant(s), pyrene-labeled water soluble polymeric, fluorescence studies, 379–386
Surfactant to monomer ratio, definition, 554
Synthesis, unimer micelles of hydrophobically modified polysulfonates, 493

T

Thermogravimetry, evaluation of weatherability of polyurethane sealants, 119–135

Thin films of polystyrene and poly(methyl methacrylate), photophysical study, 410–424

Thin polymer films
analytical techniques, 411
preparation techniques, 410–411

Three-dimensional NMR spectroscopy
description, 186,195–196
use in molecular dynamics studies, 187

Time-resolved anisotropy measurements
adsorption of water-soluble polymers at interfaces, 396
hydrophobic specie vs. conformational behavior, 391,394
interpolymer complexation, 394–396
parallel and perpendicular components of fluorescence intensity, 391,392*f*
pH vs. correlation times, 391,393*f*
poly(dimethylacrylamide) behavior studies, 365,367–375
principles, 389–391

Toluene, role in glass transition temperature of ethylene vinyl alcohol copolymer, 535–550

Trans to cis isomerization of azo probes and azo-labeled polymers, 168–169

Transmission IR spectroscopy
4,4'-hexafluoroisopropylidene dianhydride–oxydianiline on silver, 30,31*f*
(4-mercaptophenyl)phthalimide on gold, 35,36*f*
nylon 11 thin films, trichroic, *See* Trichroic transmission IR spectroscopy of nylon 11 thin films
pyromellitic dianhydride–oxydianiline on silver, 16–18

Trichroic transmission IR spectroscopy of nylon 11 thin films
electric field poling effect, 74,75*f*
experimental procedure, 72
orientation distribution of ordered hydrogen-bonded amide planes before and after poling, 76
orientation of amide groups in transverse plane, 74,76

Trichroic transmission IR spectroscopy of nylon 11 thin films—*Continued*
role of width and center of orientation distribution on IR intensities, 76
thickness direction spectrum determination, 70,72
thickness direction vs. draw direction, 72–74

Two-dimensional exchange NMR spectroscopy, description, 193–195

Two-dimensional Fourier transform IR spectroscopy
advantages, 101
concept, 101
rules for analysis, 101

Two-dimensional IR spectroscopy
advantages, 66–67
applications, 67
description, 66
experimental representation, 186,188*f*
fluoropolymer characterization
advantages, 225
{^{13}C}^{19}F heteronuclear multiple bond correlation spectroscopy, 228,231–237
{^{13}C}^{19}F heteronuclear multiple quantum coherence spectroscopy, 228,229–230*f*
{^{19}F}^{13}C heteronuclear shift correlation spectroscopy, 225,227–228
nylon 11
asynchronous correlation spectrum, 70,71*f*
experimental procedure, 67
normal IR absorbance spectrum, 67,69*f*
synchronous correlation spectrum, 70,71*f*
segment identification, 189
spectral information, 186
use
in molecular dynamics studies, 187,188*f*
in structural studies, 186–187

U

Unimer micelles of hydrophobically modified polysulfonates, spectroscopic characterization, 490–514

Unstretched low-density polyethylene film, characterization using photophysical techniques, 425–443

Urea, formation of crystalline inclusion compounds with polymers, 518

Urethane linkage breakage, weatherability of polyurethane sealants, 120–135

UV–visible spectroscopy for free volume distribution determination during multifunctional monomer polymerizations
absorbance of stilbene vs. wavelength, 173,174*f*
absorbance vs. time with and without photoinitiator, 172–173,174*f*
analytical procedure, 170–172
average deviation vs. double bond conversion, 179,180*f*
average free volume vs. double bond conversion, 179,180*f*
experimental procedure, 169–170
extent of isomerization, 173
fraction of probes that are mobile
vs. double bond conversion, 173,175,176*f*
vs. monomer size and cross-linking density, 177,178*f*
vs. polymerization temperature, 175,176*f*
previous studies, 166–167

V

Vibration
IR active conditions, 2
Raman active conditions, 2
Vibrational energy changes, theory, 90,92–93
Vibrational modes, classes, 2
Vibrational relaxation, definition, 358

Vibrational spectroscopy, applications, 137

Vinylene–phenylene polymers, photophysical behavior, polarized fluorescence spectroscopy, 446–456

Viscosity, pyrene-labeled water-soluble polymeric surfactants, fluorescence, 382–383,384*f*

Vrentas–Duda theory for diffusion, description, 256–257

W

Water-soluble polymer(s), luminescence spectroscopy, 388–407

Water-soluble polymeric surfactants, pyrene-labeled, fluorescence studies, 379–386

Wavelength effect, diffuse and specular reflectance measurements of polymeric fibrous materials, 139–141

Weatherability of polyurethane sealants
artificial weathering procedure, 119
calcium carbonate loss, 126–135
photoacoustic Fourier transform IR spectroscopic procedure, 120
plasticizer loss, 120–135
PU1, 120–125
PU2, 126,127–130*f*
PU3, 126,131–135
solvent loss, 120–135
thermal analytical procedure, 120
urethane linkage breakage, 120–135
weight loss in nitrogen, 120,121*t*

WLF equation for diffusion, form, 257

X

X-ray scattering spectroscopy, measurement of local free volume distribution, 168

Highlights from ACS Books

Good Laboratory Practice Standards: Applications for Field and Laboratory Studies
Edited by Willa Y. Garner, Maureen S. Barge, and James P. Ussary
ACS Professional Reference Book; 572 pp; clothbound ISBN 0–8412–2192–8

Silent Spring Revisited
Edited by Gino J. Marco, Robert M. Hollingworth, and William Durham
214 pp; clothbound ISBN 0–8412–0980–4; paperback ISBN 0–8412–0981–2

The Microkinetics of Heterogeneous Catalysis
By James A. Dumesic, Dale F. Rudd, Luis M. Aparicio, James E. Rekoske,
and Andrés A. Treviño
ACS Professional Reference Book; 316 pp; clothbound ISBN 0–8412–2214–2

Helping Your Child Learn Science
By Nancy Paulu with Margery Martin; Illustrated by Margaret Scott
58 pp; paperback ISBN 0–8412–2626–1

Handbook of Chemical Property Estimation Methods
By Warren J. Lyman, William F. Reehl, and David H. Rosenblatt
960 pp; clothbound ISBN 0–8412–1761–0

Understanding Chemical Patents: A Guide for the Inventor
By John T. Maynard and Howard M. Peters
184 pp; clothbound ISBN 0–8412–1997–4; paperback ISBN 0–8412–1998–2

Spectroscopy of Polymers
By Jack L. Koenig
ACS Professional Reference Book; 328 pp;
clothbound ISBN 0–8412–1904–4; paperback ISBN 0–8412–1924–9

Harnessing Biotechnology for the 21st Century
Edited by Michael R. Ladisch and Arindam Bose
Conference Proceedings Series; 612 pp;
clothbound ISBN 0–8412–2477–3

From Caveman to Chemist: Circumstances and Achievements
By Hugh W. Salzberg
300 pp; clothbound ISBN 0–8412–1786–6; paperback ISBN 0–8412–1787–4

The Green Flame: Surviving Government Secrecy
By Andrew Dequasie
300 pp; clothbound ISBN 0–8412–1857–9

For further information and a free catalog of ACS books, contact:
American Chemical Society
Product Services Office
1155 16th Street, NW, Washington, DC 20036
Telephone 800–227–5558

Bestsellers from ACS Books

The ACS Style Guide: A Manual for Authors and Editors
Edited by Janet S. Dodd
264 pp; clothbound ISBN 0–8412–0917–0; paperback ISBN 0–8412–0943–X

Understanding Chemical Patents: A Guide for the Inventor
By John T. Maynard and Howard M. Peters
184 pp; clothbound ISBN 0–8412–1997–4; paperback ISBN 0–8412–1998–2

Chemical Activities (student and teacher editions)
By Christie L. Borgford and Lee R. Summerlin
330 pp; spiralbound ISBN 0–8412–1417–4; teacher ed. ISBN 0–8412–1416–6

Chemical Demonstrations: A Sourcebook for Teachers,
Volumes 1 and 2, Second Edition
Volume 1 by Lee R. Summerlin and James L. Ealy, Jr.;
Vol. 1, 198 pp; spiralbound ISBN 0–8412–1481–6;
Volume 2 by Lee R. Summerlin, Christie L. Borgford, and Julie B. Ealy
Vol. 2, 234 pp; spiralbound ISBN 0–8412–1535–9

Chemistry and Crime: From Sherlock Holmes to Today's Courtroom
Edited by Samuel M. Gerber
135 pp; clothbound ISBN 0–8412–0784–4; paperback ISBN 0–8412–0785–2

Writing the Laboratory Notebook
By Howard M. Kanare
145 pp; clothbound ISBN 0–8412–0906–5; paperback ISBN 0–8412–0933–2

Developing a Chemical Hygiene Plan
By Jay A. Young, Warren K. Kingsley, and George H. Wahl, Jr.
paperback ISBN 0–8412–1876–5

Introduction to Microwave Sample Preparation: Theory and Practice
Edited by H. M. Kingston and Lois B. Jassie
263 pp; clothbound ISBN 0–8412–1450–6

Principles of Environmental Sampling
Edited by Lawrence H. Keith
ACS Professional Reference Book; 458 pp;
clothbound ISBN 0–8412–1173–6; paperback ISBN 0–8412–1437–9

Biotechnology and Materials Science: Chemistry for the Future
Edited by Mary L. Good (Jacqueline K. Barton, Associate Editor)
135 pp; clothbound ISBN 0–8412–1472–7; paperback ISBN 0–8412–1473–5

For further information and a free catalog of ACS books, contact:
American Chemical Society
Product Services Office
1155 16th Street, NW, Washington, DC 20036
Telephone 800–227–5558